Handbook of Industrial Chemistry

Organic Chemicals

Mohammad Farhat Ali, Ph.D.
King Fahd University of Petroleum & Minerals
Dhahran, Saudi Arabia

Bassam M. El Ali, Ph.D.
King Fahd University of Petroleum & Minerals
Dhahran, Saudi Arabia

James G. Speight, Ph.D.
CD&W Inc.
Laramie, Wyoming

McGraw-Hill

New York   Chicago   San Francisco   Lisbon   London   Madrid
Mexico City   Milan   New Delhi   San Juan   Seoul
Singapore   Sydney   Toronto

The *McGraw-Hill* Companies

CIP Data is on file with the Library of Congress

1 2 3 4 5 6 7 8 9 0 DOC/DOC 0 1 0 9 8 7 6 5 4

ISBN 0-07-141037-6

The sponsoring editor for this book was Kenneth P. McCombs and the production supervisor was Pamela A. Pelton. It was set in Century Schoolbook by International Typesetting and Composition. The art director for the cover was Anthony Landi.

Printed and bound by RR Donnelley.

This book is printed on recycled, acid-free paper containing a minimum of 50% recycled, de-inked fiber.

McGraw-Hill books are available at special quantity discounts to use as premiums and sales promotions, or for use in corporate training programs. For more information, please write to the Director of Special Sales, McGraw-Hill Professional, Two Penn Plaza, New York, NY 10121-2298. Or contact your local bookstore.

*To our wives and families, and to all scientists
and engineers who preceded us in such work*

ABOUT THE EDITORS

MOHAMMAD FARHAT ALI, PH.D., is Professor of Industrial and Petroleum Chemistry at King Fahd University of Petroleum & Minerals in Saudi Arabia. An expert in characterization studies of heavy ends, residues, and asphalt, he is also knowledgeable about crude oils and products, refining process technology, waste oil recycling, and stability characteristics of jet fuels.

BASSAM M. EL ALI, PH.D., is Professor of Industrial Chemistry at King Fahd University of Petroleum & Minerals in Saudi Arabia. His specialties include homogenous and heterogeneous catalysis using transition metal complexes in hydrocarboxylation, hydroformylation, oxidation, coupling, hydrogenation, and other important processes; investigation of the organometallic intermediates and the mechanisms of various homogenous reactions; and synthesis, characterization, and application of various supported catalytic systems in the production of fine chemicals. He has taught many industrial chemistry courses including Industrial Catalysis, Industrial Organic Chemistry, Industrial Inorganic Chemistry, and Petroleum Processes.

JAMES G. SPEIGHT, PH.D., has more than 35 years' experience in fields related to the properties and processing of conventional and synthetic fuels. He has participated in, and led, significant research in defining the uses of chemistry with heavy oil and coal. The author of well over 400 professional papers, reports, and presentations detailing his research activities, he has taught more than 50 related courses. Dr. Speight is the author, editor, or compiler of a total of 25 books and bibliographies related to fossil fuel processing and environmental issues. He lives in Laramie, Wyoming.

Contents

Contributors

Hasan A. Al-Muallem, Ph.D.
Department of Chemistry
King Fahd University of
    Petroleum & Minerals
Dhahran, Saudi Arabia

Mohammad Farhat Ali, Ph.D.
Professor of Chemistry
Department of Chemistry
King Fahd University of
    Petroleum & Minerals
Dhahran, Saudi Arabia

Bassam M. El Ali, Ph.D.
Professor of Chemistry
Department of Chemistry
King Fahd University of
    Petroleum & Minerals
Dhahran, Saudi Arabia

Manfred J. Mirbach, Ph.D.
Landis Kane Consulting
R&D Management
Fuellinsdorf, Switzerland

Ahsan Shemsi
Department of Chemistry
King Fahd University of
    Petroleum & Minerals
Dhahran, Saudi Arabia

James G. Speight, Ph.D.
CD&W Inc.
Laramie, Wyoming

Preface

The organic chemical industry is an important branch of industry and its structure usually centers on petroleum and hydrocarbon derived chemicals. The volume text of available books is generally lacking in covering other very important nonpetroleum-based organic industries such as paints, dyes, edible oils, fats and waxes, soaps and detergents, sugars, fermentation, chemical explosives, and agrochemical industries.

This book focuses primarily on the chemical processing of raw materials other than petroleum and hydrocarbons. These materials are usually converted into useful and profitable products that are, in general, used as consumer goods. The book addresses the needs of both students and practicing chemists and chemical engineers. It is intended to be a primary source of information for the young practicing professionals who wish to broaden their knowledge of the organic process industry as a whole. The book may also serve as a textbook for advanced undergraduate students in industrial chemistry.

Chapter 1 describes the development of the chemical industry and its role in welfare and employment around the world. This chapter shows how raw materials are procured and converted to consumer products.

Chapter 2 discusses safety aspects in organic industries and methods to protect the workers from hazards such as exposure to dangerous chemicals, heat, pressures, high electric fields, accelerating objects, and other sources of hazards.

Chapter 3 deals with the sources of pollution caused by raw materials, products, and wastes in petroleum, petrochemicals, pharmaceuticals, food, and other industries. The growing public concerns over the safety of chemicals in the environment, and the efforts by the governments and industries for their control, are discussed.

Chapter 4 presents the chemistry and technology of edible oil, fat, and wax processing including refining, recovery, crystallization, interesterification, and hydrogenation. The key oxidation reactions of lipids leading to quality deterioration of processed and unprocessed foods, and the mechanism of the action of the antioxidants in improving oxidation stability of foods are discussed.

Chapter 5 highlights the soap and detergent industry. The raw materials, important processes of production, and economic importance of the soap and detergent industry are elaborated.

Chapter 6 covers one of the most widely distributed and abundant organic chemicals—the sugars. The chemistry of saccharides, historical survey, and world production of sugar are presented. The sugar recovery from the two principal sources—sugar cane and sugar beets—are discussed. The chemistry and uses of nonsugar sweetening agents is also presented.

Chapter 7 describes paints, pigments, and industrial coatings. The major paint components, namely, pigments, binders, additives, and solvents are discussed in separate sections. These are followed by the principles of formulation, application techniques, durability, and testing of paints.

Chapter 8 is devoted to the industrially produced dyes with their classification, manufacture, properties, and main applications, as well as environmental and health aspects.

Chapter 9 presents an overview of modern fermentation processes and their application in food, pharmaceutical, and industrial chemical industries. The social and economic importance of fermentation processes is discussed.

The pharmaceutical industry is presented in Chapter 10 as one of the most important sectors of healthcare worldwide. The discovery, the development, and the production of drugs are covered in this chapter. The chapter also includes the correlation between the growth in the worldwide market for pharmaceuticals and the increase of the world population as a result of higher life expectancy and changes in lifestyle.

Chapter 11 presents an overview of the agrochemical industry. Beginning with the introduction and historical background, it leads to the modern trends in agriculture, chemical pest control, herbicides, fungicides, insecticides, and biological pest control agents. Social and economic aspects of pesticides use are also discussed.

Chapter 12 presents the chemistry of explosives. Chemical explosives and propellants are well-covered in this book because of their importance for peaceful uses. They are considered chemical compounds in pure form or mixtures that rapidly produce a large volume of hot gases when properly ignited. The destructive effects of explosives are much more spectacular than their peaceful uses. However, it appears that more explosives have been used by industries for peaceful purposes than in all the wars.

Chapter 13 covers the conversion of crude oil into desired products in an economically feasible and environmentally acceptable manner.

Descriptions are provided for (1) desalting and dewatering; (2) separation processes, of which distillation is the prime example; (3) conversion processes, of which coking and catalytic cracking are prime examples; and (4) finishing processes, of which hydrotreating to remove sulfur is a prime example. Descriptions of the various petroleum products (from fuel gas to asphalt and coke) are also given.

This chapter also includes a description of the petrochemical industry, and the production of the chemicals and compounds in a refinery that are destined for further processing, and used as raw material feedstocks for the fast-growing petrochemical industry.

Chapter 14 provides the basic principles of polymer science, and addresses the importance of this subject. This chapter aims to give a broad and unified description of the subject matter—describing the polymerization reactions, structures, properties, and applications of commercially important polymers, including those used as plastics, fibers, and elastomers. This chapter focuses on synthetic polymers because of the great commercial importance of these materials. The chemical reactions by which polymer molecules are synthesized are addressed along with the process conditions that can be used to carry them out. This chapter also discusses topics on degradation, stability, and environmental issues associated with the use of polymers.

This book is intended for university and college students who have studied organic chemistry, as well as for scientists and technicians who work in the organic chemical industry, and senior executives and specialists who wish to broaden their knowledge of the industrial organic processes as a whole.

At the end, we gratefully acknowledge the financial aid, facilities, and support provided by the Deanship of Scientific Research at King Fahd University of Petroleum & Minerals, Dhahran, Saudi Arabia.

Mohammad Farhat Ali, Ph.D.
Bassam M. El Ali, Ph.D.
James G. Speight, Ph.D.

Introduction: An Overview of the Chemical Process Industry and Primary Raw Materials

Mohammad Farhat Ali

1.1 The Chemical Process Industry

The chemical process industry includes those manufacturing facilities whose products result from (a) chemical reactions between organic materials, or inorganic materials, or both; (b) extraction, separation, or purification of a natural product, with or without the aid of chemical reactions; (c) the preparation of specifically formulated mixtures of materials, either natural or synthetic. Examples of products from the chemical process industry are plastics, resins, dyes, pharmaceuticals, paints, soaps, detergents, petrochemicals, perfumes, inorganics, and synthetic organic materials. Many of these processes involve a number of unit operations of chemical engineering depending on the size definition of a plant, as well as such basic chemical reactions (processes) as polymerization, oxidation, reduction, hydrogenation, and the like. The global chemical industry is valued at one and a half trillion US dollars today with more than 70,000 commercial products. The total world trade in chemicals is valued at US$400 billion, 10 percent of the value of global trade [1].

The three largest sectors within the world chemical industry are petrochemicals, pharmaceuticals, and performance chemicals. Petrochemicals dominate the global chemical industry with a share of 30 percent, followed by pharmaceuticals (16.5 percent) and performance chemicals (16 percent). The European Union (EU), the United States, and Germany are the three largest manufacturers followed by Japan, France, the United Kingdom, Italy, and other Asian countries. However, there has been a significant shift of global demand for chemicals from industrialized to developing nations, and the movement of basic chemical manufacturing from industrialized regions to Asia-Pacific and China [2].

The chemical industry as a whole makes a massive contribution to welfare and employment around the world. The European Union (EU) is the world's largest chemical producer, accounting for nearly a third of the estimated world production. Throughout the EU, about 1.7 million people are employed in some 25,000 chemical companies and the industry provides further employment in a broad range of downstream industries [2]. The U.S. chemical, petrochemical, and pharmaceutical industries together had more than 13,000 establishments, more than one million employees, and a total value of shipments worth approximately US$406.9 billion [3]. Also, many of the Asian and Latin American countries have grown rapidly and have become international competitors in the chemical industry. Consequently, the global chemical industry is among the most competitive industries in the world. Being an intermediate input industry, the chemical industry has both forward and backward linkages with other segments of the manufacturing sector thereby acting as a precursor for the good performance of the manufacturing sector as a whole.

1.2   Development of the Chemical Industry

The oldest traces of a chemical industry were found in the Middle Age and they were primarily based on the knowledge and skill in producing candles, soaps, paints, and medicaments. Manufacturing these products, at the very beginning, was a homemade affair that aimed to fulfill the needs of just one or more households. Chemical production came of age as an industry in the late 1700s but it remained small because many of the manufacturers did not have the capabilities for continual and larger production. The evolution of what we know as the modern chemical industry started more recently. Over the nineteenth and the twentieth centuries, chemists played key roles in expanding the frontiers of knowledge in advancing medicine and industry, and creating such products as aspirin, synthetic polymers, and rubbers. The discovery of the first synthetic dyestuff, mauve, in the 1860s by W. H. Perkins proved to be instrumental in the evolution of the organic chemical industry in the United Kingdom and Germany. The dawn of the twentieth century brought fundamental changes mainly as a result of the emphasis on research on the applied aspects of chemistry in Germany and the United States. Entrepreneurs took full advantage of the increasing scientific knowledge to revolutionize the chemical industry as a whole.

The organic chemical industry has grown at a remarkable rate since 1940 as a result of the development and growth of the petroleum refining and petrochemical sectors. The rapid growth in petrochemicals in the 1960s and 1970s was largely because of the enormous increase in demand for synthetic polymers. The chemical industry today is highly research and development (R&D) intensive while producing a high rate of innovation, making significant contributions to the economy. Yet, the chemical industry may be regarded as having become a mature manufacturing industry, following its rapid growth in the 1960s and 1970s, dampening the returns from the high-risk R&D investments. Moreover, many of the basic processes for producing key intermediate chemicals have lost their patent protection over the years, enabling other countries of the world, who wish to venture into this area, to buy their own manufacturing plants. As a result, petroleum-producing countries such as Korea, Mexico, Saudi Arabia, and other Middle Eastern countries have entered and rapidly expanded their production of the aromatic petrochemical intermediates together with the final polymer products such as polyethylene, polypropylene, polyesters, and epoxy resins. There is also a growing shift in the global chemical industry as a consequence of both the rapidly growing population and the industrial development of countries of southeast Asia. It is envisaged that China, with its enormous population, will become both a major market and a major producer in chemicals production during the twenty-first century.

1.3   Characteristics of the Chemical Industry

The chemical industry is essentially a science-based industry. The technologies applied in the chemical industry have their well-established scientific roots, and industry growth has been closely linked to scientific discoveries. One of the main reasons for the enormous growth of the chemical industry in the developed world has been its great commitment to the investment in R&D. This traditional investment in R&D does much to explain the outstanding growth rate of the industry in the twentieth century and its superior record of increased productivity. The industry's organized application of science to industrial problems has produced a host of new products, new processes, and new applications [4].

The chemical and drug companies in the United States now spend about $US18 billion annually on R&D. The scientific and technical research by these industries significantly contributes in making human lives safer, longer, easier, and more productive [5].

Table 1.1 shows the chemical sales and R&D spending for the global top 10 companies in the world [5].

The chemical industry produces many materials that are essential for our most fundamental needs for food, shelter, and health. It also produces products of great importance to the high technology world of computing, telecommunications, and biotechnology. The U.S. government uses the following eight standard industrial classification codes to categorize chemical companies [6].

- Industrial inorganic chemicals
- Plastics, materials, and synthetics
- Drugs
- Soap, cleaners, and toilet goods
- Paints and allied products
- Industrial organic chemicals
- Agricultural chemicals
- Miscellaneous chemical products

According to an estimate by the U.S. Environmental Protection Agency (EPA), there are 15,000 chemicals manufactured in the United

TABLE 1.1   Global Top 10 by Chemical Sales (2003–2004)

Rank	Company	Country	Chemical sales ($ millions)	R&D spending ($ millions)
1	Dow Chemicals	U.S.A.	32,632.0	981.0
2	BASF	Germany	30,768.0	904.1
3	DuPont	U.S.A.	30,249.0	1,349.0
4	Bayer	Germany	21,567.5	1,299.0
5	Total	France	20,197.3	NA
6	Exxon Mobil	U.S.A.	20,190.0	NA
7	BP	U.K.	16,075.0	NA
8	Shell	U.K.	15,186.0	NA
9	Mitsubishi Chemical	Japan	13,216.4	NA
10	Degussa	Germany	12,929.7	401.7

NA: Not available.
SOURCE: *Chemical & Engineering News*, July 19, 2004, p. 11–13.

States in quantities greater than 10,000 pounds [7]. The organic chemical industry, which manufactures carbon-containing chemicals, accounts for much of this diversity.

Chemical manufacturing has been undertaken by many different types of companies largely because of its central role in industry. Chemical products are made from raw materials supplied by a large number of different industries including petroleum, agriculture, and mines products. In turn, chemicals themselves are used as raw materials in almost all other types of manufacturing. The production of chemicals is thus attractive to companies that are seeking to upgrade their low-value feedstock (raw materials required for an industrial process) to a profitable chemical product. Petroleum companies, in particular, are increasingly acquiring leading positions in the organic chemical industry. Of the top 10 leading producers listed in Table 1.1, four are petroleum companies including Total, ExxonMobil, BP, and Shell Oil.

Mergers and acquisitions (M&As) have performed a significant role in the evolution of the chemical industry. Before World War II, the German manufacturer, IG Farben, was the largest producer of organic chemicals in the world. After the war, the Allied powers restructured German industry and IG Farben was broken into its major constituent firms: BASF, Bayer, and Hoechst. At about the same time, the British government enforced a merger of strong firms such as Brunner, Maud, and Nobel with much weaker firms—United Alkali and the British Dyes, to establish one big firm, Imperial Chemical Industries (ICI). In the United States, DuPont, during the 1980s, was involved in over 50 acquisitions, investing more than $10 billion. Total M&A activity in the United States in 1999 reached $45 billion. It is estimated that in the rest of the world M&As totaled $1.2 trillion [8]. M&As have performed a number of important roles in the chemical industry and enabled many U.S. firms to acquire foreign businesses to obtain the needed presence in the local world markets.

1.4   Raw Materials, Manufacturing, and Engineering

Industrial chemistry procures raw materials from *natural environments* to convert them into *intermediates*, which subsequently serve as base materials to every other kind of industry. There are four sources of *natural environment*:

a. The earth's crust (lithosphere)

b. The marine and oceanic environment (hydrosphere)

c. The air (atmosphere)

d. The plants (biosphere)

Raw materials derived from the above natural resources are classified as either renewable or nonrenewable. Renewable resources are those that regenerate themselves, such as agricultural, forestry, fishery, and wildlife products. If the rate at which they are consumed becomes so great that it drives these resources to exhaustion, however, these renewable resources can become nonrenewable. Nonrenewable resources are those that are formed over long periods of geologic time. They include metals, minerals, and organic materials.

The renewable resources such as agricultural materials were the main source of raw materials until the early part of the twentieth century for the manufacturing of soap, paint, ink, lubricants, greases, paper, cloth, drugs, and other chemical products. The nonrenewable feedstock based on fossil fuels was added as an alternative resource in the latter part of the twentieth century. This result was firstly because of the development of new products such as synthetic fibers, plastics, synthetic oils, and petrochemicals and then because of great advances in catalysis and polymer science. The use of petroleum gas and oil has increased during the past 30 years as a result of a complete changeover from coal-to-petroleum technology.

Recent scientific and technical developments in biotechnology, however, are beginning to shift the balance back in the direction of renewable raw materials. According to a high-level U.S. Federal Advisory Committee's suggestion, the production of chemicals and materials from bio-based feedstock will increase rapidly from today's 5-percent level to 12 percent in 2010, 18 percent in 2020, and 25 percent in 2030 [9]. It is estimated that two-thirds of the $1.5 trillion global chemical industry can eventually be based on renewable bio-sources, thus replacing petroleum and natural gas as the feedstock [9]. The renewable raw materials are being used to solve environmental problems in that the products made from them generally are more readily biodegradable and less toxic. They certainly offer a potential to contribute to emission reductions in a variety of ways. Unlike those that are petroleum-based, renewable raw materials do not contribute carbon dioxide to the atmosphere—an increasingly important concern with respect to climate change and global warming.

Because every industrial chemical process is designed to economically produce a desired product from a variety of raw materials. The economical extraction and use of exploitable raw materials are the essential prerequisites for a chemical industry. These raw materials usually have to be pretreated. They may undergo a number of steps involving physical treatment, chemical reactions, separation, and purification before their conversion into a desired product. Figure 1.1 shows a typical structure of such a process.

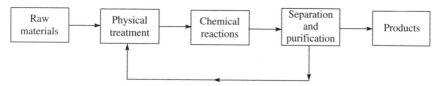

Figure 1.1   Typical chemical process structure.

As illustrated in Fig. 1.1, the organic chemical industry requires raw materials from renewable or nonrenewable resources, and sells its products either as finished materials or as intermediates for further processing by other manufacturers. The two major steps in chemical manufacturing are (1) the chemical reaction and (2) the purification of reaction products [7].

The primary types of chemical reactions are either *batch* or *continuous*. In *batch reactions*, the reactant chemicals are added to the reactor (reaction vessel) at the same time and products are emptied completely when the reaction is finished. The reactors are made of stainless steel or glass-lined carbon steel and range in size from 200 to several thousand liters. Batch reactors are provided with a stirrer to mix the reactants, an insulating jacket, and the appropriate pipes and valves to control the reaction conditions .

Batch processes generally are used for small-scale production. These processes are easier to operate, maintain, and repair. The batch equipment can be adapted to multiple uses.

In *continuous processes*, the reactants are added and products are removed at a constant rate from the reactor, so that the volume of reacting material in the reactor (reaction vessel) remains constant. Two types of reactors, either (1) a continuous stirred tank or (2) a pipe reactor, are generally used. A continuous stirred tank reactor is similar to the batch reactor described above. A pipe reactor typically is a piece of tubing arranged in a coil or helix shape that is jacketed in a heat-transfer fluid. Reactants enter one end of the pipe, and the materials are mixed under the turbulent flow and react as they pass through the system. Pipe reactors are well-suited for reactants that do not mix well, because the turbulence in the pipes causes all materials to mix thoroughly.

Because continuous processes require a substantial amount of automation and capital expenditure, this type of process is used primarily for large-scale productions.

The reaction products are often not in a pure form, usable by customers or downstream manufacturers. Therefore, the desired product must be isolated and purified by using various separation and purification

methods. Common separation methods include filtration, distillation, and extraction. Multiple methods are also used to achieve the
desired purity.

The organic chemical industry is a very high technology industry,
which uses the latest advances in electronics and engineering. Computers
are very widely used in automation of chemical plants, quality control,
and molecular modeling of structures of new compounds. Vessels for
chemical conversions and formulating and equipment for separation
processes represent the largest single expenditure in chemical plants.
The industry also buys large quantities of such generally used items as
valves, pumps, and instruments for recording and controlling processes
and product quality.

1.5   Environmental Aspects

The organic chemical industry uses and generates both large numbers
and large quantities of a wide variety of solvents, metal particulates, acid
vapors, and unreacted monomers. These chemicals are released to all
media including air, water, and land. The potential sources of pollutant
outputs by media are shown below in Table 1.2 [7].

As a result of public awareness of the dangers of chemicals in the environment, the chemical industry is one of the most highly regulated of
all industries. The regulations are intended to protect and improve the
health, safety, and environment of the public as well as the worker. The
current large expenditures for pollution control in the developed world
reflect mainly the intervention of the governments with strict laws. In
the United States, these laws are enforced by the EPA. Federal legislation on air and water pollution control in the United States provides
guidelines and training of personnel in both private industries and government agencies. The EPA has found that the U.S. industry's efforts
in finding ways to reduce both the volume and toxicity of its wastes have
resulted in a substantial decrease in the manufacturing costs and an
improvement in the production yields, while complying with government
regulations.

The biggest global organic chemical companies have been promoting
pollution prevention through various means. Some companies have cre-
atively implemented pollution prevention techniques that improve efficiency and increase profits while minimizing environmental impacts.
This is done in many ways such as reducing material inputs, reengi-
neering processes to reuse by-products, improving management practices, and substituting benign chemicals for toxic ones. Some smaller
facilities are able to actually get below regulatory thresholds just by
reducing pollutant releases through aggressive pollution prevention
policies.

TABLE 1.2   Potential Releases During Organic Chemical Manufacturing

Media	Potential sources of emissions
Air	Point source emission: stack, vent (e.g., laboratory hood, distillation unit, reactor, storage tank vent), material loading/unloading operations (including rail cars, tank trucks, and marine vessels). Fugitive emissions: pumps, valves, flanges, sample collection, mechanical seals, relief devices, tanks. Secondary emissions: waste and wastewater treatment units, cooling tower, process sewer, sump, spill or leak areas.
Liquid wastes (organic or aqueous)	Equipment wash solvent or water, lab samples, surplus chemicals, product washes or purifications, seal flushes, scrubber blow down, cooling water, steam jets, vacuum pumps, leaks, spills, spent or used solvents, housekeeping (pad wash down), waste oils or lubricants from maintenance.
Solid wastes	Spent catalysts, spent filters, sludges, wastewater treatment biological sludge, contaminated soil, old equipment or insulation, packaging material, reaction by-products, spent carbon or resins, drying aids.
Ground water contamination	Unlined ditches, process trenches, sumps, pumps, valves, or fittings, wastewater treatment ponds, product storage areas, tanks, and tank farms, aboveground and underground piping, loading/unloading areas or racks, manufacturing maintenance facilities.

SOURCE: Chemical Manufacturers Association, 1993.

The best way to reduce pollution is to study ways of preventing it at the research and development stage. At this stage, all possible reaction pathways for producing the desired product can be examined. These can be evaluated in light of yield, undesirable by-products, and their impacts on the health and environment. In general, changes made at the research and development stage will have the greatest impact.

References

1. *Chemical & Engineering News,* American Chemical Society (ACS), USA, 82 (02), 17–20, January 12, 2004.
2. European Chemical Industry Council (Cefic), *Horizon 2015: Perspectives for the European Chemical Industry, March 2004.* Available at www.cefic.org.
3. U.S. Census Board, *Economic Census,* 1997.

4. Meegan, M. K. (Ed.), *The Kline Guide to the Chemical Industry*, 5th ed., Kline and Company, Fairfield, NJ, USA, 1990.
5. *Chemical & Engineering News*, ACS, USA, 82 (29), July 19, 2004.
6. Chemical Manufacturing Association (CMA), *Statistical Handbook*, Washington D.C., p. 7, 1995.
7. U.S. Environmental Protection Agency (EPA), *Profile of the Organic Chemical Industry*, 2nd ed., Washington, D.C., November 2002.
8. Weston, J. F., B. A. Johnson, and J. A. Siu, *M&As in the Evolution of the Global Chemical Industry*, The Anderson Graduate School of Management, UCLA, USA, September 1999.
9. Wedin, R., *Chemistry on a High-Carb Diet*, *Chemistry*, ACS, USA, p. 23, Spring 2004.

2

Safety Considerations in Process Industries

Bassam El Ali and Ahsan Shemsi

2.1 Introduction

The misuse or the mishandling of a simple instrument such as a knife, hammer, or sickle may result in an injury to the holder. Workers in a factory, a manufacturing plant, or a chemical plant remain exposed to moving conveyers, machines, dangerous chemicals, heat, pressures, high electric fields, accelerating objects, and other sources of hazards. If workers are not protected from these hazards, there is the chance of incidents ranging from simple injuries to death of personnel. In addition, the damage can reach the whole manufacturing plant and its surrounding environment, causing much loss of life if the facilities or equipment are not properly controlled. These types of incidents have taken place since the beginning of the Industrial Revolution.

On December 26, 1984 at 11:30 p.m, when the people of Bhopal, India, were preparing for sleep, a worker detected a water leak in a storage tank containing methyl isocyanate (MIC) at the Union Carbide Plant. About 40 tons of MIC poured from the tank for nearly 2 hours without any preventive measures being taken. The night winds carried the MIC into the city of Bhopal. Some estimates report 4000 people were killed, many in their sleep; and as many as 400,000 more were injured or affected.

On April 26, 1986 at Chernobyl, Ukraine, a nuclear reaction went wrong and resulted in the explosion of one of the reactors in a nuclear power plant. These reactors were constructed without containment shells. The release of radioactive material covered hundreds of thousands of square kilometers. More than 3 million people in the surrounding suburbs suffered from this disaster. While 36 people died in the accident itself, the overall death toll has been estimated at 10,000.

In another incident, on January 29, 2003, an explosion and fire destroyed the West Pharmaceutical Services plant in Kinston, North Carolina, causing six deaths, dozens of injuries, and hundreds of job losses. The facility produced rubber stoppers and other products for medical use. The investigators found that the fuel for the explosion was a fine plastic powder used in producing rubber goods. Combustible polyethylene dust accumulated above a suspended ceiling over a manufacturing area at the plant and was ignited by an unknown event (Fig. 2.1).

Furthermore, on October 29, 2003, a series of explosions killed one person, severely burned another worker, injured a third, and caused property damage to the Hayes Lemmerz manufacturing plant in Huntington, Indiana. The Hayes Lemmerz plant manufactures cast aluminum automotive wheels, and the explosions were fueled by accumulated aluminum dust, a flammable by-product of the wheel production process (Fig. 2.2).

Figure 2.1   Dust explosion kills six, destroys West Pharmaceutical Services Plant, Kinston, NC (January 29, 2003). (*Source: www. chemsafety.gov/index.cfm?*)

These examples along with others show that the causes of these incidents were not only because of ergonomic factors but also because of the failure of the equipment or some other unknown reasons. The breakdown of these incidents was probably a lack of safety measures for the plant workers and also to the nearby communities.

Figure 2.2   Aluminum dust explosions at Hayes Lemmerz Auto Wheel Plant (October 29, 2003). (*Source: www.csb.gov/ index.cfm?folder=current_investigations&page=info&INV_ ID=44*)

The significance of safety measures is indicated in the proper operation of the plant, its regular checkups, overhauling, repair and maintenance, regular inspection of moving objects, electrical appliances, switches, motors, actuators, valves, pipelines, storage tanks, reactors, boilers, and pressure gauges. At the same time, the proper training of workers for running the operations and dealing with emergencies, spills, leaks, fire breakouts, chemical handling, and electrical shock avoidance should not be ignored.

2.2 OSHA (Occupational Safety and Health Administration), and PSM (Process Safety Management)

The release of toxic, reactive, or flammable liquids and gases in processes involving highly hazardous chemicals has been reported for many years. While these major incidents involving the hazardous chemicals have drawn the attention of the public to the potentials for major catastrophes, many more incidents involving released toxic chemicals have occurred in recent years. These chemicals continue to pose a significant threat to workers at facilities that use, manufacture, and handle these materials. The continuing occurrence of incidents has provided the impetus for authorities worldwide to develop or consider legislation and regulations directed toward eliminating or minimizing the potential for such events.

One such effort was the approval of the Sevaso Directive (Italy) by the European Economic Community after several large-scale incidents occurred in the 1970s. This directive addressed the major accident hazards of certain industrial activities in an effort to control those activities that could give rise to major accidents, as well as to protect the environment, human safety, and health. Subsequently, the World Bank developed guidelines for identifying, analyzing, and controlling major hazard installations in developing countries and a hazardous assessment manual that provides measures to control major fatal accidents.

By 1985, in the United States, the U.S. Congress, federal agencies, industry, and unions became actively concerned and involved in protecting the public and the environment from major chemical accidents involving highly hazardous chemicals. In response to the potential for catastrophic releases, the Environmental Protection Agency (EPA) was seriously involved in community planning and preparation against the serious release of hazardous materials.

Soon after the Bhopal incident, the Occupational Safety and Health Administration (OSHA) determined the necessity of investigating the general standards of the chemical industry and its process hazards, specifically the measures in place for employee protection from large releases of hazardous chemicals.

OSHA has introduced certain standards regarding hazardous materials, flammable liquids, compressed and liquefied petroleum gases, explosives, and fireworks. The flammable liquids and compressed and liquefied petroleum gas standards were designed to emphasize the specifications for equipment to protect employees from other hazardous situations arising from the use of highly hazardous chemicals. In certain industrial processes, standards do exist for preventing employee exposure to certain specific toxic substances. They focus on routine and daily exposure emergencies, such as spills, and precautions to prevent large accidental releases.

Unions representing employees who are immediately exposed to danger from processes using highly hazardous chemicals have demonstrated a great deal of interest and activity in controlling the major chemical accidents. The International Confederation of Free Trade Unions (ICFTU) and the International Federation of Chemical, Energy and General Workers' Union have issued a special report on safety measures.

The objectives of the process safety management of highly hazardous chemicals were to prevent the unwanted release of hazardous chemicals, especially into locations that could expose employees and others to serious harm. An effective process safety management requires a systematic approach to evaluating the whole process. The process design, process technology, operational and maintenance activities and procedures, nonroutine activities and procedures, emergency preparedness plans and procedures, training programs, and other elements that have an impact on the process are all considered in the evaluation. The various lines of defense that have been incorporated into the design and operation of the process to prevent or mitigate the release of hazardous chemicals need to be evaluated and strengthened to assure their effectiveness at each level. Process safety management is the proactive identification, evaluation and mitigation, or prevention of chemical releases that could occur as a result of failure in the procedures or equipment used in the process.

These standards also target highly hazardous chemicals and radioactive substances that have the potential to cause catastrophic incidents. This standard as a whole is to help employees in their efforts to prevent or mitigate the episodic chemical releases that could lead to a catastrophe in the workplace, and the possibility of the surrounding community to control these types of hazards. Employers must develop the necessary expertise, experience, judgment, and proactive initiative within their workforce to properly implement and maintain an effective process safety management program as envisioned in the OSHA standards.

2.3   Incident Statistics and Financial Aspects

Normally the management of any production plant is not very concerned about the safety of employees. Moreover, it is financially reluctant to engage in extensive safety planning until and unless it is very imperative or is required by some monitoring agencies that inspect the safety procedures and facilities. The situation is worse in the third world countries. There is a need to develop a culture in an organization that is safety and health oriented. The duty of the supervisors or safety managers is to realize the need for safety measures in terms of financial loss to the producer. It can be highlighted for management by bringing the information on the loss of working hours, employee injuries, property damage, fires, machinery breakdown, public liabilities, auto accidents, product liabilities, fines, costly insurance, and such to their attention.

The varying estimates of the annual cost of industrial accidents are stated in terms of millions of dollars and are usually based on the lost time of the injured worker. This is largely an employer's loss, but is far from being the complete cost to the employer. The remaining incidental cost is four times as much as the compensation and the medical payments.

2.4   Safety Decision Hierarchy

The set of commands and actions that follow a sequence of priority to reach a conclusion is called hierarchy. Hierarchy identifies the actions to be considered in an order of effectiveness to resolve hazard and risk situations. It helps in locating a problem of risk, its analysis and approaches to avoid this risk, a plan for action, and its effects on productivity.

The different sequences of a safety plan are given in Fig. 2.3.

In the first stage of risk assessment hierarchy, identify and analyze the hazard and follow up with an assessment of the risk. The alternative approaches are carried out to eliminate the hazards and risks through system design and redesign. Sometimes the risk can be reduced by substituting less hazardous materials or by incorporating new safety devices, warning systems, warning signs, new procedures, training of employees, and by providing personnel protecting equipment. A decision is normally taken after the evaluation of the various alternatives followed by the reassessment of the plan of action.

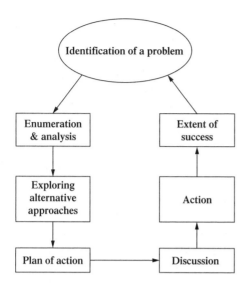

Figure 2.3 Risk assessment hierarchy.

2.5 Hazard Analysis and Risk Assessment (HARA)

The safety standards and guidelines issued from time to time are always under development regarding hazard analysis and risk assessment. The job of making a guideline becomes more difficult because of the varied nature of different industries, for example, machinery making; chemical production; manufacturing of semiconductors, pharmaceuticals, pesticides, construction materials, petroleum and refinery; and food and beverage. Each of these industries has its own hazards and risks. Therefore, it is not possible to apply a general HARA plan to all of these industries. However, this general plan can be modified for a particular process. The main features are discussed below.

- Specify the limits of the machine
- Identify the hazards and assess the risks
- Remove the hazards or limit the risks as much as possible
- Design guards and safety devices against any remaining risks
- Inform and warn the user about any residual risks of the process or machine
- Consider any necessary additional precautions

Considering all the above points, the risk management program can be started from a proper design of a machine, process, reactions, installation, operation and maintenance, and so forth.

2.6   Types of Hazards in Industries

2.6.1   Heat and temperature

In any manufacturing facility there are many sources of heat such as boilers, kilns, incinerators, evaporators, and cryogenic facilities. Extreme temperatures can lead directly to injuries of personnel and may also cause damage to the equipment. These factors can be generated by the thermal changes in the environment that lead to accidents, and therefore, indirectly to injuries and damages.

The immediate means by which temperature and heat can injure personnel is through burns that can injure the skin and muscles as well as other tissues below the skin. Continued exposure to high temperatures, humidity, or sun is a common cause of heat cramps, heat exhaustion, or heat stroke. The same degree of exposure may produce different effects, depending on the susceptibility of the person exposed.

Temperature variations affect personnel's performance. Stress generated by high temperature may degrade the performance of an employee. There are no critical boundaries of temperatures for degraded performance. Other factors that may also affect performance are the intensity of heat, duration of the exposure period, task involved, personal physical conditions, and stresses such as humidity and hot wind. There is a report indicating that the performance at high humidity is doubly lower than at high temperature. The duration of heat exposure also affects human performance. Volunteers were exposed to less than 1 hour to ambient dry bulb temperature. No significant impairment of performance by a person was observed. Long exposure to high temperature affects human performance. Other factors such as humidity and odor, fatigue and lack of sleep, smoke, dust, or temporary illness also aggravate the performance.

The effects of heat and temperature not only affect workers but also equipment and processes. For example, certain chemicals that have a low boiling point can cause an explosion at higher temperatures. In a process where these chemicals are used, they should be kept at low temperature.

The effect of excessive heat results in the degradation of the equipment by corrosion and weathering of polymer and plastic materials used in the plant. The corrosion reactions are very rapid at elevated temperatures.

The reliabilities of electronic devices are also degraded at high temperatures so that the failure of a part and thus the particular equipment becomes more frequent. The hydraulic materials or fluids generate pressures at elevated temperatures and may also cause a failure of the equipment.

The increased pressure of gas in a closed container at high temperature can cause rupture of a tank. Even a small rise in temperature of a cryogenic liquid could produce a sharp increase in vapor leading to an increase in the pressure of the container so that the container bursts.

A liquid may also expand with rise in temperature. Hence, if a tank is completely filled, the liquid will expand and overflow. An overflowing flammable liquid would then generate a severe fire hazard.

The strength of most common metals is generally reduced with increase in temperature. Most metals expand and change dimensionally on heating. This is a common cause of deformation and damage leading to the collapse of welded materials. On the other hand, reduced temperatures can cause a loss of ductility of metals and can increase their brittleness. The brittle failure of steel may seriously affect structures such as bridges causing them to collapse, ships and heavy equipment to break up, and gas transmission lines to crack. The above-mentioned facts demand a thorough inspection of the process, technical design, and regular checking of the equipment as to their safe working temperatures.

2.6.2 Pressure hazards

It is sometimes necessary to work at lower pressure to avoid serious injuries and damage. It is also commonly and mistakenly believed that injury and damage will result only from high pressures.

The damage caused by a slow-moving hurricane or wind blowing at 70 mi/h is enormous. Nevertheless, the expansive pressure exerted is in the range of 0.1 to 0.25 psi. Therefore, high pressure is a relative term. The pressures of boilers, cylinders, or compressors can be categorized in the following classes:

Low pressure	1 atmosphere (14.6 psi) to 500 psi
Medium pressure	500 to 3000 psi
High pressure	2000 to 10,000 psi
Ultra high pressure	above 10,000 psi

When the expansive force of a liquid inside a container exceeds the container's strength it will fail by rupturing. Rupturing may occur by the popping of rivets or by opening of a crack that provides a passage for fluid. When bursting is rapid and violent, the result will be destruction of the container. If employees are in the vicinity, injuries could result from impacts and from fragments. The rupture of a pressure vessel occurs when the total force that causes the rupture exceeds the vessel's strength. For example, boilers provide steam at high temperature and pressure and they are normally equipped with safety valves that permit pressures to be relieved if they exceed the set values to prevent rupturing. If the valves are not working properly, pressure from the steam may build up to a point whereby the boiler will burst.

The possibility of a rupture because of overpressurization can be minimized by providing safety valves. Possible discharges from such valves should be conducted in locations where they constitute no danger, especially if the fluid discharge is very hot, flammable, toxic, or corrosive.

Storage tanks and fermenter reactors should be pressure and temperature controlled. The high-pressure vessels should not be located near sources of heat, such as radiators, boilers, or furnaces; and if in an open area they should be covered.

Vessels containing cryogenic liquids can absorb heat from the normal environment that could cause boiling of liquids and very high pressures. Cans and other vessels used for volatile liquids should not be kept near heat or fire as they could explode violently.

The pressures in cylinders of compressed air, oxygen, or carbon dioxide are over 2000 psi. When these cylinders weigh about 200 lb, the force or thrust generated by the gas flowing through the opening when a valve breaks off a cylinder can be 20 to 50 times greater than their weight. Accidents have occurred when such cylinders were dropped or struck and the valve broke off. These cylinders sometimes took off, smashing buildings and machinery, and injuring personnel nearby. Safeguards should be used while handling, transporting, and using these cylinders.

Whipping of flexible gas lines can also generate injury and damage. A whipping line of any kind can tear through and break bones, metal, or anything else that it comes in contact with. All high-pressure lines and hoses should be restrained from possible whipping by being weighted with sand bags at short intervals, chained, clamped, or restricted by all of these means. Workers should be trained to never attempt to grab and restrain a whipping line.

A vacuum (the negative difference between atmospheric and below-atmospheric pressure) can be as damaging as the high-pressure systems. Sometimes a vacuum is more damaging to the structures that may not be built to withstand reversal stresses.

Most buildings are designed to take positive load but not to resist negative pressures. Such negative pressures might be generated on the lee side (the side opposite to the one that faces the wind) when a wind passes over. Although the actual difference is very small, the area over which the acting total negative pressure is very large so that the force involved is considerable. In most cases, the damage caused by high winds during hurricanes or tornadoes is the result of a vacuum.

The negative pressure can also be generated by the condensation of vapors that could cause a collapse of the closed containers. When vapors are cooled down to liquefy, the volume occupied by the liquid is far less than their vapors. As a result, the partial pressure inside the container decreases significantly. Vessels are designed to sustain the load imposed by the difference between the outside and inside pressures, or unless a vacuum breaker is provided.

2.6.3 Electrical hazards

The use of electricity and electrically operated equipment and appliances is so common in production and processing facilities that most persons fail to recognize the hazards involved. Electrical power is beneficial and at the same time hazardous if not properly used. The hazards involved are mainly:

1. Shock to personnel

2. Short circuiting and overheating

3. Ignition of combustible materials

4. Electrical explosions

5. Inadvertent activation of equipment

6. Electromagnetic effects on equipment and personnel

Electrical shock is initiated when a person comes into contact with a bare electrical wire and the current starts flowing through the body. This shock is a sudden and accidental stimulation of the nervous system by an electrical current. Although the potential difference determines the current flow through the body, the damaging factor and the chief source of injury and death in electrical shock is the current flow. Currents in the range of 1 to 75 mA is not damaging but above this range can be fatal.

There are many ways for a person to be shocked electrically including contact with a normally bare energized conductor or a conductor with deteriorated insulation, the equipment failure that causes an open and short circuit, static electrical discharge, and by lightning strike.

Accidents are frequent when a person is electrocuted because of lack of care near the energized bare conductor, the construction area, rooftops, or TV antennas, or working on live high-voltage lines.

Accidents may occur if a circuit is opened when an electrician begins work or if a person reenergizes the circuit by mistake. Electrical circuits shut down for repair or maintenance should be locked and tagged out after being deenergized. The circuit that uses capacitors should be discharged first by grounding. Line equipment is normally insulated, but with time the insulation deteriorates owing to many factors such as heat, elevated temperature, moisture and humidity, oxidation of insulators, chemical incompatibility, mechanical damage, high voltage, and photochemical reaction. If the insulation is defective owing to deterioration or damage, a person could be electrocuted.

Equipment failure is another cause of electrical shock. Some examples include leakage in washing machines, electrical irons, water pumps, broken energized power lines, grinding, and drilling machines. The equipment must be grounded with three wire cables.

The shock protection by these sources can be implemented in the following ways: enhanced insulation of wires and equipment, and insulation of a person who is working on a power line. Electrical equipment can be isolated. These should be properly marked by warning signs of high voltage and electrical shock.

Static charge is another hazard for electrical shock. Every moving solid, liquid, or gas carries a charge on it. Whenever there is an excess or deficiency of electrons on moving objects, it causes a potential difference between them. This is capacitive in nature because whenever two objects of different charges come close, they generate an electrical discharge. For example, a person moving on a carpet or a conveyer carrying materials that may generate static electric charge can cause a simple electric shock.

There are ways of controlling static problems. The person working in an oil refinery or in a gas station can be asked to wear cotton clothes instead of nylon or wool. A material that does not generate static electric charge can be coated on pipes and other equipment. Equipment can be sprayed by a conducting material to avoid the charge generation. Electroneutralization can generate high voltage. As a result, a gas ionizes and produces positive and negative charge species that combines opposite charges and neutralizes them. Raising the humidity above 65 percent permits the static charge to load off and dissipate.

Lightning is a massive, natural discharge of static electricity involving very high potential and high current flow. Lightning follows the path of least resistance to earth including high mountains, tall trees, T.V. antennas, light arrestor, and rods. Ground provides the path.

Lightning rods, multiple-point discharge rods, and lightning warnings are now used as protective devices. Lightning rods are placed so that their upper ends are higher than any nearby structure. Grounds are low-resistance paths to provide easy passage of current to earth. Multiple-point discharge dissipates the accumulated charges to a wider area to protect the electrical circuits and all metal equipment in a building or structure from direct passage of lightning. The lightning warning devices can detect lightning in a vast area and can be coupled with protection units. All overhead power lines are equipped with these lightning warning devices.

Keeping sparks and arc away from combustible materials or chemicals can provide protection from electrical hazards. It is also advisable to eliminate all electrical equipment from hazardous areas in which a flammable atmosphere might exist. It can also be achieved by designing inherently safe devices, explosion proof equipment with heating and overheating control, fuses, circuit breakers, reset relays, and other protection units.

2.6.4 Mechanical hazards

Most of the injuries in industrial plants are originally from mechanical causes. These industrial plants have belt-driven rotating equipment, open geared power-presses, power hammers, cutter conveyers, kilns, and incinerators. These different kinds of mechanical equipment are used in industrial plants and each has its own mechanical hazards including cutting, tearing, and breaking.

A person working in a paper plant at a manually fed paper cutter may have chances of cutting skin or body parts. Tearing of skin may occur when a sharp point or edge pierces the skin and flesh. The sharp edges of equipment and poor finishes are sometimes major causes of cutting. The equipment must be designed in such a way that it does not have sharp edges and poor finishes.

Shearing will occur when a sharp edge is in a linear motion in a direction vertical to the line of the edge. Examples include powered paper cutter and metal plates. The effect of shear can cause amputation to a person working at the machines, and can be fatal.

An impactor can crush the muscle tissues or any part of the human anatomy. Sometimes two rotating objects can cause crushing of body parts when they are moving toward each other. Common examples include meshing gears, belts running over pulleys, cables on drums, chains on sprockets, rollers on manual type washing machines, and rolls on rubber mills or paper calendars.

When a part of the body is caught between two hard surfaces it can cause a bone-shattering effect. Sometimes if an attempt is made to bend a rigid bone, a break may occur. Breakage of fragile material occurs when these are dropped or thrown on a hard surface violently.

Normally a guard is installed on a moving part of the machine, which acts as a barrier to prevent the entry of any part of the human body in the hazardous area. It is also possible that a safety device is installed that prevents or interrupts the operation if part of the operator's body is in a hazardous area or requires its withdrawal prior to machine operation. The guard or safety device itself must not constitute a hazard, must be safe, low maintenance, easy-to-use, automatically controlled, or fixed on the machines. There are different types of guards and safety devices available according to the design and demands of machines.

Total enclosure is represented by fixed covers over the pulleys, gears, shafts, and couplings to prevent access to the hazardous area. They can also be coupled with interlock devices for shutting down the machine if a portion or the whole cover is removed.

Moveable barriers or gates can also be provided that open and close easily for loading and unloading of materials. Double control devices that are operated by dual switches far apart cannot be operated by a single hand.

Mechanical feed is provided by a mechanical feeder, in which a processing material is placed over a feeding device. It moves automatically in a processing zone from which the part is ejected. There are certain safety devices such as optical sensors that monitor the light intensity of a reference source. The variation of light intensity owing to the presence of a person or a part of the body in a hazardous area prevents the activation of the machine. This can also be achieved by ultrasonic or piezoelectric detectors that produce high-frequency inaudible waves to detect the presence of any moving object in a hazardous area of the machine.

Electric field transducers can also be used. They generate a capacitive field in a hazard area. Any grounded object in the field can be detected. Operators working in that area are grounded and can be sensed by this method to stop the machine before its activation. These are different guards and safety devices that are normally used to avoid mechanical hazards.

2.6.5 Toxic materials

Many incidents caused by the release of chemicals into the environment have resulted in the loss of life and property. Fear of toxic chemicals has increased because of these incidents. The increased awareness of industrial plant workers and the general public has resulted in minimization of these releases. Highly reactive chemicals are being used more frequently in industries, agriculture, research, and defense. Many of these chemicals are found to be carcinogenic, teratogenic, and a cause of long-lasting injuries.

It is therefore necessary to provide suitable safeguards to prevent or minimize the injuries that can occur to workers in industrial plants and to the general public. There is a need to understand the ways by which these chemicals enter the human body and their physiological effects. Preventive measures should be exercised to avoid this absorption.

A material is considered toxic when a small quantity injures the body of an organism. Almost all materials are injurious to health but at different levels. The oxygen we breathe can be dangerous if taken at 100 percent without dilution. Nitrogen and carbon dioxide can be dangerous although they are present in air and lungs at high concentrations.

The concentration or the toxicity level of the substance is not the only factor of a toxic chemical. The susceptibility of the human body to toxic chemicals and their concentrations varies. Other factors that affect the severity of the injury are concentration, duration of exposure, route, and temperature.

Toxic injuries can occur at the first point of contact between the toxicant and the body, or later, systemic injuries to various organs of the body. The routes of these injuries can be through the skin, respiratory system, or the gastrointestinal tract.

The toxic materials may be solid, liquid, or gas. The solid toxic materials are radioactive substances and metals such as Pd, Cd, As, Cr, Al, and others in various forms. The chemicals are mostly in liquid and gaseous forms. For example, diethyl bromide, chlorofluoro carbons (CFCs), trichlorethane, or trichloromethane are liquids whereas phosgene, chlorine, carbon monoxide, hydrogen cyanide, and isocyanate are gases.

What happens in an industrial plant when a leak of some toxic gases such as isocyanate, ethane, or others occurs? The concentration of these gases in air increases whereas the concentration of oxygen decreases. The worker feels suffocation or asphyxia. The concentration of carbon dioxide increases; as a result, blood carbonic acid level increases, which lowers the concentration of oxygen further. The worker undergoes a condition of hypoxia (hypo: below; oxia: oxygen). The effect of hypoxia includes loss of perception, decrease in brain activity, unconsciousness, and deep breathing. It may lead to irreversible damage to the brain, paralysis, and ultimately death. Some gases alter the oxygen-carrying cells in the blood (hemoglobin). For example, the exposure to carbon monoxide (1 to 1.5 percent) decreases the oxygen-carrying capacity of blood that results in hypoxia. Some chemicals such as nitrates, nitrites, or other oxidizing agents are also harmful to the human body. Other chemicals are irritating and cause serious injuries to the body by inflaming the tissues. It may also cause inflammation of the skin, eyes, and respiratory tracts. Even a small amount of irritant can cause physiological injury to an extensive area of tissue. These may be chemical, gas, liquid, or thin particulate matter. Ammonia, acrolein, hydrazine and hydrofluoric acid, fluorosilicic acid, and asbestos can cause injuries to the upper respiratory tract, whereas chlorine, fluorine, ozone, nitric acid, and nitrogen tetroxide affect the lower portion and the alveoli.

Some chemicals are carcinogenic (cancer producing). For example, bitumen, mineral oil, aromatic compounds, vinyl chloride, benzidene, and biphenyl pyridine are the known carcinogens and their use is eliminated or replaced by noncarcinogenic chemicals. Asbestos is a particulate matter that causes asbestosis and cancer of the lungs, colon, rectum, and stomach. Therefore, OSHA has imposed a ban for zero fiber or particulate matter in the working environment.

All industrial plants are obligated to observe criteria stipulated in OSHA standards that include the exposure to different chemicals and their threshold limit for industrial workers. The preventive measures in an industrial plant depend on the type of processes involved.

Protective equipment must be used for protection from toxic gases and vapors and are required for normal hazardous operations such as working in a spray-painting plant, production and use of toxic chemicals, and fumigant use. Safe respiratory protective equipment is required for all these activities.

There are two types of respiratory protective equipments:

1. *Air purifier*: Contaminated air is purified by chemical or mechanical means. The air containing oxygen, particulate matter, gases, and vapors is first passed through a filter that removes the particles. This air is then passed through a reaction chamber that contains chemicals used for purification. For example, the removal of organic vapors and acidic gases, ammonia, carbon monoxide, and carbon dioxide is done over charcoal, silica gel, hopocalite (MnO_2:CuO [60:40%]), and soda lime, respectively.

2. *Oxygen-breathing apparatus*: The portable equipment that supplies oxygen for respiratory needs is called an oxygen-breathing apparatus. There are many types of equipment available depending on the composition of air quality supplied. They mainly consist of air or oxygen supply, face piece or helmet, and tubing for air and supply gas regulator.

The regulator controls the pressure of gas required by the user. It can supply air on a continuous or pressure demand basis. The source of air is a compressed air or liquid. They may be in closed or open circuit to reuse the air in the former case. These self-contained air breathing units have chemicals capable of generating oxygen. These are the units used for normal operations and for emergencies to protect personnel.

Special protective clothing should be provided to working personnel for protection from toxic chemicals. The clothing is made from materials resistant to acids, bases, toxic chemicals, and even to high temperatures and fire.

In an operational plant there is a need to mark the container containing chemicals with proper labeling. These chemicals and hazards have been categorized into different classes. Different colors were assigned depending on their physical or chemical hazards as shown in Table 2.1.

TABLE 2.1   Classes of Hazard Materials and Their DOT Symbols

Hazard class	Color	Symbols
Class 1: Explosives	Orange	Exploding device
Class 2: Gases	Yellow	Burning "O"
	Red	Flame
	White	Skull and cross bones
	Green	Cylinder
Class 3: Flammable liquids	Red	Flame
Class 4: Flammable solids	Red/white stripes	Flame
	Red/white/field	Flame
	Blue	Flame
Class5: Oxidizers/ organic peroxide	Yellow	Burning "O"
Class 6: Poisons/ etiologic agent	White	Skull and cross bones
	White	Sheaf of wheat with cross
	White	Broken circles
Class 7: Radioactive	Yellow/white field	Trefoil/spinning propeller
Class 8: Corrosive	Black/white field	Melting metal bar and hand
Class 9: Miscellaneous	Black stripes, white field	Black and white stripes

According to this classification an inflammable liquid or solid chemical is given a number designating its class, and a red color that indicates its physical or chemical hazard such as flammability. For toxic, corrosive, explosive radioactive material a container should be marked with different numbers and colors (Fig. 2.4).

Personnel should be informed and trained on the significance of these numbers and colors and how to handle these chemicals to avoid any incident. Clear information should be given on the pressure in a line carrying any chemical, inflammable or toxic, and at what temperature these chemicals should be pumped. Do they radiate or explode on absorbing moisture or oxygen from air? These are the technicalities that should be in the mind of personnel who are working with these chemicals.

2.6.6   Fire and explosion

Fire and explosion are common in many chemical industries. There are chances of fire breaking out in an operational plant. A fuel, an oxidizer, and a source of ignition are required to start a fire. However, fire and explosion take place only when there are appropriate conditions for it.

Many types of fuel and oxidizers are available in any industry. There are three types of fuel. They are mainly solids, liquids, or gases.

These fuels may be required for heating boilers, running engines, and for welding. Also the chemicals that are used as cleaning agents or solvents act as fuels. Lubricants, coatings, paints, industrial chemicals,

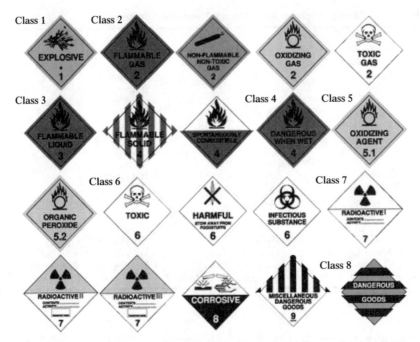

Figure 2.4 Symbols as recommended by the Department of Transportation (DOT).

refrigerants, hydraulic fluids, polymer plastics, and paper wood cartons are potential fuels.

The next element for fire is an oxidizer. The most common oxidizer is oxygen present in the air that helps in oxidizing the fuel. Sometimes a chemical can be self-ignited in the presence of an oxidizer. For example, white phosphorus catches fire as soon as it comes in contact with air. Pure oxygen is a strong oxidizer. A small leak in an oxygen cylinder may cause a fire hazard.

Fluorine is another strong oxidizer. It can react with moisture in air and catch fire. It is normally used diluted with nitrogen. Other oxidizers include chlorine, halogenated compounds, nitrates, nitrites, peroxides, and acids. These oxidizers should be handled with care and their contact with fuel should be avoided.

The source of ignition consists of materials that may initiate a fire on friction. A reaction is initiated in a mixture of fuel and oxidant. As a result of this reaction, heat is evolved in the form of flame or light that produces a fire after reaction with fuel and oxidizer. The igniter may be sunlight, an arc, or an electrical spark.

The common sources of electrical ignition in an industrial plant are the sparks of the electric motors, generators, or electrical short circuits, arcing between contacts of electrical switches or relays, discharges of charged electrical capacitors, or a discharge of static electricity accumulated on underground surfaces.

The sources of other igniters are hot plates, hot moving parts of some instruments, engines, radiators, overheated wiring, boilers, metals heated by friction, metal being welded, or sometimes a cigarette.

Fire can have a tremendous effect on human life, immediate surroundings, and even on the environment. Fire produces carbon monoxide, carbon dioxide, solid carbon particles, and smoke. Heat and high temperature make a fire highly dangerous for the employees of any industry. Death may occur as the concentration of the oxygen in air decreases in case of fire. Therefore, personnel are advised to escape before the fire expands and the temperature rises beyond 65°C.

In any industrial plant, there are devices installed to detect any kind of fire, smoke, soot, or heat. There are fire detection instruments including thermosensitive switches, thermoconductive detectors, radiant energy detectors, gas detectors, or ionization detectors.

Suppression of the increasing fire can be carried out by various methods. The very first method is to cut the supply of fuel to the fire. Fire suppression can also be achieved by blanketing a fire or by covering it with inert solid, foam, thickened water, or covering it with a nonflammable gas such as CO_2. The other available method is the dilution of the fuel, if it is a liquid fuel, by adding noncombustible liquid into it; and if it is a gas, by adding nonflammable gas.

Fire is a chain process. It can be stopped by breaking this chain. Scavengers are used to stop the free radical chain reactions and subsequently fire is extinguished. Halogenated compounds are usually good chain-reaction inhibitors.

When fire is ignited because of fuel and there is no electrical hazard nearby, water is used as a fire suppressant. This is readily available, cheap, simple to use, and effective. Normally firefighters use stream water on fuel and fire. However, water is not recommended for sodium or magnesium metals.

Water can also be used as a diluent and to stop chain reactions. The only limitation is that its effective range is very low. Sometimes thickening agents are added to the water to increase the residence time of water and its effectiveness. The thickening agents such as clays, gums, and sodium and calcium borates are used in forest fires. They act as slurries and adhere to the burning materials. The chlorides of calcium and lithium lower the freezing point to −40°C. The salts of potassium carbonate deposited on burning materials or the gas produced act as fire inhibitors.

Gas extinguishers may be used for enclosed spaces. These are largely meant for small fires or fires where electrical hazards are probable.

Carbon dioxide is widely used as a fire extinguisher. Its main function is blanketing the fire, thus lowering oxygen concentration and subsequently inhibiting the fire. It also acts as a coolant and a combustion inhibitor. When carbon dioxide is sprayed on fire it emerges as snow and lowers the temperature.

Halogenated hydrocarbons act solely by inhibiting chain reactions. The nature of halogens is very important. The least reactive would be the best fire extinguisher. However, the problem with these halogenated compounds is their toxicity, which limits their use.

Foams are also used as fire suppressants. They suppress fire by cooling, blanketing, and sealing the burning fuel from the surrounding atmosphere. They are not suitable for gaseous fuel and fuel that reacts with water.

Solid extinguishers such as sand or clay are also used to cover the oil or grease under a fire. They also suppress fire by blanketing. They are suitable for metal fires. Sodium and potassium bicarbonate are also used as solid extinguishers for liquid fuel. They act as chain reaction inhibitors. At high temperatures, they decompose to give carbon dioxide that itself is an extinguisher that suppresses fire.

The use of certain suppressants under wrong conditions may be hazardous. Water cannot be used on burning cables carrying electricity or magnesium metal.

Fire extinguishers that work automatically are available. They sense temperature, gas, or fumes and start sprinkling the extinguishing materials (CO_2 or others). There are other portable units available that are marked, A, B, or C depending on the class of fires to be extinguished. Mobile extinguishers are too heavy to be carried and therefore are often wheel-mounted. These contain potassium bicarbonate (purple-K), dry chemical, and other light water. The advantage is in their high capacity to suppress fires.

2.6.7 Accelerator and falling objects

Most of the incidents that occur in an industrial plant are because of accelerator fall or falling objects. Data have shown that nonfatal occupational injuries and illnesses involving the days away from work are more than 60 percent of the total accidents. These may be a result of getting struck by an object falling to the same or lower level. These great numbers of accidents have led to federal and state laws for corrective measures, such as provision of safeguards, safety nets, and helmets for workers. It was observed that a good number of workers fell down from heights in the fields of construction, cleaning of chimneys, and towers. Injuries also occurred when workers slipped and fell, while working on the same level. The fall may not be fatal in this case. Workers have been killed when they have struck their heads by falling from upright positions on slippery floors. The most serious damage from all of these falls is broken bones of head (skull), arms, legs, and chest. The ability of the human body to sustain an impact, such as a fall, depends on three major factors: velocity of an initial impact, magnitude of the deceleration, and orientation of the body on impact. At a free fall from a height of 11 ft, the velocity gained by the body is 18 mi/h, enough to kill a man.

During the construction and maintenance of bridges or elevated structures, numerous falls of industrial workers into water occurred. These falls resulted in various kinds of injuries such as spinal injuries, bleeding of lungs, shock, and sometimes death.

The main task is to determine the measures that should be taken to prevent these kinds of accidents. The best way to prevent a fall is by providing safeguards. Workers working at an elevation should be provided a safeguard net and fences. They may be tied with ropes as well. Their mental and physical fitness should be checked regularly to determine whether they can work at elevations and can sustain vertigo (a dizzy, confused state of mind).

A person may fall down on the floor at the same level by slipping while working or walking briskly. A person may fall because of the collapse of a piece of equipment, ladder, structural support, or hoist on which he is working. Preventive measures should be adopted while working at these places.

Workers who are not properly trained should not be allowed to work on elevated sites. A worker should be chosen for work on bridges and elevated structures depending on psychological and physiological states. Workers can be provided with emergency nets, coiled knotted ropes, ladders, fire escapes, and parachutes.

Sometimes very small objects are more damaging than bigger objects. For example, a small object thrown at a higher speed is more dangerous than bigger one. This happens when there is an explosion of gas cylinders, high pressure tanks, or gas pumps. Furthermore, the debris or fragments may travel at a very high speed and can cause bruises, tissue damage, or bone fractures. Different body parts, for example eyes, are more susceptible to an impact. While welding, grinding, tooling, spraying, or coating spray pressure, glasses should be used. These and other acceleratory effects in an industrial plant or construction site can be avoided by taking preventive measures for workers.

2.6.8 Confined space

The danger associated with working in confined spaces is not new. Since the discovery of mines, many fatalities have been reported owing to suffocation, gas poisoning, accumulated gas explosion, and asphyxiation. Workers dealing with wastewater sewage repair, cleaning, inspection, painting, and fumigation face the problems of asphyxiation, drowning, and toxicity from chemical exposure because of working in confined spaces.

A space large enough for an employee to enter and work with restricted activities or movement may have a hazardous atmosphere. The incident occurs because of failure of recognizing the hazards associated with confined spaces. The different kinds of confined spaces for a worker in a plant are tanks, silos, storage bins, vessels, hoppers, pits, and sewer lines. Big fermenters, multieffective evaporators, boilers, and wells are also included in this list.

There is another criterion called permit-required confined space such as an engulfment, entrapment, or any other recognized serious safety or health hazards.

The permit-required confined space that has a hazardous atmosphere includes chemical sludge; sewage; flammable gases or vapors; combustible, low-oxygen concentration; and higher carbon monoxide and carbon dioxide concentrations. Any recognizable environment and condition that can cause death, incapacitation, impairment of ability to rescue, injury, or acute illness is a permit-required confined space. The confined space may have a liquid, or finely divided solid substance that can be aspirated to cause the plugging of the respiratory system, or exert enough force to cause death by strangulation, constriction, or crushing.

Sometimes in a confined space the internal configuration or shape is built to have inwardly converging walls or a floor that slopes downward and tapers to a smaller cross section that could trap an entrant or contribute to asphyxiation. This is designed as a permit-required confined spaced. Examples are fermenter and digester.

The space that contains any other recognizable serious safety or health hazards is also a permit-required confined space. These hazards may be physical, electrical, mechanical, chemical, biological, radiation, temperature extremes, and structural hazards.

The atmospheric hazards are due to the presence and absence of certain gases and the presence of flammable and toxic vapors. There are three types of confined spaces:

Class A: Immediately dangerous to life that contains oxygen: 16 percent or less or greater than 25 percent and flammability of more than 20 percent and the toxicity is very high.

Class B: Dangerous but not immediately life threatening, having oxygen greater than 16 to 19.4 percent and from 21.5 to 25 percent, flammability of 10 to 19 percent and the toxicity is greater than the contamination level.

Class C: Potentially hazardous to life having oxygen 19.5 to 21.4 percent, flammability lesser than 10 percent, and the toxicity is less than the contamination level.

Physical hazards are owing to mechanical, electrical, engulfment, noise, and the size of ingress and egress-opening.

The activation of mechanical and electrical equipment, agitators, blenders, stirrer fans, pumps, and presses can cause injury to workers in confined spaces.

The chemical release into a confined space is life threatening. High-pressure liquid, falling objects, and slippery surfaces in a confined space are all potential hazards. The limited space, inadequate ventilation and light, and excessive noise are also physical hazards that increase the confined space hazards. The chemical waste and useful chemicals are also life threatening.

While working in waste streams, pools, ponds, sludge pits, sewers, or fermenters a worker is exposed to infectious microorganisms. The industrial processes that grow these infectious microorganisms for beneficial purposes can be a threat to workers in a confined space.

There should be a thorough program for confined-space working. The main points of a program are as follows:

Identifying and evaluating with respect to hazards of all confined spaces at the facility

Posting a warning sign at the entrance of all identified spaces

Performing a job safety analysis for each task at confined spaces, for example entry plan, assigned standby persons, communication between workers, rescue procedures, and specified work procedures

Testing and monitoring air quality in the confined space such as oxygen level, toxicity level, flammable materials, air pressure, and air contaminants

Preparing a confined space; for example, by isolation, lockout, tag out, purging, cleaning, and ventilation, and procuring special equipments and tools if required

The use of personnel protective equipment to protect eyes, ears, hands, feet, body, chest, and respiratory protection, harness, and mechanical lift devices

In addition to the above points, training and drill for workers, supervisors, standby personnel, and rescuers at regular intervals are absolutely needed.

2.6.9 Radiation

Since the discovery of radioactivity, some elements are classified as dangerous even if they are used for beneficial purposes. Energy is emitted by any material that travels in the form of particles or electromagnetic waves. Energy emitted by the sun reaching the earth travels in the form of electromagnetic waves and particles. Light comprises a spectrum of wavelengths that consist of high-energy cosmic rays, ultraviolet rays, visible light and low energy, infrared rays, and micro and radio waves.

The radioactive elements consist of alpha particles (helium nuclei), beta particles (positron), neutron, and gamma rays. X-rays are also emitted by elements when high-energy electrons strike a metal. The high energy of X-rays and gamma rays make them more penetrating. Beta rays have less energy than gamma rays and hence less penetration. Alpha, beta, X-rays, and gamma rays are ionizing radiation. These may cause injury by producing ionization of cellular components leading to functional changes in the tissues of the body. The energy of these radiations is great enough to cause ionization of atoms that make up the cells, producing ion pairs, free radicals, and oxidation products. The damage to the cell is mostly irreversible. These radiations have certain hazard limits in causing damage to the cell. Therefore, they are also used for diagnostic, beneficial purposes, and for the treatment of cancer cells. Radioactivity does not lose its potency by absorption or ingestion by living tissues. Thus, the radioactive material from airborne fallout on land or on grass taken up by grazing cattle ultimately passes on to human beings.

X-rays, gamma, and cosmic rays are similar except for the fact that gamma and cosmic rays are natural. They ionize matter by photoelectric effect, Compton effect, and pair production (electron and positron). These radiations are of very high energy and therefore more penetrating. They cause injury to the tissues of the whole body. Therefore, they are more damaging to the living tissues.

There are certain factors that affect the exposure and risk. These are the strengths of the source, type of radiation, and the distance. The energy order with respect to decreasing hazards is cosmic, gamma, X-rays > beta > α-particles.

The sources of ionizing radiation are nuclear power plant, nuclear material processing, and radionuclide generation for nondestructive purposes. Medical and chemical laboratories use these radionuclides—for example, iodine, thallium, and barium—as tracers. The danger of mishandling these materials could cause release of these materials into the environment. Other than medical diagnostic tests for fracture of bones and constriction of blood vessels, these are used for the treatment of cancers.

The industrial use comprises examination of welds; internal structures for the existence of cracks, voids, or contaminants; food preservations, and examination of packages and baggage for illegal articles, especially at airports.

During the last decade various nuclear power plant (NPP) accidents have made the construction and use of NPP more difficult. Among them, the Chernobyl accident was the most severe—causing damage to vegetation, animals, and property, over an area of 1000 square kilometers, and taking 36 lives immediately; but after a decade the death tolls have risen to tens of thousands. Workers engaged in the milling of uranium are also the most exposed to α-particles that can be avoided by protective clothing; however, the presence of radon gas, owing to the decay of uranium, is more dangerous.

After fission of uranium 235, the radionuclides produced in the spent fuel have cesium, strontium, iodine, and other radionuclides of very long half-lives that can be a danger. The other radio wastes include contaminated filters, wiping rags, solvents, protective clothes, hand tools, instruments and instrument parts, vials, needles, test tubes, and animal carcasses.

Precautionary and preventive measures include:

Well-trained personnel should be allowed to work, use, operate, handle, and transport the material

Safety engineers should inspect any facility producing radiation, its protective devices, and worker's protection prior to start

Access to these areas should be restricted and only an authorized person should be allowed

Suitable warning signs should be posted in the ionization equipment area

Emergency drills should be performed regularly

All instruments that use radioactive sources should be kept in a shielded enclosure and made up of lead-containing glasses, sheets, and bricks that attenuate the radiation to a permissible level: radiation going outside the area should be continuously monitored

Every personnel should be given a dosimeter or film to estimate the absorbed radiation and a record should be maintained

Keep the exposure time for personnel as low as possible

The vital parts of the body should be protected by protective clothing, glasses, gloves, masks, and shoes

Drinking, eating, and smoking in that area should be prohibited

Cleanup of any spill should be performed with the help of safety engineers that includes complete prevention of the spread, complete cleaning of the spilled area, and a thorough decontamination of the contaminated personnel

The nonionizing relations are ultraviolet, visible, infrared, and microwave. Ultraviolet radiation is the most dangerous. It is a high-energy radiation that comes from the sun naturally and is generated by human beings by electric arc wielding, Tesla lamps, plasma arc, and lasers for beneficial purposes. The danger of ultraviolet light is that it causes burning of the skin and blindness to the eye. The redness of the skin is often observed in the sun. It is as a result of the penetration of ultraviolet radiation to the dermis. The cornea and conjunctiva of the eyes absorb UV rays and become bloodshot and irritated.

The laser radiates various kinds of radiations from infrared to visible and ultraviolet. These are coherent rays with very high, focused energy. This power can be very dangerous to the human eye or skin if not used properly.

To avoid these radiations, glasses with a face UV blocker and protective coats should be worn. Goggles made of glass containing iron are more absorbing than simple glass, while quartz is nonabsorbing to UV radiation.

Visible radiations are less harming. Simple protection from visible light is beneficial. Infrared radiations are heat radiations. Any heated body radiates heat radiation that gives off thermal energy. These radiations can cause cataract to the eyes, skin burns, increased perspiration, and loss of body salt. The workers at an iron or metal casting plant, hearth, and furnace should be provided with clothing, gloves, and facemasks to protect skin against infrared radiation. Adequate cooling should be provided to personnel working near the infrared sources. It is also recommended to provide adequate salt and water in the form of juices or salt tablets to replenish the salt lost through perspiration.

Microwave radiation is emitted by dryers, ovens, and heaters normally used in the home. The high-power radars used for military purposes, communication equipment, alarm systems, and signal generators are other sources of microwave radiation. The low-power microwave radiation can cause heating and skin redness whereas high-power microwave radiation can cause inductive heating of metals and induced currents that can produce electric spark. Containers with flammable materials may catch fire if they are placed in the microwave fields. Rings, watches, metal bands, keys, and similar objects worn or carried by a person in such a field can be heated until they burn the bearers.

High-power microwave antennae should not be inspected when energized or directed toward inhabited areas. Flammable materials stored in metallic containers should not be left in microwave-induced magnetic fields. A warning device should be provided to microwave equipment to indicate when it is radiating.

Radio frequency (RF) is another kind of radiation that is used in radio, television, satellite, and mobile communications. The frequency radiated by these generators ranges from 3 KHz to 300 GHz.

The increasing use of mobile phones may have resulted in cases of brain cancer. Experiments are under way to assess the damage caused by mobile phones to the brain. Experiments are also in progress to assess the safe range of the broad spectrum of radio waves. The most restrictive limits occur between 30 and 300 MHz where absorption of RF energy by the whole body is most efficient.

2.6.10 Noise and vibrations

Vibrations, sound, and noise are other examples of common industrial hazards. The most common injury because of vibration is sound-induced hearing loss. The vibrations of machines, high-speed pumps, generators, boilers, and conveyers produce unwanted sound *noise*. The adverse effects produced by these sounds are as follows:

Loss of hearing sensitivity

Immediate physical damage (ruptured ear drum)

Interference resulting in the masking of other sounds

Destruction

Annoyance

Other disorders such as tension and mental fatigue

A normal human can hear a sound ranging from 20 to 20,000 Hz. Less than normal ability to hear a voice indicates there has been degradation. Hearing loss is an impairment that interferes with the reception of sound and with the understanding of speech in sentence form. The general loss of hearing sound in the frequency range of 200 to 5000 Hz is compensable under the Worker's Compensation Act. Degradation of hearing can also result from aging, long-term exposure to sounds of even moderately high levels, or a very high-intensity noise. Much of this degradation with age may be owing to continuous exposure to environmental noise of the modern society rather than to simple aging. Hearing losses can occur even at noise levels lower than those permitted by OSHA standards as given in Table 2.2.

OSHA has estimated a safe maximum noise level of 85 dB. The time-weighted average (TWA) is an exposure for an 8-h to a noise level not exceeding 90 dB. If this level exceeds 85 dB, OSHA requires the employer to institute a hearing conservation program (HCP). Therefore, if a company wants to avoid loss claims under worker compensation laws, it must not only meet the prescribed legal standards, but also attempt to reduce noise to the lowest possible level (< 80 dB).

Noise annoys people and causes tension among people. However, the types and levels are difficult to determine. Sometimes the noise of a different sound resonates at a frequency that masks other sounds. The noise level should be checked at all places before operations begin.

Vibration not only causes noise but also other disturbances. A vibrating instrument is difficult to handle for a long time. Vibrations also

TABLE 2.2 Permissible Noise Exposures for
Workers as Described by OSHA

Duration/day (h)	Sound level (dB)
8	90
6	92
4	95
3	97
2	100
1.5	102
1	105
0.5	110
0.25	115

SOURCE: *http://www.osha.gov/pls/oshaweb/
owadisp.show_document? p_table = FEDERAL_
REGISTER&p_id = 17368.*

cause metal fatigue of the instrument that can result in failure of rotating, moving parts, and other stressed mechanical equipment. They may also result in leakage of fluid lines, pressure vessels, containers, damage to part of the equipment, and possible injury to personnel. Vibration can cause what is known as Raynaud's phenomenon that involves paleness of the skin from oxygen deficiency owing to reduction of blood flow caused by injured blood vessels and also nerve spasms. The disease is produced by vibration directly on the fingers or hands. Vibrating tools can also cause arthritis, bursitis, injury to the soft tissues of the hands, and blockage of blood vessels. In addition to hearing loss, nervousness, psychosomatic illness and inability to relax, upset balance, and disruption of sleep are other serious effects of vibration.

The hearing conservation program (HCP) includes recording and categorizing the audiometric testing, monitoring of noise exposure, use of hearing protection devices (HPD), employee's training, and noise control engineering.

Exposure monitoring is another element of HCP. The sound level and exposure time should be measured. It is very important that sound levels measured are typical of those encountered by the worker. Proper survey techniques include sound-pressure-level (SPL) meters that should be vigorously applied at monitored workplaces. They measure the smallest pressure changes initiated by the vibrating source and transmitted through the air.

There are other instruments used for measuring noise including weighted-sound-levels and octave-band analyzers. These instruments measure the noises of different frequencies.

Ear protection can be carried out naturally and by using hearing protection devices. The ear itself has a protective mechanism that helps reduce possible effects of loud noises. The sound waves do not impinge directly on the eardrum because of the curved ear canal. Eardrum muscles are very sensitive to sudden loud noises. They contract in response to these noises by causing the ossicles to stiffen, thereby dampening the vibration transmitted.

Personnel protection devices must be used to protect the ear in an industrial plant. These are earplugs made of rubber or plastic that fit snugly in the ear canal without discomfort and effectively protect the ear. They are also available as a foam cylinder that can be compressed and twisted to be inserted into the ear canal. There are helmets available that have noise attenuating electronic components and communication features. In selecting these helmets, safety engineers must exercise caution and must take steps to ensure that the devices are properly selected and used by the workers without distraction and annoyance.

2.6.11   Ergonomics

The psychological and physiological limitations and capabilities constitute the ergonomics or human factors. It is the most important part of the occupational safety and health program. This is to evaluate personnel capabilities and improve human safety, comfort, and productivity in the workplace. Work-related musculoskeletal disorders (WMSD) are the results of ergonomics and limitations of the human body to a sudden change or continuous working on a physical job, especially where most of the jobs are carried out manually.

Efforts should be made to identify workers' complaints of undue strain, localized fatigue, discomfort, or pain that does not go away after overnight rest. Job testing that requires repetitive and strenuous exertions; frequent, heavy, or overhead lifts; awkward work positions; or the use of vibrating equipment should be identified along with the WMSD risks. The extent of the problem will determine the level of effort required to provide a reasonable prevention effort. Human factors should be an important part of a company's safety and health program. Safety efforts require the involvement and commitment of management and workers. Inputs from personnel in safety and hygiene, health care, human resources, engineering, maintenance, and human factors should be the main points of safety policy.

There are three types of control: engineering control, administrative control, and use of personal protecting instruments. The design or redesign of the job changes in a workstation layout depends on the selection and use of other tools and work procedures to take account of the capabilities and limitations of the workers. Administrative control deals with the change of job, modified rules and procedures, scheduling more rest breaks, ample supply of personnel protective equipment, use of various kinds of braces to protect from stress and strain, and the rotation of workers on physically tiring jobs. In addition, the workers should be well-trained to recognize ergonomic risk factors and techniques to reduce stress and strain while working on certain instruments.

Regular health checking of the worker can help in early detection and prompt treatment for stress. Medical care should be provided for any damage to the employee. It is supposed that an employee should follow workplace safety and health rules and work practice procedures and should report early symptoms of WMSD.

2.7   Risk Management Plan

In risk management plan, we will discuss how safety personnel organize a plan to design and modify the process to avoid any incident. The use of protective equipment and its procurement will also be discussed. The need for planning for emergencies is an important task in risk management plan.

2.7.1   The role of safety personnel

Technology is changing with time. In the past, industries often had accidents owing to mechanical and electrical failure. As industry entered new fields, new safety problems subsequently arose. Generally, inventors of these new hazards were only concerned with the utility derived from the new invention rather than with an assurance of safety. New problems arose when the laboratory equipment and processes were transformed into industrial equipment, where the safety problems involved became a concern in the process design for the plant engineers. The hazard and toxicity of chemicals, high temperatures, and pressures were tackled initially by chemists and engineers. It became necessary to have other persons responsible for accident prevention. Efforts were made to prepare trained personnel to take care of the hazards related to a particular process and the precautionary steps that should be taken to avoid them.

The job of safety personnel is much diversified and is of high skill. Safety personnel must be knowledgeable in a wide range of technical, legal, and administrative activities. It is also supposed that a safety professional has in-depth knowledge in all areas of accident prevention and is capable of solving problems that may arise.

As a result of the diverse nature of the industry, their hazards, and organizational structure, management attitudes toward a safety program and government emphasis on accident prevention have created a wide diversity of safety positions, duties, and responsibilities in industrial plants.

The safety personnel should be qualified by passing certain examinations and should be a certified safety professional. Graduate engineers who have achieved this rating by showing their knowledge in safety and accident prevention can work as safety plant engineers. Other certified personnel are certified product safety professionals, certified industrial hygienists, certified professional agronomists, and certified hazard control managers. In addition, a group of consultants may be needed to review the plants and determine their compliances with OSHA and other prevailing standards of the country. A second group may be knowledgeable in specific areas, such as flammable gases, toxic chemicals, and explosives or mines. The importance of system safety engineering grew because of efforts to evaluate hazards that might be present and potential accidents that might take place with new advanced products and processes.

With the development and specialization in the safety engineering profession, priorities regarding accident prevention have changed. The protection of personnel safety comes first followed by the protection of the environment including flora and fauna of the suburb of a plant. This includes the prevention of leakage or release of liquids, oil, chemicals, detergents, or noxious gases, metals, complex deleterious substances, and even genetically modified organisms (GMO) in the environment. Protection against damage to the environment comes right after protection of personnel and animals, before prevention of damage to equipment. Priority for rescue of equipment is last.

The responsibilities of a safety engineer are increasing with an increase in specialization. The new concern of safety engineers is the area of accidental in-process damage or loss. Avoidance of such damage usually has been the responsibility of the production manager or staff. However, accident prevention principles and methodologies are being applied more and more to process control. Lack of a simple feature, protective device, pressure regulator, and auto-trip systems are increasingly being addressed. Failure of a component of a process might cause failure in the assembled product. The expertise of a safety engineer can be beneficially applied to product safety.

2.7.2 Personal protective equipment (PPE)

The most common use of personal protective equipment is for the protection of head, eyes, ears, torso, hands, and feet. This equipment helps to protect a person from damage normally encountered in an industrial plant, a construction site, or land renovation project. PPE includes devices and clothing designed to be worn or used for the protection or safety of an individual in potentially hazardous areas or performing potentially hazardous operations.

PPE should be used as a compulsory part of the safety program and should not be considered a substitute for engineering control or work practices. The basic elements of a safety program for PPE should be an in-depth evaluation of the equipment needed to protect against the hazards at the workplace. The employee should be trained in using this equipment.

The duty of the employer and safety personnel is to assess the chances and kinds of hazards that require the use of PPE.

Head protection is an important factor where injuries are caused by falling or flying objects, or working below other employees who are working with tools that could fall down. Head injuries can occur by bumping a head against a fixed object. A helmet does resist penetration and absorb the shock of a blow. The helmet consists of a hard shell and an inner lining to absorb the shock. These also help in electrical jobs or painting. There are three classes of head protecting equipment:

Class A: General service, limited voltage protection

Class B: Utility service, high-voltage helmet

Class C: Special service, no voltage

Class A is intended for protection against head injury. Class B protects from impact and penetration by falling or flying objects and from high voltage. The materials used for the helmet must be water resistant. In case of class C, the helmet is made of aluminum or other light and strong metals and should not be used where there are chances of electricity, static charge, and microwave induction. These should be provided with an air gap between the head and the helmet by headbands.

For eye and face protection, suitable protection must be worn when there is a reasonable probability of preventing injury when dangerous equipment is used. This is also true for visitors and administrative staff if they are in the hazardous areas.

The eye protective equipment includes safety glasses, chemical goggles, face shields, welding goggles, and welding face shields. Protectors must be worn in the areas where there is a potential for injury to the eyes or face from flying particles, liquid chemicals, molten metals, acids or caustic vapors, or potentially injurious light radiation. These PPE should provide protection against particular hazards for which they are designed. They should be comfortable, fit snugly without interfering with vision and movement, durable, and cleanable. They should protect from dust, splash, chipper, welder, and cutter particles.

Loss of hearing can be a result of constant noise or a sudden noise. There is no cure for hearing loss; the only "cure" is prevention of excessive noise. The equipment for noise protection is specific for a specific kind of noise. They may be earplugs, made up of rubber, plastic, foam, wool, or earmuffs. These plugs are disposable as well as nondisposable and can be reused if working properly. Earmuffs form a perfect seal around the ear to protect the bones from sounds transducing to the ear. Certain things reduce sound protection, such as chewing, facial movement, glasses, and long hair.

The primary control to protect an employee from dust, mist, fumes, gases, and toxic vapors is the engineering control, such as enclosure or confinement of the operation, general and local ventilation, and substitution of less toxic materials. In addition, appropriate respirators should be provided to protect from occupational diseases.

Medical examination of employees should be done before posting the employee to a contaminated area. Employee fit-testing should be carried out for respirator usage. The employee should make use of PPE according to the instructions. The respirator must be used for its intended purpose, such as for toxic gases, dust particles, or mist of chemicals. The respirator should be thoroughly cleaned, disinfected, and should be kept in a clean and sanitary location after every use. The employee should be instructed and trained in using this equipment. This equipment should be routinely inspected and disinfected. Respirators for emergency use, such as self-contained devices, shall be thoroughly inspected at least once a month and after each use. The employee, who is physically able to perform the work in a hazardous environment, should be assigned tasks requiring the use of respirators. Active surveillance of working area conditions and degree of employee exposure or stress shall be maintained.

There are many dangers that threaten the torso, such as heat, splash from liquid, acids, caustic and hot metals, impacts, and cuts. Protective clothing such as vests, jackets, aprons, coveralls, and full body suits should be made available. The employee working near the hearth should be given a heat-resistant coat made of leather. Rubber and rubberized fabrics, neoprene, and plastics give protection against acids and chemicals.

For protection from cuts and bruises, a special closely woven fabric (duck) as well as any lightweight protecting cloth is helpful. Disposable suits of plastic are particularly important for protection from dusty materials. For specific chemicals, the manufacturer's guide should be consulted for effective protection.

There is a wide range of gloves available. It is important to know the characteristics of the gloves relative to the specified hazard. Hand pads, sleeves, and wristlets for protection against various hazards are also available.

To protect the feet and legs from falling or rolling objects, sharp objects, molten metal, hot surfaces, and wet slippery surfaces, a worker should be provided with appropriate foot guards, safety shoes, boots, and leggings that protect them from molten metals or welding sparks and hazards.

2.7.3   Appraising plant safety and practices

Long before the designing of a plant, facts and statistics should be collected from the same kind of facilities regarding frequencies and cause of hazards and incidents. Safety engineers are in charge of designing more accident-free plants if they raise their awareness and follow effective means of accident prevention. Many engineers would like to know whether a particular prevention action would result in improvement or degradation. Also, what economic benefits can be achieved by doing this practice? For that purpose, safety agencies, insurance companies, and OSHA-type organizations made an appraisal of plant safety using number of accidents, or resulting fatalities, or injuries. The job of a safety engineer is to minimize accidents to zero level. High safety-quality plants can achieve these accident-free periods by eliminating or minimizing the existence of unsafe conditions before accidents can occur. The corrective measure should not be taken after an accident has taken place. Safety appraisal is a means to design and construct an accident-free facility by analyzing accident frequencies and severity and by taking preventive measures to remove flaws in the design.

The job of a safety engineer is to review an old or existing plant design, future and old equipment, procedures and operation, estimates, chances of hazards, and their correction to avoid accidents in a new plant design. This appraisal can be done by the following procedures:

Any plant should include the proper marking of entrances and exits according to the local codes and must be properly maintained.

The electrification areas and their location must comply with the provisions of a standard code. They should be properly marked as hazardous areas.

Firefighting equipment should be installed and maintained by regular checking at regular intervals.

Pressure vessels should be designed and tested before operation according to a stated standard.

High-energy pressure vessels should be located at a great distance to prevent possible damage in case of their failure (explosion).

Adequate workspaces should be provided between different equipment to avoid restricted movement of the employee; and there should not be any physical interference that can cause error or accident.

Personnel protective equipment should be provided for a particular job.

Ventilation and exhaust, hoods, ducts, blowers, filters, and scrubbers should be provided and kept in order, clean, and operational, to remove air particulates or toxic chemicals.

Emergency equipment and locations for their placement or storage should be provided at the nearest readily accessible locations.

Fire lanes and other routes to locations where other emergencies could occur should be provided, marked, and maintained.

The hazardous processes are isolated so they do not constitute danger to other personnel and their activities: fuel, chemicals, electric power generators, and boilers should be isolated from other facilities to avoid danger.

Numerous problems can be avoided in plants being built or modified if plans are reviewed for safety aspects before any construction or change is initiated. Some companies now require their safety personnel to review drawings for new facilities and equipment to ensure safety.

2.7.4   Planning for emergencies

An accident is unavoidable in any industry no matter how good and flawless a safety program is. Minimizing the factors that are responsible for hazards are more important than minimizing the effects of an accident.

In any accident, the well-being of workers should be normally and morally the prime consideration. The effects of any accident can be minimized by providing emergency relief (rescue) in the shortest time possible to the victim. Normally, in an accident the person does not have time to consult any source unless he or she is previously trained and prepared for this event. In any plant each supervisor and worker should know where to call for, and how to rapidly obtain medical aid and what to do until it arrives.

Industrial plants may have their own medical staff, or a physician can be called and ambulances may be available in the nearby hospital for rapid transportation. First-aid measures can be taken until the arrival of qualified staff. In addition, medical assistance and firefighters may be requested as soon as possible, while the injured person should be given first aid and moved from the dangerous point. Normally it is not advisable to move any injured person if there is no nearby life-threatening hazard. This is to avoid aggravating any injury. If it is desirable to move a person, care should be taken that there should be no stress or strain imposed on the injured part of the body.

In any production facility the job of a safety engineer, medical personnel, and supervisor is to locate a place where first-aid equipment can be made easily accessible and without any hindrance. The workers should be given training by trained paramedical staff for first aid and other similar important practices.

A good and efficient emergency safety plan should represent good coordination between administration, engineers, supervisors, workers, and security staff. The entire program for planning for an emergency control must be a coordinated effort. The use of procedures, facilities, and personnel that would be needed in an emergency must be made a part of the plant design and operation. Although emergency planning and control is a combined effort, one person responsible should be designated for safety, security, firefighting, and medical service at the time of the emergency. Every worker should be know how to call an immediate emergency supervisor when necessary.

The main elements of a site emergency response plan (SERP) are as follows:

A list of emergency phone numbers for company team members, immediate staff personnel, management officials, medical and healthcare officials, rescue services, firefighters, organizations providing assistance to emergencies and disasters, and police should be posted in prominent working places.

Site evacuation routes and other alternative routes should be made available for reaching any site of emergency from inside and outside. Personnel accounting, procedure, assembly areas, safety zones, and exits should be known to everyone.

The location, type, and availability of equipment should be on-site from local resources or elsewhere. They should meet specific types of emergencies and be able to bring situations under control. These may include firefighting equipment, emergency medical, communication equipment, and self-contained breathing apparatus.

Means of communication must be established to alert the emergency organization personnel that their services are required. Installation of secondary communication systems for emergency use is also required in case of any failure of primary communication sources. Alarm systems should be provided for emergency.

Material safety data sheets (MSDS) on all hazardous materials should be posted at or near the location. All personnel and local response agencies should be familiarized with the hazards of the materials used on-site.

It is important to review coordination plans within the organization and with rescue-providing agencies, community and emergency officials, as well as with other neighboring industries to help during an emergency. The coordination network must ensure that all involved persons have reviewed the plan, provided their input, understood the specific functions, and agreed to those responsibilities.

Employees must be given proper training for emergencies, for example, power disconnecting, use of fire extinguishers, use of first aid, as well as search-and-rescue or emergency response procedures. Managers and supervisors must be trained as team coordinators and on-site commanders and can effectively serve as liaisons to corporate, regulatory, and local agencies.

Specific capabilities of individual team members must be kept in view and the job of emergency response activities may be assigned accordingly.

Regular drills, tests of various program elements, and response capabilities, should be carried out regularly to evaluate response procedures and corrective actions. Alarm tests, simulated drills, and mock exercises with community groups are several testing approaches. Evaluation results and proposed corrective actions must be documented and incorporated into the plan. Involvement of other agencies in the testing drill enhances relations and efficiency of the plan.

Records of past and present drills would help in improving the ESPR.

An emergency preparedness drill to deal with sabotage and terrorism must also be carried out.

References

1. Dennison, M. S., *OSHA and EPA Process Safety Management Requirements: A Practical Guide to Compliance*, Van Nostrand Reinhold, New York, 1994.
2. Ianvele, F. A., Addressing career knowledge needs, in *Innovations in Safety Management*, John Wiley and Sons, New York, 2001.
3. Della-Giustina, D. E., *Developing a Safety and Health Program*, Lewis Publishers, Boca Raton, Fl., 2000.
4. Hammer, W., and Price, D., *Occupational Safety Management and Engineering*, 5th ed., Prentice Hall, Englewood Cliffs, N.J., 2001.
5. Schnepp, R., and Gantt, P., *Hazardous Materials: Regulations, Response and Site Operations*, Delmar Publishers, New York, 1999.
6. Available at http://www.csb.gov/index.cfm? (U.S. Chemical Safety and Hazard Investigation Board).
7. Available at http://www.csb.gov/index.cfm? (U.S. Chemical Safety and Hazard Investigation Board).

Industrial Pollution Prevention

Bassam El Ali and Ahsan Shemsi

3.1 Definition of Industrial Waste

All materials produced in large amounts that are not utilizable by any means are called waste. The definition of waste can be very subjective. What represents waste to one person may represent a valuable resource to another. It must have a strict, clear, and legal definition to comply with the law.

There are different types of wastes produced by industries. Wastes are associated not only with the production of raw materials and their formation at the site of utilization but also during production, processing, and packing (Fig. 3.1) [1].

To elaborate, the pollution caused by raw material and its procurement can be shown by taking a simple example of the corn food industry, which requires growing and harvesting corn in a field and its transportation to a plant for producing food items like corn flour, corn oil, carboxymethylcellulose, and glucose. The huge amount of waste is

Figure 3.1 View of the sky across the West Coast Highway in Singapore.

also produced in the form of wastewater and corn cobs. The wastewater and corn cobs can be used in land irrigation and cattle farming, respectively. If these cannot be used properly for certain reasons, then they become wastes.

Increasing regulation of the waste management industry leads to an accurate definition of the different types of waste required for licensing of waste management facilities. In general, the nature of waste is a heterogeneous material and difficult to describe, define, and classify.

3.2 Types of Industrial Wastes

There are thousands of factories in the world that are polluting the atmosphere, water streams, and land by releasing toxic chemicals, metals, gases, particulate matter, and liquid. These toxic chemicals are released from petrochemical refineries, chemicals, metal-processing, refining, finishing, pharmaceutical, paint, pesticides, fertilizers, cement, glass, explosives, and plastic-producing plants. Examples of these chemicals are given in Table 3.1 and Fig. 3.2 [2, 3].

TABLE 3.1   List of Prescribed Substances and Major Substances Requiring Control-Environmental Protection (Prescribed Processes and Substances) Regulations

Prescribed substances

Release to air: prescribed substances
  Oxides of sulfur and other sulfur compounds
  Oxides of nitrogen and other nitrogen compounds
  Oxides of carbon
  Organic compounds and partial oxidation products
  Metals, metalloids, and their compounds
  Asbestos (suspended particulate matter and fibers), glass fibers, and mineral fibers
  Halogens and their compounds, phosphorus and its compounds, and particulate matter

Release to water: prescribed substances
  Mercury and its compounds
  Cadmium and its compounds
  All isomers of hexachlorocyclohexane
  All isomers of DDT
  Pentachlorophenol and its derivatives
  Hexachlorobenzene
  Hexachlorobutadiene
  Aldrin
  Dieldrin
  Endrin
  Polychlorinated biphenyls
  1,2-Dichloroethane

(*Continued*)

TABLE 3.1   List of Prescribed Substances and Major Substances Requiring Control-Environmental Protection (Prescribed Processes and Substances) Regulations (*Continued*)

Release to water: prescribed substances

All isomers of trichlorobenzene
Atrazine
Simazine
Tributyltin compounds
Triphenyltin compounds
Triflualin
Fenitrothion
Azinphos-methyl
Malathion
Endosulfan

Release to land: prescribed substances

Organic solvents
Azides
Halogens and their covalent compounds
Metal carbonyls
Organometallic compounds
Oxidizing agents
Polychlorinated dibenzofuran
Polychlorinated dibenzo-p-dioxin
Polyhalogenated biphenyls, terphenyls, and naphthalenes
Phosphorus
Pesticides
Alkali metals and their oxides and alkaline earth metals and their oxides

Major substances requiring control

In addition to the prevention or minimization of the release of the prescribed substances, the following substances should be considered in each application and authorization:

Particulate matter
Carbon monoxide
Hydrogen chloride
Sulfur dioxide
Oxides of nitrogen
Lead and its compounds
Cadmium and its compounds
Mercury and its compounds
Organic chemicals (trace amounts)
Dioxins
Furans

3.2.1   Classification of industrial waste

Waste is classified not only according to the type of industry producing it but also by the type of waste itself. In a broad spectrum the term waste includes the following categories [4]:

1. Inactive

2. Low activity

3. Biodegradable

Figure 3.2 Ozone is a major component of smog. (*Source: Photo courtesy U.S. EPA.*)

4. Scrap

5. Contaminated general waste

6. Healthcare waste

7. Asbestos

8. Oily waste

9. Solvents and CFCs

10. Generic types of inorganic chemical waste

11. Waste organic chemical

12. Radioactive waste

13. Explosives

14. Dust

Pollution is the most current environmental concern in waste management. The environment has been considered a sink of all wastes. Materials have been released into the atmosphere and watercourses, or dumped into landfills, which are further diluted or dispersed by natural weathering decay. Natural, biological, and geochemical processes can deal with such flows at low level without resulting in changes in the environment. However, as the levels of emissions are increasing with the rise in human activity or industrial progress, natural processes do not have sufficient turnover to prevent these changes. In some extreme cases, the overloading of the natural process of replenishment may breakdown completely, seriously affecting the environment.

Environmental pollution produced by human activity also has an effect on society through deterioration in the quality of the environment.

3.3   Public Concern over Pollution

Industrial pollution has affected not only the environment but also the human community at large. Pollution has its effect on the air quality (Fig. 3.3). The discharge of toxic gases, chemicals, and particulate matters has created problems for the people living in the cities and in the suburbs of the plants. These discharges may cause nausea, allergies, irritation to eyes, sino splash, cystic fibrosis, and other diseases. The damage done by these pollutants appears through the depletion of ozone layer and the increase in the level of CO_2, which has led to a global warming.

On March 24, 1986, at 6.30 A.M., the bungalow at 51 Clarke Avenue, Loscoe, in Derbyshire, was completely destroyed by an explosion of methane landfill gas and the three occupants of the bungalow were injured. The bungalow was situated only 70 m from the Loscoe landfill site. In fact, the site was surrounded by housing. The Loscoe landfill was

Figure 3.3   Haze over Cairo.

an old quarry, which had been worked for clay, stone, and coal from before 1879. Infilling of the quarry with waste materials commenced in 1973 and by 1979, 100 tons/day of domestic waste was being deposited in the quarry. Disposal of waste ceased in 1982 and the site was covered with a light covering of low permeable material in 1984, followed by a more extensive covering in 1986.

The identification of landfill gas as the cause of the explosion was from the gas composition evidence of 60 percent methane and 40 percent carbon dioxide, which is a characteristic of landfill gas. In addition, prior to the explosion there had been evidence of localized damage to vegetation, which was later ascribed to landfill gas. Examination of the geological characteristics of the rocks underlying the Loscoe site showed that they consisted of permeable sandstones and coal seams, allowing gas migration. In addition, blasting during quarrying operations and excavated wells may also have formed migration pathways for landfill gas. Landfill gas from the landfill site migrated through the permeable sandstone beds, resulting in a buildup of gas to form an explosive mixture with air. These and many other incidents occurred as a result of air pollution that resulted in damaging not only the property but also human lives as well. For example, the very famous Bhopal incident at Union Carbide Plant at Madhya Pradesh, India, which took more than 2000 lives resulted from a leak of isocyanate from a storage tank [5]. Liquid waste, which is discharged by the metal refining industries, has polluted the natural water resources that are essential for human beings. For example, the discharge from chromium and cadmium processing industries has polluted the nearby water sources and the human consumption of this water has resulted in defective bone fermentation, liver failure, blindness, and defective birth.

Solid waste has created more problems than the liquid wastes generated by the industries. This solid waste may come from municipal or industrial sources. The disposal of this solid waste represents a serious problem. It can be dumped as landfill or may be used as a composite. It can also be recycled into pure metals or used in the production of other useful items. In addition, the solid waste that contains no inflammable materials can be incinerated and the resulting energy can be used in the power generation and in the steam production for heating purposes.

The pollutants stored in the landfill can be leached down causing pollution of the groundwater sources.

Waste prevention and management are ways to tackle all these problems at the waste source either during its production or at the end-pipe treatment. By practicing prevention, industries can help in achieving good environmental protection and at the same time increasing its profitability and production. The industry needs to modify the methods of production to reduce waste generation at sources.

3.4 Legislation to Waste Management

Two waste disposal incidents influenced waste management and its legislation in the United Kingdom and the United States.

At Nuneaton, Coventry, Warwickshire, a series of toxic waste-dumping episodes occurred in the early months of 1972, the most serious of which was the dumping of 36 drums of sodium cyanide in disused brickworks at Nuneaton, on the outskirts of Coventry. The site was in constant use as a play area by the local children. The drums were heavily corroded and contained a total of one and a half tons of cyanide, which according to the police reports was enough to wipe out millions of people. Over the following weeks and months, further incidents of toxic waste dumping were reported extensively in the press. Drums of hazardous waste were found in numerous unauthorized sites including a woodland area and a disused caravan site. The episodes generated outrage in the population, and emergency legislation was rushed through Parliament in a matter of weeks in the form of the Deposit of Poisonous Waste Act, 1972. The new act introduced penalties of 5-year-imprisonment and unlimited fines for the illegal dumping of waste, in solid or liquid form, which is poisonous, noxious, or polluting. The basis of the legislation was the placing of responsibility for the disposal of waste on industry. Further legislation on waste treatment and disposal followed in 1974 with the Control of Pollution Act [6].

The other incident happened in the U.S.—Love Canal, Niagara Falls in New York State. This site was an unfinished canal excavated for a projected hydro-electricity project. The abandoned site was used as a dump for toxic chemical waste and more than 20,000 tons of waste containing over 248 different identified chemicals were deposited in the site between 1930 and 1952. Following the sale of the plot in 1953, a housing estate and school were built on the site. In 1977, foul-smelling liquids and sludge seeped into the basements of houses built on the site. The dump was found to be leaking and tests revealed that the air, soil, and water around the site were contaminated with a wide range of toxic chemicals, including benzene, toluene, chloroform, and trichlo-roethylene. Several hundred houses were evacuated and the site was declared a federal disaster area. There were also later reports of ill health, low growth rates for children, and birth defects among the residents. As the actual and projected clean-up costs of the site became known, Congress introduced legislation in the form of the Comprehensive Environmental Response, Compensation and Liabilities Act, 1980. This legislation placed the responsibility and cost of cleanup of contaminated waste sites back on the producers of the waste [7].

3.5   Industrial Pollution Prevention

In fact, effective industrial pollution prevention is definitely better than its cure. The first objective must be reduction in the amount of waste produced, if it cannot be avoided. The second objective is to manage the waste in a suitable way while minimizing the overall burden associated with the waste management system [8].

In order to reduce waste production, the potential waste production in an industry should be properly assessed. This step should start with the procurement of the raw materials, taking into consideration their type and nature, their conversion processing into products, their packing process, and their recycling and reuse, if possible.

Pollution prevention can ameliorate the environmental conditions by reducing the generation of waste (Fig. 3.4) [9]. This can also address the serious problems of global warming caused by ozone depletion.

In addition, pollution control prevention also has economic benefits. Although waste management and recycling in most cases increase the production cost, it pays back the initial investment in the long term. Economic benefits include the amount of reduction of waste produced or treated and disposed, and also in reduction of raw material.

Figure 3.4   Development of alternative energies (such as wind power) will decrease air pollution.

The second step is the modification of the production process, which includes the replacement of the raw material containing hazardous causalities, the optimization of the process, and the type of raw material used. The determination of the sources of leaks and spills in the process, and the separation of hazardous from nonhazardous and recyclable waste should also be considered. The third part is the management of waste including its recycling and reuse.

The modification of the plant should take into consideration the minimum or no-production of waste by installing new equipment to control the pollution. It is also possible to enhance the recovery or recycling options in the plants.

3.6 Assessment of Industrial Pollution Prevention

In order to explore all waste reduction opportunities in any process, it is desirable to have a systematic approach to consider all the important factors. These factors include the location of wastewater sources, the facility available to reduce this waste, and the determination of its economical feasibility. This requires a team of experts from among management, plant operators, engineers, analysts, environmentalists, economists and the like, who have the following clear goals:

1. Assessment of source

2. Reclamation of generated waste

3. Economic feasibility of reclamation procedure

4. Implementation by the organization

3.6.1 Assessment of waste generation

The assessment of waste generation should start with the collection of information about the plant's waste-stream process and operation. A thorough understanding of the waste-generating process and streams is considered the best option for the reduction of this waste.

The information regarding the facility's waste streams can be collected from various sources like environmental bulletins, hazardous waste manifests, waste assays, and permissible limit. The amount of waste-generating streams and their mass-balances should be made available to have a good understanding of their quantity and processing. This information gives a clear picture about the type, the nature, and the amount of each waste, its frequency of discharge, and its management cost.

After the collection of this information, priorities should be given to the hazardous waste sources by keeping in view compliance with the current environmental regulations. This should take into consideration the hazardous nature of waste, the potential of the waste minimization, the disposal cost and its volume, the facility available for the disposal, and the allocated budget.

The next step is to search for the possible ways to reduce wastes. The new potential option should have its merits over the other possible options available. The available options are published literature, conference proceedings, equipment vendors, state environmental agencies, and consultants. There may be many proposals for waste minimization.

These merits are low capital-cost requirement, operating-cost reduction, reduction of waste hazards, short recovery period, ease of implementation, and overall economical burden.

3.6.2   Feasibility of the industrial pollution prevention

The feasibility of the selected options is evaluated on the basis of three aspects. The first one is on the basis of technical evaluation, the second one is on the economical basis, and the third one is on the implementation basis.

The technical evaluation is done to assess the efficiency of this process. The new process should be compatible with the current one with a similar application and performance. It should reduce both the environmental and toxic wastes.

The economic evaluation is carried out taking into consideration the profitability and the payback for the installation of a new waste-minimizing unit. Economic evaluation is of two types: one includes the capital cost and the other the operation-running cost.

In designing, purchasing, and installing new units, capital cost is involved, whereas in running the process on a 1-year basis, the operating cost is involved. For economic feasibility of a process, both capital and operating cost should show a money-back period and profitability from the operation of the new unit in improving the product quality, in reducing waste generation, and waste-dumping cost.

The profitability of a waste minimization assessment program is important in deciding the ways to comply with the environmental regulations. Violation may ultimately result in shutting down a facility.

3.6.3   Feasibility implementation

After considering all pros and cons of the waste minimization program, the option should be qualified to be implemented. After implementation of the proposed option, the process should be monitored carefully and evaluated on a regular basis to determine its efficiency. Otherwise, modification is required to make it beneficial [10]. A successful waste minimization assessment approach has the following stages (Fig. 3.5).

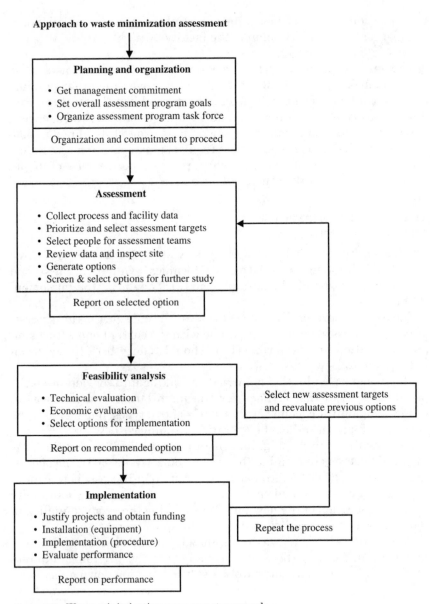

Figure 3.5 Waste minimization assessment approach.

3.7   Waste Management

Once the pollution prevention assessment program has identified the opportunities for the waste reduction, the waste management program can be implemented at that facility. This program starts with the source reduction technique, which involves the reduction of pollutant wastes at their source within the process. This technique is the most desired option in the pollution prevention hierarchy. By avoiding waste generation at its source, it is possible to eliminate the problem of waste handling and disposal. A wide variety of procedural options are available to minimize the waste generation.

Most of the source reduction options involve a change in procedural or organizational activities. For this reason, these options do not demand large capital and time investment. There are two types of product changes: procedural change and technology change.

3.7.1   Procedural change

This change involves the management, organizational, and personal functions of production. The reevaluation of plant procedures can often reveal some reduction opportunities that are relatively inexpensive and easy to implement. Many of these measures, which are used in industry mostly to improve waste reduction, can be implemented in all the areas of a plant as well. They often require little capital cost and result in a high return on investment. Procedural changes can include:

1. Good housekeeping
2. Loss prevention
3. Material handling
4. Joint personnel practices

Good housekeeping involves the careful transport and storage of the incoming raw materials with minimum spill during handling. This can be achieved by using leak-proof transport vehicles with automatic loading and unloading facilities, use of conveyor belts, and sealed storage tanks for the raw material.

Loss prevention minimizes wastes by avoiding spills and leaks from equipment during the process. The most effective way to reduce the amount of waste generated by spills is to make precautionary modifications, practices, training, and regular inspection of equipment, ensuring that spill never occurs in the first place. Cleaning of chemical spills with a typical adsorbent results in the generation of additional waste.

Several procedures can be adopted into plant design and operations to reduce the likelihood of spills, including:

1. Controlled and supervised loading, unloading, and transfer of all hazardous substances.

2. Properly designed storage tanks and containers.

3. Good physical integrity of tanks and containers.

4. Installation of overflow arms and automatic pump shutoffs.

5. Training employees for careful handling and operation of the raw material and the process.

Material handling and inventory practices include programs to reduce loss of input resulting from mishandling, expired shelf-life and improper storage conditions. A proper handling and transfer of the stored materials reduce the chances of a spill. A proper training of the employees working in the processes and the transfer of materials for an adequate spacing for stored containers and proper labeling are efficient ways to ensure the reduction of waste production.

A poor inventory control can result in the overstocking or disposal of expired material. The economic loss as a result of this malpractice can be avoided by computerizing the inventory control where one can monitor shipping, storing, raw material requests, and material tracking.

Wastes are mainly divided into two types: hazardous and nonhazardous. A proper segregation reduces the volume of hazardous waste by preventing the mixing of hazardous and nonhazardous wastes. The separation of hazardous waste from nonhazardous can significantly reduce the quantity of one kind of waste, which in turn reduces the treatment or disposal cost. Moreover, this waste can be reduced or sent to an on-site or off-site treatment plant for recovery.

The reduction of waste production can also be achieved by keeping the waste streams of wastewater separated from contaminated water or by keeping the stream of solvents separate from the hazardous materials. Cleaning solvents and wastewater can be recovered and reused, which can reduce the operational cost.

Concerning the joint personnel practices, all employees should be aware of the waste management procedures so that they can adhere efficiently to the waste reduction. Moreover, the implementation of ISO can help them in the awareness and training program, which ultimately helps in reducing the waste production. This option can be achieved by proper training and by giving incentives and bonuses and other programs to encourage employees to strive for pollution prevention.

3.7.2 Technology change

The technology change includes the process and equipment modifications to reduce waste. This option may include inexpensive, minor changes of reusing raw materials to major changes involving replacement of processes at a very high cost. After evaluating all possible procedural changes, technology changes can be considered as a last option to reduce waste, because it usually requires high capital cost. The innovations are carried out to improve product and reduce the raw material input and waste generation. Some steps involving technology change are as follows:

1. Process change

2. Equipment, piping, or layout changes

3. Changes to operation settings

4. Additional alteration

Research and development is constantly being upgraded, paving the way for the new improved processes with less raw material and energy inputs and also reducing the waste production. The development of activities should encompass a pollution prevention program with a new process resulting in a reduced volume of waste generation.

Process changes can include alteration of an existing process by addition of new unit operations or changing to a new technology to replace an outdated operation.

The inefficient equipment can be replaced to reduce waste generation. The required capital for new equipment can be justified by the higher productivity, the reduction of raw material costs, and the reduction of waste generation and its management. Many equipment changes are inexpensive and simple that make their use imperative, considering the cost of the reduced waste material.

The optimization of the operational settings is another way to reduce waste generation. This includes the adjustment of the process conditions, for example, temperature, pressure, flow rate, residence time, aeration rate, raw material input rate, and the like. These changes often represent the easiest and the least expensive ways to reduce waste generation. If a process operates at its optimum parameter, it produces less waste material. Every reaction is accompanied by other side reactions, which produce waste in a process. When a process runs optimally, the conversion rate increases, which results in a decrease of the waste material. Trial runs can be used to determine the actual optimum settings. For example, chromium coating thickness on aluminum utensils, making them corrosion proof, can be kept uniform by optimizing the time of dipping in the chromium solution, leading to a reduction of the waste production.

Additional controls can result in improved monitoring and adjustment of the operating parameters to ensure the greatest level of efficiency. Automation can reduce the human errors resulting in spills and costly downtime. The resulting increase in efficiency owing to automation can increase the product yields.

3.7.3 Input material change

Pollution prevention can also be achieved by input material change. It actually reduces or eliminates the waste material that first enters the process. Material changes can also be done to reduce the waste generation within a production process. This can be achieved by material substitution or by material purification or concentration. The new substituted material is either very pure or has a reduced amount of hazardous constituents in it, which produces less pollutants, or has lower waste generation and also satisfies the end-product specification. The best candidate for substitution are the nontoxic and nonhazardous materials.

3.7.4 Product change

The pharmaceutical industry is always challenged toward the development of new and improved medicines that have fewer side effects. The challenge for producing new products is also true for other industries. Producers are constantly seeking ways for introducing new products with the intent of reducing the resulting waste. For example, non-lead containing paints have replaced the paints containing palladium. The organic solvents used in polishes have been replaced by water-based polishes to reduce hazardous or toxic, volatile, organic emissions into the atmosphere.

Reformulation of compositional changes involves manufacturing products with lower or no composition of toxic substances to reduce the amount of hazardous waste generated during the product's formulation and end-use. However, the use of a more toxic solvent in place of a hazardous solvent can still reduce the waste but the quality of the final product should not be compromised [11].

3.8 Recycling

There are two types of recycling processes. One is preconsumer recycling and the other is postconsumer recycling. The preconsumer recycling involves raw materials, products, and by-products that have not reached a consumer for an intended end-use; but they are typically reused within an original process. The postconsumer-recycled materials are those that have served their intended end-use and have been separated from solid waste for the purpose of recycling.

Recycling through use or reuse involves returning waste material either to the original process as a substitute, or to another process as an input material. This technique allows waste material to be used for a beneficial purpose. Reclamation in the recycling process for a valuable material or for regeneration helps to eliminate waste disposal costs, reduce raw material costs, and provide income from saleable waste.

It is important to note that sometimes reducing the amount of waste generated at the site is often more economical than recycling. The effectiveness in recycling depends upon the ability to separate any recoverable waste from the wastes of other processes that is not recoverable [12].

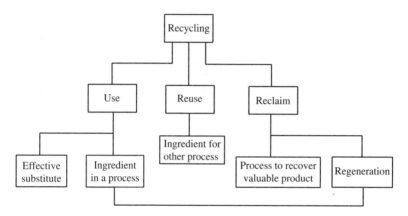

3.8.1   Options in recycling

1. On-site use/reuse

2. Off-site recovery

3. Energy recovery

On-site use/reuse.   After the recycling process, the recovered material can be used directly and carefully in a way that does not affect the quality of the product. The economic value of the waste and its recycling cost should be competitive.

On-site recycling results in less waste, leaving a facility leading to reduction in the waste. The availability and consistency of the recyclable waste determines the need for the setup of the on-site recycling process. Although on-site recycling does not involve transportation, other economical parameters, such as additional equipment, need for operators, and training can determine the feasibility of the on-site recycling process.

Off-site recovery.   Off-site recovery is preferable when the on-site waste is not available in sufficient amount to make an in-plant recovery system cost-effective. The cost of an off-site recycling process depends upon the purity of the waste and the market for the recovered material. The off-site process is also preferable when this facility exists for other plants.

Energy recovery.   Energy can be recovered through the use of waste as a fuel alternative or fuel supplement in a power generation unit for running a facility producing the waste. For example, in a cane sugar factory the waste generated from used wash-dried cane can be used for the generation of power as a fuel supplement.

3.8.2   Recycling technologies

There are many treatment technologies available for recovering a useful component from a waste. These technologies are applied to recover liquid from liquid, liquid from solid, solid from solid, solid from gas, and so on.

Vapor liquid separation.   The two important techniques used for the separation of vapor from liquid are:

1. Distillation
2. Evaporation

Distillation.   Distillation is the most widely used liquid phase separation process for recovering organic components from hazardous waste product stream. This method is based upon heating the liquid to convert it into vapors, which is condensed into liquid. This process can be done through continuous fractional distillation (CFD) or by batch distillation. CFD is used for liquids having close boiling points; whereas in a batch process, liquid having wide differences of boiling points can be separated. Distillation cannot be used to separate thick wastes such as sludges or slums.

Evaporation.   Evaporation is a technique that is conducted to remove volatile compounds and liquid solvents from slurry and sludge, suspended solids, or dissolved solids. The objective of evaporation is to concentrate a solution consisting of a nonvolatile solute in a volatile solvent. This can be achieved by heating the solutions containing solids to remove the vapors of liquid. The resulting thick liquor is collected. There are different types of evaporation technologies, including open-vat evaporation, multieffect evaporation, drum-dryer, tube-dryer, agitator-drying tube, and the like. For example, the removal of fine straws of cane from molasses in a sugar industry is an example of a drum-dryer type of removing solid from liquid.

Solid liquid separation.   The techniques available for the separation of solids and liquid include filtration, centrifugation, and sedimentation.

Filtration.   The process by which suspended solid particles are removed from liquid is called filtration. The liquid containing suspended solid particles is passed through a porous medium. A clear liquid is obtained. The porous medium may be a fabric, a canvas, a paper, a screen, or a bed of sand. For small particles a filter with very small pores is used but for larger particles sand may be used. The liquid flow may be passed through the filter under the influence of gravity, positive pressure, or vacuum. The use of medium depends upon the size of the solid particle.

Centrifugation.   Separation of solids from liquids under the influence of centrifugal force is called centrifugation. The liquid containing various particles of different densities are rotated at a very high speed in a closed system, where these get settled at the base according to their densities. This is commonly used as a preliminary step before the use of an additional recycling method.

Sedimentation.   The principle of sedimentation is almost the same as centrifugation but under the influence of gravity only. Small particles settle at the bottom of settling tanks. This technique is not energy-intensive and is mainly used for waste streams containing both liquids and solids that contain a low concentration of contaminated solid.

Liquid-liquid separation.   Liquid-liquid separation can be achieved in two methods: solvent extraction and decantation.

Solvent extraction.   In this method, organic molecules, soluble metals, and other materials are extracted from aqueous and nonaqueous streams with the help of other organic solvents. Although it is a very well-established technique, pollutants are seldom treated using this technique. Only a very few organic pollutants like CFC and phenols are removed by this technique. This is a method in which one can preconcentrate the pollutant and recover it.

Decantation.   When two immiscible liquids are mixed, the one containing the pollutant can be separated from the other on the basis of densities. The liquid is fed into a settling tank where a high-density liquid is allowed to settle at the bottom and the two layers are separated. For example, the decantation process is used to separate cutting oils from wastes.

Solute recovery.   So far we have discussed the waste management of a solute in a solvent. Now we are going to discuss solute recovery for reuse. There are different methods of solute recovery, for example, precipitation, ion exchange, ultrafiltration, and reverse osmosis.

Precipitation. This involves the alteration of ionic equilibrium to produce an insoluble precipitate. A precipitating agent such as caustic soda, lime, sulfide, sulfate, or carbonates are added to different solutions depending on the requirement to precipitate the metals of interest. Background knowledge is necessary for precipitation. Most of the metals recovered from the waste stream are precipitated through this technique. For example, chromium and cadmium are precipitated in the form of sulfides.

Precipitation is an effective and reliable treatment method. The resulting sludge can be reprocessed for metal recovery or for ultimate disposal.

Ion exchange. Ion exchange is a process in which soluble ions of metals, inorganic, and organic acid are absorbed on a solid surface containing opposite charge carrier species in a column or bed. Then the column beds are treated to specifically remove the adsorbed metals, ions, acids, and the like. Ions, which are collected through this process, may be harmless or harmful depending upon their need to be reused or disposed.

For example, the common application of ion exchange is the method of recovery of hexavalent chromium from plating waste. The other examples are the removal of copper and lead in making brass and batteries, respectively.

Ultra-filtration. A membrane with a very small pore size is used to remove solute or colloids from pressurized waste streams. It retains the larger particles and allows the solvent and small particles of interest to pass through.

Through this process, very small particles of metal ions can be filtered to be used in electrophoretic paint industry. The retained big-size particles are returned to the electropaint tank for reuse.

Reverse osmosis. Reverse osmosis is a process by which a solute form is allowed to move across a semipermeable membrane under the influence of a concentration gradient. This process can remove dissolved organic and inorganic compounds from an aqueous stream by allowing a solvent molecule to pass through the membrane and retaining the solute molecules. The membrane is micro- or mesoporous, ionic or nonionic in nature. A sufficient pressure gradient is applied to the concentrated solution to overcome the aromatic pressure and force a net flow through the membrane toward a dilute phase. This process constantly increases the solute concentration on one side of the membrane, whereas a relatively pure solvent is transported through the membrane. Ions and small molecular compounds in true solution can be separated from a solvent by this technique.

This process is widely used in a seawater desalination plant, where purified water is obtained against a high salt concentration of seawater. In the metal-making industry this purification method is used in the oil-water mixed jet-cutting tool emulsions that contain high concentration of metals. A reverse osmosis unit separates the oil from water to be reused again.

3.9   Waste Treatment

Before discussing the ultimate methods of waste disposal, other physical, chemical, and biological methods of recycling should be considered in rendering the waste nonhazardous. According to the EPA definition, treatment is "any practice, other then recycling, designed to alter the physical, chemical or biological character or composition of a hazardous substance pollutant, or contaminant, so as to neutralize said substance, pollutant, or contaminant or to render it nonhazardous through a process or activity separate from the production of a product or the providing of a service."

The physical, chemical, and biological methods are used to treat the waste just after its generation.

The physical methods are used to concentrate and reduce waste volume, and to separate the different phases of waste, whereas the chemical treatment method is used to convert hazardous waste into nonhazardous by-products.

Biological treatment is carried out with the help of microbes and enzymes to achieve the same goal as with the chemical treatment, especially for organic waste.

Although incineration is the main process used for waste disposal, it is costly. Other methods, whenever possible, are chosen as alternative ways, and are used in conjunction with the other unit process. For example, a typical process sequence might be decantation, sedimentation, biodegradation, followed by sludge agitation, and finally land-filling.

The toxicity associated with the waste is a determining factor for the selection of the different on-site or off-site physical, chemical, and biological methods.

For proper management of waste, technical consideration involved in the selection of a specific treatment process may include the characteristics of the waste including the physical form, the constituents, the concentration of each contaminant, the volume of the waste, the availability of the treatment methods, and their applicability.

The objective of the treatment procedure is another key factor in the selection of a particular treatment method. It may be desired to reduce the component or destroy it and then isolate it. The end product—waste—should be compatible with the ultimate disposal procedure to be used for the waste. For example, if the waste is to be disposed of by dumping it in the deep ocean, the end product of the waste treatment should consist of waste encapsulated in concrete.

In selecting a particular physical, chemical, and biological method of treatment, the economic feasibility of the method is an important factor as well. The cost of treatment procedure and the amount of the waste to be treated is also a strong factor. Moreover, the on-site and off-site treatments should also be considered in determining the cost of the

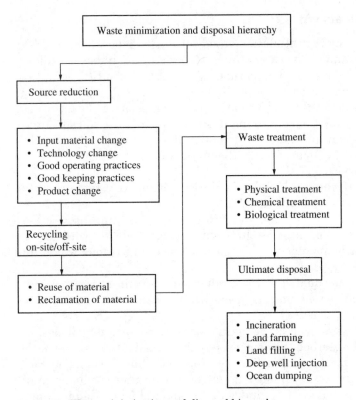

Figure 3.6   Waste minimization and disposal hierarchy.

process, the initial capital investment, and the operating cost and expenses on final waste disposal. Before discussing the waste treatment and its disposal, let us look at the waste minimization and disposal hierarchy [Fig. 3.6].

3.9.1   Physical treatment

Physical methods are employed to separate, reduce, or concentrate the waste. There are many physical methods available for waste treatment. Among them only a few have found application at the industrial level. Some methods are either in their infancy like zone-refining, freeze-drying, and electrophoresis, whereas others have found little potential application on account of their operational cost.

The most common processes today are sedimentation, filtration, flocculation, adsorption, distillation, and solvent evaporation. These treatment methods are used for phase or component separation purposes.

To reduce or concentrate the waste volume, phase separation should be carried out for sludges, slurries, and emulsions.

Separation of waste components containing large particles, filtration, centrifugation, or flotation may be used. Flocculation is carried out for colloidal systems. Removal of volatiles can be carried out by distillation or evaporation.

Ion exchange, reverse osmosis, ultrafiltration, and air stripping can also be used for separating waste components, especially for wastewater treatment.

Resin adsorption.   The resin adsorption is a good option for the selective removal of waste. This technique is normally used for the removal of thermolabile organic solutes from aqueous waste streams. The solute concentration of solution ranges from 1 to 8 percent. Moreover, synthetic cationic and anionic resins may be used to remove a hydrophobic, hydrophilic, or neutral solute, which can also be recovered by chemical methods. These resins are also used with a high concentration of dissolved inorganic salts in the waste stream. Their applications include phenol, fat, organics, and color removal from wastewater. They can be applied for the removal of pesticides, carcinogens, and chlorofluoro compounds.

Electrodialysis.   In electrodialysis, separation of an aqueous stream is achieved through the use of synthetic membranes and an electric field. The membrane allows only one type of ions to pass through and may be chosen to remove other ions that move in the opposite direction. Therefore, it produces one stream rich in particular ions and another stream depleted of those ions. The two streams can be recycled or disposed off. This technique is commonly used in the desalination of brackish water. The other uses are in acid mine drainage treatment, the desalting of sewage-plant effluents, and in sulfite-liquor recovery.

Flotation.   This method is used to remove suspended organic and inorganic solids from waste streams or slurries. This technique is basically a physical process in which solutions carrying suspended particles are agitated with the stream of air bubbles or a mechanical agitator. A froth forms at the surface of the liquid or slurry, which is then removed by skimmers or scrapers. Individual or combinations of similar materials may be removed by this method from wastewater streams.

Flotation has been used in removing oil and grease from wastewater. Selective flotation can be achieved by adding certain surfactants for the removal of metal ions, cyanides, fluorides, and carbonyl from hazardous wastes.

Air stripping.   Air stripping is used to reduce the concentration of noxious gases, for example, ammonia, and hydrogen sulfide, from biologically treated waste streams. It can also be used to remove volatile organics and other pollutant gases absorbed in waste streams. This method is also highly efficient for untreated wastewater. Wastewater containing lime for phosphate removal is first sent to a mixing tank and then to a settling tank for the calcium phosphate and calcium carbonate to settle down. The treated water is then fed at the top of a two-packed tower, while air is fed countercurrently through the tower to remove ammonia. The stripped gas is treated to decrease its concentration and wastewater is also treated.

Steam stripping.   A steam-stripping process is carried out in a distillation column, which may be either a packed or tray tower. Steam is introduced from the bottom and waste flows downward. The obtained stream, which is rich in volatile components, is further treated to remove the volatile materials. Air stripping may also be required to remove the volatile materials.

Steam stripping is similarly used to remove ammonia, hydrogen sulfide, and other volatile components from aqueous waste streams. It has also been used to recover sulfur from refinery waste and organics from industrial wastes.

Solidification.   Solidification is a physicochemical process that transforms a hazardous waste into a nonhazardous solid by fixation or encapsulation. In fixation, a solidifying agent is used to solidify the waste, whereas in encapsulation, the waste is surrounded by a binder after it has been solidified. Both processes produce durable, impermeable, and environmentally safe products.

Based on the solidifying agent, this process is divided into five categories: silicate and cement, lime, thermoplastic, organic polymer, and encapsulation techniques. All these processes are used to treat hazardous wastes.

Previously, solidification was used for the disposal of radioactive wastes. Recently, it has gained popularity for other waste treatment as a result of the strict environmental regulations.

This process is suitable for sludges from stack gas scrubbers and for fly ash from cement and iron industries as well as from plating and lead-smelting plants.

In the silicate- and cement-based process, Portland cement and other additives are added to either a wet or a sludge-like waste, which produces an impermeable rock-like solid. The degree of solidification can be achieved by controlling the composition of the waste and the type of solidifying agent. Then these solids are used for land reclamation and road construction.

The advantages of this process include the inexpensive additives and the availability of the equipment.

In the lime-based process, both lime and siliceous materials are used to solidify waste. Additives such as fly ash are also added to increase the strength of the end product. The degree of solidification depends on the reaction of the lime with the other components forming the cement. A product known as pozzolanic cement is formed, which is mainly used for landfilling, mine reclamation fill, or as a capping material.

In thermoplastic solidification, the initial waste is dried and then combined with bitumen and polyethylene at a high temperature; the mixture on cooling becomes a solid. In the second step, the solid waste is thermoplastically coated and then disposed of. This process is used for inorganic and radioactive wastes.

In the organic polymer process, a monomer and a catalyst are combined with the waste stream, and the polymer is allowed to form. The product is then containerized and disposed of. The polymer used in this technique is urea of formaldehyde (melamine). The advantages of this technique are that both solid and semiliquid waste can be processed and the weight of the processed solid is less than those produced from cement-based wastes, but they are biodegradable and must be containerized.

The encapsulation process deals with the dried wastes. The dried waste is first chemically treated and then coated with a binder, usually polyethylene. The advantages of this process are that the encapsulated waste is very durable and resistant to water and deterioration and the final product need not be stored in containers. The disadvantage is that it is an expensive process and applicable only to small volumes of wastes.

3.9.2 Chemical treatment

There are numerous chemical processes for the treatment of wastes that are used in conjunction with other methods. These methods include calcination, precipitation, catalysis, electrolysis, hydrolysis, neutralization, photolysis, cholrinolysis, oxidation, and reduction.

Calcination. Calcination is a well-established single-step process for the treatment of complex wastes, containing organic and inorganic components in slurries, sludges, tar, and aqueous solutions, by heating at higher temperatures in the absence of air to remove volatiles.

The real application of this process is the recalcination of lime sludges from water treatment plants, coking of heavy residues and tars from petroleum refinery operations, concentration and volume reduction of liquid, radioactive wastes, and treatment of refinery sludges containing hydrocarbons, phosphorus, and compounds of calcium, magnesium, iron, and aluminum.

Electrolysis. Electrolysis is a process in which oxidation or reduction reactions take place at the surface of conductive electrodes immersed in an electrolyte under the influence of an applied potential.

This method is applicable to any component carrying an electronic charge either positive or negative. The most common use of this method is for recovering heavy, toxic metals of economical importance from concentrated solutions. For example, copper, chromium, and cadmium can be recovered from waste solutions. Electrolysis is not very useful for organic waste. Pilot application includes oxidation of cyanide waste. Gas emissions may occur during the process. In case these gases are hazardous, further treatment (scrubbing) is required.

This process can also be applied to radioactive wastes that contain metals in ionic forms. The removal of collected ions from the electrodes is not difficult. These ions can be recycled and reused or can be disposed of.

Neutralization.   Neutralization is a process of adjusting the pH of a waste solution to near 7 (neutral) by adding an acid or base. The wastes might have different chemical and physical changes such as precipitation or evolution of a gas occurring as a result of a chemical reaction. The process has wide application to aqueous and nonaqueous liquids, slurries, and sledges and is widely used in waste treatment. Some applications include pickle liquor, mine drainage, acidic and basic waste, plating waste, and so on.

Neutralization is carried out in batch or continuous process reactors by mixing acidic and alkaline streams together or by passing acidic wastes through packed beds of limestone. Also, the addition of solutions of concentrated bases such as caustic or soda ash to acid liquids or compressed CO_2 to basic waste streams can be applied. The choice of acid or base depends upon the process requirement as well as on its cost. Lime and sulfuric acid are inexpensive but their use is still limited. Treatment of sulfate-bearing waste with lime produces calcium sulfate as a precipitate. However, caustic and soda ash are more expensive but widely used. Neutralization of waste containing cyanide and sulfide results in the evolution of toxic gases like hydrogen cyanide and hydrogen sulfide. These wastes require special treatment devices for neutralization such as a scrubber.

Chlorinolysis.   This process is applied to chlorinated organic wastes, which are ultimately converted into carbon tetrachloride (CCl_4). In this process, the organic feed is introduced to a reactor at 900°F along with chlorine gas under 20 atmospheres. Chlorine reacts with hydrocarbons to form carbon tetrachloride in addition to other chlorinated products, which are removed by distillation.

This process is only applicable to the waste streams containing proper organic waste. Moreover, chlorinolysis produces hydrochloric acid and phosgene gas as an effluent, which are further treated. The leakage of chlorine is another hazard associated with this process.

Oxidation.   Oxidation is a process that involves the transfer of one or more electrons. This can be carried out by adding an oxidizing agent or via electrochemistry. This process is used for the detoxification of hazardous waste; the oxidation of cyanide to cyanate, and further decomposition into CO_2 and N_2 is a good example. Metals can be oxidized to their higher oxidation state, thus making them insoluble and recoverable as a precipitate.

Some industrial applications include the oxidation of cyanide with chlorine and ozone to cyanate. Ozone coupled with ultraviolet light is used to oxidize halogenated organic compounds, which are usually resistant to oxidation with ozone alone.

Cyanide, phenol, sulfur compounds, and metal ions can be oxidized with hydrogen peroxide. Potassium permanganate ($KMnO_4$) is an excellent oxidizing agent that reacts with aldehydes, mercaptanes, phenols, and unsaturated acids. It has been used to destroy organics in waste and potable water. The reduced form of $KMnO_4$ is manganese dioxide, which can be removed from water by filtration.

Reductions.   In the reduction process, electrons are transferred from one reacting species to the chemical being reduced. As a result of this transfer, the valance state of the chemical is lowered. The resulting chemical may become less toxic or easy to precipitate. In this process, the reducing agent may be a gas, solution, or divided powder. The reduction reaction is followed by a separation step, such as precipitation, to remove the reduced compounds. For example, chromium is used in industries such as metal finishing, inorganic chemical manufacture, coil coating, corrosion proofing, aluminum forming, iron and steel manufacture, electronic manufacture, leather tanning, and pharmaceuticals. Chromium is a very toxic chemical in its Cr^{6+} state, whereas Cr^{3+} is much less toxic and can be precipitated in alkaline solution. Therefore, the reduction of chromium (Cr^{6+} to Cr^{3+}) is carried out with sodium metabisulfite and sodium bisulfite.

The other applications are mercury reduction by sodium borohydride ($NaBH_4$) and lead reduction by alkali metal hydride (MH).

3.9.3   Biological treatment

Biological processes involve chemical reactions carried out by microorganisms. The microorganisms either absorb a compound inside their cell body, decompose it with the help of enzymes, or excrete enzymes to bring it outside the cell for decomposition. The most common use of the biological process in waste treatment is the decomposition of organic compounds. These microorganisms decompose both organic and inorganic compounds.

Different biological processes used for the treatment of wastes are activated sludge, aerated lagoons, anaerobic digestion, composting enzyme treatment, trickling filter, and waste stabilization ponds.

These processes are known to be reliable and environmentally friendly. Chemical additives are usually not needed and the operational expenses are relatively low. Therefore, the biological treatment methods are expected to have an important role in the future of waste treatment facilities.

Activated sludge. Activated sludge is the most widely used treatment in the biological waste treatment processes. The process uses microorganisms to decompose organics in aqueous waste streams. These microorganisms are thoroughly mixed with the organics. These microbes take the organics into the cell, through the cell membrane, into the cytoplasm, where enzymes break down these organic compounds into smaller fragments via different reactions. Microorganisms derive energy and cellular material from these reactions. They also adsorb colloidal matter, suspended solids, and metals onto their cell surface.

There are certain controlling factors for efficient decomposition:

- The extent of mixing
- The amount of dissolved oxygen
- The concentration of toxic metals
- The type and concentration of organic compounds

Among the various organics that can be successfully decomposed by the activated sludge process include proteins, polysaccharides, fats, oils, aldehydes, alkenes, aromatics, halogenated hydrocarbons, and isoalkenes.

The advantages of this process arise from the fact that the system does not typically require chemicals and, therefore, the decomposition of organics is an environment-friendly degradation process. However, this process has limitations and cannot handle slurries, tars, or a high concentration of suspended solids. In fact, this process has been used extensively to treat waste streams from iron and steel, pulp and paper, petroleum refining, organic chemical manufacture, and pharmaceutical industries.

Aerated lagoon. The aerated lagoon is similar to that of the activated sludge process but different in a way such that biological sludge is not recycled. The aerated lagoon is an earthen basin, which is lined to make it impermeable and is artificially aerated. Aerated lagoons usually require less energy than activated sludge processes and as recycling systems are not needed, more land is required.

The aerated lagoon is fed with waste streams containing less than 1 percent solids, to avoid settling some of these solids. Therefore, the removal efficiencies are not as high as those for the activated sludge process. The residence time of an aerated lagoon is longer than that of the activated sludge process.

This process has been used for petrochemical, textile, pulp and paper mills, leather-tanning, gum, wood-processing, and some other industrial waste streams.

Composting. Composting uses aerobic digestion by microorganisms in the soil to decompose organics. This task can be accomplished by the piling of waste in the ground and aerating it occasionally by turning and moving the soil. The collection of leachate and run-off that is produced is normally required to prevent groundwater contamination.

The process, unlike some other biological processes, can tolerate some toxicants and metals. Composting is commonly used for organic wastes and a complete digestion requires 3 to 6 months depending upon the climatic conditions of the region. Composting has been successful with municipal repulse, high-concentration organic sludges, and some petroleum refineries.

Trickling filters. In place of making ponds and lagoons, the trickling filter uses microorganisms that are held in a support media in a stack and wastewater is trickled over them. Usually waste streams are sprayed over the supported filter to absorb oxygen before passing through the support media. The area, size, and number of filters are important variables to control the efficiency. Trickling filters are normally used in combination with other methods owing to their less efficiency. They are normally used for the treatment of different industrial wastes such as canneries, pharmaceuticals, petrochemicals, and refineries. The composted material is used in landfilling.

Waste-stabilization pond. For a long time, the ponds have been commonly used for sewage treatment and dilute industrial wastes. These are normally shallow basins in which wastes are fed for biological decomposition. Aeration is provided by wind, and anaerobic digestion may also occur near the bottom of the pond.

Waste-stabilization ponds are normally used as a final treatment step because they are not efficient enough to be used on their own. The industries using this process include steel, textile, oil refineries, paper and pulp, and canneries.

3.10   Waste Disposal by Incineration

The EPA pollution prevention hierarchy includes:

1. Source reduction
2. Recycling

3. Treatment

4. Ultimate disposal

So far we have covered the source reduction and recycling of wastes. In the next section, we are going to discuss their ultimate disposal including incineration and landfilling.

Incineration is a well-known process that involves the conversion of toxic and hazardous waste into a less or nontoxic waste by heating at a very high temperature to convert them into gaseous and particulate matter. Incineration is considered an attractive option after source reduction, and recycling. This method is sometimes preferred over the other treatment methods because it destroys permanently the hazardous components in the waste material.

It is also preferable to completely destroy or reduce any hazardous waste instead of keeping them in long-term land-based disposal containment.

Properly designed incineration and ancillary equipment is considered capable of the highest overall degree of destruction and control for the broadest range of hazardous waste streams. Incineration employs thermal destruction at very high temperature (>1000°C) to destroy the organic fraction through an oxidation process of the waste and converts them into inoffensive gases.

Normally all kinds of organic waste or combustible materials are potential candidates for incineration. Even contaminated water and soils are currently disposed by incineration.

The most common type of incinerators:

1. Rotary kiln

2. Liquid injection

3. Fluidized bed

4. Open-hearth (multiple-hearth) units

3.10.1 Rotary kiln incinerators

Rotary kiln is a cylindrical refractory-lined shell that is mounted at an angle from the horizontal and rotated at a certain speed. The rotation helps in mixing and moving the solid or liquid waste inside the kiln. It can handle any kind of solid or liquid waste. This incinerator is fed with feed from the top. It has different length-to-diameter ratios depending upon the requirement. The working temperature ranges from 1500°F to 3000°F and rotating speed may range from 0.2 to 2 in/s.

The slurries and liquids are injected with the help of nozzles, whereas solid feed is introduced through a pack-and-drum system.

The rotational speed and angle at which it is positioned control the residence time of the solid in the kiln. Normally solid waste is converted into CO, particulate matter, or ash. For complete oxidation of flue gases and particulate matter, the kiln is also provided with a secondary combustion chamber. The volatilized combustibles exit the kiln and enter the secondary chamber where a complete oxidation tube is placed.

A wide variety of wastes can be incinerated in a kiln simultaneously without stopping it. Numerous hazardous wastes that were previously disposed of by landfilling and deep-well injection are currently being safely and economically destroyed by the use of rotatory kiln incinerators. These include CFC, PVC, PCB, chlorinated coolant oils, and the like.

3.10.2   Liquid injection incinerators (LII)

Liquid wastes that can be pumped, injected, and converted into an aerosol under high pressure are burned in a facility called liquid injection incinerator (LII). There are three types of LII: vertical, horizontal, and tangentially fired vortex combustors. The horizontal and vertical are basically similar in operating condition. The tangentially fired unit has a much higher heat release and generally superior mixing than the other two units, making it more attractive for disposal of high water-containing wastes and poorly combustible materials. The temperature ranges for these LII are 1300°F to 3000°F.

Normally, a liquid injection incinerator consists of two stages. The primary chamber is a burner where combustible liquids and gaseous wastes are incinerated. Noncombustible liquid and gaseous wastes usually bypass the burner and are introduced downstream of the burner into the secondary chamber. These wastes are introduced in the form of an aerosol. The aerosol is brought by pressure pump, which pumps the wastes in addition to air inside the combustion chamber through an atomizer or nozzle. A good atomizer guarantees the complete burning. The liquid waste fuel is transferred from drums into a feed-tank. The tank is fed into the incinerator under high pressure. Normally a liquid fuel or a gas preheats the incinerator before waste introduction. Filtration may be required to remove solids prior to injection through the burner.

High-density liquid waste can be pumped after preheating them to decrease their viscosity, which not only helps in pumping but also in aerosol production in the incinerator.

The waste liquid, which contains alkali, organic matter, toxic substances, or a catalyst, is thermally decomposed for nonhazardous treatment and simultaneously recovered alkali salt. Such waste liquid as the above-mentioned is atomized and sprayed into the flame of a high

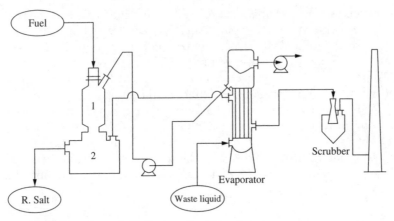

Figure 3.7 Flow sheet of alkali waste liquid treatment.

temperature by the use of a high heat-release and short-flame burner (Vortex burner) for complete decomposition (Figs. 3.7 and 3.8) [13].

The performance of this type of incinerator can be summarized:

1. Alkali waste liquid: NaCl, Na_2CO_3, or Na_2SO_4 can be recovered by oxidized roasting.

2. Organic waste liquid: Waste liquid containing amine, nitrile, or ammonium sulfate can be thermally decomposed.

3. Hazardous waste liquid: Waste liquid containing organic halogen or metal plating can be made nonhazardous.

Figure 3.8 Scheme of waste liquid treatment unit. Nittetu Chemical Engineering Co., Tokyo 174-0041, Japan.

3.10.3 Fluidized bed incinerators

The fluidized bed incinerators are used to burn finely divided solids, sludges, slurries, and liquid. The bed consists of granular material like sand, which is suspended by pressurized air in a highly turbulent state, which makes the bed act as a fluid above the combustion chamber floor. Waste is conveyed into the fluidized bed at a very high temperature, and upon direct contact it burns and gets converted into gases and ash. The gases move out of the combustion chamber, whereas ash caught in the bed material is eventually removed when the bed material is replaced.

The advantages of fluidized bed incinerators include simple compact design, low cost, high combustion efficiency, low gas temperatures, and large surface area for reaction. The disadvantages are the ash removal and carbon that build up in the bed. They are mainly used in petroleum and paper industries, wood chips, and sewage sludge disposal.

3.10.4 Multiple-hearth incinerators

Multiple-hearth incinerators consist of a series of flat hearths laid in a series from bottom to top having a central rotating shaft and supplied with rabble arms and teeth for each hearth. The hearths are lined with refractory material and also supplied with air blowers, fuel burners, an ash-removal system, and a waste-feeding system. Solid waste is fed through the roof, whereas liquids and gases are introduced from burner nozzles. The central shaft with rabble arms distributes the waste across the top of the hearth to drop holes. The waste then falls to the next hearth in a series, until discharged as ash at the bottom. Temperature ranges from 500°F at the top hearth to 1800°F in the middle hearth. An incineration tube is placed in the middle hearth. Because of higher residence time, material with low volatility can be vaporized. Evaporation of large quantities of water is an added advantage [14].

3.11 Ultimate Disposal

Another alternative option in the pollution prevention hierarchy is the ultimate disposal. This method mainly consists of land-farming, land-filling, deep-well injection, and ocean-dumping.

3.11.1 Land-farming

The oldest, simplest, easiest, and safest waste disposal method is land-farming. The wastes are dumped on a land to be stabilized by natural environmental processes. The wastes, mainly containing organics especially from food and petroleum industries, are dumped by this method.

The important criteria for land-farming of a waste are the biodegradability of the wastes containing organic materials, and they should have a neutral pH and contain moisture. These wastes should not contain materials that are capable of polluting air or contaminating ground water. Bacteria and yeast normally decompose these wastes, which result in leaching of water-soluble nutrients, volatilization, and finally incorporation into the soil.

The major steps involved in land-farming are site selection, site preparation, waste analysis, waste application, soil-waste blending, and postwaste addition care. There are several advantages of land-farming: it is effective while maintaining a low-cost disposal method; it is an environmentally safe and simple process, featuring less processing of waste; and it is also helpful in increasing the fertility and nature of the soil.

3.11.2 Landfilling

Landfilling is used for wastes in the form of sludges. There are two types of landfilling: area-filling waste disposal at the soil surface and trenching.

Wastes are subjected to pretreatments such as solidification, degradation, volume reduction, and detoxification before being landfilled. It also reduces the chances of toxic gas production and leaching out to ground water.

Area-fill can be done by mixing the waste with soil, forming a mound with the mixture, and covering it with soil. Area-fill layer is prepared by spreading alternate layers of soil and soil-waste mixture over the area and by filling a containment area surrounding the dikes with the waste and then covering it with a soil layer.

The trenching involves placing the waste in a trench and covering it with one more layer of soil. Depending upon the size of the trenches, there are two types:

1. Narrow trenches of 2 to 10 ft wide, mainly used for sludges with low solid contents.

2. Wide trenches of more than 10 ft wide used for sludges with high solid contents. In selecting an area for landfill, various factors like technical, economical, and public acceptance should be considered.

Both area-filling and trench-filling have their advantages and disadvantages. Area-filling requires less manpower and machinery and are less likely to contaminate groundwater. Trench-filling requires excavation, manpower, machinery, and constant monitoring of the site. Both methods use lime and other chemicals to control odors caused by gas production, which can lead to explosions or harming of the vegetation and contamination of the water.

3.11.3 Deep-well injection

For petroleum waste management a deep-well injection disposal process has long been used. This method transfers liquid wastes underground and away from fresh wastewater sources. It is also used to dispose of saltwater in oil fields.

For the selection of a deep-well injection process many factors are taken into account. The depth of the well is selected to avoid the contamination of fresh water and takes into consideration the nature of the underlying rock. The rock should be stronger but permeable enough to adsorb the liquid wastes. The site must be tested on a pilot scale before it is actually used.

The depth of the well is determined by the type and nature of the wastes. The more toxic the waste, the farther down the disposal zone must be.

The wastewater to be disposed of should be low in volume, high in pollutant concentration, and must be difficult to treat with other methods. The wastewater should not react in the disposal zone and should also be biologically inactive. Nuclear wastes and petroleum wastes are often disposed of by this technique.

3.11.4 Ocean dumping

The ultimate dumping method is ocean dumping. It describes two forms of waste disposal: one into shallow offshore waters and the other into deep ocean waters. Before dumping any wastes it must be treated to reduce its volume and should be less toxic to the marine environment. This dispersal actually dilutes the contaminant and biologically converts it to a nonhazardous form.

Different kinds of wastes are still dumped in the ocean without treatment along with runoff waters, but the long-term effects on the marine environment have not been well-monitored. There is a possibility of very damaging and long-term consequences [15].

There are certain advantages and disadvantages of offshore and deep-ocean dumping. The advantages of offshore dumping are that more information and experience are available, transportation costs to the sites are lower, and any resulting pollution is localized. However, the disadvantages are related to the problems with the fishing industry and ruining offshore mineral deposits. The advantages of deep-ocean dumping lie in spreading the waste over a large area; thus, diluting the contamination. However, higher transportation cost together with the difficulty in monitoring the total effects of contamination is a disadvantage.

References

1. *Industrial Pollution in Singapore: Images of the Sky by Photographs*, George P. Landow, available at http://www.scholars.nus.edu.sg/landow/post/singapore/images/ environ/sky1.jpg
2. Paul, T. W., *Waste Treatment and Disposal*, John Wiley & Sons, p. 27, 1998.
3. U.S. Environmental Protection Agency. Available at http://healthandenergy.com/ images/smog%20in%20the%20city.jpg
4. Paul, T. W., *Waste Treatment and Disposal*, John Wiley & Sons, p. 59, 1998.
5. Williams, G. M. and Aitkenhead, N., Lessons from Loscoe: The uncontrolled migration of landfill gas. *Quarterly Journal of Engineering Geology*, 24, 191–207, 1991.
6. *The Times*, Times Newspapers, London, 1972.
7. British Medical Association, *Hazardous Waste and Human Health*, Oxford University Press, Oxford, 1991.
8. Environmental Protection Agency, USA, *Waste Minimization Opportunity Assessment Manual*. EPA/625/7–88/003, 1988.
9. *Navitas Energy, Inc.* Available at http://www.windpower.com/index.cfm
10. Craig, J. W., McComas, C., and Benforado, D. M., *Measuring Pollution Prevention Progress: A Difficult Task*. AWMA Paper No. 90-41.3, Pittsburgh, PA, Annual Meeting 1990.
11. Office of Technology Assessment, Report to U.S. Congress, *Technologies and Management of Strategies for Hazardous Waste Control*, March 1985.
12. Louis, T. and McGuinn, Y. C., *Pollution Prevention*, Van Nostrand Reinhold, New York, N. Y., p. 159, 1992.
13. *Japanese Advanced Environment Equipment*. Available at http://www.nce-ltd.co.jp.
14. Louise, T. and Reynolds, J. R., *Introduction to Hazardous Waste Incineration*. Wiley Intersciences, New York, N. Y., 1988.
15. Ketchum, B. H., Keste, D. R., and Park, P. K., *Ocean Dumping of Industrial Wastes*, Plenum, New York, 1981.

Edible Oils, Fats, and Waxes

Mohammad Farhat Ali

4.1   Introduction

Fats, oils, and waxes are naturally occurring esters of long straight-chain carboxylic acids. They belong to the *saponifiable* group of lipids. Lipids are biologically produced materials that are relatively insoluble in water but soluble in organic solvents (benzene, chloroform, acetone, ether, and the like). The saponifiable lipids contain an ester group and react with hot sodium hydroxide solution undergoing hydrolysis (saponification):

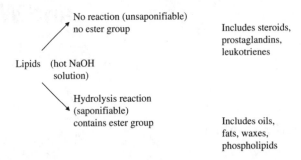

Lipids   (hot NaOH solution)

No reaction (unsaponifiable)
no ester group

Includes steroids, prostaglandins, leukotrienes

Hydrolysis reaction (saponifiable)
contains ester group

Includes oils, fats, waxes, phospholipids

Saponification is a chemical process in which an ester is heated with aqueous alkali (sodium hydroxide) to form an alcohol and the sodium salt of the acid corresponding to the ester. The sodium salt formed is called soap.

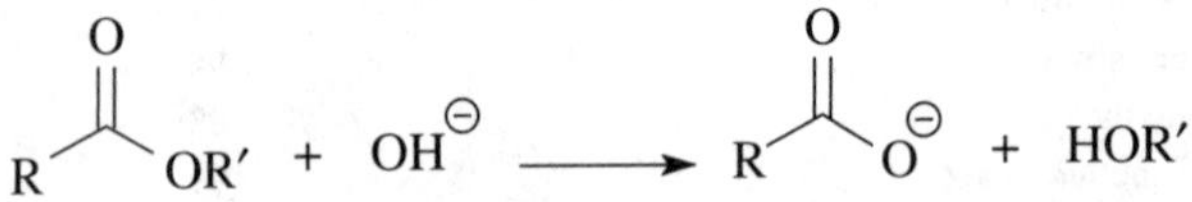

$$\underset{R}{\overset{O}{\|}}{}_{OR'} + OH^{\ominus} \longrightarrow \underset{R}{\overset{O}{\|}}{}_{O^{\ominus}} + HOR'$$

Fats and oils are esters of glycerol, the simplest triol (tri-alcohol), in which each of the three hydroxyl groups has been converted to an ester. The acid portion of the ester linkage (fatty acids) usually contains an even number of carbon atoms in an unbranched chain of 12 to 24 carbon atoms. The triesters of glycerol fats and oils are also known as triglycerides.

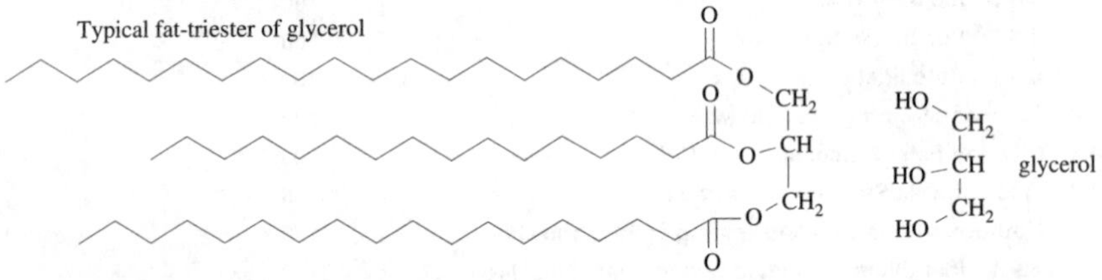

Typical fat-triester of glycerol

glycerol

The difference between fats and oils is merely one of melting point: fats are solid at room temperature (20°C) whereas oils are liquids. Both classes of compounds are triglycerides.

As glycerol is common to all fats and oils, whether animal or vegetable, it is the fatty acid part of the fat (oil) that is of interest. The differences among triglycerides (fats and oils) are because of the length of the hydrocarbon chains of the acids and the number of position of double bonds (unsaturation).

The hydrocarbon chains of the fatty acids may be completely saturated (saturated fat) or may contain one or more double bonds. The geometric configuration of the double bond in fats and oils is normally *cis*. If the chain includes more than one double bond, the fat is called polyunsaturated. The presence of a double bond puts a *kink* in the regular zigzag arrangement characteristic of saturated carbons. Because of this kink in the chains, the molecules cannot form a neat, compact lattice and tend to coil, so unsaturated triglycerides often melt below room temperature and are thus classified as oils.

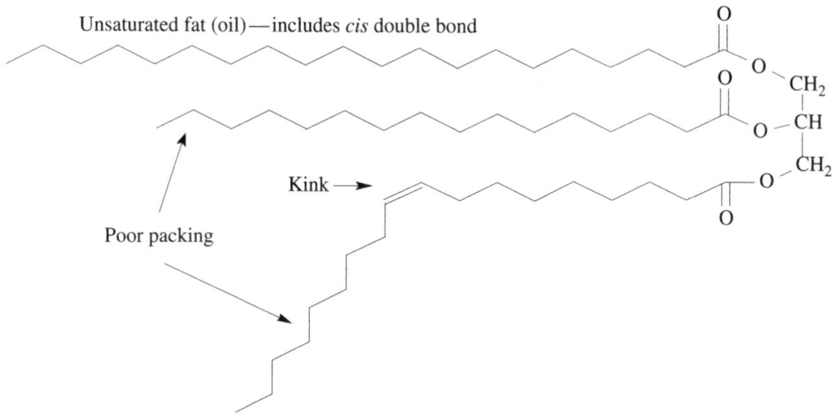

Unsaturated fat (oil)—includes *cis* double bond

Kink →

Poor packing

Fats and oils are the most concentrated source of energy. They provide approximately 9 kcal of energy per gram, compared to 4 kcal/g for proteins and carbohydrates. They are carriers of fat-soluble vitamins and essential fatty acids. They also contribute to food flavor and mouth-feel as well as to the sensation of product richness. They are used as frying fats or cooking oils where their role is to provide a controlled heat-exchange medium as well as to contribute to color and flavor. They are also used in many other commercial applications, including soaps, detergents, and emulsifiers, printing inks, protective coatings, and feeds for domesticated animals.

Waxes are monoesters of long-chain fatty acids, usually containing 24 to 28 carbon atoms, with long-chain fatty primary alcohols. A fatty alcohol has a primary alcohol group ($-CH_2OH$) attached to a

16-to-36 carbon atoms long chain. Waxes are normally saturated and are solids at room temperature.

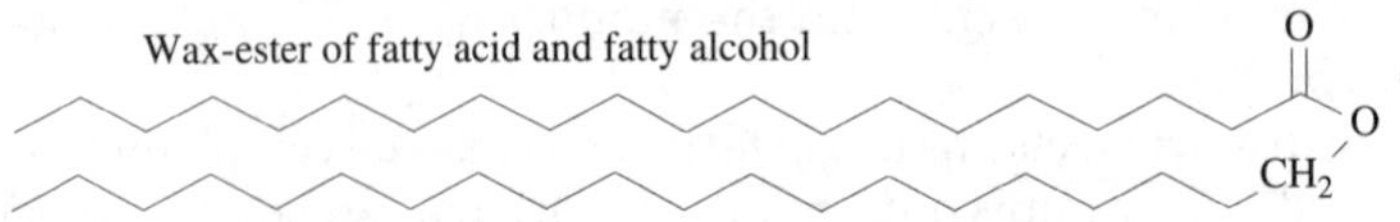

Wax-ester of fatty acid and fatty alcohol

Plant waxes are usually found on leaves or seeds. Thus, cabbage leaf wax consists of the primary alcohols C_{12} and C_{18}—C_{28} esterified with palmitic acid and other acids. The dominant components are stearyl and ceryl alcohol ($C_{26}H_{53}OH$). In addition to primary alcohols, esters of secondary alcohols, for example, esters of nonacosane-15-ol, are present:

$$H_3C - (CH_2)_{13} - CH - (CH_2)_{13} - CH_3$$
$$|$$
$$OH$$

Waxes can be classified according to their origins as naturally occurring or synthetic. The naturally occurring waxes can be classified into animal, vegetable, and mineral waxes. Beeswax, spermaceti, wool grease, and lanolin are important animal waxes. The vegetable waxes include carnauba, ouricouri, and candelilla. Petroleum waxes are the most prominent mineral waxes. Paraffin wax is petroleum wax consisting mainly of normal alkanes with molecular weights usually less than 450. Microcrystalline wax is another type of petroleum wax containing substantial amounts of hydrocarbons other than normal alkanes, and the components have higher molecular weights than paraffin wax components. Compositions of significant commercial waxes from natural sources are given in Table 4.1 [1].

4.2   Fatty Acids

The carboxylic acids obtained from the hydrolysis of fat or oil are called fatty acids. They are the building blocks of the triglycerides and the fats and oils are often named as derivatives of these fatty acids. For example, the tristearate of glycerol is named tristearin and the tripalmitate of glycerol is named tripalmitin.

Normal saturated fatty acids have a long, unbranched hydrocarbon chain having a general formula $CH_3(CH_2)_nCOOH$, where n is usually even and varies from 2 to 24.

The unsaturated fatty acids may have one double bond (monosaturated) or have more than one cis-methylene interrupted double bond (polyunsaturated) as illustrated in Fig. 4.1.

TABLE 4.1   Sources and Compositions of Normal Waxes

Type	Melting point °C	Main components
Animal waxes		
Beeswax	64	Myricyl palmitate
Chinese	82–84	Isoheptacosyl isoheptacosanoate, ceryl lignocerate
Shellac	81–82	Ceryl lignocerate, ceryl cerotate
Spermaceti	—	Cetyl palmitate
Wool (anhydrous lanolin)	36–42	Cholesteryl estolidic esters, alcohol esters of iso- and anteiso acids
Mineral waxes		
Montan	86	Tricontanyl esters of C_{28-30} acids
Petroleum waxes		
Microcrystalline	71–88	Hydrocarbons (490–800 molecular weights)
Paraffin	54–57	Hydrocarbons (350–420 molecular weights)
Vegetable waxes		
Bayberry	43–48	Trimyristin, tristearin
Candelilla	70–80	C_{29-33} hydrocarbons, simple esters and lactones
Carnauba	80–85	Esters of C_{26-30} alcohols and C_{26-30} ω-hydroxy acids
Esparto	69–81	Hydrocarbons, esters of C_{26-32} acids and alcohols
Japan	51–62	Tripalmitin
Jojoba (a liquid wax)	11–12	Docosenyl eicosanoate
Ouricury	79–85	Myricyl cerotate and hydroxycerotate
Sugarcane	79–81	Myricyl palmitate stigmasteryl palmitate

SOURCE: *Riegel's Handbook of Industrial Chemistry*, 9th ed., 1992.

The following three systems of nomenclature are in use for naming fatty acids.

a. The common (trivial) names

b. The number of carbon atoms in the chain

c. The International Union of Pure and Applied Chemistry (IUPAC) system

The trivial names that indicate the initial source of fatty acids are used more often than the IUPAC names in the industry. For example, butyric acid is a major component of butter flavor, palmitic acid comes from palm kernel, and oleic acid from olives.

The number of carbon atoms in the chain, followed by a colon and additional numbers indicating the number of double bonds, are used as an abbreviation to designate fatty acids. Thus in the 18-carbon series, C18:0, C18:1, C18:2, and C18:3 represent stearic, oleic, linoleic, and linolenic acids, respectively. One or two letter abbreviations are also used, and these four acids sometimes are designated by St, O, L, and Ln, respectively.

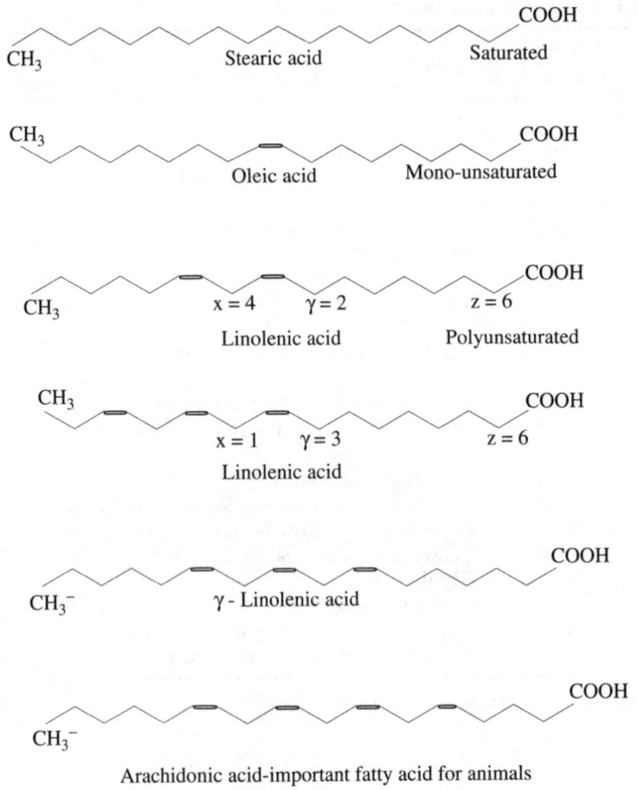

Figure 4.1 The structure of some fatty acids.

The IUPAC name of fatty acid is that of the alkane parent with the –*e* changed to –*oic* acid. The carboxyl carbon is carbon 1:

$$CH_3.CH_2.CH_2.CH_2.CH_2.CH_2.CH_2.CH_3 \text{ (octane)}$$

$$8 \quad 7 \quad 6 \quad 5 \quad 4 \quad 3 \quad 2 \quad 1$$

$$CH_3.CH_2.CH_2.CH_2.CH_2.CH_2.CH_2.COOH \text{ (octanoic acid)}$$

The suffix *dioic* is used if the acid contains two carboxyl groups.

The double bonds in fatty acids differ in (a) number, (b) location, (c) geometrical configuration, and (d) conjugation. Conjugation is a special case of location in which two double bonds are separated only by a single carbon-carbon bond. Unsaturated fatty acids may have one or as many as six double bonds. Those containing multiple double bonds usually have a methylene (CH_2) group between the double-bond sequence, so the system is not conjugated.

When double bonds are present, the suffix *anoic* is changed to *enoic*, *dienoic*, or *trienoic* to indicate the number of bonds present. The location of the first carbon in the double bond is indicated by a number preceding the IUPAC systematic name. The geometric configuration of double bonds is indicated by the Latin prefixes *cis-* (both hydrogens on one side) and *trans-* (hydrogen across from each other). Unsaturation between the 9 and 10 carbons with *cis* orientation is most common in polyunsaturated fatty acids. Accordingly, oleic, linoleic, and linolenic acids are called 9-octadecenoic, 9,12-octadecadienoic and 9,12,15-octadecatrienoic acids, respectively.

Table 4.2 lists some common examples of fatty acids, their sources, common names, and systematic names [1]. Many additional terms are used to distinguish unsaturated fatty acids by the location of the first double bond relative to the omega (ω) or $-CH_3$ carbon. Thus oleic acid is both Δ^9 and a C18:1 ω-9 acid. Linoleic acid is a $\Delta^{9,12}$ and C18:2 ω-6 acid. Linolenic acid is both $\Delta^{9,12,15}$ and a C18:3 ω-3 acid.

TABLE 4.2   Some Important Fatty Acids, Their Names and Common Sources

Carbon atoms	Common name	Systematic name	Common sources
Saturated fatty acids			
3	Propionic	Propanoic	Bacterial fermentation
4	Butyric	Butanoic	Milk fats
5	Valeric	Pentanoic	Bacterial fermentation
5	Isovaleric	3-Methylbutanoic	Dolphin and porpoise fats
6	Caproic	Hexanoic	Milk fats, some seed oils
8	Caprylic	Octanoic	Milk fats, Palmae seed oils
10	Capric	Decanoic	Sheep and goat milk, palm seed oils, sperm head oil
12	Lauric	Dodecanoic	Coconut oil
14	Myristic	Tetradecanoic	Palm and coconut oils
16	Palmitic	Hexadecanoic	Palm oil
18	Stearic	Octadecanoic	Animal fats
19	Tuberculostearic	10-Methylstearic	Tubercle bacillus lipids
20	Arachidic	Eicosanoic	Some animal fats
22	Behenic	Docosanoic	Peanut and various other seed oils
24	Lignoceric	Tetracosanoic	Minor amounts in some seed oils
26	Cerotic	Hexacosanoic	Plant waxes
28	Montanic	Octacosanoic	Beeswax and other waxes
30	Mellisic	Triacontanoic	Beeswax and other waxes
Unsaturated fatty acids			
10	Caproleic	9-Decenoic	Milk fats
10	Stillingic	2,4-Decadienoic	Stillingia oil
12	Lauroleic	2-Dodecenoic	Butterfat
18	Linolenic	9,12,15-Octadecatrienoic	Linseed oil

SOURCE: *Riegel's Handbook of Industrial Chemistry*, 9th ed., 1992.

4.3   Glycerides

Glycerol can be esterified commercially with one, two, or three fatty acids to produce mono-, di-, or triglycerides. Fats and oils are naturally occurring triglycerides, the distribution of which varies in different plants and animals.

$$
\begin{array}{l}
CH_2OH \\
|\\
CHOH \\
|\\
CH_2OH \\
\text{(Glycerol)}
\end{array}
\quad
\xrightarrow[\text{R-COOH}]{\text{Fatty acids}}
\quad
\begin{array}{l}
\overset{\displaystyle O}{\overset{\displaystyle \|}{CH_2O-C}}-R' \\
\overset{\displaystyle O}{\overset{\displaystyle \|}{CHO-C}}-R'' \\
\overset{\displaystyle O}{\overset{\displaystyle \|}{CH_2O-C}}-R'''
\end{array}
$$

(A fat with three different
carboxylic acid-triglyceride)

The properties of triglycerides depend on the fatty acid composition and on the relative location of fatty acids on the glycerol. As accurate methods for determining the composition are available, several conventions have been developed to specify arrangements of fatty acids on the glycerol molecule. Natural fats and oils are designated as the triglyceride type in terms of saturated and unsaturated acids and isomeric forms.

Table 4.3 illustrates the triglyceride types and isomeric forms of some natural fats. GS_3 in the table refers to a fully saturated glyceride and GS_2U refers to a glyceride composed of two saturated acids and one unsaturated acid. Distinguishing between the 1, 3, and 2 positions permits identification of the SUS and SSU isomers of GS_2U and the USU and UUS isomers of GSU_2.

A stereospecific numbering system (Sn) is used to indicate the location of specific fatty acids in triglyceride molecules such as in 1-stearoyl-2-oleoyl-3-myristoyl-Sn-glycerol; the respective fatty acids are indicated in the 1, 2, and 3 positions. This kind of information is very valuable in relating properties of certain fats to compositional data. Table 4.4

TABLE 4.3   Triglyceride Types and Isomeric Forms of Natural Fats

	Types (% wt)				Isomers (% wt)			
	GS_3	GS_2U	GSU_2	GU_3	SUS	SSU	USU	UUS
Pig fat (lard)	2.5	22.4	55.7	19.4	1.0	21.4	46.9	8.8
Peanut oil	0.1	9.9	42.5	47.5	9.3	0.6	0.7	41.8
Beef fat (tallow)	12.6	43.7	35.3	8.4	30.6	13.1	3.4	31.9
Cocoa butter	7.1	67.5	23.3	2.1	65.0	2.5	0.2	23.1
Soybean oil	0	3.7	31.0	65.3	3.7	0	0	31.0

TABLE 4.4 Fatty Acid Composition of Some Edible Oils and Fats

Source	<14:0	14:0	16:0	16:1	18:0	18:1	18:2	18:3	20:0	20:1	22:0	22:1	24:0	24:1
Almond oil	—	0.0	6.5	0.6	1.7	69.4	17.4	—	—	—	—	—	—	—
Avocado oil	—	—	11.0	3.4	0.7	71.5	12.0	1.5	—	—	—	—	—	—
Butter fat	23.8	8.2	21.3	1.8	9.8	20.4	1.8	1.2	1.1	1.5	0.3	0.1	0.1	—
Canola oil	—	—	4.8	0.5	1.6	53.8	22.1	11.1	—	0.0	—	—	—	—
Cocoa butter	—	0.1	25.4	0.2	32.2	32.6	2.8	0.1	—	—	—	—	—	—
Coconut oil	58.7	16.8	8.2	—	2.8	5.8	1.8	—	—	—	—	—	—	—
Corn oil	0.0	0.0	10.9	—	1.8	24.2	58.0	0.7	—	—	—	—	—	—
Cotton seed oil	—	0.8	22.7	0.8	2.3	17.0	51.5	0.2	—	—	—	—	—	—
Fish (manhaden) oil	—	9.6	20.5	12.6	3.3	11.0	0.7	1.6	0.3	—	—	0.8	—	—
Lard	0.5	1.3	23.8	2.7	13.5	41.2	10.2	1.0	—	1.0	—	—	—	1.9
Mustard seed oil	—	0.1	1.9	0.3	0.1	17.7	9.1	0.5	0.6	3.91	1.8	55.1	0.2	—
Olive oil	—	0.0	11.0	0.8	2.2	72.5	7.9	0.6	—	—	—	—	—	—
Palm oil	0.1	1.0	43.5	0.3	4.3	36.6	9.1	0.2	—	0.1	—	—	—	—
Palm kernel oil	54.2	16.4	8.1	—	2.8	11.4	1.6	—	—	—	—	—	—	1.6
Peanut oil	—	0.1	9.5	0.1	2.2	44.8	32.0	—	—	1.3	—	—	1.8	—
Rapeseed oil	—	—	1.7	—	0.9	12.3	12.7	7.6	1.2	5.8	0.9	59.4	0.5	—
Rice bran oil	—	0.7	16.9	0.2	1.6	39.1	33.4	1.6	—	—	—	—	—	—
Safflower oil	—	0.1	6.2	0.4	2.2	11.7	74.1	0.4	—	—	—	—	—	—
Sesame oil	—	—	8.9	0.2	4.8	39.3	41.3	0.3	—	0.2	—	—	—	—
Soybean oil	—	0.1	10.3	0.2	3.8	22.3	51.0	6.8	—	—	—	—	—	—
Sunflower oil	—	—	5.4	0.2	3.5	80.6	8.4	0.2	0.3	—	—	—	—	—
Tallow	0.9	3.7	24.9	4.2	18.9	36.0	3.1	0.6	—	0.3	—	—	—	—
Walnut oil	—	—	7.0	0.1	2.0	22.2	0.4	52.9	10.4	—	—	—	—	—

SOURCE: *Riegel's Handbook of Industrial Chemistry*, 9th ed., 1992.

summarizes general distribution for the major edible fats and oils, and Table 4.5 that for industrial fats and oils [1].

4.4 Physical Properties of Triglycerides

The physical properties, such as melting points, specific heat, viscosity, density, and refractive index depend on the type of fatty acids present in the triglyceride and their location, chain length of fatty acids, number and location of *cis* and *trans* double bonds on the fatty acid chains as well as compatibility of the different triglycerides in the mixture and the type of crystal present.

4.4.1 Melting point

The melting point of fats usually occurs over a wide range of temperature. The melting range of fats depends on the triglyceride composition. Among the fatty acids, melting points increase with chain length. *Trans* fatty acids always have higher melting points than their *cis* counterparts for any chain length. Melting point data are useful in animal fats and processed fats but are of little value for vegetable oils because most oils are liquids at ambient temperatures.

4.4.2 Specific heat

The specific heat of fats is defined as the ratio of the heat capacity of a fat to the heat capacity of water; or the quantity of heat required for a one-degree temperature change in a unit weight of fats. Although the specific heat of most triglycerides in a given physical state are similar, specific heat does increase with increasing unsaturation of fatty acids in both the liquid and solid states of a fat. Liquid fats have almost twice the specific heat values than those of solid fats. Knowledge of the specific heat of fats and oils is useful in processing operation.

4.4.3 Viscosity

Viscosity of an oil or a fat is a measure of its resistance to flow. Viscosity is an important factor in designing systems for handling oils and fats. Among the fatty acids, viscosity increases with chain length and decreases with increasing unsaturation. Viscosity is thus a function of molecular size and orientation of the molecules. There is an approximately linear relationship between log viscosity and temperature. The viscosity of oils usually increases in prolonged heating as a result of polymerization (gum formation).

TABLE 4.5 Fatty Acid Composition of Some Industrial Oils and Fats

Source	<14:0	14:0	16:0	16:1	18:0	18:1	18:2	18:3	20:0	20:1	22:0	22:1	24:0	24:1
Castor oil	—	—	1.1	0.2	1	3.3	3.6	0.32	0.4	—	—	—	—	—
Chinese tallow	1.3	2.1	65	—	4.4	22.5	0.8	—	—	—	—	—	—	—
Croton oil	2.5	5.4	6.2	0.2	3.2	15.8	49.4	3	2.9	8.9	0.2	0.6	—	—
Jojoba oil	—	—	1	—	—	9	—	—	—	70.7	—	16.3	—	3
Linseed oil	—	—	5.3	—	4.1	20.2	12.7	53.3	—	—	—	—	—	—
Rapeseed oil	—	0.1	2.6	0.3	0.9	11.2	12.8	8.6	—	7.5	—	48.1	—	—
Tall oil	—	—	—	—	—	50	7	41	—	—	—	—	—	—
Tung oil	—	—	3.1	—	2.1	11.2	14.6	69	—	—	—	—	—	—
Whale oil	—	3.3	8.1	26.9	1.1	33.3	—	—	—	10.9	—	2.2	—	—

SOURCE: *Riegel's Handbook of Industrial Chemistry*, 9th ed., 1992.

4.4.4 Density

The density of fats and oils is an index of the weight of a measured volume of the material. This property is important not only for designing of equipment but also for the estimation of the solid fat index (SFI). The SFI is related, approximately, to the percentage of solids in a fat at a given temperature. When determined at a number of specified temperatures, it can be especially useful to margarine manufacturers or other processors who need to control the characteristics of their manufactured products by blending.

4.4.5 Refractive index

The refractive index of fats and oils is sensitive to composition. The refractive index of a fat increases with increasing chain length of fatty acids in the triglycerides or with increasing unsaturation. This makes it an excellent spot test for uniformity of compositions of oils and fats. Further, the refractive index is an additive as well as constructive property, thus it can be used as a control procedure during hydrogenation processes.

4.4.6 Polymorphism

The existence of a substance in two or more forms, which are significantly different in physical or chemical properties, is known as polymorphism. The difference between the forms arises from different modes of molecular packing in the crystal structure of certain triglycerides. Certain pure or mixed fatty acid triglycerides may show as many as five different melting points. Each crystal system has a characteristic melting point, x-ray diffraction pattern, and infrared spectrum. For example, tristearin can exist in three polymorphic forms with melting points of 54.7, 63.2, and 73.5°C.

An awareness of crystal-packing characteristics and polymorphism helps one to understand incompatibility problems of different fats. Polymorphism has several industrial implications in use of fats as shortenings, margarines, and cocoa butter.

4.4.7 Other physical properties

Other physical properties such as the smoke, flash, and fire points of oils and fats are measures of their thermal stability when heated. The smoke point is important for the oils and fats used for deep-frying. The flash point and fire points are a measure of residual solvent in crude and refined oils and are also a safety requirement.

Table 4.6 shows some physical properties of oils and fats [1].

TABLE 4.6 Physical Properties of Oils and Fats

Properties	Rapeseed oil	Peanut oil	Cottonseed oil	Soybean oil	Sunflower oil	Coconut oil	Palm oil	Olive oil	Beef tallow	Lard
1) Density @15°C g/cm^3	0.910	0.991	0.917	0.922	9.20	0.919	0.921	0.914	0.936	0.914
2) R.I. @20°C	1.472	1.460	1.472	1.470	1.474	1.448	1.453	1.467	1.454	1.458
3) Smoke Pt. °C	218	207	223	213	209	200	223	—	—	—
4) Flash Pt. °C	317	315	322	317	316	300	314	—	316	—
5) Fire Pt. °C	344	342	342	342	341	—	341	—	344	—
6) Viscosity @50°C n.m.pa.s	30	23	27	25	26	19	28	30	25	25
7) Sp. Heat J/g	—	—	2.18 @79.6°C	2.06 @80.4°C	2.50 @175°C	—	2.40 @140°C	—	—	—

4.5   Chemical Properties of Triglycerides

The most important chemical reactions for triglycerides (fats and edible oils) are hydrolysis, methanolysis, and interesterification. The other reactions, such as hydrogenation, isomerization, polymerization, and autoxidation that are primarily relevant to the processing of edible oils and fats are also discussed in this section.

4.5.1   Hydrolysis

The fat or oil can be hydrolyzed into fatty acids and glycerol by treatment with steam under elevated pressure and temperature. The reaction is reversible and is catalyzed by inorganic catalysts (ZnO, MgO, or CaO) and an acid catalyst (aromatic sulfonic acid).

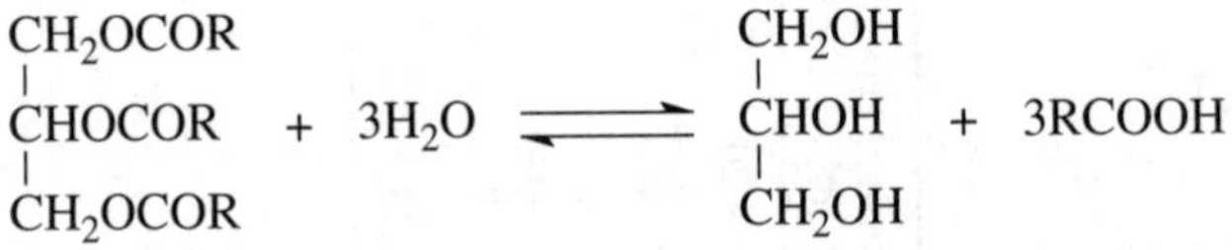

$$
\begin{array}{l}
CH_2OCOR \\
| \\
CHOCOR \\
| \\
CH_2OCOR
\end{array}
+ 3H_2O \rightleftharpoons
\begin{array}{l}
CH_2OH \\
| \\
CHOH \\
| \\
CH_2OH
\end{array}
+ 3RCOOH
$$

Glycerides can also be hydrolyzed by treatment with alkali (saponification). After acidification and extraction, the free fatty acids are recovered as alkali salts (soaps).

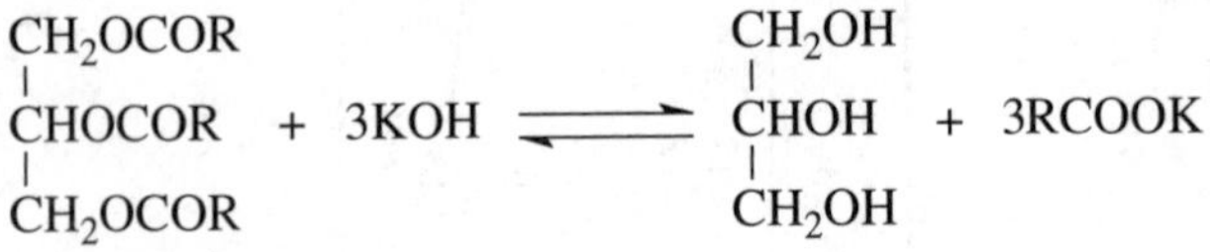

$$
\begin{array}{l}
CH_2OCOR \\
| \\
CHOCOR \\
| \\
CH_2OCOR
\end{array}
+ 3KOH \rightleftharpoons
\begin{array}{l}
CH_2OH \\
| \\
CHOH \\
| \\
CH_2OH
\end{array}
+ 3RCOOK
$$

4.5.2   Methanolysis

The fats and oil react with methanol to produce fatty methyl esters. Inorganic alkali, quaternary ammonium salts, and enzymes (lipase) have been used as catalysts for methanolysis in commercially practiced processes for soap manufacture.

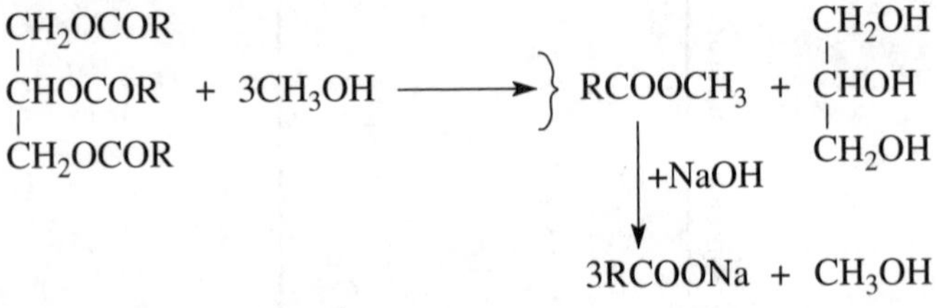

$$
\begin{array}{l}
CH_2OCOR \\
| \\
CHOCOR \\
| \\
CH_2OCOR
\end{array}
+ 3CH_3OH \longrightarrow \left.\begin{array}{l} \\ \\ \\ \end{array}\right\} RCOOCH_3 +
\begin{array}{l}
CH_2OH \\
| \\
CHOH \\
| \\
CH_2OH
\end{array}
$$

$$
\downarrow +NaOH
$$

$$
3RCOONa + CH_3OH
$$

4.5.3   Interesterification

Interesterification causes a fatty acid redistribution within and among triglyceride molecules, which can lead to substantial changes in the physical properties of fats and oils or their mixtures without altering the chemical structure of the fatty acids. Intermolecular acyl groups exchange triglycerides in the reaction until an equilibrium is reached, which depends on the structure and composition of the triglyceride molecules. The reaction is very slow even at 200 to 300°C, but the rate of reaction can be accelerated by using sodium methylate.

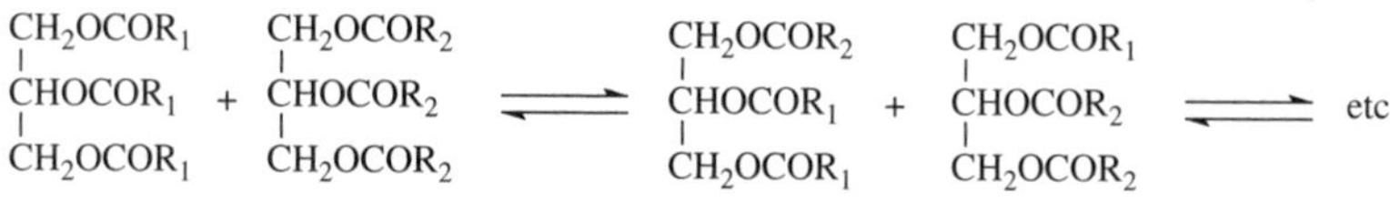

$$
\begin{array}{l}
CH_2OCOR_1 \\
CHOCOR_1 \\
CH_2OCOR_1
\end{array}
\; + \;
\begin{array}{l}
CH_2OCOR_2 \\
CHOCOR_2 \\
CH_2OCOR_2
\end{array}
\; \rightleftharpoons \;
\begin{array}{l}
CH_2OCOR_2 \\
CHOCOR_1 \\
CH_2OCOR_1
\end{array}
\; + \;
\begin{array}{l}
CH_2OCOR_1 \\
CHOCOR_2 \\
CH_2OCOR_2
\end{array}
\; \rightleftharpoons \; etc
$$

Interesterification may be either random or directed. In *random interesterification* the acyl groups are randomly distributed as demonstrated by the following example where equal proportions of tristearin (S—S—S) and triolein (O—O—O) are allowed to interesterify.

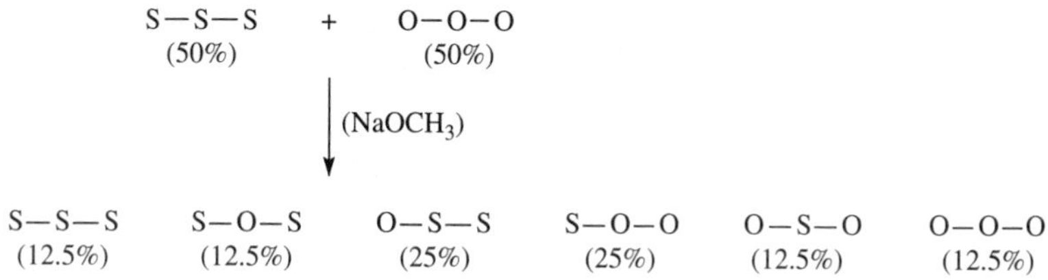

$$
\underset{(50\%)}{S-S-S} \quad + \quad \underset{(50\%)}{O-O-O}
$$

$$\downarrow (NaOCH_3)$$

$$
\underset{(12.5\%)}{S-S-S} \quad \underset{(12.5\%)}{S-O-S} \quad \underset{(25\%)}{O-S-S} \quad \underset{(25\%)}{S-O-O} \quad \underset{(12.5\%)}{O-S-O} \quad \underset{(12.5\%)}{O-O-O}
$$

In *directed interesterification*, the reaction temperature is lowered until the higher-melting and least-soluble triglyceride molecule in the mixture crystallizes. In this way, a fat can be separated into higher and lower melting fractions.

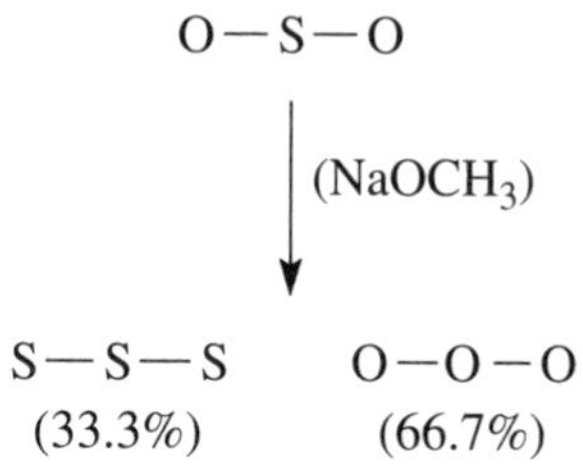

$$O-S-O$$

$$\downarrow (NaOCH_3)$$

$$
\underset{(33.3\%)}{S-S-S} \qquad \underset{(66.7\%)}{O-O-O}
$$

The directed interesterification is of much industrial significance because it can be used to convert oils into more and/or less saturated fractions of original fat or oil or blend of two oils.

4.5.4 Hydrogenation

The unsaturated double bonds in a fatty-acid chain are converted to saturated bonds by addition of hydrogen. The reaction between the liquid oil and hydrogen gas is accelerated by using a suitable solid catalyst such as nickel, platinum, copper, or palladium. Hydrogenation is exothermic, and leads to an increase in melting point and drop in iodine value. Partial hydrogenation can lead to isomerization of *cis* double bonds (geometrical isomerization).

Polyunsaturated fatty acids such as linolenic acid (C18:3) are hydrogenated more quickly to linoleic (C18:2) or oleic acid (C18:1) than linoleic to oleic acid or oleic acid to stearic acid (C18:0). The conversion steps can be represented as follows:

$$\text{C18:3} \xrightarrow[\text{+H}_2]{K_1} \text{C18:2} \xrightarrow[\text{+H}_2]{K_2} \text{C18:1} \xrightarrow[\text{+H}_2]{K_3} \text{C18:0}$$

4.5.5 Isomerization

The configuration of the double bond in naturally occurring fatty acids, present in oils and fats, is predominantly in the *cis* form. Isomerization can occur if oils and fats are heated at temperatures above 100°C in the presence of bleaching earths or catalysts such as nickel, selenium, sulfur, or iodine.

Two types of isomerization spontaneously occur during hydrogenation: geometrical and positional. The extent to which isomerization occurs can be affected by processing conditions and catalyst selection.

4.5.6 Polymerization

Under deep-frying conditions (200 to 300°C) the unsaturated fatty acids undergo polymerization reactions forming dimeric, oligomeric, and polymeric compounds. The rate of polymerization increases with increasing degree of unsaturation: saturated fatty acids are not polymerized. In thermal polymerization polyunsaturated fatty acids are first isomerized into conjugated fatty acids, which in turn interact by Diels-Alder reactions producing cyclohexene derivatives. The cyclohexene ring is readily dehydrogenated to an aromatic ring; hence compounds related to benzoic acid can be formed.

On the other hand, oxidative polymerization involves formation of $C-O-C$ bonds. Polymers with ether and peroxide linkages are formed in the presence of oxygen. They may also contain hydroxy, oxo, or epoxy groups. Such compounds are undesirable in deep-fried oil or fat because they permanently diminish the flavoring characteristics of the oil or fat and also cause a foaming problem.

4.5.7 Autoxidation

Fats and oils often contain double bonds. Autoxidation of a fat or oil yields a mixture of products that include low molecular weight carboxylic acids, aldehydes, and methyl ketones.

Drying oils such as linseed oil contain many double bonds. These oils are purposely allowed to undergo air oxidation leading to a tough polymer film on the painted surface.

The autoxidation reactions involve three steps: initiation, propagation, and termination. The initiation step leads to the formation of a hydroperoxide on a methylene group adjacent to a double bond; this step proceeds via a free-radical mechanism:

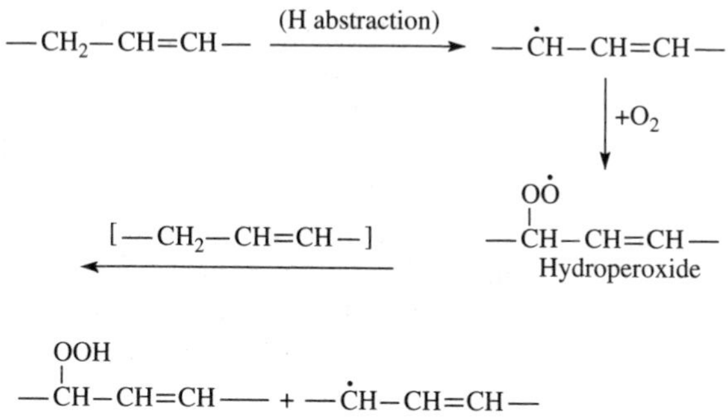

$$-CH_2-CH=CH- \xrightarrow{\text{(H abstraction)}} -\overset{\cdot}{C}H-CH=CH-$$

$$\downarrow +O_2$$

$$[-CH_2-CH=CH-] \quad \overset{O\overset{\cdot}{O}}{\underset{\text{Hydroperoxide}}{-\overset{|}{C}H-CH=CH-}}$$

$$\overset{OOH}{-\overset{|}{C}H-CH=CH-} + -\overset{\cdot}{C}H-CH=CH-$$

The second step, which is also a reaction in the propagation cycle, is the addition of another molecule to the hydroperoxide radical to generate new free radicals.

The chain length of these two radical-reaction steps is about 100. When the radical concentration has reached a certain limit, the chain reaction is gradually stopped by mutual combination of radicals, the termination step.

Numerous compounds result from these reactions. For example, Table 4.7 lists a series of aldehydes and methyl ketones derived preferentially when tristearin is heated in air at 192°C [2].

TABLE 4.7   Oxidation Products of Tristearin

Class of compounds	% Portion	C-number	Major compounds
Alcohols	2.7	4–14	n-Octanol, n-Nonanol, n-Deanol
v-Lactones	4.1	4–14	v-Butyroactone, v-Pentalactone, v-Heptalactone
Alkanes	8.8	4–17	n-C_7, nC_{-9}, nC_{-10}
Acids	9.7	2–12	Caproic, Butyric
Aldehydes	36.1	3–17	n Hexanal, n Octanal
Methyl ketones	38.4	3–17	2-Heptanone, 2-Decanone

4.6   Sources of Edible Oils and Main Fats

Several hundred plants and animals produce fats and oils in sufficient quantities to warrant processing into edible oils; however, only a few sources are commercially significant. Table 4.8 summarizes the major sources in the world and the method of processing.

The organ fats of domestic animals, such as cattle and hogs, and milk fat are important raw materials for fat production. Edible oils are mostly of plant origin. Olive oil and palm oil are extracted from fruits. All other oils are extracted from oilseeds. The world production of oilseeds and other crops has significantly increased in recent years to meet the growing needs for oils and fats in the world.

World oilseed production in 2003 was 335.9 million tons [3]. Figure 4.2 shows world production of the different oilseeds. Soybeans constitute the largest share (56 percent) and the United States is the main crop-grower. Malaysia grows mainly palm. The Philippines grows coconut. China, Europe, India, and Canada grow rapeseed (canola). Sunflower is grown in the United States, Australia, Europe, and Argentina. The cottonseed market is dominated by the United States, China, Pakistan, India, and the former Soviet Union.

World vegetable oil consumption in 2003 was 87.2 million tons. U.S. consumption was 9.91 million tons. In the U.S. market, animal fats (tallow and lard) have a relatively small share (2 percent) compared to vegetable oils. The consumption of four oils—soybean (80 percent), corn (4 percent), canola (4 percent), and cottonseed (3 percent) has grown rapidly over the past 30 years compared to the traditional oils and animal fats. Figure 4.3 shows U.S. consumption of edible fats and oils in 2003 [3].

TABLE 4.8   Major Edible Fats and Oils in the World and Methods of Processing

Source	Oil content (%)	Prevalent method of recovery
Soybean	19	Direct solvent extraction
Corn (germ)	40	Wet or dry milling and prepress solvent extraction
Tallow (edible tissue)	70–95	Wet or dry rendering
Canola	42	Prepress solvent extraction
Coconut (dried copra)	66	Hard pressing
Cottonseed	19	Hard pressing or prepressing or direct solvent extraction
Lard (edible tissue)	70–95	Wet or dry rendering
Palm	47	Hard pressing
Palm kernel	48	Hard pressing
Sunflower	40	Prepress solvent extraction
Peanut (shelled)	47	Hard pressing or prepress solvent extraction

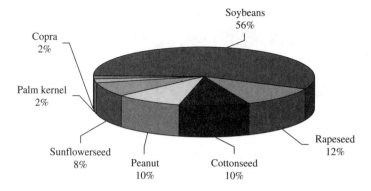

	Million short tons	Million metric tons
Soybeans	209.5	190.1
Rapeseed	43	39
Cottonseed	38.7	35.2
Peanut	35.4	32.1
Sunflowerseed	28.7	26
Palm kernel	8.9	8.1
Copra	5.9	5.4
Total	370.1	335.9

Source: USDA

Figure 4.2   World oilseed production 2003.

4.7   Oils and Fats: Processing and Refining

Crude fats and oils consist primarily of glycerides. However, they also contain many other lipids in minor quantitites. Corn oil, for example, may contain glycerides plus phospholipids, glycolipids, many isomers of sitosterol and stigmasterol (plant steroids), several tocopherols (vitamins E), vitamin A, waxes, unsaturated hydrocarbons such as squalane and dozens of carotenoids and chlorophyll compounds, as well as many products of decomposition, hydrolysis, oxidation, and polymerization of any of the natural constituents.

All crude oils and fats obtained after rendering, crushing, or solvent extraction, inevitably contain variable amounts of nonglyceridic coconstituents like fatty acids, partial glycerides (mono- and diglycerides), phosphatides, sterols, tocopherols, hydrocarbons, pigments (gossypol, chlorophyll), vitamins (carotene), sterol glucosides, protein fragments as well as resinous and mucilaginous materials, traces of pesticides, and

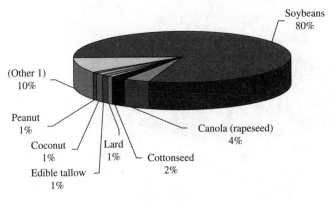

	Million pounds	Million metric tons
Soybeans	17,471	7.92
Canola (rapeseed)	857	0.39
Cottonseed	482	0.22
Lard	205	0.09
Edible tallow	236	0.11
Coconut	310	0.14
Peanut	185	0.08
(Other 1)	2,112	0.96
Total	21,858	9.91

Source: USDA

Figure 4.3   U.S. fats and oils edible consumption, 2003.

"heavy" metals. Table 4.9 shows some of the most important coconstituents found in some major oils [4].

Some of these materials are highly undesirable and must be removed to provide satisfactory processing characteristics and to provide desirable color, odor, flavor, and keeping qualities in the finished products.

TABLE 4.9   Minor Components in Some Major Oils

Component	Soybean oil	Canola	Palm oil
FFA	0.3–0.8%	0.3–1.0%	2.0–5.0%
Phosphatides	1.0–3.0%	0.5–3.5%	0.03–0.1%
Sterols/triterpenic alcohols	0.04–0.07%	0.04–0.6%	0.1–0.2%
Tocopherols/tocotrienols	0.06–0.2%	0.06–0.1%	0.05–0.1%
Carotenoids	40–50 ppm	25–70 ppm	500–800 ppm
Chlorophyll/pheophytine	1–2 ppm	5–30 ppm	Traces
Peroxides meg O_2/kg	<10	<10	1–5
Fe	1–3 ppm	1–5 ppm	4–10 ppm
Cu	0–30 ppb	0–30 ppb	0–50 ppb

These objectionable coconstituents are removed during the refining process in such a way that the glyceride yield and the desirable constituents in the oil are not affected.

Figure 4.4 shows the different stages in seed preparation and a classical chemical refining process. The general methods employed to produce edible oils suitable for human consumption consist of (a) seed preparation, (b) extraction, (c) degumming, (d) neutralization, (e) bleaching, (f) deodorization, (g) hydrogenation, and sometimes winterization.

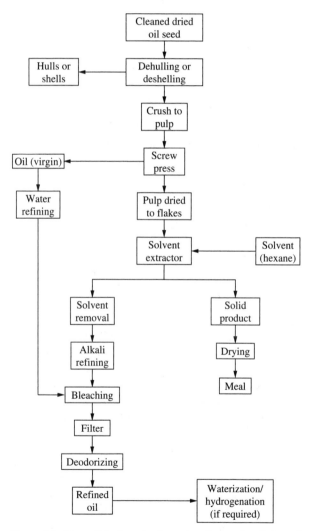

Figure 4.4 Essential steps in the extracting and refining of edible oil from oilseeds.

Seed preparation. When oilseeds are received at the oil mill, they still contain plant residues, damaged seeds, dust, sand, wood, pieces of metal, and foreign seeds. The oilseeds are carefully cleaned of these materials using magnets, screens, and aspirator systems. The cleaned seeds are dried to remove moisture. Next, the dried oilseeds are usually decorticated to remove the hull that surrounds the oilseed *meat* before being further processed. Hulls always contain much less oil than kernels or meats. Further, removing the hull reduces the amount of material that must be handled, extracted, and refined. Dehulling is normally preformed very carefully, to ensure that the meat is not broken into too many small pieces. Corrugated roller mills or bar mills are used to cut or break the hull to free the meat. The hulls are separated by screening and air clarification. Hulls may be blended back with meat to control protein level, sold as a cattle feed, or burned in boilers to generate steam and electrical energy.

Extraction. The raw material for the fat and oil industry comes from animals (hogs, sheep, and fatty fish); fleshy fruits (palm and olive); and various oilseeds. Most oilseeds are grown specifically for processing of oils and protein meals.

The purpose of oil extraction is twofold: first, extracting the maximum amount of good quality oils and then getting maximum value from the residual press cake or meal. The following three methods, with varying degrees of mechanical simplicity, are used: (1) rendering, (2) pressing with mechanical presses, and (3) extracting with a volatile solvent.

Rendering. The rendering process is applied on a large scale to the production of animal fats, such as tallow, lard, bone fat, and whale oil. The fatty tissues are chopped into small pieces and are boiled in steam digesters. The fat is gradually liberated from the cells and floats to the surface of the water, where it is collected by skimming. A similar method is used in the extraction of palm oil from fresh palm fruits.

Pressing. Oil seeds do not have fat cells like those of animals for storing fats. Instead oil is stored in microscopic globules throughout the cells. In these cases, rendering will not liberate the oil from the cellular structures and the cell walls are broken only by grinding, flaking, rolling, or pressing under high pressures to liberate the oil. The general sequence of modern operations in pressing oilseeds and nuts is as follows: (1) Preparation of the seed to remove stray bits of metals and removal of hulls; (2) Reduction of particle size of the kernels (meats) by grinding; and (3) Cooking and pressing in hydraulic or screw presses.

Efficient oil extraction by a mechanical press is highly dependent on having the correct preparation before pressing. The extraction stage

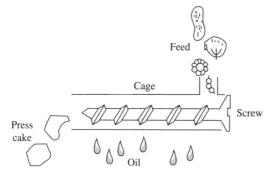

Feed

Cage

Screw

Press
cake

Oil

Figure 4.5   Screw press.

itself is carried out using a screw press (Fig. 4.5). The press is fed by means of a variable speed conveyor, within the feeder unit. The feeder regulates the flow of material into the press and thereby controls the loading on the press main motor. Oil released along the length of the cage is allowed to drain into the base of the press where it is collected. The solid material (press cake) remaining within the press is finally discharged into conveyors to be removed for subsequent processing or storage [5].

Oil expressed without heating contains the least amount of impurities and is often of edible quality without refining or further processing. Such oils are known as cold-drawn, cold-pressed, or virgin oils. The expressed oil from cooked seeds contains greater quantities of nonglyceride impurities such as phospholipids, color bodies, and unsaponifiable matter. Such oils are highly colored and are not suitable for edible use.

Solvent extraction.   The press cake emerging from a screw press still retains 3 to 15 percent of residual oil. More complete extraction is done by solvent extraction of the residues obtained from mechanical pressing. The greater efficiency obtained in the solvent extraction process encouraged the industry for direct application to oilseeds. In the United States and Europe, continuous extractor units are used in which fresh seed flakes are added continuously and are subjected to a counterflow of solvent by which intimate contact is achieved between the seeds and solvent. The common solvent for edible oil is commercial hexane or heptane, commonly known as petroleum ethers, boiling in the range of 146 to 156°F (63.3 to 68.9°C). After extraction, maximum solvent recovery is necessary for economical operation. The solvent is recovered by distillation and is reused. The extraction oil is mixed with prepress oil for refining. The extracted meals contain less than 1 percent of residual oil.

In large-scale operations, solvent extraction is a more economical means of recovering oil than mechanical pressing.

Refining (degumming and neutralization). The usual refining of vegetable oils involves degumming and alkali refining. Degumming mainly reduces the phosphatides and metal content of the crude oil by mixing it with an acid and water. The phosphatides are present in free hydratable form (HP) or in nonhydratable form (NHP), mostly in combination with some Ca^{++}, Mg^{++}, or Fe^{++}. In alkali refining, the NHP that remains behind in the oil after acid treatment, and the free fatty acids formed during the hydrolysis (lipolysis) of the HP, are further removed by neutralization.

Degumming consists of treating the oil with a small amount (0.05 percent) of concentrated phosphoric acid and water, followed by centrifugal separation of coagulated material (lecithin). The process is applied to many oils (e.g., soybean oil) that contain phospholipids in significant amounts. Refining with alkali removes free fatty acids that are formed during the lipolysis of the fat or oil before rendering or extraction. The oil is treated with an excess of 0.1 percent caustic soda solution and the mixture is heated to about 75°C to break any emulsions formed. The mixture is allowed to settle. The settlings, called "foots," are collected and sold as "soapstock." In the continuous system, the emulsion is separated with centrifuges. After the oil has been refined, it is usually washed with water to remove traces of alkali and soapstock. After water washing, the oil may be dried by heating in a vacuum or by filtering through a dry filter and material.

A whole variety of processes have been developed to improve the removal of the nonhydratable phosphatides. The best known are the uni- and superdegumming processes (unilever) and the TOP degumming process (vandemoortele). They are principally based on a special acid treatment of the NHP. Over the last few years, several new technological approaches have been developed, which guarantee very low levels of phosphorus (less than 10 ppm).

In the enzymatic degumming process, part of the hydratable phosphatides is enzymatically modified by removing the fatty acid on the C-2 position of the glycerol, using a phospholipase A2 enzyme as biocatalyst. These modified phosphatides facilitate the removal of the remaining NHP. Table 4.10 shows the results of an enzymatic degumming

TABLE 4.10   **Example of Enzymatic Degumming**

Rapeseed oil	Crude oil	Degummed oil
Phospholipids	0.60%	0.01–0.02%
P	240 ppm	4–8 ppm
H_2O	0.10%	0.30%

NOTE: P-content = phosphatide content $\times 10^4 t/25$.

of crude rapeseed oil. The phosphorus levels are reduced from 240 ppm to 4–8 ppm [4].

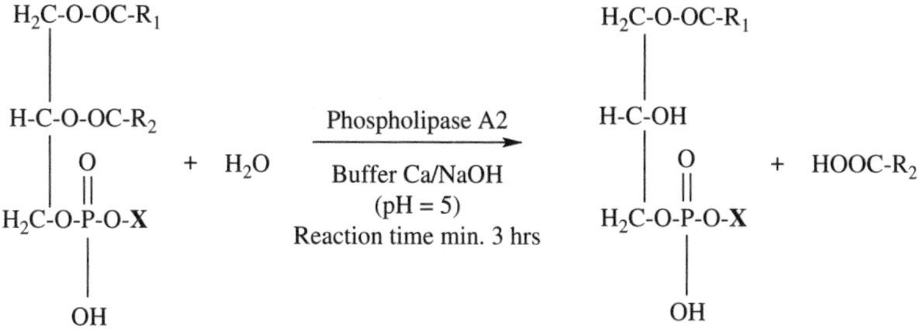

$$H_2C\text{-}O\text{-}OC\text{-}R_1$$

$$H\text{-}C\text{-}O\text{-}OC\text{-}R_2 \quad \xrightarrow[\substack{\text{Buffer Ca/NaOH} \\ (pH = 5) \\ \text{Reaction time min. 3 hrs}}]{\text{Phospholipase A2}} \quad H\text{-}C\text{-}OH$$

$$\underset{H_2C\text{-}O\text{-}P\text{-}O\text{-}\mathbf{X}}{\overset{O}{\underset{\parallel}{}}} + H_2O$$

$$OH$$

Phosphatide

$$H_2C\text{-}O\text{-}OC\text{-}R_1$$

$$\underset{H_2C\text{-}O\text{-}P\text{-}O\text{-}\mathbf{X}}{\overset{O}{\underset{\parallel}{}}} + HOOC\text{-}R_2$$

$$OH$$

Lyso-lecithin

In the soft-degumming process, a chelating agent (EDTA) is added to the oil to remove the cations from the nonhydratable phosphatides, thereby making them hydratable again (Table 4.11) [4].

$$(\text{phosphatide})_2^{-}M^{++} + EDTA \xrightarrow{\text{"catalyst"}} \text{phosphatide} + EDTA\text{-}M^{++}$$

$$M^{++}: Ca^{++}, Mg^{++}, Fe^{++}$$

In the improved acid degumming (IMPAC De Smet), special additives are mixed to improve the wetting of the NHP and hence their solubility in the aqueous phase [4]. Furthermore, an intimate contact between the reaction products (acid/alkaline) and the phosphatides is ensured to improve hydratability of the NHP and hence the efficiency of their removal. However, in all dugumming processes the efficiency of the degumming treatment depends to a great extent on the quality of the crude oil. A good-quality fresh oil is easier to degum than an adulterated oil. This is a direct consequence of the NHP to HP ratio: the lower the NHP content, the easier the degumming procedure and the better the degummed oil yield. Figure 4.6 summarizes the different steps of degumming processes in the form of a process diagram [4].

TABLE 4.11   Example of Soft Degumming

Soybean oil	Crude oil	Degummed oil
Phospholipids	0.30%	0.01%
P^*	120 ppm	3–5 ppm
Ca	59	0.1
Mg	35	0.02
Fe	1.6	0.04

NOTE: P-content = phosphatide content $\times 10^4/25$.
*Represents laboratory data.

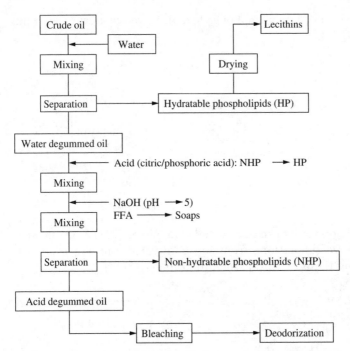

Figure 4.6 Generalized flow sheet for degumming process. (*Source*: Courtesy of DeSmet Group, Antwerp, Belgium.)

Bleaching. The refined oils are usually dark in color owing to the presence of some pigmented materials such as chlorophyll or carotenoids and minor impurities like residual phosphatides, soaps, metals, and oxidatin products. Bleaching reduces the color by absorbing these colorants on bleaching earth (bentonite clays), or activated charcoal, or both. In addition to decolorization, bleaching clay also absorbs suspended matter and other minor impurities.

The bleaching process comprises three stages:

- Initial mixing of oil with bleaching earth.
- Oil heating under a vacuum with sparge steam to ensure complete contact between the oil and the earth.
- Filtration on hermetic leaf filters, followed by polishing filtration. The spent cake is dried by steam blowing and the recovered oil is recycled.

Natural bleaching clays are aluminum silicates (bentonite, attapulgite, and montmorillonite), containing relatively high amounts of Mg, Ca, or Fe. The clays are generally activated by heat treatment. The high metal content, however, limits the adsorptive activity of these clays. The metals can be removed from the reactive spots by means of acid treatment, yielding clays with much higher adsorptive capacity. In some cases, active carbon is added in the course of bleaching to improve the removal of blue and green pigments as well as polycyclic aromatic hydrocarbons. As a result of high cost and high-oil retention, active carbon is used only in specific cases (e.g., coconut and palm kernel oil), in combination with bleaching clays, mostly in a ratio of 1/10 to 1/20.

Bleaching is by far the most expensive process in refining in terms of utilities cost. The relatively high cost of the bleaching clays as well as the oil-in-earth losses and bleaching adsorbent disposal costs largely affect the operating cost of a bleaching plant. Moreover, the ever stronger environmental regulations are now forcing oil refiners to drastically reduce mainly the solid wastes as these are the most difficult to treat.

Several bleaching processes have been developed over the years to limit the consumption of bleaching earth. Figure 4.7 shows the flow sheet of a double-batch bleaching process. This is by far the oldest bleaching method, still in use in many refining plants [4].

In order to reduce bleaching earth consumption, alternative new multistep bleaching processes have been developed. Figure 4.8 shows the flow sheet of a two-stage counter-current bleaching with a prefiltration process. The main function of prefiltration is to remove all solid impurities as well as to adsorb most of the phosphatides and soaps. This increases bleaching efficiency in the second stage [4].

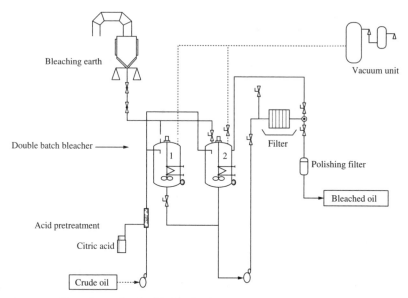

Figure 4.7 Flow sheet of a double batch bleaching process with acid pretreatment. (*Source*: Courtesy of DeSmet Group, Antwerp, Belgium.)

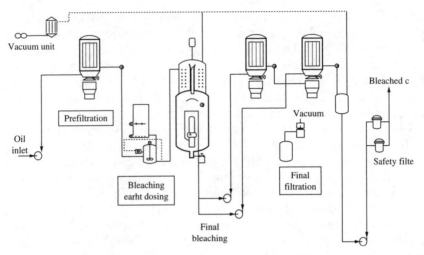

Figure 4.8   Two-stage counter-current bleaching with prefiltration. (*Source*: Courtesy of DeSmet Group, Antwerp, Belgium.)

Deodorization.   Most fats and oils, even after refining, have characteristic flavors and odors owing to the presence of minor amounts of free fatty acids, aldehydes, ketones, and other compounds. The concentration of these undesirable substances, found in most oils, is generally low, between 0.2 and 0.5 percent. The efficient removal of undesirable substances depends on (a) the vapor pressure of the different minor compounds, (b) the deodorizing conditions (temperature, pressure, residence time), (c) the amount of stripping steam, and (d) the geometry of the vessel (e.g., sparge-steam distribution, oil depth, stainless steel, and so on.).

Deodorization is usually conducted at a temperature between 220 and 260°C, at a pressure between 2 and 4 mbar, and under injection of 0.5 to 3 percent steam in a stainless steel vessel.

The different stages in the deodorization process are as follows:

- Deaeration
- Heating
- Deodorization or steam stripping
- Heat recovery or cooling
- Final cooling

The volatiles, evaporated during deodorization, are condensed and are usually recovered in a direct condenser or scrubber for fatty substances.

Both batch and continuous deodorization processes are used. The batch deodorizer is the earliest and most simple type of process. It generally

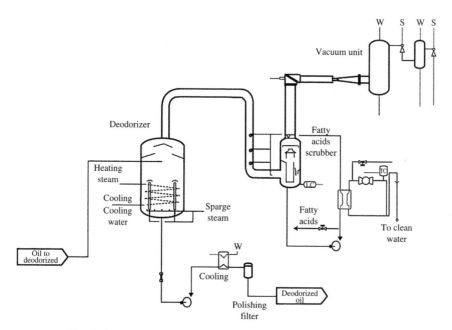

Figure 4.9   Batch deodorization process. (*Source*: Courtesy of DeSmet Group, Antwerp, Belgium.)

uses a single-shell vertical cylindrical vessel to perform all desired functions (Fig. 4.9). These units are used for small capacities (less than 50 TPD) and irregular production or to process small batches of multiple oils that demand minimum intermixing [4].

The continuous deodorizer performs the same basic functions but is designed for a larger operation, requiring few feedstock changes (Fig. 4.10). It is generally the best choice for modern integrated refineries [4].

About 0.01 percent of citric acid is commonly added to deodorized oils to inactivate trace-metal contaminants such as soluble iron or copper compounds that would otherwise promote oxidation and the development of rancidity.

Hydrogenation.   Hydrogenation is used to convert liquid fats to plastic fats, thereby making them suitable for the manufacture of margarine or shortening. Hydrogenated oils and fats also exhibit improved oxidative stability and color. Close control of hydrogenation results in highly specific results. For example, salad and cooking oils can be improved by controlled hydrogenation.

In the hydrogenation process, the hydrogen is added directly to double bonds of fatty acids. The most unsaturated fatty acid groups are most easily hydrogenated and thus react first with the hydrogen if conditions are

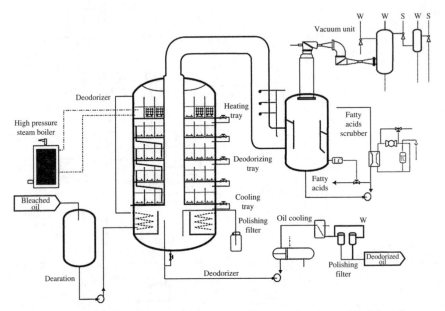

Figure 4.10   Continuous deodorization process. (*Source*: Courtesy of DeSmet Group, Antwerp, Belgium.)

somewhat selective, that is, to add hydrogen to the linolenic (three double bonds) and linoleic (two double bonds) acid radicals before adding to the oleic (one double bond) acid radicals. The reaction may be generalized:

$$\underset{R}{\overset{H}{\diagdown}} C = C \underset{R'}{\overset{H}{\diagup}} \; + \; H_2 \quad \overset{B}{\longrightarrow} \quad H\!-\!\underset{R}{\overset{H}{\diagdown}} C - C \underset{R'}{\overset{H}{\diagup}}\!-\!H$$

Raney nickel-type and copper-containing catalysts are normally used. Variables affecting hydrogenation include the catalyst, temperature, hydrogen pressure, and amount of agitation. If very hard fats with low amounts of unsaturation are derived and selectivity is unimportant, higher temperatures and pressures are employed to shorten the reaction time and to use the partially spent catalyst that would otherwise be wasted. The catalytic hydrogenation is frequently accompanied by isomerization with a significant increase of melting point, caused, for example, by oleic (*cis*) isomerizing to olaidic (*trans*) acid. The *trans* isomers are much higher melting than natural *cis* forms. After hydrogenation, hot oil is filtered to remove the metallic catalyst for either reuse or recovery.

Winterization.   Winterization is an operation that consists in removing from certain oils the components that solidify at low temperatures and therefore are a source of turbidity or settling in the bottle. The process consists in filtering cooled oil under strict control.

Winterizing is not practiced so widely in hot countries and its application is restricted mainly to sunflower, maize, cotton, olive, ricebran, and partially hydrogenated soybean oils. The feedstock of the winterizing plant is usually bleached oil, sometimes neutralized or deodorized oil. The winterizing process is conducted in four steps:

- the oil is precooled in heat exchangers;

- the filter aid is added just before the first crystallizer and the oil is cooled slowly, in a minimum of 6 h, a prerequisite for gentle crystallization;

- the crystallized form stays 6 h in the maturator, at final cooling temperature;

- the oil is then filtered in horizontal hermetic leaf filters. The oil temperature is often raised just before filtration, to reduce viscosity and hence facilitate filtration.

Normally, a winterized oil should remain clear for at least 24 h at 0°C. This corresponds to a wax content of below 50 ppm.

4.8   Fats and Oils Stability and Antioxidants

The capability of a fat, oil, or fatty food to retain a fresh taste and odor during storage and use is referred to as its stability. It is related to the ester linkages of triglycerides, which are either subjected to hydrolysis liberating free fatty acids (lipolysis), or may undergo oxidation forming peroxides and aldehydes (oxidation). Lipolysis can occur from chemical, enzymatic, or thermal stress actions. Lipolytic spoilage is known as *lipolytic rancidity,* or *hydrolytic rancidity.* Fats and oils also can become rancid as a result of oxidation and this oxidative spoilage is known as *oxidative rancidity.* Lipolytic rancidity usually poses less of a flavor problem than oxidative rancidity because the former develops off-flavor only in those fats, which contain short-chain fatty acids (less than C_{12}).

Oxidation of fats and oils is thought to be the major cause because of which the ester linkages of triglycerides in polyunsaturated oil deteriorate. The reaction of oxygen with unsaturated fatty acids in fats and oils proceeds through a free-radical chain-reaction mechanism involving three stages: (1) initiation-formation of free radicals; (2) propagation-free-radical chain reaction; (3) termination-formation of nonradical products. Hydroperoxides are the major initial-reaction products of fatty acids with oxygen. Subsequent reactions control both the rate of reaction and the nature of products formed. The initiation step is the removal of weakly held methylene hydrogen to produce an allylic radical $RH \rightarrow R^{\bullet} + H^{\bullet}$, while oxygen adds to the double bond to form a diradical.

$$R-\underset{\underset{H}{|}}{C}=\underset{\underset{H}{|}}{C}-R' + O_2 \longrightarrow R-\underset{\underset{\bullet}{|}}{\underset{\underset{H}{|}}{C}}-\underset{\underset{O-O^{\circ}}{|}}{\underset{\underset{H}{|}}{C}}-R'$$

Alternatively, oxidation is thought to be induced by oxygen in the singlet-state interposing between a labile hydrogen to form a hydroperoxide directly $(RH + O_2 \rightarrow ROOH)$.

During propagation, the chain reaction is continued by $R^{\bullet} + O_2 \rightarrow ROO^{\bullet}$ and $ROO^{\bullet} + RH \rightarrow ROOH + R^{\bullet}$ and hydroperoxides, and new hydrocarbon radicals. The new radical formed then contributes to the chain by reacting with another oxygen molecule.

Termination occurs when the radicals interact as follows:

$$R^{\bullet} + R^{\bullet} \rightarrow RR$$
$$ROO^{\bullet} + R^{\bullet} \rightarrow ROOR \qquad \text{(Nonradical products)}$$
$$ROO^{\bullet} + ROO^{\bullet} \rightarrow ROOR + O_2$$

Trace metals, especially copper, catalyze autoxidation by reacting with hydroperoxides to create new free radicals and initiate new chain reactions.

$$M^{+} + ROOH \rightarrow RO^{\bullet} + OH^{-} + M^{2+} \text{ (metal ion is oxidized)}$$
$$\underline{M^{2+} + ROOH \rightarrow ROO^{\bullet} + H^{+} + M^{+} \text{ (metal ion is reduced)}}$$
$$2ROOH \rightarrow RO^{\bullet} + ROO^{\bullet} + H_2O : \text{Net reaction}$$

In order to avoid the autoxidation problem it has become a common practice in the edible oil industry to construct reaction vessels of stainless steel. Citric acid is often added as a metal sequestrant to effectively inactivate trace metal ions.

Antioxidants are used to stabilize fats and fat-containing products against oxidation and thereby prolong their stability and storage time. An antioxidant (AH) acts as a free radical trap to terminate oxidation chain reactions:

$$R^{\bullet} + AH \rightarrow RH + A^{\bullet}$$
$$RO^{\bullet} + AH \rightarrow ROH + A^{\bullet}$$
$$ROO^{\bullet} + AH \rightarrow ROOH + A^{\bullet}$$
$$R^{\bullet} + A^{\bullet} \rightarrow RA$$
$$RO^{\bullet} + A^{\bullet} \rightarrow ROA$$

The major food grade synthetic antioxidants used include butylated hydroxytoluene (BHT), tertiary butylhydroquinone (TBHQ), propyl gallate (PG), and 2,4,5-trihydroxybutylophenone (THBP). The structures of some of the most important synthetic antioxidants are shown below [5]:

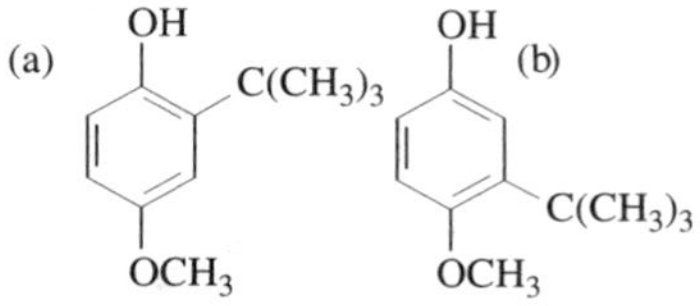

(a) BHA
Butylated hydroxyanisole
(a) 2-teritary-4-hydroxyanisol
(b) 3-tertiary-4-hydroxyanisol

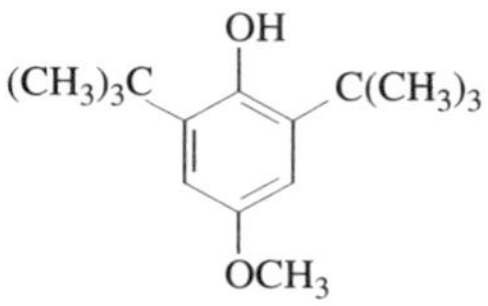

BHT
Butylated hydroxytoluene
2,6-ditertiary-butyl-4-methylphenol

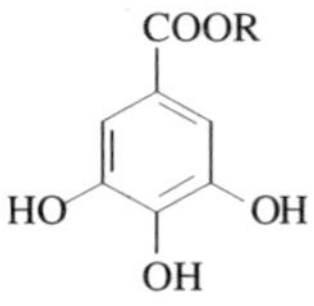

Gallates
Esters of 3,4,5-tri-hydroxybenzoic acid
R = C_3H_7 (Propyl)
= C_8H_{17} (Octyl)
= $C_{12}H_{25}$ (Dodecyl)

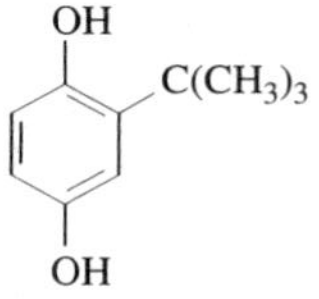

Ethoxyquin

TBHQ
Tertiary butylhydroquinone

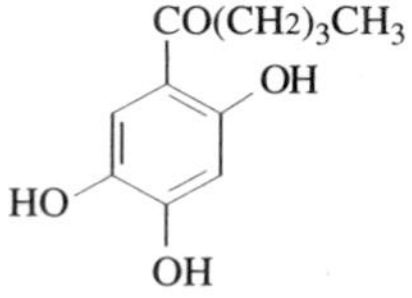

THBP
Trihydroxybutylophenone
2,4,5-trihydroxybutyrophenone

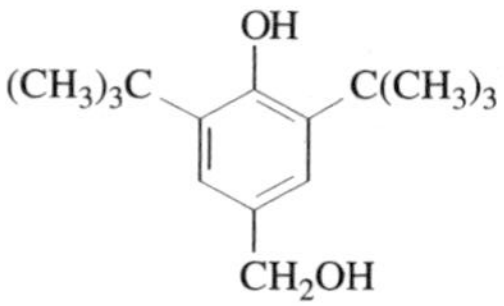

Ionox-100

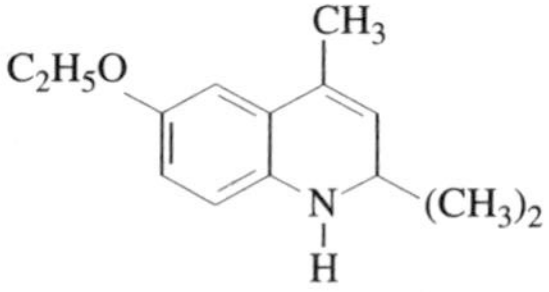

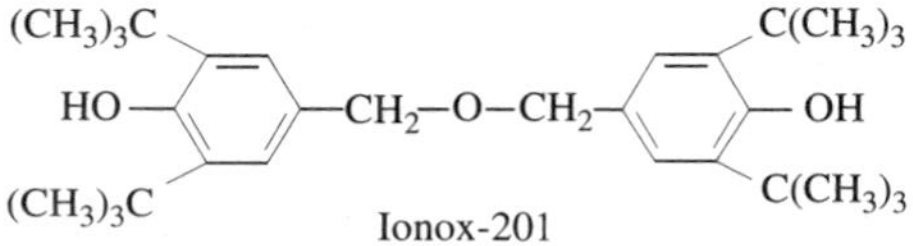

Ionox-201

Structures of some synthetic antioxidants

117

The concentration of an antioxidant in a food fat is important for reasons of cost, safety, sensory properties, and functionality, and generally they are allowed in food products at 0.01 percent level. Substances such as citric acid, isopropyl acid, phosphoric acid, ascorbic acid, and tartaric acid are sometimes added as synergists to enhance the effectiveness of antioxidants.

4.9   Methods of Analysis and Testing of Fats and Oils

The main organizations that develop or validate methods for fats and oil analysis include: International Union for Pure and Applied Chemistry (IUPAC), International Organization for Standardization (ISO), Association of Official Analytical Chemists (AOAC), American Oil Chemists' Society (AOCS), American Society for Testing and Materials (ASTM), Association Francaise de Normalization (AFNOR), British Standards Institution (BSI), Deustsches Institute für Normung (DIN), and the Federation of Oils, Seeds and Fats Association (FOSFA). In addition, there is an increasing trend for the various national standard institutions to develop their own standard methods based on the standard ISO methods; these are generally adopted as official methods.

In the United States most of the analytical and test methods are described by the AOCS. These well-established test procedures have afforded an adequate quality assessment of raw materials or end products or both. Selected analytical methods adopted by AOCS for identification of the type, determination of the composition, detection of additives, physical properties, and stability of fats and oils, are summarized below. These methods are subject to the use of prescribed equipment and techniques of sampling and testing as specified by the AOCS [6].

4.9.1   Identification and compositional analysis

The following older, well-established methods based on the estimation of the selected functional groups or calculation of oil or fat contents are still used to differentiate oils or fats. However, more accurate and fast analytical methods, such as gas chromatography of fatty acids and the HPLC of triglycerides have reduced the significance of these older methods.

Saponification number (SN).   This is the weight of KOH (in mg) needed to hydrolyze a 1-g sample of an oil or fat. The higher the SN, the lower the average molecular weight of the fatty acid in the triglycerides. Some SN of various oils and fats are presented in Table 4.12 [2].

TABLE 4.12   Iodine (IN) and Saponification
Numbers (SN) of Various Edible Fats
and Oils

Oil/Fat	IN	SN
Coconut	256	9
Palm kernel	250	17
Cocoa	194	37
Palm	199	55
Olive	190	84
Peanut	192	156
Rapeseed	225	30
Sunflower	190	132
Soya	192	134
Butter	225	30

Acid value (AV).   This value is important for determination of free fatty acids (FFA) in crude and refined oils and fats. It is the number of milligrams of KOH needed to neutralize the organic acids present in 1 g of oil or fat. FFA is calculated as free oleic acid and reported as a percentage. The AV is determined by multiplying percent FFA with a factor of 1.99.

Iodine value (IN).   This value measures the unsaturation of the oils and fats in terms of the number of grams of iodine absorbed per 100-g sample. The method is applicable to all normal oils and fats not containing conjugated systems. Some IN of various oils and fats are presented in Table 4.12.

Hydroxyl number (OHN).   This number reflects the content of hydroxy fatty acids, fatty alcohols, mono- and diacylglycerols, and free glycerol.

Color reactions.   Some oils give specific color reactions caused by particular ingredients. The Halphen test for detecting cottonseed oil is one example. This test estimates the presence of cottonseed oil in vegetable or animal fats or oils as the result of a pink color formed between the reagent (sulfur and carbon disulfide) and cyclopropenoic fatty acids normally present in cottonseed oil.

Composition of fatty acids.   The saturated and unsaturated fatty acids with 8 to 24 carbon atoms in animal fats, vegetable oils, marine oils, and fatty acids are quantitatively determined by gas chromatography (GC) after conversion to their methyl ester forms. However, free fatty acid analysis is also possible by using specially selected stationary solid phases. A capillary gas liquid chromatographic method is also used to measure fatty acid composition and levels of *trans* unsaturation and *cis, cis* methylene-interrupted unsaturation of vegetable oils.

Coupled HPLC-GC methods are used for the detection of adulteration of oils and fats. The identification is done by analysis of unsaponifiable minor constituents specific to a particular oil or fat as shown below:

Minor Constituent (analysis)	**Identification**
Squalene	Olive or rice oil and fish liver oil
Campesterol/stigmasterol	Cocoa butter substitutes
Carotene	Raw palm oil
γ-/β-Tocopherol	Corn oil
γ-Tocopherol	Wheat germ oil
α-/γ-Tocopherol	Sunflower oil
γ-/δ-Tocopherol	Soybean oil
Cholesterol	Animal fat

Physical tests for identification.   Specific density, index of refraction, color, viscosity, and melting point tests are used to identify fats and oils. The onset, flow point, and the temperature range over which melting occurs are indicative of specific numbers in fats. They are determined by standardized procedures.

The Solid Fat Index (SFI) estimates the percent of solids in a semi-solid fat on the basis of changes in volume with temperatures. This is of importance in fat hydrogenation and interesterification processes.

4.9.2 Quality control tests

A number of analytical methods are used for assessing the quality and deterioration of oils or fat during refining and subsequent storage conditions. These methods are summarized below:

Lipolysis.   Free fatty acids (FFA) result from lipolysis (hydrolysis) of oils and fats. This is determined by the FFA or acid value. The crude oils and animal fats usually have a FFA content exceeding 1 percent. The FFA content is lowered to less than 0.1 percent by the refining of oil or fat.

Peroxide value.   The oxidation of oils and fats leads to the formation of hydroperoxides. The hydroperoxides readily decompose to produce aldehydes, ketones, and other volatile products, which are characteristic of oxidation rancidity. The method for determination of peroxide concentration is based on the reduction of the hydroperoxide group with HI (or KI) to liberate free iodine, which may be titrated. The peroxide value is expressed in terms of a milliequivalent of iodine formed per kg of fat.

Shelf stability test. Shelf life is predicted by the active oxygen method (AOM). The fat or oil is subjected to an accelerated oxidation test under standardized conditions so that the signs of deterioration are revealed within several hours or days. The sample is heated at 97.8°C while air is blown through it. The AOM value is reported as the number of hours to reach a peroxide value of 100 meq/kg.

References

1. Kent, J. A., *Riegel's Handbook of Industrial Chemistry*, 9th ed., Chapman & Hall, New York, 1992.
2. Belitz, H. D. and Grosch, W., Food chemistry, *Lipids*, Chap. 3, Springer, London, 1999.
3. U.S. Census Bureau, 2003, *U.S. Fats & Oils Edible Production and Consumption* 2003, 2004.
4. Kellens, I. M., *Current Developments in Oil Refining Technology*, De Smet Group, Antwerp, Belgium.
5. Akoh, C. C. and Min, D. B., *Food Lipids*, Marcel Dekker Inc., New York, 1998.
6. Firestone, D. (Ed.), *Official Methods of American Oil Chemist's Society*, 4th ed., AOCS, Champaign, 1989.

Soaps and Detergents

Bassam El Ali

5.1 Soap

5.1.1 Introduction

According to an ancient Roman legend, soap got its name from Mount
Sapo, where animals were sacrificed. Rain washed a mixture of melted
animal fat or tallow and wood ashes down into the clay soil along the

Tiber River. It was found that this clay mixture cleaned the washed clothes with much less effort [1–6].

The term soap is a class name for the sodium and potassium salts of fatty acids. These fatty acids were found in animal fats and in plant oils such as coconut oil, palm oil, olive oil, castor oil, or cottonseed oil [1–6].

5.1.2 History

Records show that ancient Egyptians bathed regularly. The Ebers Papyrus, a medical document, describes the combining of animal and vegetable oils with alkaline salts to form a soap-like material used for treating skin diseases, as well as for washing.

The early Greeks bathed for aesthetic reasons and apparently did not use soap. Instead, they cleaned their bodies with blocks of clay, sand, pumice, and ashes, then anointed themselves with oil, and scraped off the oil and dirt with a metal instrument known as a strigil. They also used oil with ashes. Clothes were washed without soap in streams.

The roman civilization showed a more positive progression in cleanliness than did the early Greeks. As Roman civilization advanced, so did bathing. The first of the famous Roman baths was supplied with water from their aqueducts. The baths were luxurious, and bathing became very popular [1–6].

In Europe, by the 7th century, soap making was an established craft. Soap-maker guilds guarded their trade secrets closely. Vegetable and animal oils were used with ashes of plants along with fragrance. Gradually more varieties of soap became available for shaving and shampooing, as well as bathing and laundering.

Italy, Spain, and France were early centers of soap manufacturing, because of their ready supply of raw materials such as oil from olive trees. The English began making soap during the 12th century. The soap business was so good that in 1622, King James I granted a monopoly to a soap maker for US$100,000 a year. Well into the 19th century, soap was heavily taxed as a luxury item in several countries. When the high tax was removed, soap became available to ordinary people and cleanliness standards improved.

A major step toward large-scale commercial soap making occurred in 1791 when a French chemist, Nicholas Leblanc, patented a process for making soda ash, or sodium carbonate, from common salt. Soda ash in the alkali obtained from ashes combines with fat to form soap. The Leblanc process yielded quantities of good quality, inexpensive soda ash.

Some soap manufacture took place in Venice and Sauona in the 15th century and in Marseilles in the 17th century. By the 18th century, manufacture was widespread throughout Europe and North America and by the 19th century the making of soap had become a major industry [1–6].

TABLE 5.1   Fatty Acid Distribution [1, 6]

Fatty Acid	Coconut	Stripped coconut	Palm kernel	Palm	Tallow
Caprylic ($C_8H_{16}O_2$)	7	1	4		
Capric ($C_{10}H_{20}O_2$)	6	1	5		
Lauric ($C_{12}H_{24}O_2$)	48	55	50		
Myristic ($C_{14}H_{28}O_2$)	19	22	15	2	2.5
Palmitic ($C_{16}H_{32}O_2$)	9	11	7	42	27
Stearic ($C_{18}H_{36}O_2$)	2	2	2	5	20
Oleic ($C_{18}H_{32}O_2$)	8	7	15	41	42
Linoleic ($C_{18}H_{32}O_2$)	1	1	1	10	5
C_{20} or C_{22} isomers			1		2.5

5.1.3 Raw materials [1–6]

Soaps are manufactured from a renewable source. The triglycerides (or triesters of fatty acids) are the raw material for the production of soap. The triglycerides occur widely in plants and animals.

Tallow and coconut oil are the principal fatty materials in soap making in the United States. The palm oils, palm kernel oil, and their derivatives are used in soap manufacture in many other parts of the world. Greases, obtained from hogs and smaller domestic animals, are the second most important source of glycerides of fatty acids. Coconut oil has long been important in soap making. The soap from coconut oil is firm and lathers well. It contains a large amount of the desired glycerides of lauric and myristic acids (Table 5.1) [1–6].

The soap maker represents one of the larger consumers of chemicals, especially caustic soda, salt, soda ash, caustic potash, sodium silicate, sodium bicarbonate, and trisodium phosphate. Builders are inorganic chemicals added to soap. In particular, tetrasodium pyrophosphate and sodium tripolyphosphate were usually effective soap builders [1–6].

5.1.4 Chemistry of soaps

Soaps are water-soluble sodium or potassium salts of fatty acids containing 8 to 22 carbon atoms. The fatty acids are generally a mixture of saturated and unsaturated moieties.

Saturated soap: $CH_3(CH_2)_nCOOM$

Mono-unsaturated soap: $CH_3(CH_2)_nCH_2CH\!=\!CHCH_2(CH_2)_m\!-\!COOM$

Poly-unsaturated soap: $CH_3(CH_2CH\!=\!CH)_xCH_2(CH_2)_y\!-\!COOM$
[$M\!=\!Na, K, R_4N^+$]

The basic chemical reaction in the making of soap is saponification:

$$3NaOH + C_{17}H_{35}COO_3C_3H_5 \longrightarrow 3C_{17}H_{35}COONa + C_3H_5(OH)_3$$

Sodium   Glyceryl stearate   Sodium stearate   Glycerin
hydroxide

The other method for making soap comprises fat splitting followed by the neutralization process with sodium hydroxide.

$$(C_{17}H_{35}COO)_3C_3H_5 + 3H_2O \xrightarrow[\text{splitting}]{\text{Fat}} 3C_{17}H_{35}COOH + C_3H_5(OH)_3$$

Stearic acid   Glycerin

$$C_{17}H_{35}COOH + NaOH \xrightarrow{\text{Neutralization}} C_{17}H_{35}COONa + H_2O$$

Stearic acid   Sodium
hydroxide

Sodium hydroxide, potassium hydroxide, sodium carbonate, and triethanolamine are the most commonly used alkaline moieties in both the processes.

Recently, soap manufacture by the saponification of fatty methyl esters has been developed in Japan and Italy. The methanolysis of triglycerides takes place in the presence of enzymes (lipase) as catalysts to produce fatty methyl ester and glycerin. The fatty methyl ester undergoes the saponification and forms the final product [1–6].

$$(C_{17}H_{35}COO)_3C_3H_5 + 3CH_3OH \xrightarrow{\text{Catalyst}} 3C_{17}H_{35}COOCH_3 + C_3H_5(OH)_3$$

Methylstearate   Glycerin

$$C_{17}H_{35}COOCH_3 + NaOH \xrightarrow{\text{Saponification}} C_{17}H_{35}COONa + CH_3OH$$

Methyl stearate   Sodium   Sodium stearate
hydroxide

5.1.5 Classification of soaps

The two main classes of soaps are toilet soap and industrial soap. Toilet soap is usually made from mixtures of tallow and coconut in the ratios 80–90:10–20. The bar soap includes regular and super fatted toilet soaps, deodorant and antimicrobial soaps, floating soaps, and hard water soaps. The super fatted soaps are also made from a mixture of tallow and coconut oil in ratios 50–60:40–50. All soaps practically contain 10 to 30 percent water and also contain perfume that serves to improve the original soap odor. Toilet soaps usually contain only 10 to 15 percent moisture; they have little added material, except for perfume and a fraction of a percent of titanium dioxide used as a whitening agent. Shaving soaps, in contrast, contain a considerable amount of potassium soap and an excess of stearic acid; the combination gives a slower drying lather [1, 2, 5].

Milled toilet soap is another type of bar soap. Because of the milling operation, the soap lathers better and generally has an improved performance, especially in cool water.

Laundry soap bars are precursors of the chip and the powder forms. They are generally made from tallow or a combination of tallow and coconut oil. Borax and builders, such as sodium silicate and sodium carbonate, are included to improve performance and help soften water [1, 2, 5].

5.1.6 Manufacturing of soaps

Before the 1950s, soap was manufactured in the saponification process. Soap was prepared in large kettles in which fats, oils, and caustic soda were mixed and heated. After cooling, salt was added to the mixture forming two layers: soap and water. The soap was pumped from the top layer to a closed mixing tank where builders, perfumes, and other ingredients were added. Finally, the soap was rolled into flakes, cast, or milled into bars, or spray-dried into soap powder [1, 2].

After the 1940s, an important modern process for producing soap is based on the direct hydrolysis of fats by water at high temperatures. The chart of the continuous process is shown in Fig. 5.1. The process involves splitting (or hydrolysis) where fatty acids are neutralized to soap [1, 2, 5].

The saponification of triglycerides with an alkali is a bimolecular nucleophilic substitution (SN2). The rate of the reaction depends on the increase of the reaction temperature and on the high mixing during the processing. In the saponification of triglycerides with an *alkali*, the two reactants are immiscible. The formation of soap as a product affects the emulsification of the two immiscible reactants, which causes an increase in the reaction rate [1, 2, 5].

$$\text{Triglyceride} + 3\ NaOH \rightarrow 3\ RCOONa + \text{Glycerin}$$

The flow diagram of the continuous process of converting fatty acids into soap (Fig. 5.1) includes an important component—the hydrolyzer—to which the fats and the catalyst were introduced after mixing and preheating in the blend tank. At the same time, deaerated-demineralized hot water is fed to the top contacting section of the hydrolyzer. The fatty acids are discharged from the top of the splitter and the glycerin is

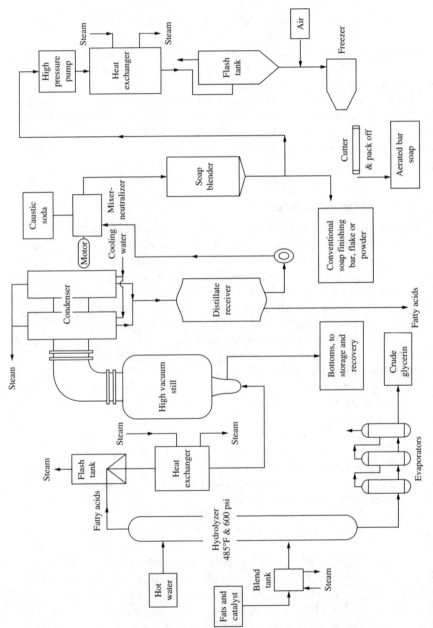

Figure 5.1 Continuous process for fatty acids and soaps.

eliminated from the bottom of the hydrolyzer. The fatty acids are sent to a flash tank, where the water is separated or flushed off.

$$\text{Triglyceride} \xrightarrow[\text{Fat splitting}]{+H_2O} \text{Fatty acids} + \text{Glycerin}$$

The resulting crude mixtures of fatty acids may be used, but a separation into more useful components is made.

The fatty acids are deaerated and distilled in a high-vacuum still. The fatty acids are deaerated to prevent darkening by oxidation during the process. The acids are charged at a controlled rate to the bottom of the high-vacuum still unit. The hot fatty acids are cooled down to room temperature into two parallel condensers prior to neutralization with 50% caustic soda in a high-speed mixer neutralizer. The fatty acids are converted into their corresponding sodium salts that form the soap.

$$\text{Fatty acid} + \text{NaOH} \xrightarrow{\text{Neutralization}} \text{Soap}$$

The amount of caustic soda (NaOH) required for neutralizing a fatty acid blend can be calculated as follows:

$$\text{NaOH} = [\text{weight fatty acid} \times 40]/M \text{ fatty acid}$$

The average molecular weight of a fatty acid is calculated from the following equation:

$$M_{\text{fatty acid}} = 56.1 \times 1000/AV$$

where AV = acid value of fatty acid blend = mg of KOH required to neutralize 1 g of fatty acid.

Neat soap (60 to 63 percent total fatty matter) is discharged into a slowly agitated blender to assure a complete neutralization. The neat soap at this stage may be extracted for conventional soap production (bar, flake, or powder) or undergoes further treatment at high pressure and high temperature in a high-pressure steam exchanger. The heated soap is sent to a flush tank for partial drying. The resulting viscous, pasty soap is then mixed with air in a heat exchanger, where the soap is cooled from 105 to 65°C. The soap is continuously extracted in strip form and is cut into bars. The operation is completed by further cooling, stamping, and wrapping to produce the aerated bar soap.

The advantages of soap manufacturing by this process include the color improvement of the soap, the excellent glycerin recovery, and the need for less space and labor [1, 2, 5].

5.1.7 Environmental aspects

The main atmospheric pollution problem in soap manufacturing is odor. The sources of this odor are from the storage and handling of liquid ingredients such as sulfonic acids and salts, and also from sulfates. The raw materials and product storage, the ventilation lines, the vacuum exhausts, and the waste streams are all potential odor sources. The control of these odors may be achieved by scrubbing exhaust fumes, or by incinerating the remaining volatile organic compounds. The odors originating from the spray dryer may also be controlled by scrubbing with an acid solution. Blending, mixing, drying, packaging, and other operations may involve dust emissions. The dust emission from the finishing operations can be controlled by dry filters such as bag houses. The large sizes of the particulate from the soap powder dryer mean the high-efficiency cyclones installed in a series can achieve a satisfactory control [1, 2, 5, 6, 7].

5.2 Detergent

5.2.1 Introduction and history

It is not easy to determine when the detergent industry came into existence. The problem was to define exactly what is being referred to as a synthetic detergent. For example, in the United States the term *surfactants* was used, while in Europe the term *tenside* was applied to point out the detergent industry.

Many general definitions of a detergent have been suggested [4]:

Detergent. A product that after formulation is devised to promote the development of detergency.

Surface active agent. A chemical compound, which when dissolved or dispersed in a liquid, is absorbed at an interface giving rise to a number of important chemical properties. The compound includes in its molecule one group that has an affinity for polar surfaces, ensuring solubilization in water, and a group that has little affinity for water.

Amphiphilic product. A product that contains in its structure one or more hydrophilic groups and one or more hydrophobic groups.

In the last 1000 years, soap has been used as a general-purpose washing and laundry agent. Soap remained a luxury until the beginning of the 20th century when the first self-acting laundry detergent was introduced in Germany (Persil, 1907). Soap took its place as an ingredient in the multicomponent systems for the routine washing of textiles. Soap was combined with the so-called builders, such as sodium carbonate, sodium silicate, and sodium perborate. The first practical substitutes for soap were fatty alcohol sulfates, discovered in Germany by Bertsch and coworkers in 1928 [4]. The availability of synthetic alkyl sulfates based on natural fats and oils made the introduction of the first neutral detergent for delicate fabrics (Fewa) possible in Germany in 1932. Fatty alcohol sulfates and their derivatives (alkyl ether sulfates, obtained by tracking fatty alcohols with ethylene oxide and subsequent sulfation) still retain their importance in many applications: heavy-duty detergents, dishwashing agents, cosmetics, and toiletries. In 1946, Procter and Gamble introduced the synthetic detergent Tide in the United States. By the 1950s, the widespread availability of tetra propylenebenzene sulfonate (TPS), a product of the petrochemical industry, had largely displaced soap as the key surfactant from the detergent.

In 1977, the German firm Hankel patented the use of synthetic zeolites as a partial replacement for phosphates. The sodium aluminum silicates as zeolites have a particular lattice structure capable of absorbing heavy metal cations through ion exchange process. The role of zeolites that were added to TPS was to soften water by rapid reaction with calcium at normal temperature [4].

Nitrilo triacetic acid (NTA) also represents a partial replacement of phosphate. However, the US Surgeon General issued a report stating that NTA caused birth defects in rats. The use of NTA was immediately phased out in the United States [4].

An important new criterion, the so-called biodegradability of detergents, appeared for surfactants. The insufficient biodegradability of TPS and nonylphenol ethoxylates caused great masses of stable foam to build up in the vicinity of weirs, locks, and other constructions in waterways. Subsequently, the first German Detergent Law was adopted in 1961 [4]. The German precedent was soon followed by the adoption of a similar legislation in France, Italy, and Japan. In the United Kingdom and the United States, voluntary agreement toward the transition to biodegradable surfactants took place between the industry and government [4].

The world production figures of different countries showed that the per capita consumption of detergents varies quite largely (Fig. 5.2). For example, it amounted to 2 to 3 kilograms per year (kg/a) in countries such as Brazil, China, and Russia, and to more than 10 kg/a in Mexico and some European countries in 1997 [8, 9].

The world production of laundry detergents amounted to 21.5×10^6 t in 1998. The total produced volume remained constant in the 1990s worldwide (Table 5.2).

The use of enzyme additives was the most important and revolutionary trend in the detergent industry in the past decade. The lower-temperature washes of the new high-efficiency machines are expected to give a boost to the demand for enzymes. As catalysts, enzymes are efficient and effective at low doses and at low temperatures where surfactant

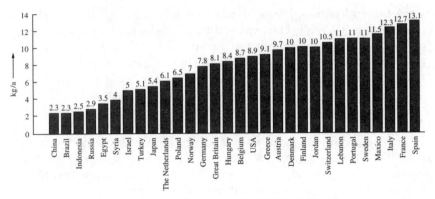

Figure 5.2   Per capita consumption of detergents in 1996 (kg/a) [8, 9].

activity is challenged. Enzymes continue to be popular additives for stain removal and fabric care.

The Freedonie group pegged the US demand for detergent enzymes at US$155 million in 1996 and anticipated that it will grow to US$260 million by 2001 [10].

5.2.2 Principle groups of synthetic detergents

Detergents are complex formulations that contain more than 25 different ingredients that can be categorized into the following main groups:

1. Surfactants
2. Builders
3. Bleaching agents
4. Additives

TABLE 5.2   Worldwide Production Detergent Industry [4]

Production 10^3 t	1994	1998
Soap bars	4579	3651
Detergent bars	1251	968
Powders	13,310	14,137
Pastes	496	505
Liquids	2052	2372
Total	21,688	21,633
World average Per capita consumption, kg/a	4.5	3.9

SOURCE: Ciba Specialty Chemicals Ltd., Switzerland.

Each group of detergents has its own specific functions during the washing process. They have synergistic effect on one another [1–4].

5.2.3 Surfactants

Surfactants represent the most important group of detergent components. They are present in all types of detergents. In general, surfactants are water-soluble surface-active agents comprising a hydrophobic group (a long alkyl chain) attached to a hydrophilic group.

The hydrophilic group is usually added synthetically to a hydrophobic material to produce a compound that is soluble in water. However, this solubilization does not necessarily produce a detergent, because detergency depends on the balance of the molecular weight of the hydrophobic portion to that of the hydrophilic portion.

Therefore, there are four main groups of surfactants: anionic, cationic, nonionic, and amphoteric[1–4].

Anionic surfactants. These are compounds in which the detergency is realized in the anion that has to be neutralized with a basic material before the full detergency is developed.

$$R - SO_3^- Na^+$$

Alkylsulfonates (anionic surfactants)

Cationic surfactants. These are compounds in which the detergency is in the cation, and although in the manufacturing process no neutralization takes place, the material is in effect neutralized by a strong acid.

$$R_2 N^+(CH_3)_2 Cl^-$$

Dialkyl dimethylammonium chlorides (cationic surfactants)

Nonionic surfactants. These contain nonionic constituents.

$$RO - (CH_2 - CH_2 - O)_n H$$

Alkyl poly(ethylene glycol)ethers (non-ionic surfactants)

Amphoteric surfactants. They include both acidic and basic groups in the same molecule.

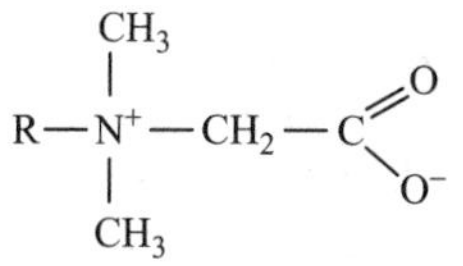

$$R - \overset{\overset{\displaystyle CH_3}{|}}{\underset{\underset{\displaystyle CH_3}{|}}{N^+}} - CH_2 - C\underset{O^-}{\overset{O}{\diagdown}}$$

Betain (amphoteric surfactants)

Anionic surfactants

Alkylbenzene sulphonates (LAS and TPS).   Alkylbenzene suofonates represent
the largest class of synthetic surfactants and until the mid-1960s tetra
propylene benzene sulfonate (TPS) was the most prominent detergent.

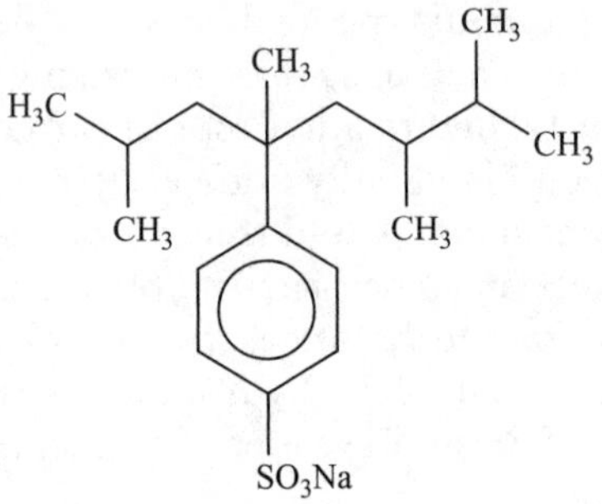

TPS

TPS has largely replaced soap as an active ingredient in laundry
detergents in Europe, the United States, and Japan. However, the
branched side present in TPS prevents the compounds from undergoing sufficient biodegradation. The replacement of TPS by more degrad-
able straight-chain homologues was an urgent task for scientists in the
1950s. Economic circumstances have permitted the straight-chain or
linear alkylbenzenesulfonates (LASs) to take the lead in the detergent
industry in Europe, the Americas, and Asia [1–4].

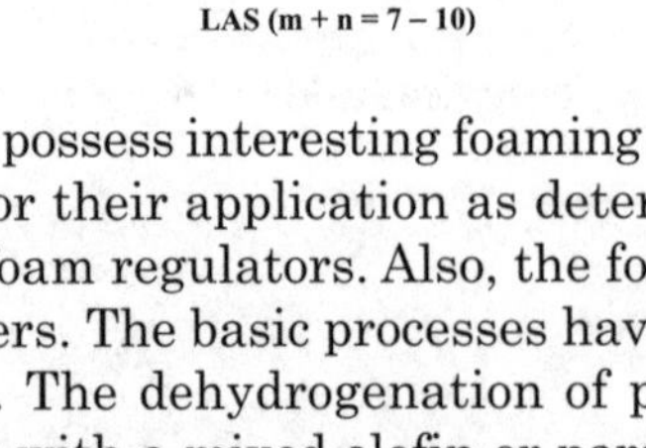

LAS (m + n = 7 – 10)

LASs were found to possess interesting foaming characteristics, which
are very significant for their application as detergents. However, LAS
can be controlled by foam regulators. Also, the foam produced is stabilized by form stabilizers. The basic processes have been applied for the
manufacture of LAS. The dehydrogenation of paraffins, followed by
alkylation of benzene with a mixed olefin or paraffin feedstock, represents the most important route for the production of LAS. This process
is catalyzed by hydrogen fluoride (HF) [1–4].

The partial chlorination of paraffins, followed by alkylation of the chloroparaffin or paraffin feedstock using aluminum chloride ($AlCl_3$) as a catalyst, represents another important route toward LAS. The third process implicates partial chlorination, but includes a dehydrochlorination to olefins prior to alkylation, with $AlCl_3$ or HF as a catalyst.

Universal Oil Products Company (UOP) offers processes, catalysts, adsorbents, and equipment for the production of linear alkylbenzenes (LAB) from kerosene or normal paraffins. The UOP LAB processes consist of a combination of several UOP processes, including the kerosene prefractionation and detergent alkylate. The LAB technology is the most economical technology available today, and over 70 percent of the world's LAB is produced using UOP technologies. Until 1995, alkylation used hydrofluoric (HF) acid as the catalyst. In 1995, the first commercial Detal process unit using a solid best catalyst alkylation process was commissioned. This revolutionary technology abolished the use of liquid acid in the plant, reducing capital investments, maintenance costs, and waste treatment [1–4].

The continued worldwide demand for LAB creates new growth and development of new technologies to improve the process. LAS made from sulfonation of LAB is the most cost-effective surfactant available for use in detergent formulation. Environmentally proven LAS has the largest volume of existing surfactants [1–4].

Sulfonation of LAB. The sulfonation of alkylbenzenes leads to sulfonic acid tyre product, which is then neutralized with a base such as sodium hydroxide to produce sodium alkylbenzene sulfonate. The sulfonation reaction is highly exothermic and instantaneous. An efficient reactor heat removal system is used to prevent the decomposition of the resultant sulfonic acid. The sulfonation reaction takes place by using oleum ($SO_3H_2SO_4$) or sulfur trioxide (SO_3). Although, the oleum sulfonation requires relatively inexpensive equipment, the oleum process has major disadvantages compared to sulfur trioxide. The need for spent acid stream disposal and the potential corrosion owing to sulfuric acid generation increased the problems related to oleum process [1].

Sulfonation

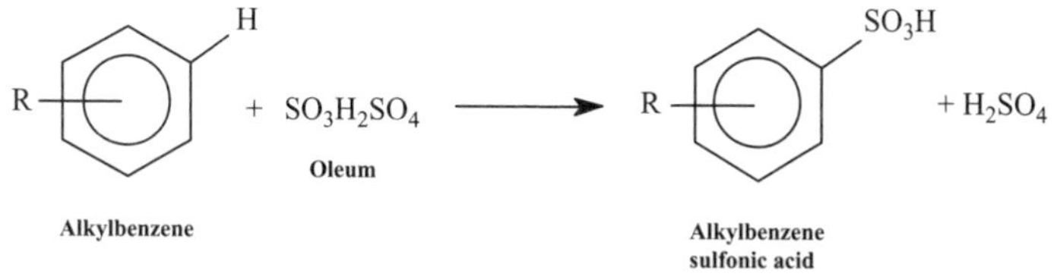

Alkylbenzene   Alkylbenzene
                                sulfonic acid

Neutralization

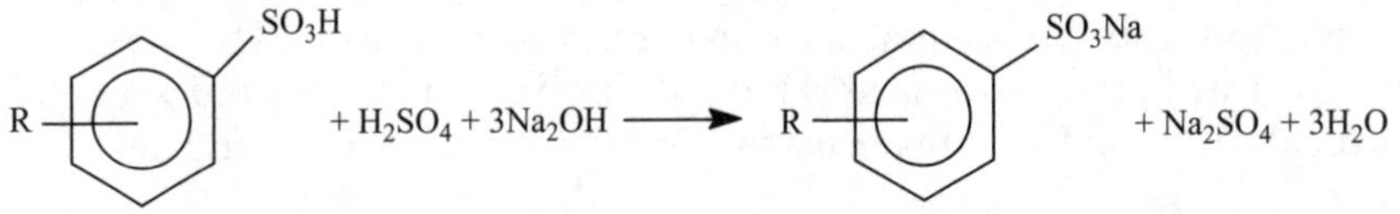

The gaseous air or SO_3 sulfonation process leads to high yields of sulfonic acid (95 to 98 percent). This process comprises three major steps. The sulfonation of alkylbenzene with air or SO_3 forms the alkylbenzene sulfonic acid and anhydride. The latter is decomposed into the alkylbenzene sulfonic acid by hydration. The neutralization of the sulfonic acid into the corresponding sodium salt represents the last chemical step in the process of formation of detergents [1].

Sulfonation

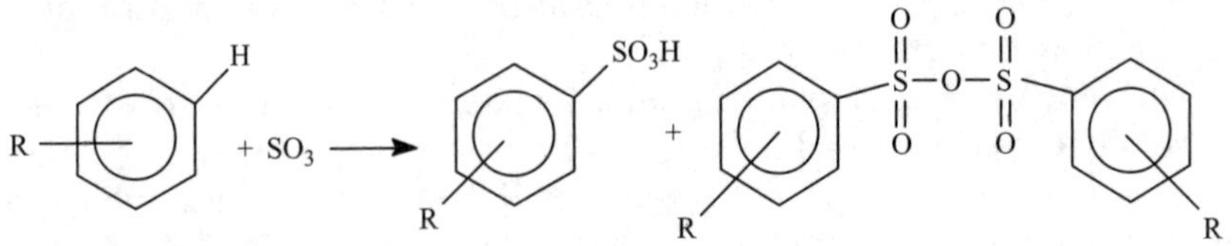

Anhydrides

Hydration

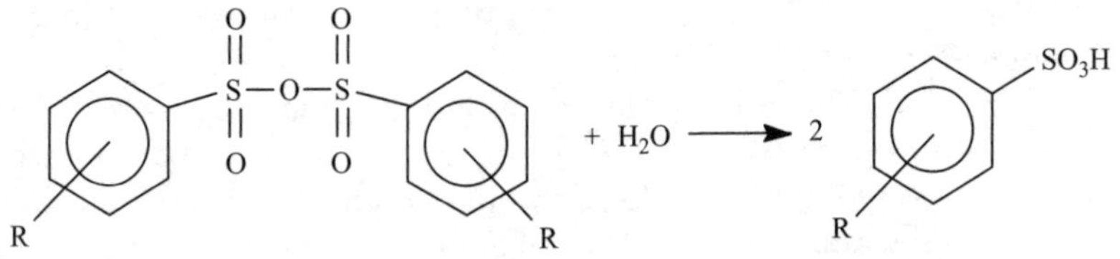

Neutralization

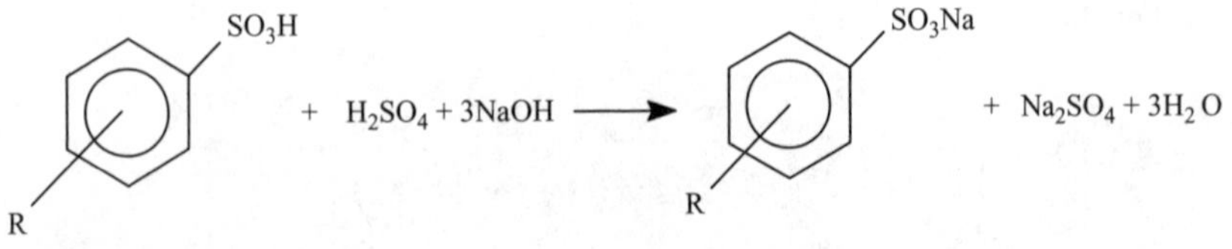

Secondary alkanesulfonates (SAS). The large-scale production of sodium alkanesulfonates (SAS) began in the late 1960s.

These sulfonates are still valued as anionic surfactants for consumer products. The secondary alkanesulfonates are known to have high solubility, fast wetting properties, chemical stability to alkali, acids, and strong oxidants such as chlorine.

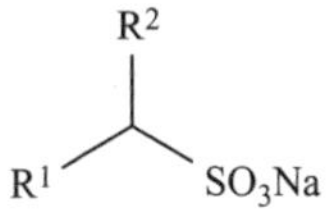

$$R^1 \overset{R^2}{\diagup\!\diagdown} SO_3Na \qquad R^1, R^2 = C_{11\text{-}17}$$

Sodium alkanesulfonates are produced by photochemical sulfoxidation or sulfochlorination of suitable C_{12}–C_{18} paraffin cuts. SAS can largely be substituted for LAS in formulations because of the similarity in terms of solubility, solubilizing properties, and wetting power [4].

Sulfonated olefins

$$R—CH=CHCH_2SO_3Na \qquad \text{and} \qquad R—CH—(CH_2)_nSO_3Na$$

$$R=C_8 \quad C_{12} \qquad\qquad\qquad\qquad \underset{OH}{|}$$

$$R=C_7—C_{13}$$

Alkenesulfonates   Hydroxyalkanesulfonates

The first commercial sulfonation of olefins with SO_3 involved the use of an SO_3-organo compound complex. Nowadays best results are obtained by sulfonating with uncomplexed diluted SO_3 in a film reactor [3, 4].

The reaction between α-olefins and SO_3 is not straightforward because of the formation of mixtures of alkene sulfonic acids, sultones, alkene disulfonic acids, and sultone sulfonic acids. Alkaline hydrolysis of sultone intermediates results in Ca 60 to 65 percent alkenesulfonates and Ca 35 to 40 percent hydroxyalkanesulfonates. The materials are sold as α-olefinsulfonates (AOS) because of the use of olefinic precursors. AOS has not made great strides in the heavy-duty laundry field but is being used successfully for light-duty detergents, hand dishwashing, shampoos, bubble baths, and synthetic soap bars [3, 4].

α-Sulfo fatty acid ester (MES)

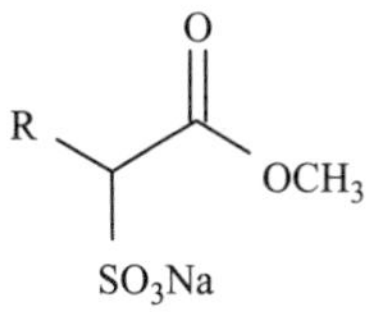

$$R \overset{\overset{O}{\|}}{\underset{\underset{SO_3Na}{|}}{\diagup\!\diagdown}} OCH_3 \qquad R=C_{12\text{-}16}$$

The α-sulfo fatty acid esters represent another class of anionic surfactants. Methyl ester sulfonates (MES) are surfactants that are derived from a variety of methyl ester feedstocks such as coconut, palm kernel, palm stearin, beef tallow, and soy. Palmitic and stearic acid derivatives lead to good detergency because of the long hydrophobic residues [3, 4]. The sensitivity of MES to water hardness is similar to AOS and small compared to LAS and SAS. MES have exceptional dispersion power with respect to lime soap. They have only been used in a few Japanese detergents [3, 4].

Alkyl ether sulfates (AES)

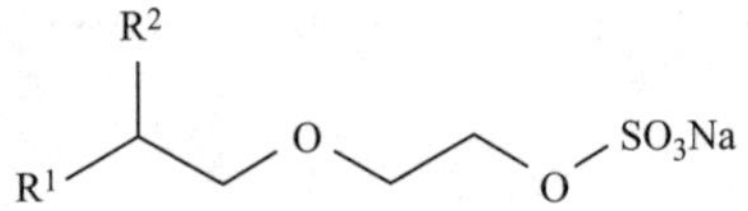

$R^1 =$ H, $R^2 = C_{10\text{-}12}$ Fatty alcohol ether sulfates

$R^1 + R^2 = C_{11\text{-}13}$ Oxo alcohol ether sulfates

Alkyl ether sulfates (AES) are anionic surfactants obtained by ethoxylation and subsequent salfation of alcohols derived from feedstock or synthetic alcohol. AES, also known as alcohol ether sulfates, have low sensitivity to water hardness (Fig. 5.3), high solubility, and good storage stability at low temperature [4, 11].

The main components of the commercial AES are alkyl ether sulfates and alkyl sulfates. Other by-products such as unsulfated alcohols, alcohol ethoxylates, inorganic salts, and polyethylene oxide sulfates are also present in the commercial product [4, 12].

Sodium $C_{12\text{-}14}$ n-alkyl diethylene glycol ether sulfates demonstrate increased detergency performance as the hardness increases. This is a result of the positive electrolyte effects attributable to calcium or magnesium ions.

AES are very intensively foaming compounds, which increased their use in high-foam detergents for vertical-axis washing machines. AES are suitable components of detergents for delicate or wool washables, as well as foam baths, hair shampoos, and manual dishwashing agents because of their specific properties. The optimal carbon chain length has been established to be $C_{12\text{-}14}$ with 2 mol of ethylene oxide.

AES have been used in the United States and Japan because of their critical micelle concentration, which is lower than LAS, resulting in very satisfactory washing power. In Europe, the use of alcohol ether sulfates has been largely restricted to specialty detergents [3, 4].

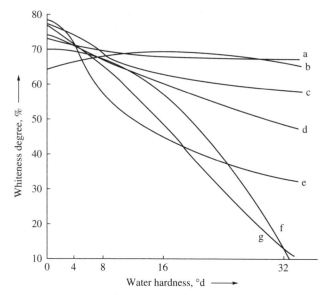

Figure 5.3 Detergency performance on wool by various surfactants as a function of water hardness [4]. Time: 15 min; temperature: 30°C; bath ratio: 1:50; concentration: 0.5 g/L surfactant + 1.5 g/L sodium sulfate; (a) Nonylphenol 9 EO; (b) C_{12-14} alcohol 2 EO sulfates; (c) C_{15-18} α-olefinsulfonates; (d) C_{16-18} a-sulfo fatty acid esters; (e) C_{12-18} Alcohol sulfates; (f) C_{10-13} alkylbenzenesulfonates; (g) C_{13-18} alkanesulfonates.

Nonionic surfactants. The majority of nonionic surfactants are condensation products of ethylene oxide with a hydrophobe. This hydrophobe is invariably a high molecular weight material with an active hydrogen atom, and the nonionic material can be one of the following reaction products [1–4]:

1. Fatty alcohol and alkylphenol condensates
2. Fatty acid condensates
3. Condensates of ethylene oxide with an amine
4. Condensates of ethylene oxide with an amide

Fatty alcohol and alkylphenol condensates

$$R - O - CH_2 - CH_2 \, (OCH_2CH_2)_{n-1}OCH_2CH_2OH$$

R = alkyl or phenyl

The alcohol ethoxylates and alkylphenol ethoxylates are produced by the reaction of alcohol with an excess of ethylene oxide. The reaction of one molecule of ethylene oxide is not sufficient to produce a water-soluble detergent product.

$$R-O-CH_2CH_2OH + n\,CH_2\!\!-\!\!\underset{O}{\diagdown\!\!/}\!\!-\!\!CH_2 \;\longrightarrow\; R-OCH_2CH_2-(OCH_2CH_2)_{n-1}-OCH_2-CH_2OH$$

R = alkyl or phenyl; n = 16–50

The ether linkage is a strong bond, therefore, the material cannot be hydrolyzed and there is no possibility of ionization and subsequently it is not affected by metal ions. In practice, the optimum detergency is found in the range of 10 to 15 molecules of ethylene oxide per molecule of fatty alcohol. The source of the alcohol varies greatly, but natural, Ziegler and OXO alcohols of various molecular weights are being used. Another source of alcohols is the one used by Union Carbide. These alcohols are obtained from n-paraffins that are oxidized to secondary alcohols. The ethoxylation of these alcohols is done in two stages [3]. The product containing 1 mol of ethylene oxide is produced first with an acid catalyst. After this stage the catalyst and the unreacted alcohol are removed and ethoxylation takes place as usual [3].

The alkylphenols behave in the same manner as fatty alcohols. The nonyl (or octyl) phenol is widely used with 8 to 12 molecules of ethylene oxide. Nonylphenol is completely soluble in water at room temperature and exhibits excellent detergency. Dodecylphenol ethoxylate is used in certain agriculture emulsifiers and dinonylphenol as low or nonfoaming ingredients of household washing machine powders [3, 4].

The alkylphenol ethylene oxide condensates have been the most widely produced nonionic detergent. These condensates are solubilized by the ethylene oxide units forming hydrates with water. Compared with all other materials that dissolve in water, these products show an apparent anomaly: their solutions are completely clear in the cold, but when heated they become turbid and, if the temperature is raised sufficiently, separation into two phases can take place. This is explained by the fact that at an elevated temperature the hydrates are destroyed.

The important use of polyester fibers in clothing and bedding increased the nonionic constituent in laundering as it was found that nonionics remove soil from these fibers better than the anionics.

Fatty acid condensates.   Fatty acid condensates are another type of nonionic surfactants that are prepared by the reaction of fatty acids with ethylene oxide.

$$RCOOH + n(CH_2OCH_2) \;\longrightarrow\; R\overset{\displaystyle O}{\overset{\|}{C}}OCH_2CH_2-(OCH_2CH_2)_{n-2}-OCH_2-CH_2OH$$

These condensates can also be produced by the esterification of a fatty acid with a polyethylene glycol, $HOCH_2(CH_2CH_2)_nCH_2OH$. Esters are materials that have some disadvantages compared to other nonionic detergents. In general, nonionic detergents are not affected by metallic ions, acids, or alkalis. However, the fatty acid condensates of ethylene oxide are readily hydrolyzed by acids or alkaline solutions to the corresponding fatty acid and polyethylene glycol. In strong alkaline solutions the esters are no better than the soap of the corresponding fatty acid, and in strong acid solution they are much affected. However, they do perform well as components of household detergent powders [3, 4].

Condensates of ethylene oxide with an amine.   The condensation of alkylamines with ethylene oxide leads to secondary or tertiary substituted amines depending on the concentration of ethylene oxide.

$$R-NH-CH_2-CH_2(OCH_2CH_2)_{n-2} - OCH_2-CH_2OH$$

$$R-N\begin{cases} CH_2CH_2(OCH_2CH_2)_{n-2} - OCH_2CH_2OH \\ CH_2CH_2(OCH_2CH_2)_{n-2} - OCH_2CH_2OH \end{cases}$$

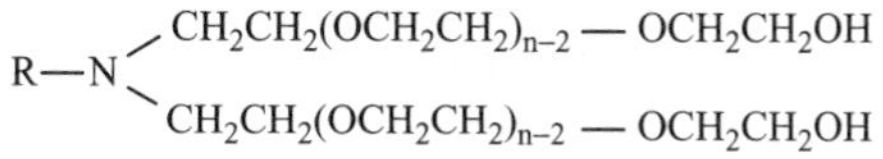

$$R = 11\text{-}17$$

This class of nonionic surfactants has not been used largely in cleaning detergents. However, it is noteworthy that these materials in acidic solution can exhibit cationic characteristics, whereas, in neutral or alkaline solution, they are nonionic [3, 4].

Condensates of ethylene oxides with an amide.   Fatty acid alkanolamides are amides of alkylolamines and fatty acids. Certain members of this class exhibit detergency and others do not. The nondetergent materials are converted into detergents by condensation with ethylene oxide.

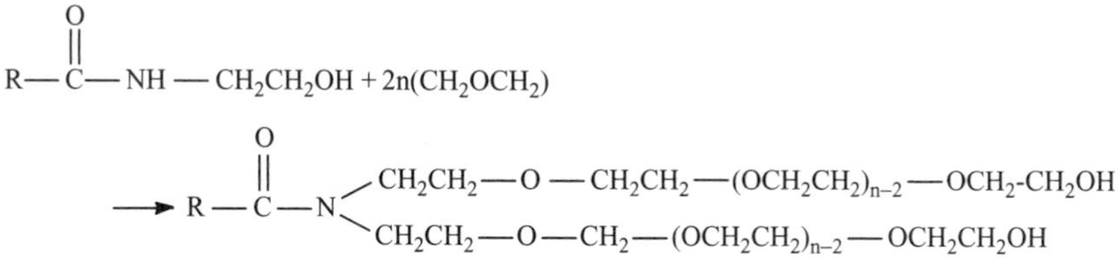

$$R-\overset{\overset{\textstyle O}{\|}}{C}-NH-CH_2CH_2OH + 2n(CH_2OCH_2)$$

$$\longrightarrow R-\overset{\overset{\textstyle O}{\|}}{C}-N\begin{cases} CH_2CH_2-O-CH_2CH_2-(OCH_2CH_2)_{n-2}-OCH_2-CH_2OH \\ CH_2CH_2-O-CH_2-(OCH_2CH_2)_{n-2}-OCH_2CH_2OH \end{cases}$$

$$R = C_{11\text{-}17}$$
$$n = 1, 2$$

Fatty acids alkanolamides find only a little application in laundry detergents. They are used as foam booting, that is, adding desired stability to the foam produced by detergents. Small amounts of these materials as cosurfactants are capable of enhancing the soil removal properties of the classical detergent components at low washing temperatures [3, 4].

Cationic surfactants. Long chain cationic surfactants such as disteral diamethyl ammonium chloride (DSDMAC) exhibit extraordinarily high sorption power with respect to a wide variety of surfaces [13–15]. These surfactants are very strongly absorbed by the surface of natural fibers, such as cotton, wool, and linen [4].

Cationic surfactants are mainly employed for the purpose of achieving certain special effects, which include applications in rinse-cycle fabric softeners, antistatic agents, and microbicides [4].

Nonionic surfactants are more tolerant of the presence of cationic surfactants than anionic surfactants. Mixtures of the two are sometimes used in the production of specialty detergents that are powerful antistatic products.

The cationic detergents invariably contain amino compounds. The most widely used would be the quaternary ammonium salts, such as cetyltrimethylammonium chloride, a well-known germicide and distearyldimethylammonium chloride (DSDMAC), known as fabric softener for cotton diapers and as a laundry rinse-cycle fabric softener [4].

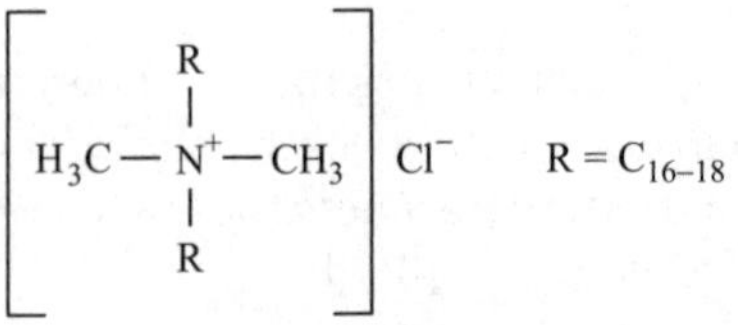

$$\left[\; H_3C - \overset{\overset{\displaystyle R}{|}}{\underset{\underset{\displaystyle R}{|}}{N^+}} - CH_3 \;\right] \; Cl^- \qquad R = C_{16-18}$$

DSDMAC

Imidazolinium salts, such as 1-alkylamidoethyl-2-alkyl-3-methylimidazolinium methylsulfate have been used as rinse-cycle softening agents, but in less significance compared to DSDMAC [4].

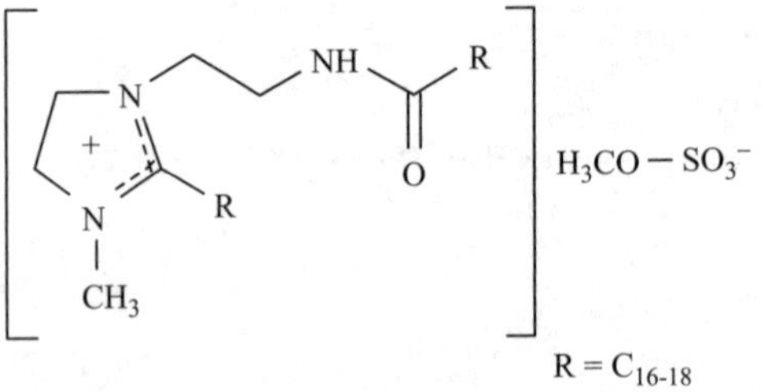

$$H_3CO - SO_3^-$$

$$R = C_{16-18}$$

The compounds based on alkyldimethylbenzylammonium chloride type have been used as laundry disinfecting agents. They show limited fabric softening character [4].

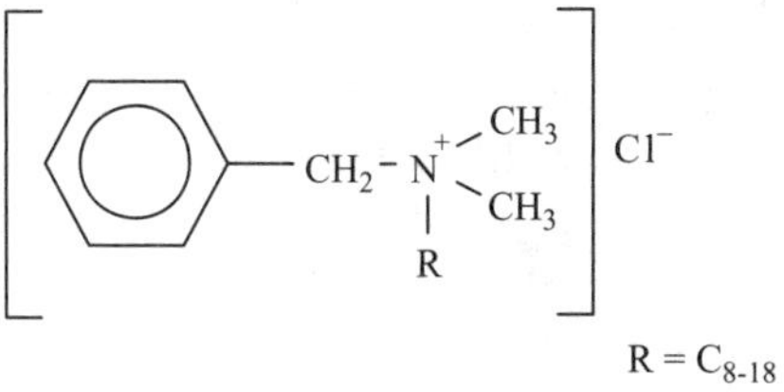

$$R = C_{8\text{-}18}$$

Esterquats represent a new generation of fabric softeners that have been developed in the 1980s or 1990s and have replaced DSDMAC [4, 16].

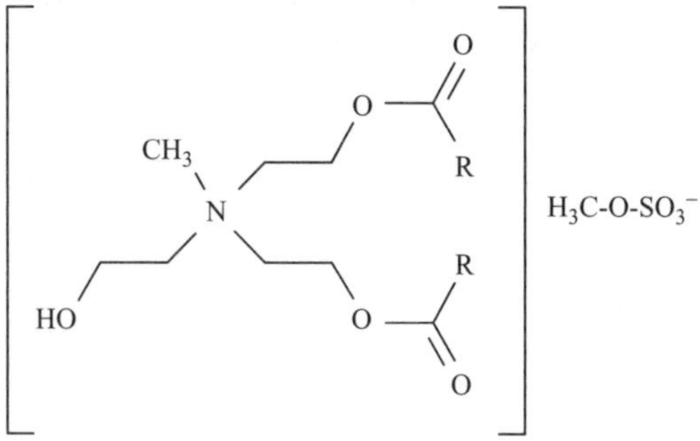

$$H_3C\text{-}O\text{-}SO_3^-$$

N-Methyl-N,N-bis[2-($C_{16\text{-}18}$-acyloxy)ethyl]-N-(2-hydroethyl) ammonium methosulfate.
(Esterquat) EQ

Amphoteric surfactants.   These compounds have the characteristics of both anionic detergents and cationic fabric softeners. They tend to work best at neutral pH, and are found in shampoos, skin cleaners, and carpet shampoos. They are very stable in strong acidic conditions and are favorably used with hydrofluoric acid. For example, compounds of alkyl-betain or alkylsulfobetaine type possess both anionic and cationic groups in the same molecule even in aqueous solution. These surfactants are rarely employed in laundry detergents because of their high costs [3, 4].

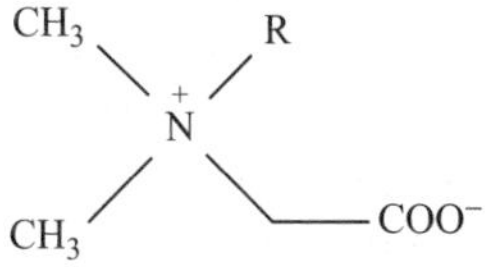

$R = C_{12\text{-}18}$ **Alkylbetaines**

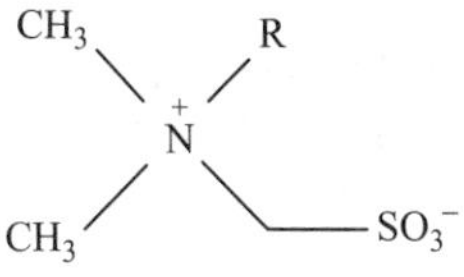

$R = C_{12\text{-}18}$ **Alkylsulfobetaines**

5.2.4 Inorganic builders

The use of the detergents mentioned so far as end products, would be costly, and so water is used to dilute these ingredients into the more consumer useable form. Sodium sulphate that is used as a diluting agent in powder detergents is cheap and user friendly.

Builders enhance the detergency action so that less can be used of the more expensive detergents of high activity. The combination of builders and surfactants exhibits a synergistic effect to boost total detergency and cleaning efficacy. Detergent builders should have the ability to control water hardness and other metal ions by eliminating calcium and magnesium ions that arise from the water and soil. Builders should also be compatible with other formulation ingredients and detergent additives. Consumer safety and environmental acceptability are important characteristics of builders [3, 4].

The inorganic constituents fall into the five following groups:

1. Phosphates

2. Silicates

3. Carbonates

4. Zeolites

5. Bleach-active compounds

Phosphates.   There are two classes of phosphates—orthophosphates and complex phosphates. The orthophosphates used in the detergent industry are trisodium phosphate in hydrated and anhydrous forms (Na_3PO_4 and $Na_3PO_4 \cdot 12H_2O$), disodium phosphate, another form of orthophosphates in anhydrous form (Na_2HPO_4), and crystalline $Na_2HPO_4 \cdot 12H_2O$. In the past, trisodium phosphate was used as a soap builder, but it is seldom used in detergent formulations nowadays because phosphates cause eutrophication of water. It has the property of softening water by precipitating metallic ions, as a gelatinous precipitate [3,4].

Trisodium phosphate can cocrystallize with sodium hypochlorite to produce the material called commercially *chlorinated trisodium phosphate* that has a formula approximating to ($Na_3PO_4 \cdot 12H_2O$) NaOCl. This material is added to the cleaning detergents that require the bleaching and sterilizing effect of NaOCl together with the high alkalinity and water-softening of trisodium phosphate [3, 4].

The detergent industry has introduced another form of phosphate the so-called *condensed phosphates*. These materials have a higher proportion of P_2O_5 and a lower proportion of Na_2O in the molecule. These phosphates have a lower alkalinity than trisodium phosphate. The commonly used complex phosphates are:

Tetrasodium pyrophosphate, $Na_4P_2O_7$ (TSPP)

Sodium tripolyphosphate, $Na_5P_3O_{10}$ (STP)

Sodium tetraphosphate, $Na_6P_4O_{13}$

Sodium hexametaphosphate, $(NaPO_3)_6$

Both sodium tetraphosphate and sodium hexametaphosphate are hygroscopic and are unsuitable for formulation into dry powders. Tetrasodium pyrophosphate (TSPP) is now used for specialized purposes only, but its potassium analogue is used in liquids. The combination of TSPP and sodium carbonate was more effective than either ingredient used separately [3, 4].

Sodium tripolyphosphate (STP) was commercially available in the mid-1940s and had replaced TSPP because of its superior detergent processing, solubility, and hardness ion-sequestering characteristics. Sequestration is defined as the reaction of a cation with an anion to form a soluble complex. The sequestration of Ca^{2+} and Mg^{2+} ions leads to softened water and is the most important function of any detergent builder [3, 4].

STP is the major builder ingredient in heavy duty laundry detergents, automatic dishwashing compounds, and industrial and institutional cleaners [3, 4].

Silicates.   The addition of sodium and potassium silicates to synthetic detergent has proved very beneficial. They have in solid or solution form important characteristics such as emulsification, buffering, deflocculation, and antiredeposition ability.

Sodium silicate is prepared by the fusion of sand that contains a high amount of silica with soda ash according to the following equation:

$$Na_2CO_3 + SiO_2 \rightarrow Na_2SiO_3 + CO_2$$

The proportion of silica, sand, and soda ash is important in providing a variety of alkalinity. The molecular formula of silicates has been adopted according to the method of Berzlins and is written as $Na_2O:SiO_2$ [3].

Silicates are produced in liquid, crystalline, or powdered form. The ratio of $Na_2O:SiO_2$ is usually selected to meet the product requirement and its application. The sodium metasilicate ($Na_2O/SiO_2 = 1/1$) is used in dry blending. The liquid silicates having $Na_2O:SiO_2$ ratio of 1:2 or higher are used in laundry and automatic dishwashing applications.

Soluble glass and soluble powders are the two forms of detergents that are prepared in the ratios of $Na_2O:SiO_2$ of 1:2 and 1:33, respectively. A ratio of 1:2.4 is commonly used in making detergent powders and is produced by blending the neutral soluble powder and the alkaline soluble glasses or by adding caustic soda to neutral glass [3].

A *wet method* for the production of soluble detergents of up to 40 percent disilicate by the reaction of fine sand and caustic soda (3 to 50 percent) is as follows.

$$2 \; SiO_2 + 2NaOH \rightarrow Na_2O.2SiO_2 + H_2O$$

Potassium silicate, available commercially in colloidal ratios, is used nowadays for specialized liquid detergents. It is available in weight ratios of 1:1.5 to 1:2.5. Silicates soften water by the formation of precipitates that can be easily rinsed away. Silicates are corrosion inhibitors of stainless steel and aluminum that can be caused by synthetic detergents and complex phosphates [3].

Carbonates.   Carbonates are being used because of the restriction in the use of phosphates in certain areas of the United States. Sodium carbonate (Na_2CO_3) or a combination of Na_2CO_3 and zeolite has replaced sodium tripolyphosphate (STP) as a builder in granular laundry products [3].

Sodium carbonate provides high alkalinity because only sodium hydroxide is high on a weight per volume basis. Na_2CO_3 softens water by the precipitation of calcium and magnesium carbonates, provided the pH of the solution is greater than 9.

Synthetic soda is of superior quality compared to mined natural soda ash. There are two important grades of carbonates: light soda ash and dense soda ash. Light soda ash can absorb large amounts of liquid material onto its surface and remains dry [3].

Sodium carbonate is commonly used in powdered laundry detergent, automatic dishwashing compounds, and hand surface cleaners.

Sodium carbonate is produced by the Solvay process, which uses sodium dichloride, carbon dioxide, and ammonia. Carbon dioxide obtained from burning limestone to lime is introduced in countercurrent to the solution of sodium chloride (known as brine) saturated with ammonia. Sodium bicarbonate, which is almost insoluble in solution and precipitates, is separated and roasted to sodium carbonate [3].

$$CaCO_3 \longrightarrow CaO + CO_2$$

$$CO_2 + NaCl + NH_4OH \longrightarrow NaHCO_3 + NH_4Cl$$

$$2NaHCO_3 \longrightarrow Na_2CO_3 + CO_2 + H_2O$$

The carbon dioxide released from the conversion of the bicarbonate to carbonate is reused in the absorption tower in addition to CO_2 obtained from the burning of the lime.

Zeolites.   Silicon is the second most abundant element in the earth's crust after oxygen. Silicon, which is never found free, is invariably associated with oxygen (silica, quartz) and with metallic oxides.

Zeolites, also known as molecular sieves, are important alternative builders for powdered laundry detergents and replaced phosphate salts that were banned for legislative reasons. Zeolites exist in the form of calcium, sodium, magnesium, potassium, and barium salts.

The most widely used form of zeolites is type A—which are hydrated sodium aluminum silicates of empirical formula $Na_2O \cdot Al_2O_3 \cdot x\ SiO_2 \cdot yH_2O$.

The crystalline material of type-A zeolites has a three-dimensional lattice structure of a simplified formula:

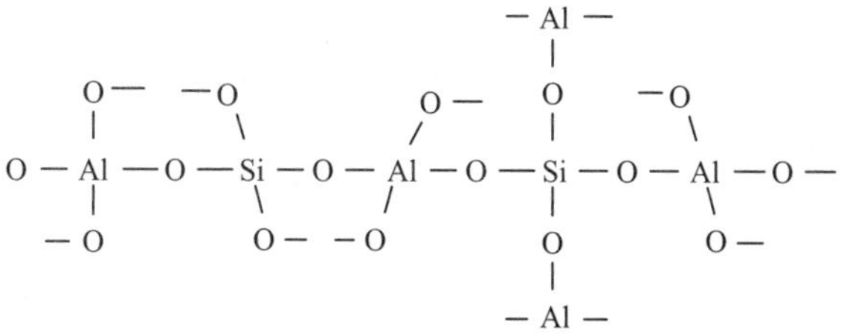

$$
\begin{array}{c}
- \text{Al} - \\
|
\end{array}
$$

Type-A zeolites are the most widely used form for laundry detergents.

The main advantage of zeolites compared to phosphates is nonsolubility in water, and they subsequently remove heavy metal ions such as manganese and iron readily and rapidly from the solution. However, magnesium ions are not totally removed by zeolites because of the size of magnesium; therefore, zeolites are then used in association with other builders such as sodium carbonate [3, 4].

Bleach-active compounds. Oxygen-releasing compounds are added to detergent powders as bleach-active materials. Peroxide-active compounds are the most used bleaches in Europe and many other regions of the world. Among these peroxides, hydrogen peroxide (H_2O_2) is converted by alkaline medium into hydrogen peroxide anion as active intermediate [4].

$$H_2O_2 + OH^- \rightleftharpoons H_2O + HOO^-$$

HOO^- oxidizes bleachable soils and stains. The most used source of hydrogen peroxide is sodium perborate, known as sodium peroxoborate tetrahydrate, $NaBO_3 \cdot 4H_2O$, that in crystalline form contains the peroxodiborate anion [4, 17].

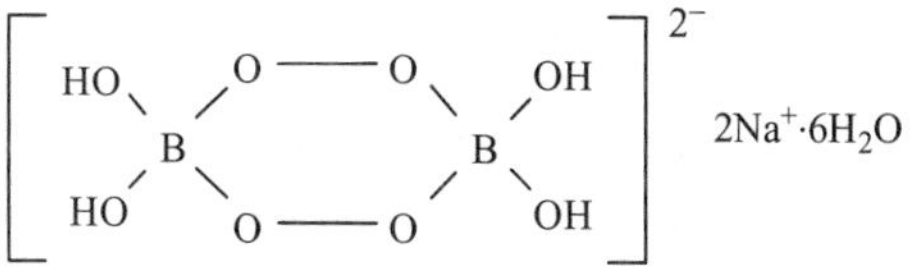

Peroxodiborate anion

Peroxodiborate is hydrolyzed in water to form hydrogen peroxide.

$$(NaBO_2H_2O_2)_2 \cdot XH_2O \longrightarrow 2NaBO_2 + 2H_2O_2 + XH_2O$$

In solution, sodium perborate monohydrate is similar in action to that of hydrogen peroxide. At elevated temperatures, active oxygen is released and has a bleaching effect but does not affect animal, vegetable, and synthetic fibers [3, 4].

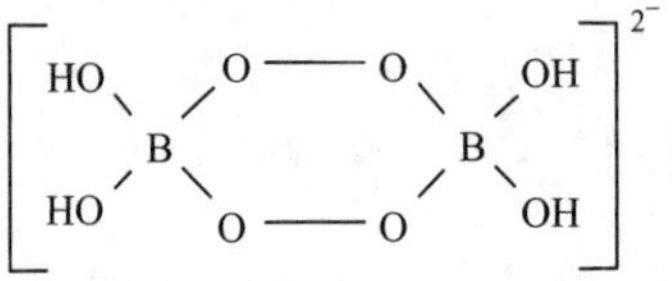

Sodium perborate

Sodium perborate can be used as the universal laundry bleach and has been used in Europe for the past 50 years [3, 4].

Sodium perborate is a stable material when mixed with other dry ingredients. However, the presence of traces of water and certain heavy metals will catalyze the decomposition of the perborate. Therefore, magnesium sulphate or silicate, or tetrasodium pyrophosphate is added to adsorb traces of water and metal to prolong the storage life of the powders [3, 4].

Hypochlorite is another effective bleaching compound at normal temperature. The use of liquid hypochlorite has gained importance in many countries where laundry habits, such as cold water washing, made the sodium perborate less effective. Hypochlorite reacts with an alkaline to produce hypochlorite anions [4, 18, 19].

$$HOCl + OH^- \rightleftharpoons ClO^- + H_2O$$

The aqueous solution of sodium hypochlorite (NaOCl) is used as a source of active chlorine that can be used in either the wash or the rinse cycle.

Powdered sodium perborate has some advantages over liquid sodium hypochlorite (NaOCl). NaOCl must be added separately in either the wash or the rinse cycle whereas perborate can be included directly in the powder laundry product. Also, a high dosage of NaOCl may cause significant damage to laundry and colors. In addition, sodium hypochlorite solutions have limited storage stability especially in the presence of impurities such as heavy metal ions. Sodium hypochlorite has a high reactivity and oxidation potential and may cause problems with textile dyes and most fluorescent whitening agents.

Studies showed that chlorine bleach is used predominantly in the Mediterranean countries whereas peroxide bleach use dominates in Europe. However, hypochlorite bleach in either the wash or the rinse cycle is still a preferred bleaching agent in a large part of the world (Table 5.3) [4].

TABLE 5.3   Washing Habits in Different Regions of the World [4]

Washing conditions	United States	Japan	Europe
Washing machine	vertical axis or agitator type	vertical axis or pulsator type	horizontal axis drum type
Heating equipment	no	no	yes
Fabric load, kg	23–25	4–8	2–5
Amount of water, L	medium loads: approx. 50–60	low: 30	low: 10–16
	large loads: approx. 65–75	high: 65	high: 20–25
Total water consumption, L (regular cycle)	120	120–150	55–90
Wash liquor ratio	approx. 1:25	approx. 1:10 to 1:15	approx. 1:4 to 1:10
Washing time, min	8–18	5–15	50–60 (90°C) 20–30 (30°C)
Washing rinsing, and spinning time, min	20–35	15–35	100–120 (90°C) 50–70 (30°C)
Washing temperature, °C	hot: 55 (130°F) warm: 30–45 (90–110°F) cold: 10–25 (50–80°F)	5–25	90 60 40
Water hardness (average), ppm CaCO$_3$	relatively low, 100	very low, 30	relatively high, 250
Detergent dispenser	mostly no	mostly no	yes
Recommended detergent Dosage*, g/L g/kg laundry	1.0–1.2 15–30	0.5–0.8 5–7	4.5–8.0 15–35
Peroxide bleach	mostly added separately	mostly dosed separately	mostly incorporated in detergent
Chlorine bleach	added separately	dosed separately	predominantly in the Mediterranean countries

*In the United States and Japan mostly without bleaching components.

5.2.5   Sundry organic builders

Antiredeposition agents. The detergent has an important characteristic that allows the removal of the soil from textile fibers during the washing process. Furthermore, the redeposition of displaced soil can be prevented by the addition of special antiredeposition agents. The role of these agents is to be adsorbed on the surface of the textile, there by creating a protective layer that strictly inhibits redeposition of the removed soil. For example, the adsorption behavior of gelatin on glass, both in the presence and in the absence of sodium n-dodecyl sulfate, clearly showed the effect of the

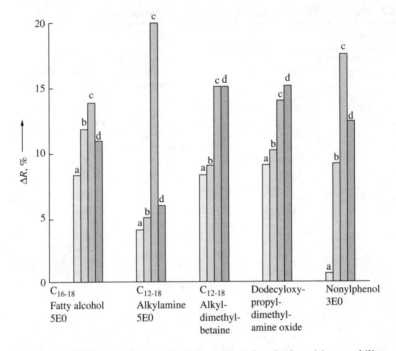

Figure 5.4 Influence of specific surfactants on the oil antiredeposition capability of detergents [20]. ΔR indicates the improvement in the soil antiredeposition capability against the additive [20]. (a) Cotton, 95°C; (b) Permanent-press cotton, 95°C; (c) Polyester, 60°C; (d) Polyester or cotton, 60°C. 5g/L detergent; 0.2 g/L surfactant; number of wash cycles = 3; time = 30 min [20].

addition of the surfactant. The presence of n-dodecyl sulfate eliminates the adsorption of the macromolecules onto the substrate [3, 4].

The carboxymethyl cellulose (CMC) derivatives and carboxymethyl starch (CMS) are effective antiredeposition agents that are cellulose-containing fibers such as cotton and blends of cotton and synthetic fibers. However, CMC has virtually no effect on pure synthetic fibers. Other effective antiredeposition agents and soil repellents have been developed (Fig. 5.4) [20].

The nonionic cellulose ethers have been found suitable for use with synthetic fibers [3, 4].

$$R = CH_3, C_2H_5, CH_2CH_2OH,$$
$$CH_2CH(OH)CH_3,$$
$$CH_2CH_2CH(OH)CH_3$$

Nonionic cellulose ethers

The modern detergents are provided with mixtures of anionic and nonionic polymers (e.g., carboxymethylcellulose-methylcellulose) and also with polymers from polyethylene glycol and terephthalic acid [3, 4].

Thickening agents. Carboxymethyl cellulose (CMC) is also used as a thickening agent in addition to its characteristic as a soil antiredeposition agent. Modified nonionic celluloses, methyl cellulose, hydroxyethyl cellulose methylhydroxy propyl cellulose are being used as thickening agents. All these modified celluloses are soluble in cold water and insoluble in hot water and most organic solvents [3, 4].

Optical brighteners. Optical brighteners are an integral part of laundry products, whether laundry powders or laundry liquids. They are organic compounds capable of converting a portion of the invisible ultraviolet light into longer wavelength visible blue light. They are dyestuffs, which are absorbed by textile fibers, but are not easily rinsed off. The reflection of the blue light makes the clothes look brighter than they actually are.

Optical brighteners are usually derivatives of coumarin, stilbene, distyrylbipheny, and bis(benzoxazole). Examples of chemical structures of some important optical brighteners are given:

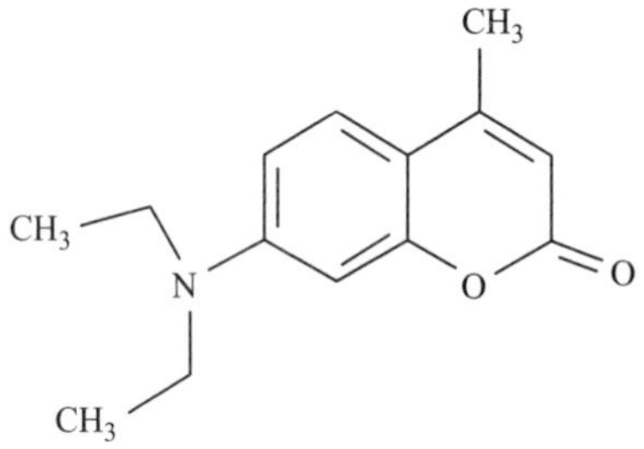

7-diethylamino-4-methyl coumarin

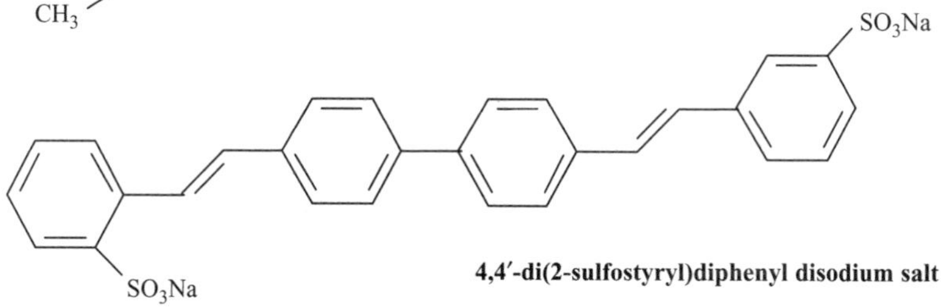

4,4′-di(2-sulfostyryl)diphenyl disodium salt

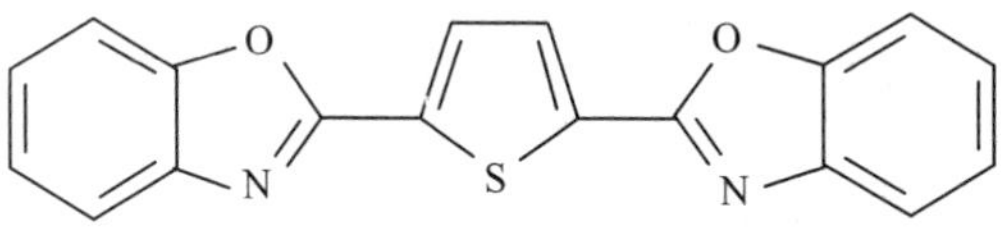

2,5-bis(benzoxazol-2-yl)thiophene

The optical brighteners binding occur, in the case of cotton and chlorine-resistant materials, through the formation of hydrogen bond to the fibers. Whitening effects achieved with polyamide brighteners are largely because of the diffusing power of whitening agent molecules present at the fiber surfaces [3, 4].

Chelating agents.   The role of the chelating agents is to block the polyvalent ions and to make them undetectable and ineffective. Sodium salts of ethylene diamine tetraacetic acid (EDTA) and of nitrilo triacetic acid (NTA) are members of the group of chelating agents and their use has increased in the last few years.

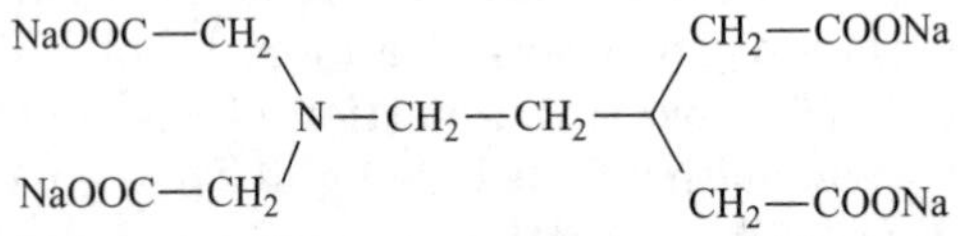

$$NaOOC-CH_2 \diagdown \atop NaOOC-CH_2 \diagup N-CH_2-CH_2- \diagup CH_2-COONa \atop \diagdown CH_2-COONa$$

Sodium salt of EDTA

The sodium salt of EDTA reacts with calcium ions to give a complex in which Ca is bidentated to nitrogen atoms of EDTA. Two sodium ions are released from the reaction [3, 4].

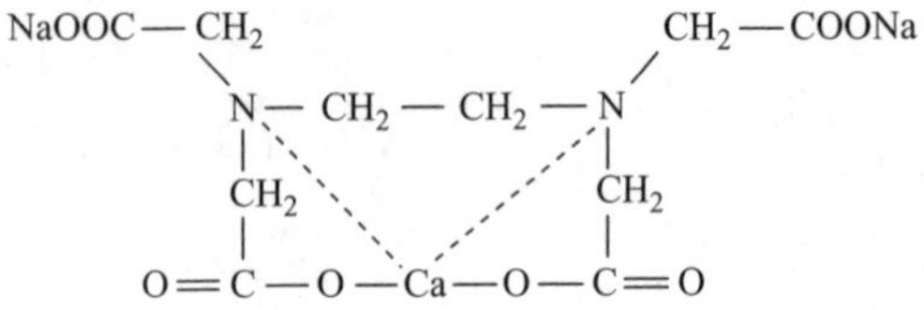

$$NaOOC-CH_2 \diagdown \atop N-CH_2-CH_2-N \diagup CH_2-COONa$$

$$\underset{O=C-O-Ca-O-C=O}{\overset{CH_2 \qquad\qquad CH_2}{|\qquad\qquad\quad|}}$$

There are three main groups of the chelating agents. Aminocarboxylic acids:

a. Ethylene diamine tetraacetic acid

b. Nintrilo triacetic acid

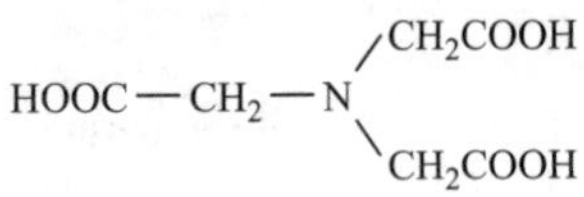

$$HOOC-CH_2-N \diagup CH_2COOH \atop \diagdown CH_2COOH$$

c. Diethylene triamine pentacetic acid (DTPA)

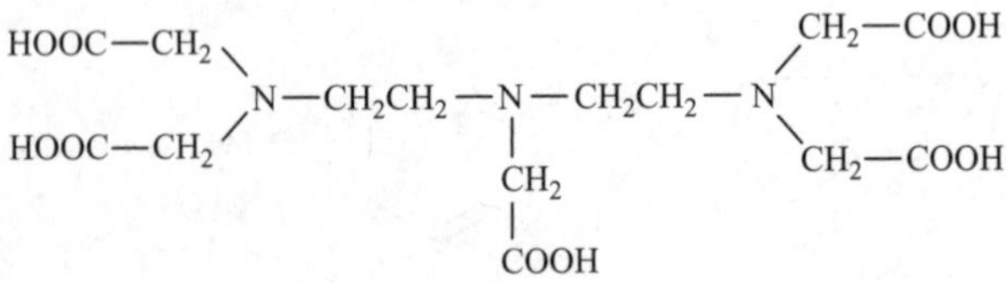

$$HOOC-CH_2 \diagdown \atop HOOC-CH_2 \diagup N-CH_2CH_2-N-CH_2CH_2-N \diagup CH_2-COOH \atop \diagdown CH_2-COOH$$

$$\underset{COOH}{\overset{CH_2}{|}}$$

NTA sequestrates more calcium ions per unit weight than EDTA because of its lowest molecular weight. However, both will sequester more calcium ions than magnesium because the molecular weight of magnesium is smaller than that of calcium. EDTA and NTA are used in laundering formulations to chelate trivalent ions, thus preventing iron stains in laundering [3, 4].

Enzymes. Enzymes are defined as organic catalysts that promote specific chemical reactions in the body upon which all life depends. Naturally occurring enzymes are related to proteins and are largely composed of amino acids.

The most significant development in the detergent industry in recent years is the introduction of enzyme additives.

Enzymes catalyze destruction and removal of stubborn protein-aceous stains and specific types of soils by detergent. Chocolate and starch-based food stains as well as greasy stains that are difficult to remove in low-temperature washing, are eliminated by detergent-enzymes [1–4].

There are four types of enzymes of interest to the detergent industry:

1. Proteases: act on protein to form amino acids

2. Amyloses: convert starches into dextrins

3. Lipases: attack fats and oils

4. Celluloses: hydrolyze cellulose of broken surface fibers and remove micro-pills from cotton and restore color

The mechanism of the enzymes is to cleave the protein into smaller peptide fraction that is soluble in water. Even if the conversion is not complete, the protein is degraded into a product that can be easily removed by the detergent. The activity of enzymes depends on temperature and pH value. At a certain temperature of the wash bath (mostly above 55°C) the activity of enzymes decreases. The proper choice of the type of enzyme and the appropriate formulation has led to detergent-grade enzymes that are not significantly damaged during the storage or in the wash liquor [1–4].

5.2.6 Manufacturing of detergents

Production of alkylbenzene sulfonates [1, 2]. The alkylbenzene sulfonates, used as liquid surfactants in making the detergent slurry, are produced by the sulfonation of linear alkylates followed by the neutralization step with a caustic solution containing sodium hydroxide (NaOH). The process of sulfonation of alkylbenzenes with oleum takes

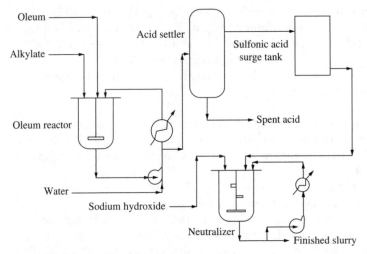

Figure 5.5 Oleum sulfonation process.

place in a batch system where five basic processing operations are used (Fig. 5.5):

1. Sulfonation
2. Digestion
3. Dilution
4. Phase separation
5. Neutralization

The sulfonation stage includes mixing of alkylate with oleum that leads to an exothermic reaction. The removal of heat is done by reactor jacketing. The key parameters that control the reaction of sulfonation are temperature, acid strength, reaction time, and oleum-to-alkylate ratio. The reaction was completed at the digestion stage where the product from the sulfonation zone is aged for 15 to 30 min. The mixture of sulfonic acid and sulfuric acid is diluted with water to quench the reaction. The reaction mixture is sent to a separator to allow gravity settling of the spent sulfuric acid from the lighter sulfonic acid. The lower spent acid layer contains, approximately, 75 to 80 percent sulfuric acid. The upper layer contains, approximately, 88 to 91 percent sulfonic acid and 6 to 10 percent sulfuric acid [1, 2].

The linear alkyl sulfonates can be neutralized with aqueous solutions of base such as NaOH, KOH, NH_4OH, or alkanolamines. The sodium salts are used in the formulation process to produce spray-dried detergents

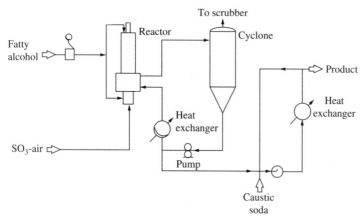

Figure 5.6   Fatty alcohol sulfation process.

for household laundry. However, ammonium and alkanolamine neutralized salts are usually employed in light duty liquid detergents [1, 2].

Fatty alcohols sulfation [1, 2].   The sulfation of fatty alcohols takes place in falling film reactors but cooling water and sulfation temperatures are adjusted to lower values. The sulfonic acids obtained are neutralized immediately to minimize degradation and side reactions in storage [1, 2].

$$R-OH + SO_3 \longrightarrow R-O-SO_3H$$

A typical process for the sulfation of fatty alcohols is shown in Fig. 5.6. A posthydrolysis step includes bleaching to remove color before neutralization. The neutralization step of the sulfonic acid is similar to the case of oleum sulfonation process [1, 2].

The surfactant slurry, builders, and other miscellaneous additives are introduced in the crutcher. A considerable amount of water is removed, and the paste is thickened by the tripolyphosphate (used as a builder) hydration reaction:

$$Na_5P_3O_{10} + 6H_2O \longrightarrow Na_5P_3O_{10} \cdot 6H_2O$$

Spray-drying process.   The first step in the spray-drying process includes the preparation of slurry of thermally stable and chemically compatible ingredients of the detergent (Fig. 5.7)[1–4]. The slurry, the builder, and other additives are mixed in a crutcher. The blended slurry is transferred and held in a stirred storage vessel for continuous pumping to a spray dryer. The slurry is sprayed through nozzles into the tower at pressures of 4.1 to 6.9 kilopascals (kPa) in single-fluid nozzles and at

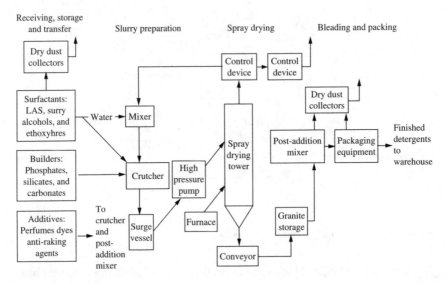

LAS = Linear alkyl salfonate

Figure 5.7   Spray-drying process.

pressures of 340 to 690 kPa in 2-fluid nozzles. Steam or air is used as the atomizing fluid in the 2-fluid nozzles. The slurry is sprayed at high pressure into a vertical drying tower having a steam of hot air at 315 to 400°C. The dried powder flows off the tower at a temperature of 90 to 100°C. An airlift is used for cooling to prevent lumping. The most common towers designed for detergent production are countercurrent. The slurry is introduced at the top of the tower and heated air is introduced at the bottom. The detergent granules are conveyed mechanically or by air from the tower to a mixer to incorporate additional ingredients, such as a perfume, and finally to packaging and storage [1–4].

5.3   Environmental Aspects

5.3.1   Emissions and controls

The exhaust air from detergent towers contains two types of contaminants:

1. Fine detergent particles
2. Organics vaporized in the higher zone of the tower

Dust emissions are generated at scale hoppers, mixers, and crutchers during the batching and mixing of fine dry ingredients to form slurry. Conveying, mixing, and packaging of detergent granules can also cause dust emissions. For this process, fabric filters are used, not only to reduce or eliminate dust emission, but also to recover raw materials [1–4].

Dry cyclones are the primary collection equipment used to capture the detergent dust in the spray dryer exhaust and recycle it back to the crutcher. The dry cyclones can remove up to 90 percent by weight of the detergent product fines from the exhaust air. Fabric filters have been used after cyclones but have limited applicability, especially on efficient spray dryers, because of condensing water and organic aerosols binding the fabric filters [4, 21].

In addition to particulate emissions, volatile organic compounds (VOCs) may also be emitted when the slurry contains organic materials with low vapor pressures. The surfactants included in the slurry represent the origin of the VOCs. The vaporized organic materials condense in the tower exhaust air stream into droplets or particles. Paraffin alcohols and amides in the exhaust stream can result in highly visible plume that persists after the condensed water vapor plume has dissipated [4, 21]. Some of the VOCs identified in the organic emissions are: hexane, methyl alcohol, 1,1,1-trichloroethane, perchloroethylene, benzene, and toluene [1–4].

A method for controlling visible emissions would be to remove by substitution the offending organic compounds from the slurry.

5.3.2 Wastewater and the environment

Wastewater is the by-product that has the potential for causing a number of undesirable effects in sewage treatment plants in the environment. In fact, the clean water, which was brought into the process, is later released to the sewage system in the form of contaminated wastewater containing additional energy, soil from the laundry, lint, dyes, finishing agents, and detergents. Detergents are released as the products of reaction with other material during the washing cycle or in unchanged form [4].

The variations in laundry soil levels affect the concentration and the composition of laundry wastewater considerably. Laundry wastewater is generally a heavy source of contamination; therefore, it should not be returned to receiving waters in untreated form [4].

Fortunately, the laundry wastewater is mixed with general water in the public sewage and in the sewage treatment plants. As a result of dilution, both temperature and the high pH value of wastewater are considerably decreased. In addition, if laundry wastewater was treated separately, major problems would be encountered in dealing with the load of organic pollutants introduced by household and commercial laundry operations [4].

5.3.3 Biodegradation

The removal of organic compounds from sewages, surface waters, and soils can be done by the biodegradation process. This is a stepwise process that is mainly affected by aerobic microorganisms (Fig. 5.8). The first step involves the transformation of the sodium sulfonate to a first degradation product (primary degradation). The subsequent degradation to the second, third, and so on, intermediate steps takes place with the decreasing of molecular mass and structural complexity. The ultimate biodegradation represents the total decomposition of the total organic structure into carbon dioxide, water, inorganic salts, and at the same time, partly into bacterial biomass.

Since 1973, the regulation in Europe requires a minimum of 80 percent biodegradability for the anionic and nonionic surfactants in a package detergent. Anionic surfactants are determined as *methylene blue active substance* (MBAS), that is, materials forming a chloroform soluble complex with cationic dye methylene blue. Nonionic surfactants are defined as *bismuth active substance* (BiAS), that is, materials forming an insoluble complex with the bismuth-containing Drangendorff reagent [4]. The primary biodegradation of anionic and nonionic surfactants is determined in standardized tests by measuring the removal of MBAS and BiAS, respectively. The ultimate biodegradation of chemicals can be followed in the tests by means of nonspecific analytical parameters such as carbon dioxide evolution (BOD) or the removal of dissolved organic carbon (DOC).

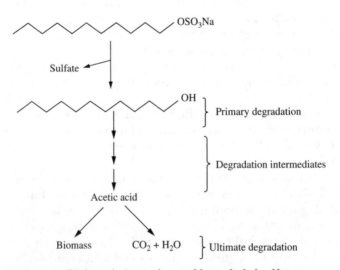

Figure 5.8 Biodegradation pathway of fatty alcohol sulfates.

Primary and ultimate biodegradability of test substances is normally evaluated by applying standardized and internationally used (OECD, ISO, EU) test procedures [4].

5.4 Economic Aspects

The role of detergents represents an important factor in the worldwide economy. The increase of the standards of living in the last 40 years was accompanied with a tremendous growth in the world consumption of the detergents. The absolute quantity has increased from approximately 10×10^6 t in the late 1950s to 21.5×10^6 t worldwide in 1988 [4]. For example, the total consumption of laundry, dishwashing, and cleaning detergents in Europe was 6.4×10^6 t in 1991. The quantity has increased to 6.8×10^6 t in 1999 (Table 5.4) [4, 22].

TABLE 5.4   The Total Consumption of Surfactants Worldwide (10^3 t) [4, 22]

End use	1992	1998
Household		
North America	866	1115
Western Europe	1061	1119
Asia	1208	1599
Other regions	998	1235
Total	**4133**	**5068**
Personal care		
North America	204	210
Western Europe	143	163
Asia	119	194
Other regions	86	155
Total	**552**	**722**
Industrial and institutional cleaners		
North America	263	311
Western Europe	155	192
Asia	60	75
Other regions	42	53
Total	**520**	**631**
Industrial process aids		
North America	1191	1316
Western Europe	644	676
Asia	858	1085
Other regions	355	364
Total*	**3048**	**3441**
Grand total	**8253**	**9862**

*Does not include soap.

TABLE 5.5   The Major Surfactant Consumption by Type
Worldwide (10^3 t) [4, 22]

Surfactant type	1992	1998
Linear alkylbenzene sulfonates	2385	3027
Branched alkylbenzene sulfonates	411	148
Alcohol sulfates	466	479
Alcohol ethoxy sulfates	511	911
Alcohol ethoxylates	742	849
Alkyl phenol ethoxylates	652	701
Quaternaries	312	434
Others	2774	3263
Total	**8253**	**9862**

The two largest end uses, which are household products and industrial processing aids, together account for 86 percent of the total production of surfactants [4, 22]. Household products include heavy-duty powder and liquid detergents, light-duty liquids, and fabric softeners.

Industrial processing materials are covered by the following major end uses: plastics and elastomers, textiles, agricultural chemicals, leather and paper chemicals, and other miscellaneous uses.

The anionic surfactants, such as linear alkyl-benzene surfactants, represent the largest consumed type of surfactants worldwide followed by alcohol ethoxylate, and alkylphenol ethoxylates (Table 5.5) [4, 22].

For legal restrictions, the use of sodium triphosphate as a builder has sharply decreased mainly in the United States, Europe, and Japan. The application of sodium triphosphate in detergents dropped by more than 50 percent within 10 years in these countries [4, 23].

Zeolite A, the most important phosphate substitute, became the highly demanded builder worldwide. The worldwide production of zeolite A increased in western Europe in the 1990s. Detergent builder zeolites represent the largest application field for zeolite. Almost 90 percent of zeolites produced worldwide (or ~ 215,000 tons/year) in 2003 were used for detergents. Meanwhile, production capacities for detergent-grade zeolites have largely surpassed the demand [4].

References

1. Kent, J., *Riegel's Handbook of Industrial Chemistry*, Van Notrand Reinhold, New York, N.Y., 1992.
2. Austin, G., *Shreve's Chemical Process Industries*, 5th ed., McGraw-Hill Book Company, New York, N.Y., 1985.
3. Davidsohn, A. S. and B. Milwidsky, *Synthetic Detergents*, Longman Scientific and Technical, Burnt Mill, Harlow-England, 1987.
4. Smulders, E., *Laundry Detergents*, Wiley-VCH, Verlag-Germany, 2002.
5. The Soap and Detergent Association *Soaps and Detergents*, New York, N.Y., 1981.

6. Gupta, S., Chemistry, chemical and physical properties and raw materials, *Soap Technology for the 1990's*, L. Spitz (Ed.), American Oil Chemists Society, 1990.
7. Danielson, J. A., *Air Pollution Engineering Manual*, 2nd ed., AP-40, U.S. Environmental Protection Agency, Research Triangle Park, NC, May 1973 (from Organic Chemical Process Industry).
8. Smulders, E., and M. O. Hernandez, *Comunicaciones presentadas a las Jornadas del Comite Espanol de la Detergencia*, 28, 13–32, 1998.
9. Smulders, E., P. Krings, and H. Verbeek, *Tenside, Surfactants, Detergents* 34, 386–292, 1997.
10. Kirshner, E. M., Soaps & Detergents, *Chemical and Engineering News*, American Chemical Society, January 26, pp. 39–54, 1998.
11. Andree, H., and P. Krings, *Chemiker Zeitung*, 99, 168–174, 1975.
12. Smulders, E., and Krings, P., China Surfactant Detergent & Cosmetics, 1, 18–34.
13. Brüer, K., and Fehr, H. R., Puchta, *Tenside Surfactants Detergent*, 17, 281–287, 1980.
14. Linfield, W. M., Cationic Surfactants, *Surfactant Science Series*, 4, 49, 1970.
15. Berenbold, H., *Tenside, Surfactants, Detergent*, 27, 58–61, 1990.
16. Puchta, R., P. Krings, and P. Sandkühler, *Tenside Surfactants Detergent*, 30, 186–191, 1993.
17. Kuzel, P., *Seifen Ocle Fette Wachse*, 105, 423–424, 1979.
18. Church, J. A., Hypochlorite bleech, in U. Zoller and G. Broze (Eds.) *Handbook of Detergents: Properties, Surfactant Sci. Ser.*, 82, 619–629, 1999.
19. Julemont, M., Application of hypochlorite in U. Zoller and G. Broze (Eds.) *Handbook of Detergents: Properties, Surfactant Sci. Ser.*, 82, 631–637, 1999.
20. Jakobi, G. *Tenside-Taschenbuch*, 2nd ed., H. Stache, (Ed.), Hanser Verlag, München-Wien 253–337, 1981.
21. Buonicore, A. J. *Air Pollution Engineering Manual*, W. T. Davis (Ed.)., Van Nostrand Runhold, New York, N.Y., 1992.
22. A. Colin Houston & Associates, Inc. *High Alcohols Market Forecast*, Pound Ridge, New York, N.Y., 2000.
23. Roland E., P. Kleinschmit, *Zeolites in Ullmann's Encyclopedia of Industrial Chemistry*, 6th ed., Wiley-VCH, Weinheim, 2001.

Chapter

6

Sugar

Mohammad Farhat Ali

6.1 Introduction

Carbohydrates, sugars, and starches are the most widely distributed and abundant *organic chemicals* on earth. They have a central role in the metabolism of animals and plants. These serve as a basic food, accounting for a large portion of total nutrient intake. They act as sweetening agents, gel- or paste-forming and thickening agents, stabilizers, and also precursors for aroma and coloring substances generated within the food by a series of reactions and during handling and processing.

Historically, carbohydrates were thought to be hydrates of carbon on the basis of their molecular formulas $C_n(H_2O)_m$. More realistically, carbohydrates are now considered polyhydroxy aldehydes, polyhydroxy ketones, or substances that yield such compounds on acid hydrolysis.

6.2   The Chemistry of Saccharides

Saccharum is the Latin word for *sugar* and the derived term *saccharide* is the basis of a system of carbohydrate classification. The simplest sugars belong to the carbohydrate class, monosaccharide; they include fructose and glucose. Glucose and fructose are structural isomers. The molecular formula is as follows:

$$C_6H_{12}O_6$$

The structural formula of glucose contains a ring having six atoms (C_5O, *pyranose form*) and an aldehyde group (*aldoses*). However, the structural formula of fructose contains a ring containing five atoms (C_4O, *furanose form*) and a ketone group (*ketoses*):

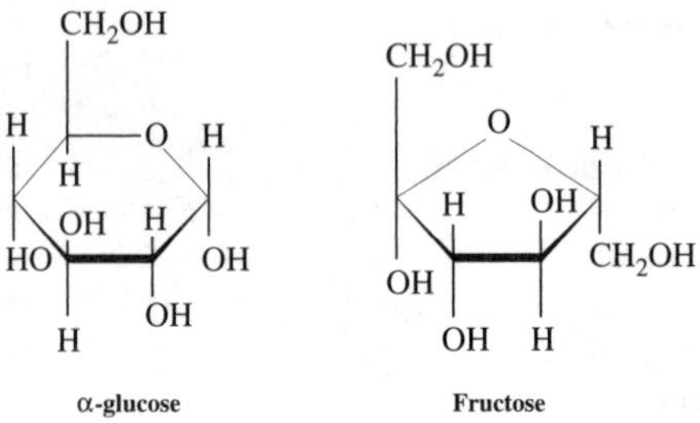

α-glucose   Fructose

The formulas of D-glucose and D-fructose using simplified Fischer projections may also be represented as follows:

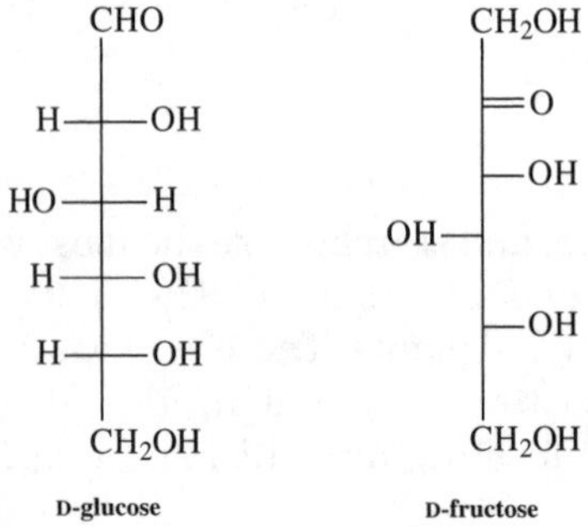

D-glucose   D-fructose

Disaccharides are formed by the union of two monosaccharides with the loss of one molecule of water. Disaccharides include lactose, cellobiose, maltose, and sucrose. The molecular formula for sucrose is as follows:

$$C_{12}H_{22}O_{11}$$

The structural formula of sucrose contains two units of monosaccharide (glucose and fructose) joined together by an atom of oxygen, a glycoside link, from carbon 1 of one unit (glucose) to an OH of the other unit (fructose).

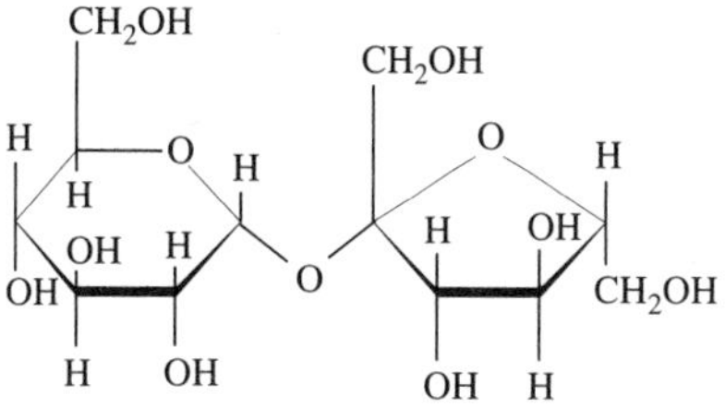

Sucrose

(glucose (α1 → 2) fructose)

Hydrolysis of sucrose yields D-glucose and D-fructose; the process is called inversion and the sugar mixture produced is known as invert sugar because, although sucrose itself rotates plane-polarized light to the right, the mixture inverts this light by rotating to the left.

The carbohydrate class, polysaccharide, represents compounds in which the molecules contain many units of monosaccharides joined together by glycoside links. Upon complete hydrolysis, a polysaccharide yields monosaccharides. Starch is the most valuable polysaccharide. The starch molecules (amylose and anylopectin) are tree-like, containing 250 to 1000 or more glucose units per molecule joined together through alpha linkages.

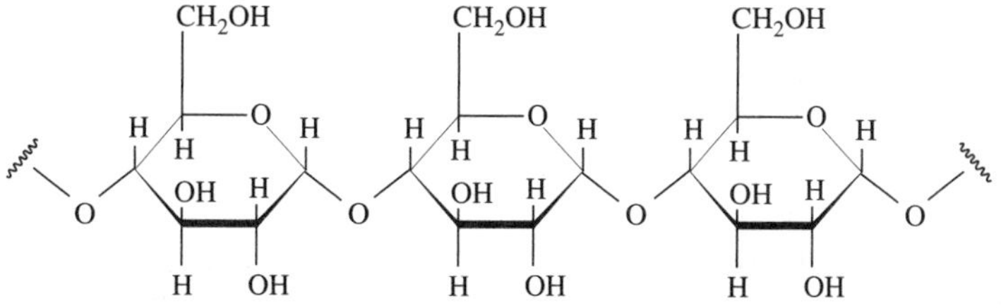

Amylose

poly (1,4'-O-α-D-glucopyranoside)

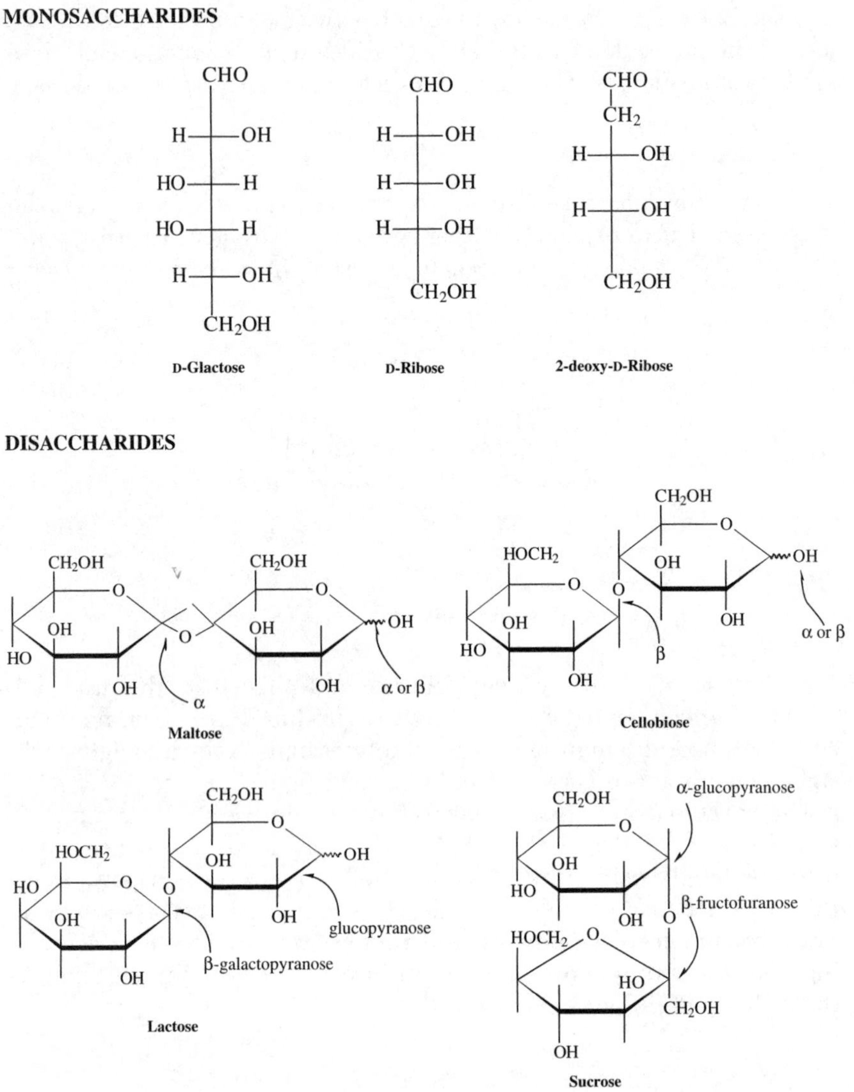

MONOSACCHARIDES

D-Glactose   D-Ribose   2-deoxy-D-Ribose

DISACCHARIDES

Maltose   Cellobiose

Lactose   Sucrose

Figure 6.1   The structures of some important mono- and disaccharides.

Cellulose is the most abundant polysaccharide. It is the fibrous component of a plant cell wall (e.g., cotton). Cellulose molecules are chains of molecules of up to 14,000 units of D-glucose linked together by beta linkages:

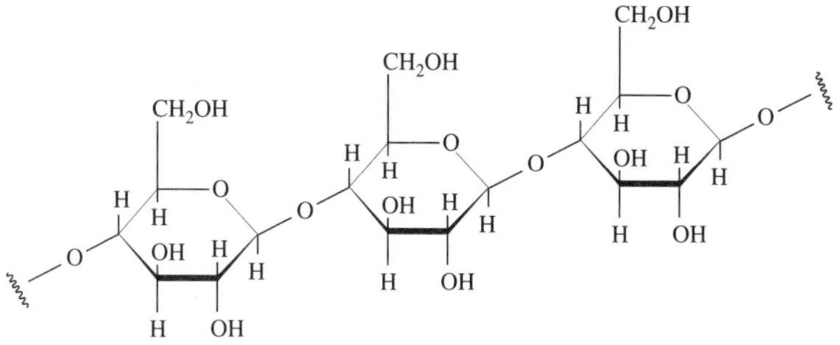

Cellulose
poly (1,4'-O-β-D-glucopyranoside)

6.3 Properties of Sucrose

In commercial usage, the term sugar usually refers to sucrose. Sucrose is a disaccharide sugar that occurs naturally in every fruit and vegetable. It is a major product of photosynthesis, the process by which plants transform the energy of the sun into food. Sugar occurs in greatest quantities in sugarcane and sugar beets from which it is separated for commercial use. Fully refined sugar, whether made from cane or beet, is pure sucrose and the consumer cannot tell from which of the two plants it is derived. Despite the identical end product, the sugarcane and the sugar beet industries differ greatly in methods of production and organization and each has its own distinctive history and geography.

Sucrose crystallizes from aqueous solutions as monoclinic, hemimorphic crystals. The presence of raffinose or dextran impurities, if present in quantity, produces long needle-like shapes.

The melting point of sucrose, approximately 188°C, is rather indefinite, because it appears to depend upon the solvent used for its crystallization. The density of sucrose is 1.5879 g/cm^3. The linear expansion coefficient ranges from 0.0028 to 0.0050 percent, depending on the axis. Characteristic infrared (IR) absorption bands occur at 1010, 990, 940, 920, 870, 850 cm^{-1} (sharp), and at 680, 580 cm^{-1} (broad). The specific heat of crystalline sucrose is 415.98 J/mol at 20°C. The dipole moment is 2.8×10^{-29} cm (8.3D). Sucrose is readily soluble in water and the solubility increases with the rise in temperature. It is sparingly soluble in alcohol but moderately soluble in organic solvents, such as dimethyl formamide, pyridine, and dimethyl sulfoxide.

An important property of sucrose in solution is its polarization. It is well-known that sucrose in solution rotates a polarized light to the right in equal proportion to the quantity of sucrose present. The great utility of this property is in the use of the saccharimeter, an instrument which directly reads the percentage of sucrose present in sucrose-bearing materials. Therefore, the sucrose is purchased, and various stages of manufacturing processes are monitored, by polarization measurements.

6.4   Historical Survey and World Production

Lack of data makes it difficult to establish just when sugar first became known to humankind, but it probably was first used by man in Polynesia from where it spread to India. In 510 B.C. Emperor Darius of what was then Persia, invaded India, where he found sugarcane growing on the banks of the River Indus. He called it the reeds, which produce honey without bees.

Much later, in the sixth century A.D., it was grown in Persia and the Arab Muslims took it to Egypt. The word sugar is derived from an Arabic word, *sukkar*. By the end of the tenth century A.D., the early Islamic movement established sugar production throughout the Mediterranean area. The subsequent centuries saw a major expansion of western European trade with the east, including the importation of sugar. It is recorded, for instance, that sugar was available in London at *two shillings a pound* in 1319 A.D. This equates to about US$ 100 per kg at today's prices, so it was very much a luxury.

By the middle of the fifteenth century there were plantations in Madeiva, the Canary Islands, and St. Thomas, and they supplied sugar to Europe. In the same century Columbus sailed to the Americas called the *New World*. It is recorded that in 1493 A.D. he took sugarcane plants to grow in the Caribbean. There the climate was so advantageous for the growth of the cane that an industry was quickly established. By 1600, the sugar industry was the largest in tropical America.

At this stage, sugar was still a luxury and vast profits were made, to the extent that sugar was called *white gold*. It was only after 1700 A.D. that sugar was transformed from a luxury product into one of everyday use, even by the poor. This took place as Brazil and the new West Indian colonies began producing sugar from sugarcane in such large quantities that the price was significantly reduced. Lower prices led to increased consumption, which, in turn, fueled demand, leading to the expansion of the sugar industry in the Americas and later elsewhere in the tropical world.

The other main source of sugar, *sugar beet*, although known as a sweet vegetable, was first developed as a commercial source of sugar in the second half of the 18th century by two German chemists, Margraf and his pupil Achard. Its further development occurred in large part because of the Napoleonic wars at the start of the 19th century when Britain blockaded sugar imports to continental Europe. By 1880, sugar beet had replaced sugarcane as the main source of sugar in continental Europe.

Presently the United States, Brazil, and the European Union (EU) are the leading producers and exporters of sugar, whereas India is relatively a newcomer to the ranks. Indian production began to increase about 16 years ago. India is also one of the largest consumers of sugar in the world. For the first 8 years most of its production was consumed domestically, but in the last 8 years, average production has increased substantially (45 percent) over its previous 8-year average, resulting in a significant increase in Indian sugar export. The United States ranks among the top four sugar producers and is one of the few countries with significant production of both sugar beets and sugarcane. The United States is also the largest consumer of sweeteners, including high fructose corn syrup. By virtue of the price support loan program and the tariff rate quota, the United States is one of the largest global sugar importers.

World centrifugal production for 2003–04 was over 142 million tons. About 60 percent of this came from cane, and about 40 percent from beet. World sugar consumption for 2003–04 is estimated at 140 million tons [1].

The data in Table 6.1 taken from Economic Research Service, U.S. Department of Agriculture (2004), list the annual production and

TABLE 6.1 Annual Production and Consumption of Refined Sugar for Some Countries in the World

Country	Year	Production ×1000 tons	Consumption ×1000 tons
Brazil	1995–96	13,700	8100
	2003–04	25,530	10,050
Canada	1995–96	164	1209
	2003–04	98	1,431
India	1995–96	18,225	14,820
	2003–04	16,670	20,500
Pakistan	1995–96	2643	3090
	2003–04	4,047	3,600
Russia	1995–96	2060	5000
	2003–04	1,930	6,480
USA	1995–96	11,510	14,016
	2003–04	8,132	8,946
W. Europe	1995–96	135	499
	2003–04	16,707	15,064

consumption of refined sugar for the years indicated for some countries in the world [2].

6.5   Cane Sugar

6.5.1   Raw sugar manufacture

White sugar is essentially pure sucrose derived from sugarcane. The cane plant is a coarse-growing member of the grass family *saccharum*. Several species of saccharum are found in Southeast Asia and neighboring islands, South America, and the West Indies. There are three basic species, *s. officinarum*, *s. robustum*, and *s. spontaneous*. Many forms of these species interbreed, making a highly diverse genus. Sugarcane plants are propagated by planting sections of the stem having a bud at the base of each leaf. The buds sprout into shoots from which several other shoots arise below the soil level to form a clump of stalk, or *stool*. The mature sugarcane stalk (stem) may vary from 4 to 12 ft or more in height, and in commercial varieties are from 0.75 to 2 in. in diameter. From 12 to 20 months are required for crops to mature from new plantings. As the plants become tall, lower leaves along the stems are spaded and die. These ultimately drop off, so leaves toward the top remain green and active. Between the nodes the stems have a hard, thin, outer tissue, or *rind*, and softer center. The high-sugar-containing juice is in this center. More than one crop is harvested from a single planting. After the first crop is removed two or more so-called stubble crops are obtained, which are the results from a growth of new stalks from the bases of stalks cut near the ground level in harvesting [3].

The bulk of all sugarcane harvested in the world today is still cut by hand with a cane knife or machete. However, harvesting of cane in Hawaii and Louisiana is highly mechanized. Machines top the canes at a uniform height, cut them off at ground level, and deposit them in rows. Infield roads are then made alongside the rows, and the cane is picked up by large grabs installed on three-quarter yard cranes especially designed for this operation. A second system is to burn the leaves from the standing cane, after which it is cut and taken directly to the mill. Delay between cutting and milling in either case should be as short as possible because delay results in loss of sugar content.

The mechanically harvested cane picks up field mud, sand, trash, and fine dirt during transportation into the factory. All of this creates problems in grinding cane and clarification of juices, as well as further processing in the factory. To avoid these problems the cane first goes through a washer. Washing systems vary from a simple system of spraying warm water on a table to a very elaborate system consisting of conveyors with water jets, stripping rolls, and baths for removal of stones and mud.

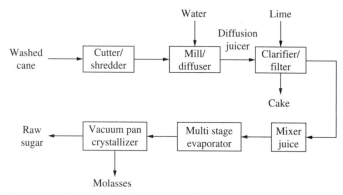

Figure 6.2 Typical flow diagram for the manufacture of raw sugar from sugarcane.

Figure 6.2 shows different stages for the manufacture of raw sugar. These steps are described as follows.

Extraction of juice. The juice is extracted from the cane either by *milling*, in which case the cane is pressed between heavy rolls, or by *diffusion*, in which case the sugar is leached out with water. In either case, the cane is chopped into short segments (8 to 12 in. long) and these segments are passed through two sets of rotating knives. The first set of knives cuts the cane into small pieces and also acts as a leveler to distribute the cane more evenly on the carrier. The second set of knives acts as a shredder and thoroughly cuts up and shreds the cane into a fluffy mat of pieces a few centimeters in length. In case of diffusion, the cane is put through an even finer shredder called a buster. No juice is extracted in the shredders.

Milling. The prime objective in sugarcane milling is to extract the greatest possible amount of sucrose from the sugarcane in juice form, and to make the final bagasse (fibrous residue from the cane) as dry as possible so that it will burn readily in the boilers. The cut pieces from the shredder then pass through a series of mills, called a tandem or milling train. The classical combination of three rollers arranged in triangular form (Fig. 6.3) is the standard milling unit of the sugar industry [4]. The top roller rotates counterclockwise and the bottom two rollers clockwise. From three to seven sets of such three-roller units, described as a 9-roller to 21-roller mill, are in use. After passing through the mill, the bagasse is carried to the next mill by bagasse carriers and is directed from the first squeeze in a mill to the second by a turn plate. In order to achieve good extraction, a process called *compound imbibition* is used

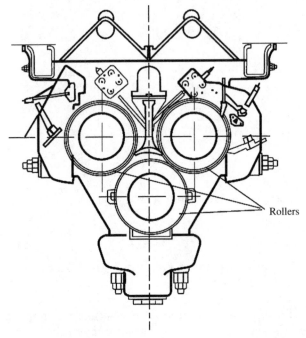

Rollers

Figure 6.3 Cross section of the three-roller mills. (This material is used by permission of John Wiley & Sons, Inc.)

to reduce the sucrose in the fiber by repeated dilution and milling. The juice is collected from the first mill and is mixed with the juice from the crusher. The mixed juice is then passed through perforated metal screens, with openings about 1 mm in diameter, for clarification.

Diffusion.   Diffusers are universally used for the extraction of sucrose from sugar beets but their installation for sugarcane factories has been relatively recent. The process in cane diffusion is mostly lixivation (washing) with only a little true diffusion (osmosis). There are two main systems in cane diffusion—diffusion of cane or diffusion of bagasse. In the first, the prepared whole cane passes through the diffusion unit. In the second, the shredded cane passes through one or two sets of mills where most of the juice is extracted, leaving only some 30 percent residual juice in the bagasse to go through the diffuser.

Operation of the diffuser is based on systematic countercurrent washing of the cane or bagasse by means of imbibition of hot water (65 to 75°C). In practice, this is achieved by forming a bed of shredded cane or first-mill bagasse on a conveyor. Water is added at the discharge end of the conveyor and percolates through the bed of bagasse and the perforated slot of the conveyor. The water dissolves the sugar in the bagasse and the thin juice thus formed is collected in a hopper. This juice is moved forward one stage by pumping and the process is repeated until the juice reaches maximum concentration at the feed end of the diffuser. The diffuser may be conditioned either for single-flow or for parallel-flow juice circulation.

There are many advantages of diffusion over conventional milling. The diffusion plants are simple, more efficient, low in cost, and take much less energy to run. However, the bagasse from diffusion contains much more water and must be dried in a mill or some sort of bagasse press before burning in a boiler. The bagasse from the conventional mill process will burn directly. Most bagasse is burned in the boilers that run the factories.

Purification of juice. The raw juice leaving the mill is slightly acidic (pH., 5.5 to 6.5), turbid, and colored. It carries in suspension cane fiber, field soil, protein, fats, waxes, gums, coloring matter, and soluble salts. Under slightly acidic conditions, the sucrose in the juice is gradually inverted (hydrolyzing to glucose and fructose) under the influence of the native inverter's enzyme. The process designed to stop inversion and remove impurities is called clarification or defecation. The inversion reaction is stopped by adding the milk of lime to the cold juice in amounts sufficient to raise the pH to the range of 7.5 to 8.5. The limed juice is then gradually heated to nearly 100°C to inactivate the enzyme and stop microbiological action. At the same time, a large fraction of the suspended material is removed by settling. Clarification by heat and lime, a process called defecation, is the oldest and in many ways the most effective means of purifying the juice. There are many modifications to this basic process with differing geographical conditions and juice composition. Phosphate is added to juice deficient in phosphate to increase the amount of calcium phosphate precipitate, which makes a floc that helps clarification. When mud settles poorly, polyelectrolyte flocculants such as polyacrylamides are sometimes used.

The precipitated mud is removed in a settling device known as a clarifier. The better known clarifiers in the market are Rapi Dorr 444, the Graver clarifier, and Prima sap, Bach clarifier, Poly-cell, and BMA clarifier. Basically, a clarifier consists of a vertical cylindrical vessel composed of a number of trays with conical bottoms stacked one over the other. The limed raw juice enters near the center of each tray and flows toward a circumference. A sweep arm in each tray turns quite slowly and sweeps the settled mud toward a central mud outlet. The clear juice from the top circumference overflows into a collection compartment. With more uniform juice takeoff, the potential stagnant pockets are eliminated.

The clarification process divides the whole juice into two portions: (1) the clarified juice and (2) the precipitated settlings, which are the scums or mud waters. The clarified juice goes to evaporators without further treatment. The mud is filtered and subsequently washed for removal of sugar in continuous rotary drum vacuum filters with the addition of fine screening of extracted bagasse. The mud mostly consists of field soil and very finely divided fiber. It also contains nearly all the proteins and cane wax. The mud is returned to the fields as a fertilizer.

Clarified juice is usually darker (dark brown) than the combined raw juice from crusher, mills, and subsequent diffusers because of heating (100°C) during lime treatment. The raw sugar produced from this juice is brown in color and contains about 98 percent sucrose. In order to produce plantation white sugar, or at least very light brown raw sugar, it is necessary to resort to processes more elaborate than simple defecation. The procedures using SO_2 or CO_2 in conjunction with lime are used by some sugar producers.

The use of sulfur dioxide in addition to lime makes a better clarification. In this method (sulfitation process), lime is added as usual, but then sulfur dioxide from a sulfur burner is bubbled through the juice. The precipitate is settled as in the ordinary clarification process. The bleaching effect of sulfur makes a lighter-colored sugar. The extra cost of sulfitation, increased sealing in heaters and evaporators, and higher ash in raw sugars are reasons for the discontinuance of this process.

The purification of cane-juice by an excess of lime and by removing the surplus by carbonic acid originated in Java in 1880. In this process (carbonation process), a large excess of lime is added to produce a juice of pH greater than 10.5. Then, carbon dioxide from the lime kiln is bubbled through to precipitate out calcium carbonate along with other impurities, which are removed by filtration. Although the juices from carbonation are much better clarified and somewhat decolorized, this process is not general in modern practice.

There are other white sugar processes such as: (1) treatment with vegetable carbons, (2) treatment with ion exchange resins, and (3) Talodura process. These processes are in fact supplemental clarification methods, adopted by some manufacturers with specific objectives.

Evaporation and heating.   Clarified juice contains about 85 percent water and requires evaporation of water from the sugar solution to yield a final crystalline product. The evaporation is done in two stages: first, the solution is concentrated in an evaporator, and second, the sugar is crystallized in a vacuum pan. Evaporation is carried out in multiple-effect evaporators where the evaporation is divided into a number of stages, in only one of which live steam is used, the heat required for the remaining stages being supplied by the vapor of the liquid itself.

The usual single-effect evaporator, in its simple form, is a closed, vertical-body pressure vessel divided into two sections by a pressure-tight divider. One section is connected to a steam source and the other is partially filled with juice. The steam is kept at a higher temperature than the boiling point of juice resulting in the transfer of heat to the juice, thus driving off water vapor from the juice. The juice concentrate is made in several evaporators connected in series, called a multiple-effect evaporator. The juice travels from one vessel to another because of the gradual increase in vacuum. The vapor obtained in each vessel of multiple-effect evaporator serves to heat the juice in the following vessel, thus driving off additional water. The basic requirement to ensure operations is that the boiling temperature difference of the liquid in each succeeding vessel should be lower than the temperature of steam entering the vessel. This temperature difference provides the driving force for heat transfer from steam to liquid and in turn is provided by a lower liquid-side vapor pressure than steam-side pressure. A typical multiple effect evaporator (Fig. 6.4) consists of three to five evaporator vessels in series [4].

The evaporation is carried on to a final 65 to 68 percent sugar concentrate (syrup). The syrup thus obtained is very dark brown and turbid. The concentration of sugar solution (juice) is measured on the Brix scale, which is a density scale for sugar (sucrose) solutions. The degrees on the Brix scale are numerically equal to the percentage of sucrose in the solution (e.g., 65 to 68 degrees Brix is 65 to 68 percent sucrose in solution).

Crystallization (sugar boiling). The clarified juice (syrup) from the multiple-effect evaporator is transferred to a vacuum pan—a vessel in which syrup is boiled under vacuum to form a heavy mixture of crystals and the mother liquor, called *massecuite.* The vacuum pan is a single-effect evaporator varying in size up to 14 ft. (4.270 m) diameter, and sometimes even larger. The function of the vacuum pan is to produce and develop satisfactory sugar crystals from the syrup or molasses fed to it. In general, two types of pans are used: coil pans, which operate satisfactorily on live steam, and Calandria pans, which are usually a chest of vertical tubes using low-pressure exhaust steam from a preevaporator (concentrator). A typical pan is shown in Fig. 6.5 [4].

Any of the different methods of crystallization of sugar in a vacuum pan is known as *sugar boiling.* Each batch of finished massecuite in a batch-sugar boiling process is a *strike.* Ideally, maximum crystal yield and maximum exhaustion of final molasses should be obtained, with the fewest possible strikes.

The vacuum pan is a horizontal cylinder with compartments in its lower part. It has a very large discharge opening: typically 1 m in diameter. At the end of a strike, the massecuite contains more crystals than syrup and is therefore very viscous. This large opening is required to empty the pan in a reasonable time. The control of pan-boiling is achieved by *seeding,* which supplies crystal nuclei (sugar grains of about 0.35 mm

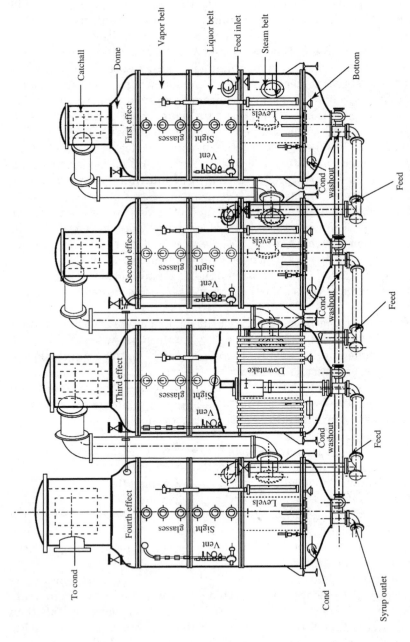

Figure 6.4 A typical multiple-effect evaporator. (This materials is used by permission of John Wiley & Sons Inc.)

First effect
Second effect
Third effect
Fourth effect

Catchall
Dome
Vapor belt
Liquor belt
Feed inlet
Steam belt
Bottom

Sight glasses
Vent
Levels

Cond washout
Feed
Downtake
To cond
Cond
Syrup outlet

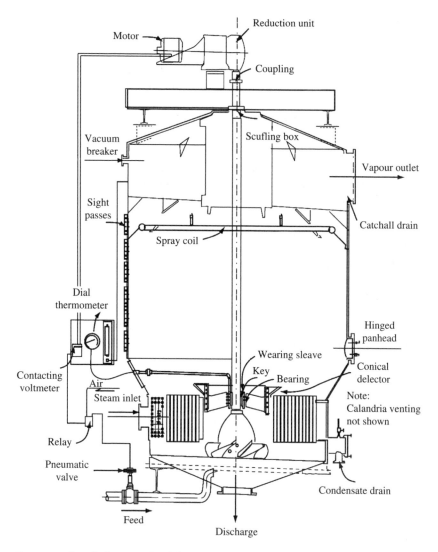

Figure 6.5 A typical conventional Calendria pan. (This material is used by permission of John Wiley & Sons Inc.)

size) to the supersaturated sucrose solution (massecuite) in numbers equal to the desired crystal population of the finished massecuite. The quantity of seed used for each strike is called a *footing*. Seed is added to the first compartment and the resulting strike moves progressively through the compartments of the pan. Additional syrup is added to each compartment to control the viscosity of the massecuite. Because a single crystallization does not recover all of the sucrose from the syrup, the mother liquor from a *strike* is recycled for recovery of additional sugar. The most commonly used methods of crystallizing sugar and exhausting the final molasses (run-off from a low-grade massecuite) are two-, two-and-a-half-, three-, and four-strike boiling systems. A three-boiling system, the Talisman system (Fig. 6.6), is the most popular [5]. In order to distinguish among the boilings, a letter is assigned to each boiling and its products. For example, the first boiling (syrup massecuite) is assigned the letter A, and its products after centrifugation are A sugar and A molasses. The second and third boilings are assigned the letters B and C, respectively.

The Talisman system, developed by Diago, is a three-strike boiling system in which all the commercial sugar is obtained from a high-purity A strike. The sugar from B massecuite is made into seed magma with high-purity syrup and used as footing for A strike, and the C massecuite completes the exhaustion of the final molasses. All C sugar is melted in the hot clarified juice (Fig. 6.6) [5].

The B strikes are boiled with a footing of fine-grained sugar (massecuite), resulting from seeding A molasses, and whole massecuite is built up with A molasses. The amount of seed used must be sufficient to boil four B and four C massecuites. When the footing strike is completed it is cut into another pan. This footing is fed with A molasses and enough syrup to maintain purity, and is properly cut to produce four B massecuites for each footing. The same is done for C strikes, though the C massecuites are fed with B molasses. After purging B massecuites, the

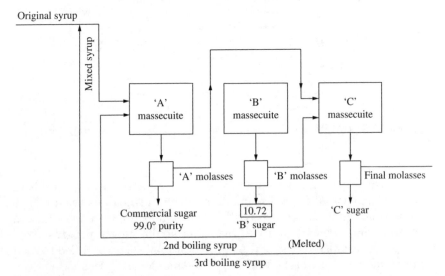

Figure 6.6 Talisman boiling system.

B sugar is mingled with syrup to make a high-purity seed magma, and B molasses is fed into C strikes. The B sugar grain is double the size of C sugar grain, and therefore massecuite A does not require as many cuts as it does in other systems of boiling with C sugar magma of much lower purity. The advantage of this system of boiling is that the commercial A sugar produced is of very high quality while there is maximum exhaustion of final molasses. The dissolved C sugar raises the Brix of clarified juice and therefore requires less steam and capacity for evaporation. Also, melted C sugar increases the purity of the clarified juice.

Centrifuging. The massecuites from the vacuum pans or from a crystallizer first go to a holding tank called a mixer that has revolving arms to prevent the crystals from settling. The crystals in the massecuite are separated from the surrounding molasses or syrup by the centrifugal force in batch-type centrifuges. A high-speed, direct-motor-driven, typical batch centrifuge is shown in Fig. 6.7 [3].

The essential part of the centrifuge is a perforated drum or a *basket* attached directly to a vertical driving motor. The basket is made of steel and has perforated vertical sides lined toward center for good drainage. The complete assembly is dynamically balanced. The loading gate is opened and the hot massecuite is immediately added to the revolving basket. The massecuite rises in the basket and forms a vertical layer on the screen lining. At the operational speed of 1000 to 2000 rpm, the syrup flows through the perforated lining and is removed through an outlet at the bottom of the casing. The sugar is sprayed with wash water to expel molasses. A final spin then allows further purging of molasses and drying of raw sugar. The dried raw sugar is removed using an automatically operating plough and is sent to the warehouse for storage.

The continuous centrifuges are used to save a greater amount of energy and time. The massecuite is fed into the spinning conical basket as in the batch process. The centrifugal force moves the massecuite up the wall of the cone over a perforated stainless steel screen. As the massecuite is moving up the cone, the sugar crystals and the molasses are being separated and the sugar is washed with steam, water, or both. The molasses passes through the screen into a receiving trough while sugar is discharged over the rim into the annular space surrounding the molasses receiving compartment.

Packing and storing. The raw sugar from centrifuges and driers goes for packaging. The packaging in bags is no longer practiced by the large sugar manufacturers. The sugar producers worldwide have adopted the practice of shipping raw sugar in bulk. Proper facilities and equipment are provided for storing and handling of bulk sugar. The sugar is moved into and out of the warehouses using dump trucks, railroad cars, or any

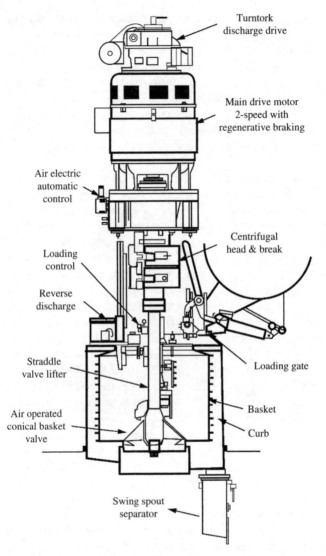

Turntork
discharge drive

Main drive motor
2-speed with
regenerative braking

Air electric
automatic
control

Centrifugal
head & break

Loading
control

Reverse
discharge

Straddle
valve lifter

Loading gate

Air operated
conical basket
valve

Basket

Curb

Swing spout
separator

Figure 6.7   A typical batch centrifuge. (This material is used by permission of John Wiley & Sons Inc.)

other inexpensive means. Unloading and loading of ships may be done by gantries of elaborate design or by clamshell buckets similar to those used for grain and other bulk materials.

6.5.2   Refining of raw sugar

Raw sugar is light to dark brown in color, slightly sticky, and contains about 1 to 2 percent of ash, starch, and coloring matter. The purpose of

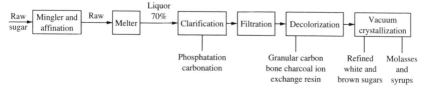

Figure 6.8   Principal steps in refining of raw sugar.

refining raw sugar is to remove these impurities and produce refined sugar of nearly 100 percent purity. There are several stages in the refining procedure (Fig. 6.8), and in each a certain amount of impurities and color is removed [6]. The several steps in the raw sugar refining are described in the following paragraphs.

Affination.   The first step in refining is removing the adhering film of molasses from the crystals of raw sugar by a washing process known as affination. This is done by mingling the raw sugar with a hot raw syrup in a large trough containing a mixer paddle and scroll. The raw syrup, which has dissolved some of the adherent molasses film from the crystal faces, is spun off and the residual sugar is washed with a jet of hot water to remove residual syrup. The affination process yields a pale-colored sugar, which is discharged by ploughs into continuous-melter tanks.

Melting.   The washed sugar is melted in hot water in the melter tank where it meets a stream of hot sweet waters from the process. The dark brown sugar liquor coming from the melter is adjusted to a density of about 65 degrees Brix. The melter liquor is strained through a plain screen to remove insoluble debris.

Clarification.   The raw washed sugar liquor from the melter contains particulate matter coming from all sources, for example, field soil and fiber, yeast, molds, gums, colloids, and foreign contaminants. The object of clarification is the complete removal of these undesirables. The raw liquor is also acidic. One of three processes is then used: *carbonation, phosphatation,* or *filtration.*

Carbonation and phosphatation are chemical treatment processes, which form a precipitate in the liquor; filtration uses inert filter aids that permit filtration under pressure.

The carbonation process is similar to that used in the purification of juice for raw sugar processing except for the high liquor density compared to the juice in the cane. The process involves two stages. In the first stage, the high density liquor is heated from 60 to 80°C in a saturated tank and is limed to about pH 10. The carbon dioxide gas from the flue is bubbled through the limed liquor until the pH drops to between 8.4 and 9.0. Several saturation tanks are used during processing. A voluminous and gelatinous precipitate of calcium carbonate along with coprecipitated salts and entrapped color bodies are formed. In the second stage, the precipitate formed in the first stage is conditioned to improve its filter-ability. The precipitate carries down with it most of the coloring matter present and is removed by filtration.

In phosphatation, the high-density liquor is heated to 60 to 70°C and treated with small amounts of phosphoric acid and sufficient milk of lime to bring the pH to 7.2 to 7.8. After the addition of lime, the temperature of the liquor is raised to 90°C and time is allowed for floc formation. Some air is injected (blown) into the system for entrapping air bubbles into the floc. The precipitate then floats to the surface as a scum and is scraped off without any filtration. The phosphate clarifiers have many sizes and shapes. Essentially all the present-day systems are based on the Williamson patents of 1918 in which the liquor is impregnated with air, then heated in a suitable vessel so that the insoluble material floats and does not depend on filtration. Some of the best-known clarifiers are Williamsons, Bulkley-Dunton, Sveen-Pederson, and Talo clarification systems.

Filtration is the most straightforward, probably the oldest form of sugar liquor clarification process. Filtration is effected with plate and frame pressure-filters or some type of leaf pressure-filter such as Sweetland filters with stationary suspended circular leaves covered on both sides with filter cloth. A filter aid of some sort (diatomaceous earth, paper pulp, or kieselghur) is essential to the operation. The precipitated calcium carbonate serves as a filter aid in the carbonation process. The liquor is mixed with the filter aid and forced under pump pressure through the fabric, which retains the cake and allows the clear liquor to flow through. The retained sugar in the cake can be washed out with hot water. Filtration is no longer used as the sole means of clarification. The process is used for further clarification of the liquor from a carbonation or phosphatation process.

Decolorization.   The clarified liquor, although brilliantly clear of contaminants, is very dark brown in color. Colorants may originate from plant pigments, which have remained in the sugar solution; may have resulted from the reaction of amino acids as reducing sugars (melanoidins); or caramels resulting from the thermal decomposition of sucrose or reducing sugars. Some degree of decolorization occurs at every step in refining; but color is also being formed in process, specially in hot, concentrated liquors. Decolorization is, therefore, an important step because the presence of any colorant in the sucrose is highly objectionable in the use of the product in food manufacturing. The color has an aesthetic appeal for the consumers and is the main property that distinguishes refined sugar from raw sugar.

Carbon adsorbents have traditionally been used for removal of coloration in the clarified effluent liquor. The general carbon adsorbents, which may be bone char, granular carbon, or powdered carbon products, are used in either a fixed-bed operation or a moving-bed process. These carbons are regenerated at intervals to maintain their effectiveness.

The use of ion exchangers for decolorization is also more widely applied these days in sugar refining. The chloride form of a strong anionic resin decolorizes the liquor, and the sodium form of a strong cationic resin softens the liquor. The significant advantages of ion exchange are in situ regeneration without the use of heat, short contact time, and small size of equipment.

The choice of the decolorization method depends on many factors and each refiner must analyze his own requirements. The principal factors to be considered are the cost of adsorbents, cost of process equipment, labor, water, energy, and waste disposal. Multiple treatment by several different decolorizing methods is common in the sugar industry and is an economical way of ensuring good quality.

Crystallization and finishing.   The decolorized liquors going into the crystallization and finishing processes range from water that is white-to-pale yellow with a solid content of 55 to 65 degree Brix (55 to 65 percent). In many refineries, this liquor is further polish-filtered to make sure that it is sparkling clear with no turbidity. In many cases, this liquor is preevaporated to bring the Brix to ≥ 68 (68 percent). The evaporated liquor is then drawn into the vacuum pan in which the massecuite is boiled. The vacuum pans are the Calendria type, just as those described earlier under raw sugar manufacture, and their operation is the same. The controlled crystallization of sugar to produce various sizes of crystals is a complex process depending largely on the boiling rate throughout and proper seeding. The proper moment for seeding is determined by instruments, or by stretching a string of evaporated liquor (massecuite) between the thumb and index finger, to approximately 4 to 5 cm without breaking it. The seed is finely pulverized sugar dispersed in isopropyl alcohol or sugar liquor. Addition of the seed induces an immediate formation of crystal nuclei throughout the supersaturated syrup (massecuite). Once the nuclei are grown to a significant grain size, rapid crystallization known as a *strike* takes place. Ordinarily, three, four, or five strikes of refined sugar are obtained. The syrup from the fourth strike is sent back to decolorization and recycled.

The crystallized sugar is transferred to centrifuges for removal of free liquor. The sugar from centrifuges contains about 1 percent moisture. They are conveyed to holding bins located above dryers. These dryers are called granulators because they rotate and keep the crystals from sticking together, thereby keeping them granular.

The final handling step is packaging and storing. The granulated sugar is stored in silos, bins, or bags made from paper, jute, or plastic. The unbagged, loose, or bulk sugar is distributed to wholesalers in bins, or trucks, or rail freight cars.

6.6   Beet Sugar

Sugar beets, *Beta vulgaris*, is a biennial plant that is grown in colder climates. The sugar beet differs from the ordinary table beet in that it is a white-rooted, much larger beet, and is not red. The sugar beets are harvested for sugar in their roots. During the last 70 years there has been an impressive increase in sugar yield from 4 to 5 percent to 15 to 20 percent sucrose as a result of crossing and prolonged selection. This fact, plus the relative simplicity of the process of manufacture, has resulted in an increase in sugar beet production accounting for nearly 50 percent of total world production.

The sugar beet plants reach maturity normally in the fall (October and November in northern latitudes). It has been established that the sugar accumulation in beets peaks in October and because sugar decomposition as a result of respiration occurs during subsequent storage of beets, they are rapidly harvested and processed from the middle to the end of September. Therefore, the manufacture of beet sugar is very much a seasonal operation.

The basic steps in the manufacture of beet sugar (Fig. 6.9) consist of (1) washing, (2) slicing, (3) diffusion, (4) juice purification, (5) evaporation, and (6) crystallization.

Washing and slicing.   The sugar beets are transferred from factory storage to the processing area in water flumes. These flumes are equipped with rock and stone removers, vegetation and trash catchers, and beet washers. The beets are rewashed, weighed, and sliced into long, narrow strips called *cossettes*.

Diffusion.   The cossettes are fed into the diffuser by means of a continuous weigher. Countercurrent hot water is introduced at the upper end of the diffuser. The extracted sugar leaves from the lower end of the diffuser as a grayish-black 10 to 15 percent sucrose solution called diffusion or raw juice. The extracted pulp is discharged at the top. The pulp contains residual sugar of approximately 0.2 percent of the beet dry weight. The pulp is pressed, dried, and enriched by the addition of 2 to 3 percent molasses. It serves as cattle feed.

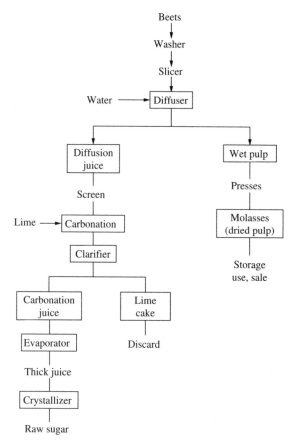

Figure 6.9   Flow sheet of a sugar beet.

Today, continuous diffusers are in use. Two types, RT-4 and BMA tower diffuser, are preferred choices.

The RT-4 consists of a horizontal rotating cylinder with an internal helix that separates the drum into moving compartments. The cossettes move forward with the rotation of a cylinder whereas the juice moves in the opposite direction. In this way, a countercurrent extraction is achieved.

The BMA tower diffuser consists of a vertical tank with a rotating vertical central shaft with conveying arms that move the cossettes upward against a downward flow of juice.

Juice purification.   The beet sugar juice is turbid and contains about 17 percent sucrose, 1 to 2 percent organic nonsucrose matter including proteins, aminoacids, amides, saponius, and pectic substances. It also contains up to 1 percent inorganics including sodium, potassium, calcium, phosphorus, and numerous trace metals. It also contains phenolic acids and numerous enzymes.

Three types of purification treatments are in general use: *carbonation,* *sulfitation,* and *ion exchange.*

In the carbonation process, the juice is first screened to remove small particles and then heated to 80 to 90°C. A small amount of lime is added under controlled conditions up to pH 10.8 to 11.9. The carbon dioxide is bubbled through the mixture in a carbonator. A number of organic acids and phosphate are precipitated as calcium salts and colloid flocculate. The other nonsucrose products are removed through coagulation by calcium or hydroxyl ions. The substances coagulated are proteins, saponius, and various vegetable coloring matters. In addition, small amounts of iron, aluminum, and magnesium are precipitated as hydroxides.

The carbonation process can be either continuous or batch processes and is performed in two steps. In the first step, the sludge is separated. In the second step, further addition of carbon dioxide removes the excess lime, which is left in solution. The resulting solution is called *thin juice.*

In subsequent sulfitation, a small amount of sulfur dioxide is frequently added to the thin syrup to lighten and stabilize the color during the evaporation process. After sulfitation, the thin juice is again clarified by filtration and is sent to the evaporators.

The use of ion exchangers is also becoming very important in juice purification. In cation exchange, the substitution of Ca^{2+} and Mg^{2+} for the alkali ions (Na^+ and K^+) softens the thin syrup and prevents the formation of hardness scale on the evaporator coil. The larger pore ion exchangers also help in bleaching the thin juice because of the binding of pigments, mainly by adsorption in the ion exchangers.

Evaporation.   The thin juice is heated and pumped from the filter presses to the multiple-effect evaporators. The dissolved solid concentration is raised from an initial concentration of about 15 percent to 50 to 65%. The concentrated juice is now known as *thick juice.* A decolorizing adsorbent (granular carbon) is added countercurrently in towers to purify and decolorize the thick juice, followed by a tight filtration process.

Crystallization.   The filtered thick juice is now sent to the vacuum-pan for boiling. It is essential to use low temperatures (65 to 80°C) in this process to avoid sucrose inversion and caramelization. Pan boiling and crystallization practices are very similar to the same unit process in the sugarcane refinery described earlier for cane sugar. Crystallization is continued until the crystals have reached the required size. This mixture of crystals and mother liquor, known as massecuite, is discharged into a mixer tank and from there it is sent to centrifugal separators. In centrifugal separators, the sugar is spun free of syrup (mother liquor) and after brief washing, is sent to a drier and finally to the cooler. The syrup is reboiled to yield a further crop of massecuite from which the sugar and second syrup are again separated by centrifuges. The syrup remaining after several crystallizations, called beet molasses, is still quite high in sugar (up to 60 percent dry basis). It is sold for cattle-feed directly or desugared commercially by the following process.

Sugar recovery from molasses. The black syrup that remains after repeated centrifugal separations, and the sugar it contains which cannot be removed economically, is commonly known as molasses. A typical analysis of final molasses from sugar beets includes sucrose (34.1 percent), reducing sugars (16.5 percent), ash (11.3 percent), water (21.8 percent), and various sugars, gums, and acids (16.3 percent). The ash includes calcium, magnesium, potassium, silicon, iron, phosphorus, and other elements in the form of inorganic salts.

Sucrose can be partly recovered from molasses by means of lime, strontia, or byryta processes. These processes are all based on the formation of insoluble saccharates. The process using dry lime as the precipitant is known as the Steffen process. About 95 percent of the sucrose is recovered from beet molasses. The diluted molasses is cooled to about 6°C and dry lime is added. The precipitated calcium saccharate is carbonated to reduce the lime content, filtered, and concentrated. About 90 percent of the sucrose can be recovered by the Steffen process. The calcium carbonate precipitated can be roasted and reconverted to quick lime.

Steffen process is not applicable for extraction of sugar from cane molasses (black strap) because of the large content of invert sugar in black strap. Chromatographic methods using ion exchange resins have been developed for separation of sucrose, invert sugars, and nonsugars from both beet and cane molasses. The separation of sucrose, invert sugars, and nonsugars is accomplished in special columns filled with a special cation exchanger, which acts as an adsorbent for the fractions. By selecting the resin and appropriate conditions, three different fractions, namely sucrose, reducing sugars, and nonsugars may be withdrawn.

Sugar crops NES. The Food and Agricultural Organization (FAO) lists three primary sugar crops. The two traditional sources are sugar cane and sugar beet. The third source is listed under the name *sweeteners*. This includes products used for sweetening that are derived from minor sugar crops of local importance such as cereals, fruits, or milk, or that which are produced by insects. This category includes a wide variety of monosaccharides (glucose and fructose) and disaccharides (sucrose and saccharose). They exist either in a crystallized state as sugar, or in thick liquid-form as syrups. In recent years, ever-larger quantities of cereals (mainly corn) have been used to produce sweeteners derived from starch. Sugar and syrups are also produced from the sap of certain species of maple trees, from sweet sorghum, and from sugar palm.

Starch sugars (syrup).   Starch and cellulose could be used as a source of saccharides, but currently, only starch hydrolysis is of economic importance. The starch (mainly corn) can undergo various hydrolytic processes to yield mixtures of various saccharides in the form of syrups or crystalline products. Such saccharification is achieved by either acidic or enzymatic hydrolysis. Controlled processing conditions yield products of widely different compositions to suit various applications. The extent of starch conversion into sugars is generally expressed as dextrose equivalents (DE value), that is, the amount of reducing sugars produced, calculated as glucose (DE value: glucose = 100, starch = 0).

Acid hydrolysis of starch is conducted with hydrochloric acid or sulfuric acid, mainly in a continuous process, yielding syrups with 20 to 68 DE. The process consists of the acidification of a starch slurry with hydrochloric acid to a pH of about 1.8 to 1.9, and pumping the suspension into a converter (autoclave) where live steam is gradually admitted to a pressure of 30 to 45 psi. Converted liquids are neutralized with sodium carbonate to a pH of 5 to 7. Proteins, lipids, and colloidal matter are separated as sludge. Pigments are eliminated with activated carbon and minerals with ion exchangers. The raw juice, thus obtained, is evaporated under a vacuum (falling-film evaporator) up to a solids content of 70 to 85 percent.

During acid hydrolysis, 5 to 6 percent of glucose undergoes side reactions to produce reversion products. These are predominantly isomaltose and gentiobiose, and, in addition, other di- and trisaccharides.

In enzymatic processes, α-amylases, β-amylases, glucoamylases, and pullulanases are used. First, starch is liquefied and hydrolyzed with acid, with α-amylase, or with a combination acid or enzyme process.

The enzyme most commonly used is α-amylase. Optimal pH and temperature are 6.5 and 70 to 90°C, respectively. Hydrolysis can be carried out to obtain a product consisting mostly of maltose with small amounts of glucose or vice versa. The sweet-taste intensity of starch hydrolysates depends on the degree of saccharification and ranges from 25 to 50 percent of that of sucrose. Table 6.2 provides an average composition on some hydrolysis products. The wide range of starch syrups starts with those with a low DE value of 10 to 20 (maltotriose) and ends with a high DE value of 96 (dextrose)[7].

TABLE 6.2   Average Composition of Starch Hydrolysates*

DE-value	Glucose (Dextrose)	Maltose	Maltotroise	Higher oligo-saccharides
Acid hydrolysis				
30	10	9	9	72
40	17	13	11	59
60	36	20	13	31
Enzymatic hydrolysis				
20	1	5	6	88
45	5	50	20	25
65	39	35	11	15
97	96	2	—	2

*All values expressed as % of starch hydrolysate (dry weight) basis.

Glucose-fructose syrup (high fructose syrup).   The next development in starch chemistry involved enzyme catalyzed isomerization of dextrose (α-D glucose) to fructose, which has 1.3 times the sweetening power of sucrose. The first commercial shipment of high fructose corn syrup (HFCS) took place in 1967. The fructose content of the syrup was around 15 percent. Further research enabled the industry to develop a higher conversion and the first commercial shipment of HFCS42 or 42-percent fructose syrup took place a year later. Further refinements in the process were developed in the late 1970s and by the mid-1980s, HFCS became the sweetener of choice for the soft drink industry in the United States.

Glucose syrup of about 94 DE is filtered, treated with activated carbon to remove pigments, and deionized with ion exchange resins to remove excess of calcium ions. The pH is adjusted to a slightly alkaline medium (8.2) and at 35 to 60°C a different immobilized enzyme (glucose-fructose isomerase) converts glucose into fructose. Under appropriate conditions up to 40 to 50 percent of glucose in the syrup can be converted to fructose. In some modified processes, chromatographic separation techniques are used when products up to 90 percent fructose are obtainable. Figure 6.10 shows a flow diagram for the production of HFCS from glucose syrup [8]. HFCS are used as invert sugars derived from sucrose.

6.7   Other Sugars

Sugar industries all over the world produce four basic types of sugar products: granulated, brown, liquid sugar, and invert sugar.

Granulated.   Granulated sugar is pure crystalline sucrose. It can be classified into seven types of sugar based on the crystal size. Most of these

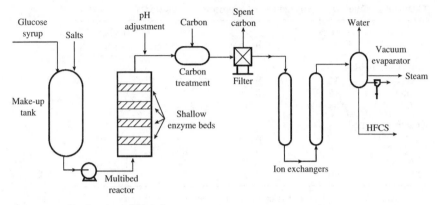

Figure 6.10 Production of high fructose corn syrup.

are used only by food processors and professional bakers. Each crystal size provides unique functional characteristics that make the sugar appropriate for the food processor's special need. The different types of granulated sugar and their suitability for usage are as given in Table 6.3.

Brown sugar. It is used in the home and food industry to develop the rich molasses-type flavor in cookies, candies, and similar products. It

TABLE 6.3   Different Types of Granulated Sugars and Their Special Characteristics

Type	Usage	Special characteristic
Regular sugar	Household use, food processing	Easier for bulk handling, not susceptible to caking
Fruit sugar	Dry mixes like gelatine desserts, pudding mixes, and drink mixes	Uniformity of crystal size prevents separation or settling of smaller crystals at the bottom of the box
Bakers special	Baking industry	
Superfine sugar	Sweetening fruits and iced-drinks	Dissolves easily
Confectioners or powdered sugar	Icings, confections, and whipping cream	
Coarse sugar	Making of fondants, confections, and liquors	Highly resistant to color change or inversion at high temperatures
Sanding sugar	To sprinkle on top of baked goods	Large crystals reflect light and give the product a sparkling appearance

consists of sugar crystals coated in a molasses syrup with natural flavor and color.

Many sugar refiners produce brown sugar by boiling a special molasses syrup until brown sugar crystals are formed. A centrifuge spins the crystals dry. Some of the syrup remains giving the sugar its brown color and molasses flavor. Other manufacturers produce brown sugar by blending a special molasses syrup with white sugar crystals.

Liquid sugar. Liquid sugars were developed before today's methods of sugar processing made transport and handling of granulated sugars practical. Liquid sugar is essentially liquid, granulated sugar and can be used in products wherever dissolved granulated sugar might be used.

Invert sugar. Inversion or chemical breakdown of sucrose results in invert sugars, an equal mixture of glucose and fructose. Available commercially only in liquid form, invert sugar is sweeter than granulated sugar. It is used in the carbonated beverage industry and in food products to retard crystallization of sugar and retain moisture.

6.8 By-Products of the Sugar Industry

The sugar industry produces many by-products along with sugar. These include molasses, bagasse, beet pulp, alcohol, pulp for paper industry, press-mud fertilizer, and a power source from burning bagasse.

Molasses. Molasses is the final effluent obtained in the preparation of sugar by repeated crystallization. The sugar it contains cannot be removed economically. The molasses from cane sugar is most commonly known as blackstrap and that from beet is called beet molasses. Molasses is mainly used for the manufacture of ethyl alcohol (ethanol), yeast, and cattle feed. Another use of sweet molasses is in cooking, spreads for bread, topping for pancakes, and in the manufacture of alcoholic beverages.

Bagasse and beet pulp. Bagasse is a fibrous residue of the cane stalk that is obtained after crushing and extraction of juice. It consists of water, fiber, and relatively small quantities of soluble solids. Bagasse is usually used as fuel in the furnaces to produce steam in sugar factories. It is also used as a raw material for production of paper and as feedstock for cattle.

The pulp, leaving the diffuser after the extraction of sugar from the beet cossettes, is pressed in screw-type presses to remove water. The pressed pulp is enriched by the addition of molasses or concentrated Steffen filtrate and is dried in rotary driers. Molasses-dried beet pulp is excellent cattle feed and its sale contributes appreciably to the profits of the beet-sugar factory.

Mud. The material removed from the filters during clarification of the juice contains the settled insoluble solids. The mud is returned to the fields as fertilizer. Some sugar factories extract a crude wax from the filtered mud, which is used in the manufacture of polishes.

6.9   Other Sweeteners

Traditionally, sugar (sucrose) or honey was used to sweeten food. In modern food processing, however, a number of other sweeteners are used, both bulk sweeteners that are used in amounts similar to the amounts of sugar they replace, and sugar substitutes that are many times sweeter than sugar and are used in very small amounts.

High fructose syrups, prepared from starch, are commonly used instead of sugar. They are generally known as high fructose corn syrup (HFCS) containing 30 percent fructose, 35 percent glucose, and 6 percent higher saccharides. They are only 74 percent as sweet as sucrose, therefore more has to be used to achieve the same sweetness.

Sugar alcohols, such as sorbitol, mannitol, and xylitol are also used as sweeteners. They are derivatives of sugars that occur naturally in some fruits, and are manufactured by chemical reduction of the parent sugars. Sugar alcohols have a lower calorie yield than the sugars from which they are derived, and are commonly used in the preparation of jams and sweets used by diabetics. They are considered safe and can be used in foods in any required amount; however, an intake greater than about 20 to 50 g/day may cause gastrointestinal discomfort and have a laxative effect.

Fructose, prepared from hydrolysis of sucrose and isomerization of glucose is used in the manufacture of candy, soft drinks, and other processed foods. As fructose is 124 percent as sweet as sucrose, substitution of fructose for sucrose in foods permits a reduction in the calorie content of a food.

Sugar substitutes (sometimes called artificial sweeteners or intense sweeteners) are synthetic compounds that are many times sweeter than sucrose. They are commonly used instead of sugar (sucrose) in the manufacture of low-calorie soft drinks and other foods. Unlike sucrose, which is metabolized by the human body and ultimately changed into fat, intense sweeteners are nonnutritive and low calorie, providing a means of calorie-controlled yet good-tasting foods and beverages. They usually do not metabolize and go through the human digestive system without being digested at all. They also have the added benefit of improving the palatability of foul-tasting drugs, helping people who suffer from diabetes.

Artificial sweeteners encompass a broad group of chemicals that are available as sugar substitutes and as ingredients in many food products aimed at consumers interested in body weight management and in *diet* foods. The first artificial sweetener was saccharin. It was discovered in 1879 by a British chemist, Constantine Fahlberg, during experiments with toluene derivatives as food preservatives. It was used worldwide as a sweetener for years until concerns that it could cause cancer arose. In the USA, the Food and Drug Administration (FDA) determined that the studies linking saccharin and cancer were inconclusive and did not ban it. Its use, however, was banned in Canada in 1977.

Cyclamate was another major artificial sweetener, marketed from 1950 to 1969 in the United States. Mixtures of cyclamate and saccharin were especially well-accepted from taste consideration, and met stability and compatibility requirements. In late 1969, the U.S. FDA enforced a ban on the use of cyclamates in foods. Currently, the use of cyclamate is permitted in low-calorie foods in about 40 countries including Canada. Still, the U.S. FDA has not approved the use of cyclamates in foods.

The increasing market demand for sweeteners resulted in the development of a number of chemicals. The major artificial sweeteners in the present market include acesulfame-K, alitame, aspartame, cyclamate, saccharin, and sucralose. Sweetness-intensity factors of several sweeteners compared with sucrose are given below:

Sweetener	Sweetness intensity*
Sucrose	1
Fructose	1.73
Glucose	0.74
Invert sugar	1.30
Maltose	0.32
Lactose	0.16
Sodium cyclamate	35
Acesulfame-K	200
Alitame	2500
Aspartame	200
Saccharin	550
Sucralose	600–800
Thaumatin	3500

*Commonly cited relative sweetness-intensity values are listed; however, the concentration and the food or beverage matrix may greatly influence actual values.

6.9.1 Acesulfame-K

Acesulfame-K is the potassium salt of 6-methyl-1,2,3-oxathiazin-4(3H)-one-2,2-dioxide. This sweetener was discovered in Germany and was first approved by the FDA in 1988 for use as a nonnutritive sweetener. The complex chemical name of this substance led to the creation of the trademark common name, acesulfame-K, which is based on its following relationships to acetocetic acid and sulfanic acid, and to its potassium salt nature [9]:

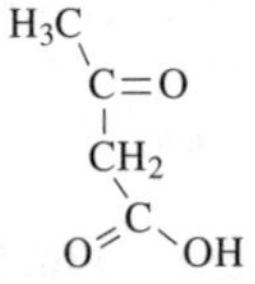

| Acetoacetic acid | Sulfamic acid | Acesulfame K |

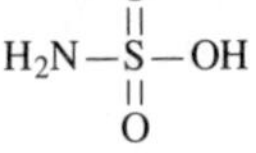

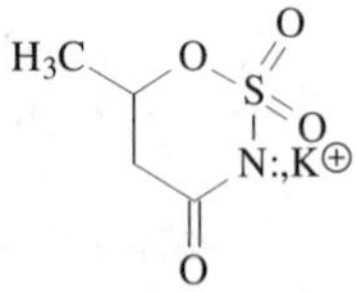

Acesulfame-K is 200 times as sweet as sugar and is not metabolized and is thus noncaloric. It is exceptionally stable at elevated temperatures encountered in baking, and it is also stable in acidic products, such as carbonated soft drinks. It has a synergistic effect when mixed with other low-calorie sweetners, such as aspartame. Common applications of acesulfame-K are table uses, chewing gums, beverages, foods, bakery products, confectionary, oral hygiene products, and pharmaceuticals.

6.9.2 Alitame

Alitame [L-α-aspartyl-N-(2,2,4,4-tetramethyl-3-thietanyl)-D-alaninamide] is a sweetener based on an amino acid. It is a very intense sweetener, possessing a sweetening power of about 2000 times that of sucrose. It also exhibits a clean sweet taste similar to sucrose. Although it is metabolized, so little is needed that its caloric contribution is insignificant. Alitame is prepared from the amino acids, L-aspartic acid, D-alamine, and a novel amine [9].

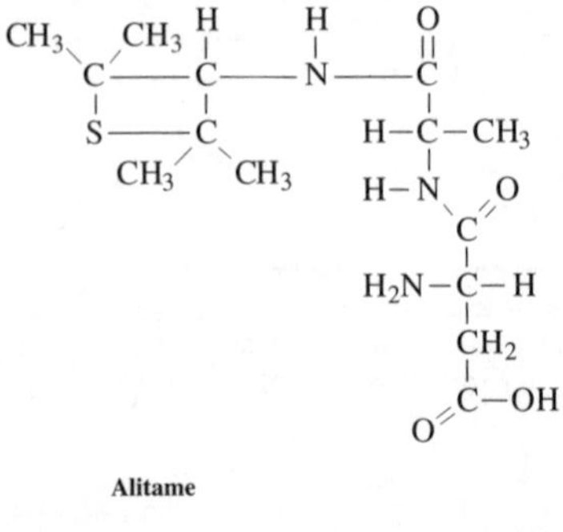

Alitame

6.9.3 Aspartame

The chemical name for aspartame is L-aspartyl-L-phenylalamine methyl ester.

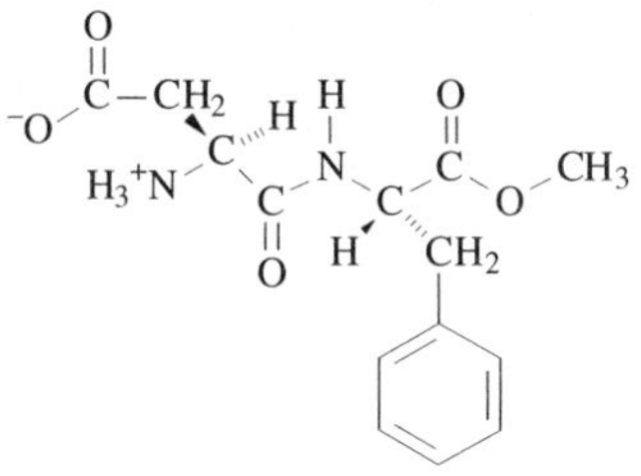

It is a white crystalline powder and is about 200 times as sweet as sucrose. It is noted for a clean, sweet taste that is similar to that of sucrose.

Aspartame is the most widely used artificial sweetener in the world. It was approved by the FDA for use in the USA in 1981, and now is approved for use in several other countries of the world. One of the drawbacks of aspartame is its instability to heat and acid. Under acidic conditions aspartame slowly hydrolyzes leading to a loss of sweetness, chemical interaction, and microbial degradation. The shelf life of the aspartame-sweetened products with high water content is limited to about 6 months, after which it breaks down into its constituent components and loses its sweetening abilities. At elevated temperatures, solid aspartame slowly releases methanol to form aspartyl phenylalamine and the dioxopiperazine. This reaction is especially favored at neutral and alkaline pH values. Because of this reason, aspartame cannot be used in hot baking foods.

Another disadvantage of aspartame was noticed in the human digestive system. When the body ingests aspartame, it breaks down into its three constituent components: phenylalamine, aspartate, and methanol. The phenylalamine and aspartate are handled by enzymes in the stomach and in the small intestine, while the methanol is transported to the liver for detoxification. The metabolism of phenylalamine requires an enzyme that is not produced by a small proportion of the population having a genetic disorder called phenyl keton uria (PKU). Aspartame should be avoided by persons suffering from PKU. A warning to PKU sufferers on aspartame-containing products is required in many countries.

6.9.4 Cyclamate

Cyclamate (cyclohexylsulfamic acid and its calcium and sodium salts) were discovered in the United States in 1937. They are 30 to 80 times as sweet as sucrose and were widely used until late 1969, when it was banned by the FDA because of questions on safety. It is not banned in Canada and the European Union.

Cyclamate is produced by reacting cyclohexylamine with sulfonating agents, followed by reactions with sodium or calcium hydroxides to produce cyclamates and free cyclohexylamine as follows:

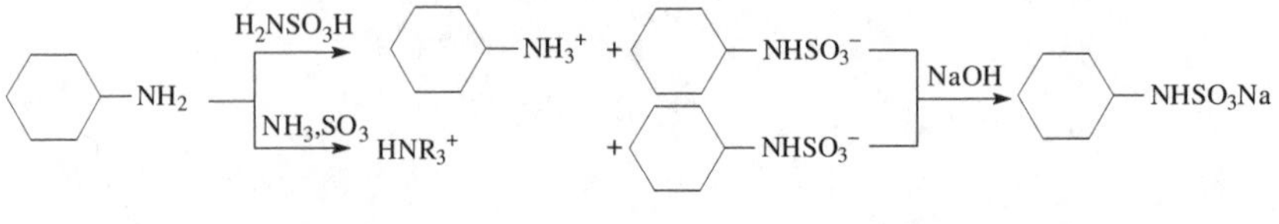

Owing to their good stability, cyclamates are suitable for all applications of intense sweeteners without a significant interfering taste sensation, and are heat stable. The main application of cyclamates is in blends with saccharin in a 10:1 ratio by weight. The mixture is more than twice as sweet as either component alone, making them an important sweetener in countries approving the use of both sweeteners [10].

6.9.5 Saccharin

Saccharin (3-oxo-2,3-dihydro-1,2 benzisothiazole-1,1 dioxide) has been available as a nonnutritive sweetener since 1900. It is 300 to 500 times as sweet as sucrose. It leaves a bitter, metallic aftertaste, especially to some individuals. Commercially, saccharin is usually produced by the Remsen-Fahlberg Process [11] according to the following reactions:

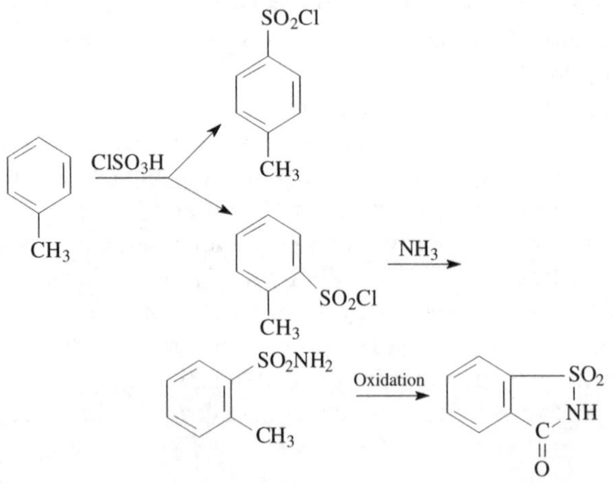

The calcium and sodium salts of saccharin are used as table-top sweeteners, soft drinks, and deserts. Saccharin is blended with other sweeteners for improving taste and sweetness, and sometimes combined with sucrose to reduce the sugar level in the preparation of diet deserts. In oral hygiene products, for example, toothpastes and mouthwashes, and other pharmaceutical products, saccharin is used to mask undesirable tastes of the other ingredients [11].

6.9.6 Sucralose

Sucralose (1,6-dichloro-1,6-dideoxy-β-fructofuranosyl-4-chloro-α-D-galactopyra- noside) is a nonnutritive sweetener based on sucrose. It is selectively chlorinated and the glycoside link between the two rings is resistant to hydrolysis by acid or enzymes, so it is not metabolized. It has 400 to 800 times the sweetness of sucrose, is very soluble in water, and is stable in heat. It can be used in food products that are baked or fried.

Sucralose is produced by the selective chlorination of the sucrose molecule using a patented process by Tate and Lyle that replaces the three hydroxyl groups (OH) with three chlorine (Cl) atoms.

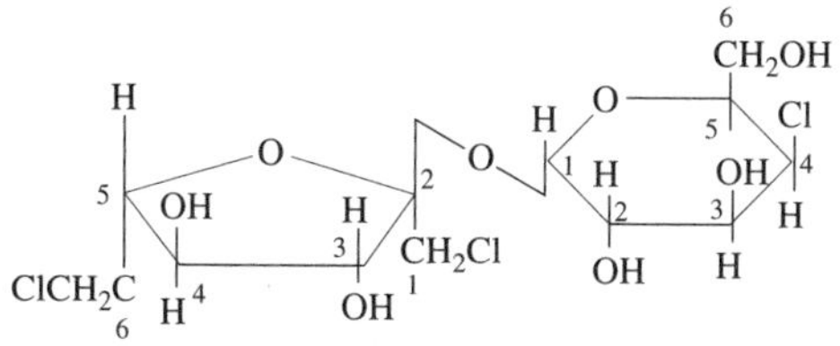

This modified sugar is minimally absorbed by the body and passes out unchanged. It was approved for use in foods and beverages in 1999 in the United States.

6.10 Sugar Analysis

The chemical and technical control of a sugar factory is the responsibility of the laboratory. All sugar must be accounted for through the whole manufacturing process from the time it enters as sugarcane until it emerges as raw or refined crystals. The primary purpose of laboratory control is to guide the operations to optimize process conditions for the best practical results, and then to estimate the extent of sugar lost and suggest the ways to reduce these losses. In order to keep an accurate account of the products in circulation and especially of sugar entering and leaving the factory, weighings and analyses are made on the delivered sugarcane, extracted raw juice from the mills or diffusion, and finished raw sugar. The sum of sugar in the mixed juice and of sugar left in the bagasse is the total sugar in the cane. If the sugar balance cannot be obtained, the reason must be determined and cause of loss eliminated.

Most large sugar manufacturers have developed their own control test methods and manuals. The International Society of Sugar Cane Technologists (ISSCT) undertook efforts toward codifying and standardizing the terms, definitions, control figures, and methods of calculating these control figures, throughout the cane sugar industry. The ISSCT methods were first published by the society in book form in 1942. In 1971, the third edition of this publication, edited by Claylon, was published in Australia [5].

For proper control in sugar manufacture, the laboratory must be well-equipped with accurate scales, saccharimeter (polariscope), and the Brix spindle or refractometer. The laboratory control of factory operations, from milling to final product, basically depends on the following three operations.

1. Sampling

2. Weighing and measuring

3. Pol and Brix determination

The system of sampling must be uniform, representative, and agreed upon between the operations and laboratory departments.

All the scales for weighing and measuring must be properly calibrated and checked periodically.

Pol and Brix test methods are used by a greater number of sugar factories in the world. *Pol* is the value determined by direct or single polarization of the normal weight solution in the saccharimeter. The term is used in calculation of the polarization balance percent of cane. In this balance the polarization loss in bagasse, final molasses, mud, and undetermined losses are represented. The polarization in sugarcane is determined by adding the polarization in the above losses to the polarization in sugar recovered.

Brix is the percentage by weight of the solids in a pure sucrose solution. By general acceptance, the Brix represents the apparent solids in a sugar solution as determined by the Brix hydrometer or other densimetric measurement converted to the Brix scale.

The method using the refractometer is also used to calculate Brix or solids in molasses solutions.

References

1. Foreign Agricultural Services (FAS), U.S. Department of Agriculture, *World sugar situation brief*, May 2004.
2. Economic Research Service (ERS), U.S. Department of Agriculture, *Annual production* and Consumption of Refined sugar for Some Countries of the World, 2004.
3. Sugar, *Kirth-Othmer Encyclopedia of Chemical Technology*, 3rd ed., Vol. 21, pp. 865–920, John Wiley & Sons, New York, 1983.

4. Meade, G. P. and Chen J.C.P., *Cane Sugar Handbook*, 10th ed., John Wiley & Sons, New York, 1977.
5. Baikow, V. E., *Manufacture and Refining of Raw Cane Sugar, Sugar Series 2*, 2nd ed., Elsevier Scientific Publishing Co., New York, 1982.
6. Kent, J. A., Sugar and Other sweeteners, *Riegel's Handbook of Industrial Chemistry*, 9th ed., Chap. 9, Chapman & Hall, New York, 1992.
7. Belitz, H. D. and Grosch W., Sugars, sugar alcohols, honey, *Food Chemistry*, 2nd ed., Chap.19, Springer, New York, USA, 1999.
8. Austin, G. T., Sugar and starch industries, *Shreve's Chemical Process Industries*, 5th ed., Chap. 30, McGraw-Hill Book Co., New York, 1984.
9. Fennema, O. R., Food additives, *Food Chemistry*, 3rd ed., Chap. 12, Marcel Dekker, New York, 1996.
10. Sweeteners, *Kirth-Othmer Encyclopedia of Chemical Technology*, 3rd ed., Vol. 22, pp. 448–464, John Wiley & Sons, New York, 1983.
11. Sweeteners, *Ullmann's Encyclopedia of Industrial Chemistry*, 6th ed., Vol. 35, pp. 407–429, Wiley-Tech., New York, 2003.

Paints, Pigments, and Industrial Coatings

Mohammad Farhat Ali

7.1 Introduction

Background. The word *paint* covers a whole variety of decorative and protective coatings that are used to impart a high degree of protection to engineering, building, and other materials. The range of substrates to which paints are applied includes a vast range of materials such as metals, wood, plaster, cement, concrete, paper, leather, and the like. The most commonly used protective coatings in household and industry are diverse materials such as lacquers, varnish, plastic resin solutions, pigmented liquids, metal powders, shellacs, and stains.

The paints and coating industry is divided into two distinct subsectors—architectural and industrial. The architectural coatings subsector depends heavily on the performance of the construction sector, whereas industrial coatings are closely associated to the automotive, major appliance, and industrial equipment sectors.

Architectural coatings include interior and exterior house paints, primers, sealers, varnishes, and stains. Industrial coatings include automotive paints, can coatings, furniture finishing, and road-marking paints.

World market.   Worldwide, paint makers shipped 5 billion gal valued US$ 80 billion in 2002. North American paint markers shipped 1.4 billion gal at US$21.2 billion. European producers sold 1.7 billion gal valued at US$23.9 billion. Asian paint makers shipped 1.1 billion gal valued at US$14.8 billion, and producers in the rest of the world—including South America, Africa, and the Middle East—shipped 900 million gal with a value of US$10.6 billion [1, 2].

Definitions.   The following are some of the common terms used in the paint industry.

Binder.   A resinous or resin-forming constituent of a paint that binds together the pigment particles and holds them on the surface.

Chalking.   Paint failure that is characterized by a layer of loose pigment powder on the surface of a weathered film. Chalking is often a desirable failure because of its self-cleaning action.

Checking.   Slight fine breaks on the surface of a film that do not extend to the substrate and that are visible to the naked eye.

Coating.   A generic term of paints, lacquers, enamels, and the like. Also a liquid composition, which is converted to a solid protective, decorative, or functional adherent film after application as a thin layer.

Cracking.   Breaks that extend from the film surface to the underlying substrate.

Drier.   A composition that accelerates the drying of oil paint, printing ink, or varnish. Driers are usually metallic compositions and are available in both solid and liquid forms.

Drying Oil.   An oil that possesses, to a marked degree, the property of readily taking up oxygen from the air and changing to a relatively hard, tough, elastic substance when exposed to a thin film of air.

Enamel.   A paint that is characterized by an ability to form an especially smooth film.

Extender.   A carbonate or silicate pigment that has little hiding power but which is included in paints for other useful purposes, for example, flattening, color dilution, or rheology control. It is usually considered chemically inert.

Filler.   A pigmented composition for filling the pores or irregularities in a surface, preparatory to application of other finishes.

Glaze.   A very thin coating of a paint product, usually a semitransparent coating tinted with pigment, applied on a previously painted surface to produce a decorative effect.

Hiding Power.   The ability of a paint to obscure underlying color varies with different pigments. The difference between the index of refraction of the vehicle and that of the pigment determines the hiding power.

Japan.   A varnish yielding a hard, glossy, dark-colored film. Japans are usually dried by baking at relatively high temperatures.

Lacquer.   A coating composition that is based on synthetic thermoplastic film-forming material dissolved in organic solvent that dries primarily by solvent evaporation. Typical lacquers include those based on nitrocellulose, other cellulose derivatives, vinyl resins, acrylic resins, and the like.

Lake.   A special type of organic pigment essentially consisting of an organic soluble coloring matter combined more or less definitely with an inorganic base or carrier. It is characterized generally by a bright color and a more or less pronounced translucency when made into an oil paint.

Paint.   A classification sometimes employed to distinguish pigmented drying oil coatings (*paints*) from synthetic enamels and lacquers.

Pigment.   The fine solid particles used in the preparation of paint or printing ink and substantially insoluble in the vehicle.

Primer.   The first of the two or more coats of paint, varnish, or lacquer.

Printing Ink.   A colored or pigmented liquid or paste composition that dries to a solid film after application as a thin layer by printing machinery.

Putty.   A dough-like material consisting of pigment and vehicle, used for sealing glass in frames and for filling imperfections in wood or metal surfaces.

Sealer.   A liquid composition to prevent excessive absorption of finish coats into porous surfaces. It is also a composition to prevent bleeding.

Shellac.   Orange-colored resin, which is a secretion of the lac beetle found in great quantities in India and Indochina. Shellac is ordinarily dissolved in denatured ethyl alcohol.

Stain.   A penetrating composition that changes the color of a surface, usually transparent and leaving practically no surface film.

Thinner.   A hydrocarbon solvent used to reduce the viscosity of paints to appropriate working consistency usually just prior to application.

Tinting Strength.   The relative capacity of a pigment to impart color to a white base.

Toner.   An organic pigment that does not contain inorganic pigment or an inorganic carrying base, and is insoluble in the pure form.

Varnish.   A liquid composition that is converted to a transparent or translucent solid film after application as a thin layer.

Vehicle.   The liquid portion of a paint or printing ink. Anything that is dissolved in the liquid portion of a paint or printing ink is a part of the vehicle. It is composed of a binder and a thinner.

7.2   Constituents of Paints

The range of substrates to which paints are supplied differ markedly not only in their physical and chemical characteristics but also in the severity of the service environment to which a painted surface is to be exposed. This requires a multiplicity of paint materials that are used to coat a very wide variety of surfaces. Despite the apparent complexity of substrates that require coating, all paints are basically similar in composition in that they contain a suspension of finely ground solids (pigments) in a liquid medium (vehicle) consisting of a polymeric or resinous material (binder) and a volatile solvent. During the drying of paint, the binder forms a continuous film with the necessary attributes of adhesion, flexibility, toughness, and durability to the substrate (the surface being coated).

Paints also contain additives, which are added in small quantities to modify some properties of the pigments and binder constituents.

The four broad fundamental constituents: (1) pigments, (2) binders, (3) solvents, and (4) additives will be discussed in greater detail as follows.

7.2.1   Pigments

Pigments are insoluble, fine particle-size materials that confer on a paint its color and opacity. Pigments are used in paint formulation to carry out one or more of the following tasks:

1. Provide color
2. Hide substrates and obliterate previous colors
3. Improve the strength of the paint film
4. Improve the adhesion of the paint film
5. Reduce gloss
6. Reduce cost

Pigments should be insoluble in the medium in which they are used, chemically inert, free of soluble salts, and unaffected by normal temperatures. They should be easily wetted for proper dispersion, nontoxic, noncorrosive, and have low oil-absorption characteristics. They should be durable and fast to light (as much as possible).

In general the following properties of pigments are important in selecting a pigment for any particular product:

a. Hiding power
b. Tinting strength
c. Refractive index
d. Light-fastness
e. Bleeding characteristics
f. Particle size and shape

Hiding power. The ability of paint to completely obliterate any underlying color is defined as the hiding power and usually expressed as the number of square meters of a surface covered by 1 L of paint. To obliterate, the pigments used must prevent light from passing through the film to the previous colored layer and back to the eye of the observer. In general, dark pigments, because they are more opaque, are more effective than light pigments in this respect. Hiding power depends upon the wavelength and the total amount of light that a pigment will absorb, on its refractive index (RI) and also on particle size and shape. The difference between the RI of the vehicle and that of the pigment is used by the paint formulators as an indicator for the hiding power. The greater the difference the greater is the hiding power of the paint. The indices of refraction of some common paint materials are given in Table 7.1 [3]. It can be seen that the pigment rutile (TiO_2) with the highest RI is the most effective white pigment for hiding power.

Tinting strength. During application the majority of white pigments are tinted to the appropriate pastel of mid-shade with colored pigments. The amount of colored pigment required to tint (color) a given weight of a

TABLE 7.1   Indices of Refraction of Some Common
Paint Materials

Material	Refractive index
Rutile titanium dioxide	2.76
Anatase titanium dioxide	2.55
Zinc sulfide	2.37
Antimony oxide	2.09
Zinc oxide	2.02
Basic lead carbonate	2.00
Basic lead sulfate	1.93
Barytes	1.64
Calcium sulfate (anhydrite)	1.59
Magnesium silicate	1.59
Calcium carbonate	1.57
China clay	1.56
Silica	1.55
Phenolic resins	1.55–1.68
Malamine resins	1.55–1.68
Urea-formaldehyde resins	1.55–1.60
Alkyd resins	1.50–1.60
Natural resins	1.50–1.55
China wood oil	1.52
Linseed oil	1.48
Soya bean oil	1.48

white pigment to produce a given shade is described as the tinting strength of a paint. Tinting strengths are always relative to a standard sample of the pigment under test, and for two samples of the same pigment, the tinting strength is a measure of the difference in particle size and distribution. Comparative tinting strengths of white pigments in a standard blue pigment (Table 7.2) show that the tinting strength of rutile titanium pigment far exceeds that of all the other listed pigments

TABLE 7.2   Comparative Tinting Strengths of
Common Pigments

Material	Tinting strength
Rutile titanium pigment	1850
Anatase titanium pigment	1350
Zinc sulfide	900
Antimony oxide	400
Lithopone	300
Zinc oxide	200
White lead	100

and is one of the reasons for the wide use of rutile pigments throughout the paint industry [4].

The tinting strength of a pigment is independent of its hiding power, because the comparison of shades is done at film thickness that completely hides the substrate. Relatively transparent pigments can have high tinting strengths.

Refractive index.   When light falls on a pigmented paint film, a part is reflected back and some part enters the film. The light, which is reflected back, interacts with the pigment on its way back through the film. The black and strongly colored pigments absorb the light to obliterate any surface, whereas the white pigments confer opacity solely through scattering of light. White pigments have a higher RI than most colored pigments, with consequently greater scattering power. In particular, the RI of titanium dioxide pigments is so much higher than those of film-formers (binders) that they possess excellent hiding power (Table 7.1).

Light fastness.   Light-fastness of a paint is its ability to resist deterioration under the action of sunlight and industrial fumes. Pigment stability during exposure to sunlight and environment is of considerable importance. Many pigments fade, darken, or change shade badly in light. This is because the ultraviolet rays of the sunlight are sufficiently energetic to break certain chemical bonds and thus change molecules. This change in chemical structure leads to an absorption of light in the visible region of the spectrum resulting in a loss of color or variation of hue. On the other hand, if the pigment can absorb ultraviolet rays without breakdown, it will protect the binder. Color changes can also occur in pigments by a chemical attack of the environment to which they are exposed, for example, blackening of lead pigments in sulfur-rich environment and the discoloration of Prussian blue on alkaline substrates. The chemical composition of the pigment is therefore an important factor in determining its chemical resistance and color or light-fastness.

Bleeding characteristics.   Some pigments (organic type) are soluble in aromatic solvents and are slightly soluble in alcohols and other aliphatic solvents. This solvent solubility results in the phenomenon of bleeding, whereby organic pigments in paint films can be solubilized and carried through subsequently applied paint coats by the solvents used in the paint formulation. The bleeding results in discoloration of paint films.

Particle size and shape.   The particle size, shape, and distribution of a pigment influence the rheological properties, shade, gloss, weathering characteristics, and ease of dispersion.

Pigment particles can occur in three different forms: primary particles, aggregates, and agglomerates. Primary particles in a single *piece* of pigment can be identified as an individual by microscopic examination. Aggregates are primary particles that are firmly *cemented* together at crystalline areas. Agglomerates are comparatively loosely bound primary particles and aggregates that are joined at crystal corners and edges. In general, *particle size* refers to primary particle size. The particle size of the dispersed pigment agglomerates or primary particles is of great importance in determining the performance of paint systems. No sample of pigment contains particles of an identical size; rather there is a mixture of sizes with an average diameter. The size of particles of pigments may range between 1 μ and 60 μ diameters.

Most pigments and extenders used in paints are crystalline in nature. Particles may be tetragonal, rhombic, cubic, nodular, rod-like, or plate-like. Noncrystalline pigments such as the carbon blacks are also used in the paint industry. As particle shape affects pigment packing, it also affects its hiding power.

The classification of pigments.   The materials used to impart color may either be pigments or dyestuffs. The difference between pigments and dyes is their relative solubility in the liquid media (solvent plus binder) in which they are dispersed; dyes are soluble, whereas pigments are insoluble. This solubility or insolubility is the reason a surface colored with an insoluble pigment is opaque with their good light-fastness. A dye, on the other hand, may impart an intense color to the surface but remain transparent, and generally, their light-fastness is fairly poor.

Pigments, which can be organic or inorganic in origin, have been classified in a variety of ways, such as color, natural or synthetic, and by chemical types. There is a further class of solid materials that are also insoluble in the paint medium but which impart little or no opacity or color to the film into which they are incorporated. These materials are known as extenders and they are all inorganic in origin. Extenders are incorporated into paints to modify the flow properties, gloss, surface topography and the mechanical and permeability characteristics of the film.

The classification of pigments and extenders used for further discussion is shown in Table 7.3 and certain description and properties of each class are generalized as follows:

Many inorganic pigments are found in nature as minerals and are dug out of the earth's crust, crushed, washed, and graded by size. The light stability, degree of opacity, and chemical resistance of natural inorganic pigment is normally very high. Frequently, inorganic pigments are chemically prepared from inorganic raw materials. The synthetic inorganic pigments are apparently the same chemically as

TABLE 7.3   Major Pigment Classifications

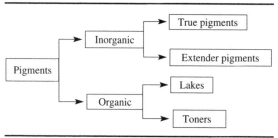

the naturally occurring pigments, but often quite different in properties. The texture of synthetic inorganic pigments is much finer and this renders them more readily dispersible than the naturally occurring inorganic pigments during paint preparations. Moreover, natural pigments may be contaminated by some impurity, such as silica, which is uneconomical to remove; the synthetic products on the other hand are pure.

The naturally occurring organic pigments are mainly of historical interest and are no longer used. There are now far more synthetic organic pigments and dyes than inorganic ones. In the manufacture of organic pigments certain materials become insoluble in the pure form, whereas others require a metal or an inorganic base to precipitate them. The coloring materials, which are insoluble in the pure form, are known as *toners* and those which require a base are called *lakes*. Synthetic organic pigments are very finely textured and they provide clean, intense colors; however, they do not provide a high level of opacity. Both light-fastness and heat stability of organic pigments are generally lower than that of inorganic pigments. The brilliance and clarity of hue for organic pigments is much superior. The most attractive, cleanest colors can only be obtained with organic pigments.

7.2.2 Inorganic pigments

Inorganic pigments can conveniently be subdivided by color. The extender pigments, although generally white in color, will be discussed separately. The categorization of inorganic pigments and extenders is shown in Table 7.4 [5].

White pigments.   White pigments are the major contributors in paint formulation. White pigments are used not only in white paints, but also in a substantial fraction of other pigmented paints to give lighter colors than would be obtained using color pigments alone. All white pigments are inorganic compounds of titanium, zinc, antimony, or lead. Presently,

TABLE 7.4   Classification of Inorganic Pigments

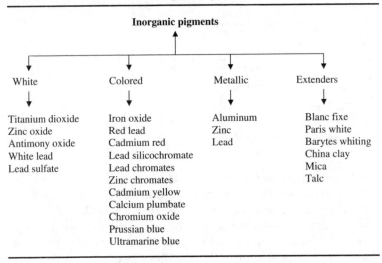

White	Colored	Metallic	Extenders
Titanium dioxide	Iron oxide	Aluminum	Blanc fixe
Zinc oxide	Red lead	Zinc	Paris white
Antimony oxide	Cadmium red	Lead	Barytes whiting
White lead	Lead silicochromate		China clay
Lead sulfate	Lead chromates		Mica
	Zinc chromates		Talc
	Cadmium yellow		
	Calcium plumbate		
	Chromium oxide		
	Prussian blue		
	Ultramarine blue		

the most important white pigment used in paints is titanium dioxide. Formerly, white lead and zinc oxide were widely used. Table 7.5 compares some characteristics of white pigments [6].

Titanium dioxide. Titanium dioxide is the most important white pigment produced commercially. Titanium dioxide exists in three crystal forms: rutile, anatase, and brookite. Only anatase and rutile are important as pigments. Anatase and rutile differ in their chemical structures. The rutile crystal has a more compact structure than anatase and hence a higher density, higher RI, greater pacifying power, and greater exterior durability. Rutile is used in larger volumes primarily because it gives about 20 percent greater hiding power than anatase. However, rutile is not perfectly white and absorbs a certain amount of radiation in the

TABLE 7.5   Summary of the Characteristics of White Pigments

Characteristics	Titanium dioxide		Zinc oxide	White lead Basic lead carbonate
	Anatase	Rutile		
Refractive index	2.55	2.70	2.08	2.0
Average particle size, μm	0.2	0.2–0.3	0.2–0.35	1.0
Density, g/cm^3	3.8–4.1	3.9–4.2	5.6	7.8–6.9
Oil absorption, grams of oil/100 g pigment	18–30	16–48	10–25	11–25
Relative hiding power	100	125–135	20	15

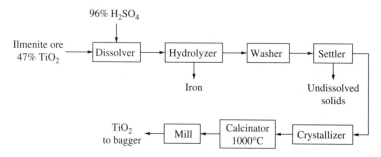

Figure 7.1   Titanium dioxide manufacture by the sulfuric acid process.

400- to 500-nm region, giving a yellowish undertone, whereas anatase absorbs almost no light. The color of rutile coatings can be adjusted by tinting with a violet pigment.

There are two major processes for the manufacture of titanium dioxide pigments, namely (1) sulfate route and (2) chloride route. In the sulfate process, the ore limonite, $FeOTiO_2$, is dissolved in sulfuric acid and the resultant solution is hydrolyzed by boiling to produce a hydrated oxide, while the iron remains in solution. The precipitated titanium hydrate is washed and leached free of soluble impurities. Controlled calcinations at about 1000°C produce pigmentary titanium dioxide of the correct crystal size distribution; this material is then subjected to a finishing coating treatment and milling. The process flow sheet is shown in Fig. 7.1 [4].

The chloride process uses gaseous chlorination of mineral rutile, followed by distillation and finally a vapor phase oxidation of the titanium tetrachloride. By adjusting the oxidation conditions, both the crystal form and the particle size of the pigment can be controlled. A flow diagram for the chloride process is shown in Fig. 7.2 [6].

Titanium dioxide pigments are used in all types of paint systems where the inclusion of white pigment is needed. The major markets for

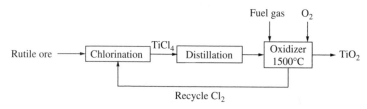

Figure 7.2   Titanium dioxide manufacture by the chloride process.

titanium pigments are paints (>60 percent), and to a lesser degree plastics, paper, rubber ceramic, textile masonry products, and cosmetic industries.

Other white pigments. The range of available white pigment is wide and includes white lead (basic lead carbonate, $2PbCO_3 \cdot Pb(OH)_2$) lithopone (mixed $ZnS/BaSO_4$); zinc oxide (ZnO); antimony oxide (Sb_2O_3); and titanium dioxide (TiO_2).

White lead used to be the most widely used pigment until the late 1930s. However, because of high toxicity of lead salts, the lead content of any paint sold to retail consumers in the United States is limited to 0.06 percent of the dry weight. Because of this restriction and also easy availability of titanium dioxide pigments, which are known for their relatively high hiding power, usage of white lead dropped rapidly and is no longer permitted as a constituent of most paints.

Lithopone is a mixed zinc sulfide-barium sulfate pigment available in two types; one containing 30 percent zinc sulfide and one containing 60 percent zinc sulfide. Coprecipitation is achieved by reacting an aqueous solution of zinc sulfate with barium sulfide. The barium sulfide solution is prepared by reducing barite ore ($BaSO_4$) with carbon. The equations are as follows:

$$BaSO_4 + 4C \longrightarrow BaS + 4CO$$

$$ZnSO_4 + BaS \longrightarrow BaSO_4 \downarrow + ZnS \downarrow$$

After TiO_2, zinc sulfide is the strongest white pigment on account of its brilliant white color, extremely fine texture, and relative cheaper cost. The lithopones are correspondingly weaker, depending on their zinc sulfide content. Lithopones are remarkably unreactive and particularly well-adapted to interior coatings.

Zinc oxide, ZnO, is a reactive white pigment prepared by vaporizing metallic zinc at a temperature of about 900°C in the presence of oxygen. As a pigment, ZnO is basic in nature and can react with certain types of acidic paint resins resulting in the formation of a brittle film on drying. Formation of such films leads to premature failure of paint. For this reason as well as because of its low RI, 2.02, ZnO cannot compete for the hiding power of TiO_2. Consequently, ZnO is rarely used as the sole pigment in modern coatings, although it finds some use in admixture with other pigments. ZnO is used in exterior house paints as a fungicide and in some can linings as a sulfide scavenger.

Antimony oxide, Sb_2O_3, is a nonreactive white pigment prepared from metallic antimony using a similar technique to that used for the preparation of zinc oxide. Antimony oxide is widely used in the preparation of fire retardant paint in conjunction with chlorine containing resins. On exposure to fire, the chlorine gas liberated by decomposition of the resin component of the paint film reacts with the antimony oxide to produce a vapor of antimony chloride that blankets the flames. Antimony oxide is also used to modify the heavy chalking characteristics of anatase form of titanium oxide.

Color pigments.   The area of color pigments is far too wide, so that only an overview can be given. The chemistry, properties, economics, and uses are discussed in Refs. [5], [7], and [8]. Color pigments can be divided into inorganic and organic products. The inorganic pigments are chemically inert, very light-fast products based on oxides and sulfides of the elements iron and chromium in particular, and of zinc, molybdenum, and cadmium to a smaller level. The color pigments may be either of natural or synthetic origin. Synthetically produced pigments are preferred by the paint formulators, because only they fulfill today's requirements for color consistency and uniformity. The considerations involved for selecting color pigments include color strength, opacity, or transparency, ease of dispersion, exterior durability, heat resistance, chemical resistance, solubility, and cost. The most important organic color pigments include azo compounds, carbonyl colorants, and phthalocyanins, as well as their salts and metal complexes. These will be discussed under the heading organic pigments.

The most significant inorganic color pigments are classified by color tint and discussed as follows.

Yellow and orange pigments.   Yellow iron oxides (FeO(OH)), lead chromates ($PbCrO_4$), zinc chromates ($ZnCrO_4$), and cadmium yellow (CdS) belong to standard pigments among the yellow pigments.

Yellow iron oxides are of both natural and synthetic origin. The naturally occurring forms are composed of hydrated iron oxides and they range in shade from a dull but clean yellow to a dark yellow-brown. The synthetic iron oxides are available in a wider range of shades than the naturally occurring varieties. Yellow iron oxides give opaque films with good hiding and high exterior durability; chemical and solvent resistance is excellent. The pigments are generally easily dispersed and are comparatively inexpensive.

Chrome yellow pigments are an important class of synthetic inorganic pigments. They are used to impart color and opacity to a wide range of decorative and industrial undercoats and bright finishes. The lead chromate ($PbCrO_4$) is medium yellow in color. Primrose chrome and lemon chrome are cocrystals of lead chromate with lead sulfate. The color of these pigments is very pale yellow becoming redder in shade. The crystals of lead chromate with lead oxide (chrome oranges) are redder yellow in color. Scarlet chrome pigments are crystals of lead chromate, lead molybdate, and lead sulfate. Chrome yellows are relatively low-cost pigments with good light-fastness, high tinting strength, and opacity. Despite the good light-fastness of this class of pigments, bleaching by sulfur dioxide results in a gradual loss of color in films containing lead chromates on prolonged exposure in an industrial atmosphere. Because of their lead content, their use is not permitted in consumer paints in many parts of the world. Their use in industrial applications is also declining because of their toxic nature. Their major current use is in yellow traffic stripping paint.

Zinc chromates are used for decorative, and as anticorrosive, yellow paints. This pigment has the advantage of being nontoxic and, furthermore, its color does not change by exposure to sulfur containing atmospheres. It is characterized by excellent light-fastness, but its use is restricted because of poor opacity and poor tinting strength. The color stability of zinc chromate when exposed to lime permits its use as a pigment in paints for plaster and concrete. Zinc yellow has the composition $4ZnCrO_4 \cdot K_2O \cdot H_2O$. Two other yellow chromate pigments are strontium chromate and barium chromate, both used as corrosion inhibitors.

Red pigments.   Red iron oxide (Fe_2O_3) is an inorganic pigment of either natural or synthetic origin. It is a low chroma red with excellent durability and low cost. Synthetic pigment is made by heating iron sulfate with quicklime in a furnace. The second preparatory technique involves calcining iron sulfate in the presence of air at high temperatures.

Natural and oxides of iron are mined either as the mineral hematite (Fe_2O_3) or as hematite in its hydrated form. Indian red is a naturally occurring mineral whose ferric oxide content may vary from 80 to 95 percent, the remainder being clay and silica. It is made by grinding hematite and floating off the fines for use.

A range of synthetic inorganic pigments can be obtained from compounds containing various ratios of cadmium sulfide and cadmium selenide. These cadmium colors range in shade from orange to deep maroon. The pigment is prepared by passing hydrogen sulfide gas through solutions of cadmium sulfate and sodium selenide, and the pigment is obtained as a precipitate. The final shades of the pigment are developed by the calcination process.

Red lead (Pb_3O_4) is a brilliant red-orange colored synthetic inorganic pigment used mainly as a protective priming coat for steel work rather than a coloring pigment in paints. The toxic nature of this pigment restricts its use in modern coating systems.

Blue and green pigments.   Ultramarine blue is a complex sodium aluminum silicate and sulfide, made by calcining an intimate mixture of sodium carbonate, china clay, sulfur, and silica together with some organic resinous material such as rosin. The color of the pigment is attributed to the presence of sulfur. The pigment has good light-fastness, heat resistance, and is unaffected by alkalies but it has poor tinting strength in paints. Ultramarine is widely used as bluing in laundering to neutralize the yellowish tone in cotton and linen fabrics.

Prussian blue ($KFe(Fe(CN)_6)$) is an intense reddish blue pigment with fairly good properties. It is used as a coloring pigment in many types of paint systems and is also used in the production of lead chrome greens.

Lead chrome greens ($PbCrO_4$: $KFe(Fe(CN)_6)$), are synthetic inorganic pigments varying in shades from grass green to deep green. The pigment is prepared by dry grinding lead chrome yellow and Prussian blue. The pigments have good opacity in films, but they tend to deepen in color upon atmospheric exposure. The use of lead chrome green is, however, limited because of the toxicity of lead.

Chromium oxide (Cr_2O_3) is a dull green synthetic inorganic pigment, which can be used in all types of paint systems where high chemical resistance and outstanding light-fastness are required.

Inorganic blues and greens are, however, increasingly supplanted by phthalocyanine blues and greens (see Sec. 7.2.3, "Organic Pigments"), which have greater color strength.

Black pigments.   Black iron oxide (Fe_3O_4) is a synthetic inorganic pigment, produced by oxidation of iron (II) hydroxide obtained from the action of alkali on iron (II) sulfate solution. The composition of the pigment corresponds with that of magnetite. The pigment is of fine texture, but has only low tinting strength in paint and other coatings. It is mainly used as a colorant in fillers, primers, and undercoats. The major black pigments in the paint industry are carbon blacks (see Sec. 7.2.3).

Metallic pigments.   Metallic pigments are used on the surfaces for luster and brilliance finishes that are normally not produced by conventional pigments. For many applications, a metallic effect is highly desirable and can be achieved by adding aluminum, zinc, bronze, stainless steel, or pearlescent pigments.

Aluminum powder.   Aluminum powder is available in two forms: Leafing grade and nonleafing grade. Both grades are manufactured from pure aluminum (99.3 to 99.7 percent purity) and the particle is lamellar in shape (0.1 to 2 μm in thickness and 0.5 to 200 μm in diameter). In the milling process, stearic acid is used to give the leafing grade having a bright silvery appearance. These pigments orient themselves in a parallel overlapping fashion in coating films providing protective coating systems for metal and wood surfaces. In such applications, leafing aluminum is normally used at relatively low levels of addition, either as the sole pigment or as admixture with other pigments. The nonleafing grade aluminum types are manufactured using oleic acid. They do not possess the required surface tension to leaf and consequently the aluminum platelets are randomly located within the paint film. The nonleafing grade is primarily used in automotive topcoats where they impart an aesthetically pleasing sparkle to the finish.

Zinc dust.   There are two forms of zinc: powder and dust. Only zinc dust is used in coatings. Zinc dust is prepared by distilling zinc oxide (ZnO) in the presence of carbon. The main use of zinc dust is in the preparation of zinc dust primers for steelwork where they are used as the sole pigment. Zinc dust primers provide a high degree of corrosion protection to the steel substrate because of the tendency of the metallic zinc to corrode preferentially to steel.

Bronze powders.   These are like aluminum powders and are used to obtain copper and gold metallic effects. They are largely used by printing ink manufacturers, especially for labeling and packaging printing. They are available as rich gold and pale gold.

Stainless steel flake.   Stainless steel has proved very popular for exterior coatings. It is used in place of aluminum wherever absolute chemical inertness is required. Stainless steel coatings are specially used for the interior coating to the food storage tanks and bins.

Pearlescent pigments.   Mica pigments coated with titanium dioxide, or iron oxide, or both are industrially important pigments. They are safe, stable, and environmentally acceptable for use in coatings. The pearlescent effect is produced by the behavior of incident light on the oxide-coated mica; partial reflection from the partial transmission through the plates creates a sense of depth.

Lead powder.   The oil-suspension of metallic lead powder is used in the preparation of protective primers for steelwork where it is normally present as the sole pigment. Lead pigments act as a barrier and provide a degree of sacrificial protection similar to that of zinc dust.

Extender pigments.   Extenders or extender pigments are white inorganic minerals that are relatively deficient in both color and opacity and are commonly used as partial replacement for the more expensive prime pigments. These pigments are also referred to as inert pigments because of their optically inert behavior in surface coatings. The principal function of most extenders is to provide bulk and to adjust density and flow properties. They can also affect physical properties such as hardness, permeability, and gloss. They improve a coatings' resistance to corrosion and degradation by UV light. The extenders commonly used by the surface coatings industry include, for the most part, the following: calcite (whiting), silica, kaolin (clay), talc and baryites.

Calcite and whitings (RI: 1.5–1.7) are naturally occurring calcium carbonate deposits. The lowest cost grades are ground limestone or the mixed calcium magnesium carbonate ore, dolomite. Synthetic calcium carbonate is also used as an extender, but it is more expensive.

Calcium carbonate is the most widely used of the extender pigments. It is employed as a bright yet inexpensive extender for titanium dioxide and is available in many grades varying in average particle size and particle size distribution. It is used throughout the range of water and solvent-based paints for both interior and exterior application. In some applications, the reactivity of calcium carbonate with acids makes carbonate pigments undesirable, especially in exterior applications.

Silica (SiO_2) (RI: 1.48) is mined from deposits of diatomaceous soft chalk-like rock (keiselghur). This is an important group of extender pigments, which is used in a variety of particle sizes. They are used as a flatting agent to reduce gloss of clear coatings and to impart shear thinning flow properties to coatings. They are relatively expensive.

Kaolin or china clay (RI: 1.56) is hydrated aluminum silicates of very fine colloidal dimensions in the natural state. Clays are used in the paints because of their extremely good dispersibility in water-based systems, good suspension properties, and good brushability and opacity. However, they have poor weather resistance.

Bentonite and attapulgite clays are used to modify viscosity of coatings. Mica clay has a platelet structure and can be useful in reducing permeability of paint films.

Talcs (RI: 1.55) are hydrated magnesium silicates. It is extremely soft and is characterized by its slippery feel. Talc is used in paint to assist the dispersion of titanium dioxide and to improve the application properties of the paint film.

Baryite (RI: 1.64) is a naturally occurring mineral ($BaSO_4$). It has the highest RI and density (4.5) among the extender pigments. It is used in automotive primer formulations because of its nodular particle shape allowing densely packed films with good holdout in undercoats and primers. It also shows good wear resistance in traffic paints.

7.2.3 Organic pigments

Naturally occurring dyestuff, for example, indigo shrub, madder root, cochineal insects, Persian berries, and the like, have been known and used for centuries. These soluble dyes were rendered insoluble in water by treatment with suitable precipitating agents and used as lakes. With the increase in the variety of synthetic dyes available and multiplicity of methods of preparing these pigments, a number of pigments of organic origin have been developed. Chemically, there is little difference between organic pigments and certain dyes. The technical difference is their relative solubility; dyes are soluble, whereas pigments are essentially insoluble in the liquid media in which they are dispersed. The organic coloring materials, which are insoluble in the pure form, are known as *toner pigments*, and those which require a base are referred to as *lakes*. Compared with inorganic pigments, organic pigments generally are brighter in color, more transparent (lower hiding power), considerably greater in tinting strength, and poorer in heat and light-fastness. A large number of organic pigments are available in the market. A few of the more common ones are discussed in the following sections:

Red pigments.   Toluidine red, barium lithol red, and BON red are the three widely used organic red pigments.

The toluidine reds are a class of organic compounds known as insoluble azo dyes. It is an azo derivative of B-naphthol [9] (Fig. 7.3). Toluidine red is bright red of moderate light-fastness, good chemical resistance, and good hiding power. Toluidine red is soluble in some solvents and gives coatings that are likely to bleed.

Barium lithol red (Fig. 7.4) is bright red in color and is suitable for interior use only because of its relatively poor light-fastness and poor chemical resistance [10].

2-hydroxy-3-naphthoic acid (BON) is coupled with diazo compounds and their calcium salts (Fig. 7.5) are bright red bleed resistance pigments [10]. The somewhat higher cost manganese salt shows better exterior durability than the calcium or barium salts. The BON red pigments are characterized by an extremely high degree of color stability, resistance to acids and alkalies and are nontoxic.

Yellow pigments.   The most common organic yellow pigments are members of the insoluble azo class of pigments and they belong to four main classes: monoarylide yellows, diarylide yellows, benzimidiazolone yellows, and heterocyclic yellows.

The Hansa yellow (Fig. 7.6) is a bright monoarylide often used in trade sales and emulsion paints. They have low opacity in paint films and are soluble in aromatic solvents [10].

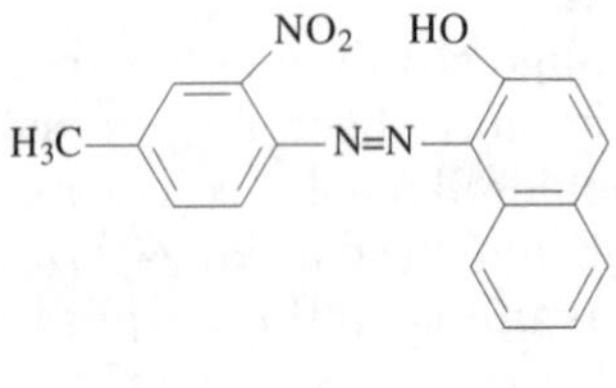

Figure 7.3   Structure of toluidine red.

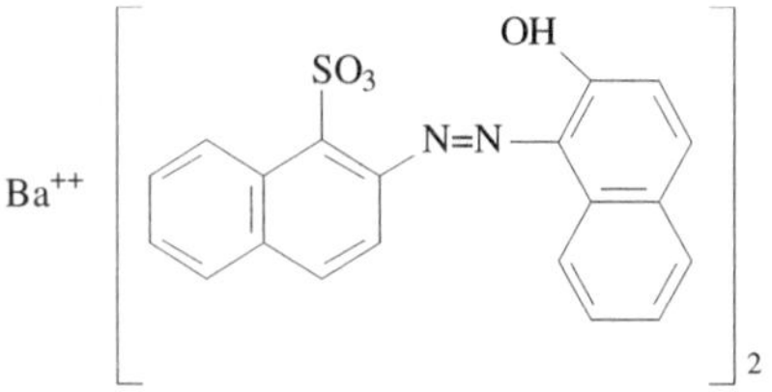

Figure 7.4   Structure of barium lithol red.

Diarylide yellows are bisazo pigments derived from 3,3′-dichloroben-zidene (Fig. 7.7). They have high color strengths and a high chroma. They are of low cost and have reasonable heat and chemical resistance and improved bleed resistance when compared to the monoarylide yellows. However, they have poor light-fastness [9].

Benzimidiazolone yellows have good opacity, very good heat resistance, good solvent resistance (little tendency to bleed) and very good light-fastness. Pigment yellow 151 (Fig. 7.8) is an example, which is used as replacements for the lead chrome pigments.

There are numerous heterocyclic yellow pigments such as nickel azo yellow (Fig. 7.9). They are high-cost, high-performance yellow pigments with excellent light-fastness.

Green pigments.   The most common organic green pigments are phthalocyanine greens. They have achieved significance for water-based paints, especially in the form of color pastes. Phthalogreens are made by halogenating copper phthalocyanine (CPC) to produce mixtures of isomers in which many of the 16 hydrogen atoms of CPC have been replaced with chlorine or mixtures of chlorine and bromine (Fig. 7.10). The pigments can go from a blue green to yellow green, depending on the ratio of bromine to chlorine. The yellowish green are obtained with nine to ten bromine atoms per molecule. The phthalocyamine greens are economical and have good light-fastness. The excellent stability of these pigments permits their use as colorants in all forms of decorative and industrial coating systems [9].

Blue pigments.   The most common organic blue pigments in the coating industry is copper phthalocyanine (Fig. 7.11) [10]. This is a bright,

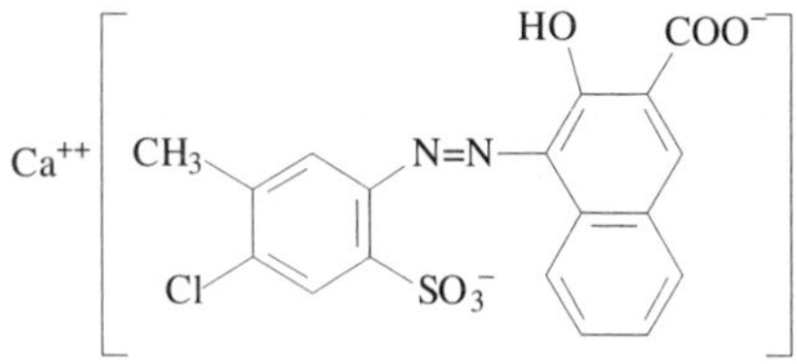

Figure 7.5   Structure of calcium BON red.

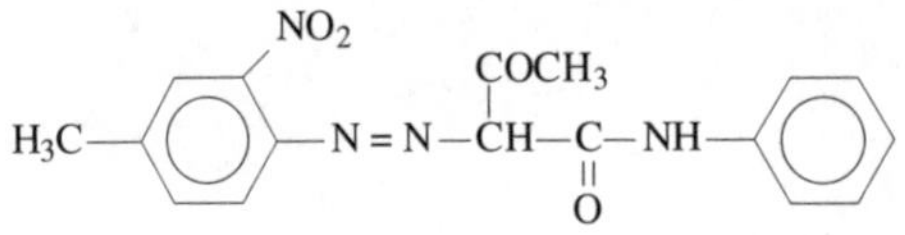

Figure 7.6   Structure of pigment yellow 1.

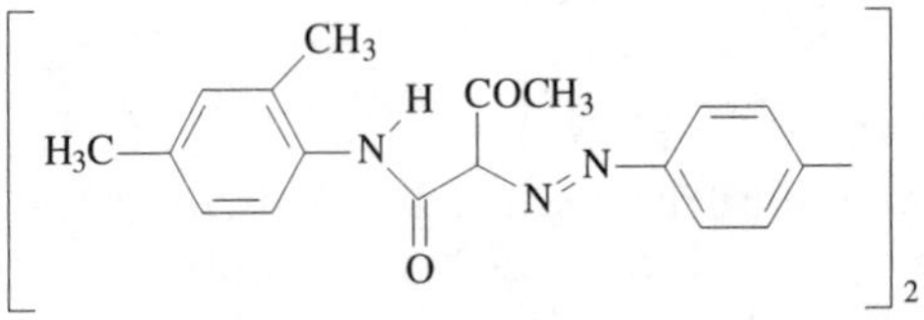

Figure 7.7   Structure of a diarylide yellow.

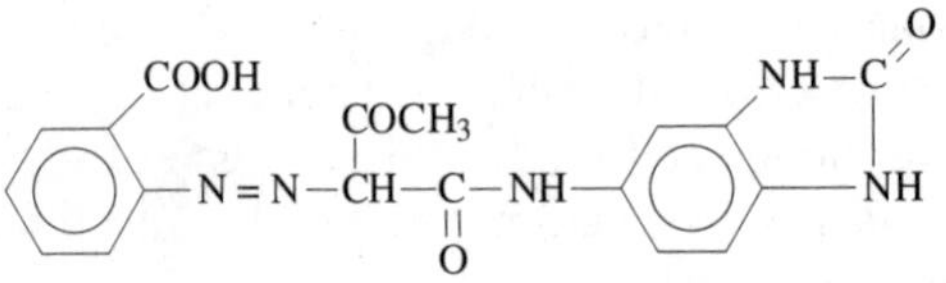

Figure 7.8   Structure of pigment yellow 151.

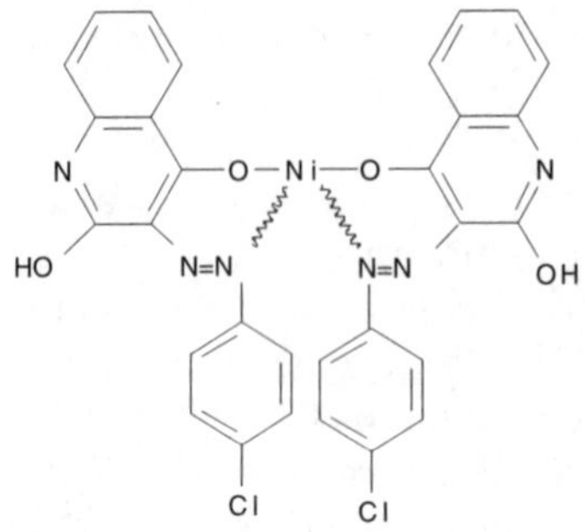

Figure 7.9   Structure of a nickel azo yellow.

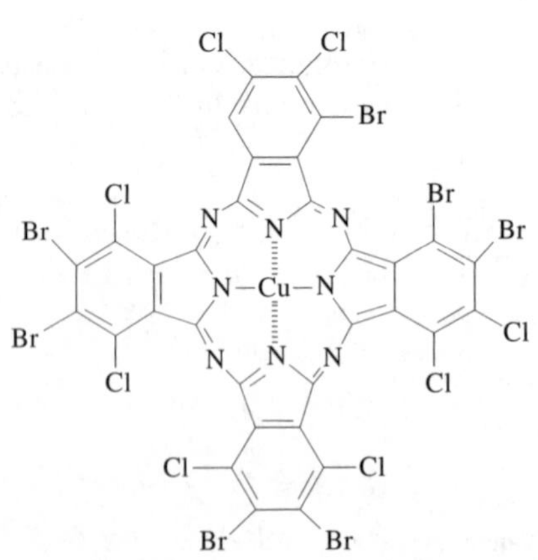

Figure 7.10   Representative phthalocyanine pigments.

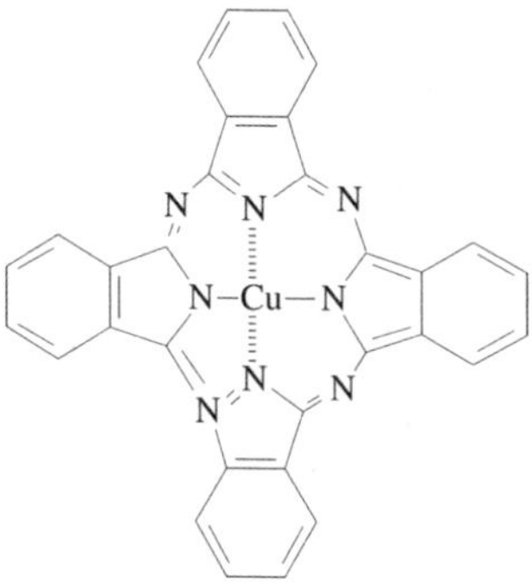

Figure 7.11   Structure of copper phthalocyanine blue.

versatile pigment of outstanding light-fastness. Phthalo blues are available commercially in three crystal forms: alpha, beta, and the seldom-used epsilon. The beta form is the most stable. Phthalocyanine pigments are characterized by a high tinting strength and opacity together with excellent color stability on exposure to light. As a class, the pigments are nontoxic, heat stable up to 500°C, and are resistant to most chemicals. These pigments are also insoluble in most solvents used in paints and hence are not prone to bleeding. Phthalo blues show a tendency to *chalk-fading*. Chalking arises from the erosion of the paint surface, resulting in a dull white surface causing *chalk-fade*. This is not true fading and can be eliminated by wiping or polishing. Chalking is also strongly dependent on the type of vehicle and pigment to binder ratio.

Black pigments.   Carbon blacks are organic pigments produced by partial combustion of petroleum products or natural gas. The particle size and intensity of blackness depends on the process and the raw materials used. For example, carbon black pigment prepared from vegetable oils or coal-tar distillates are inferior in color and opacity compared with the high carbon blacks prepared from the petroleum products or natural gas.

As a class, carbon blacks are insoluble in solvents, stable to acids and alkalies, and have excellent light-fastness. They are used as coloring pigments in all types of decorative and industrial paints.

7.2.4   Binders

The second basic constituent of a paint is a *binder*, which binds together the pigment particles and holds them on to the surface. Until the early 1950s, the binders used in paints were principally natural polyunsaturated oils (drying oils) such as tung, fish, and linseed oils; or natural resins, and exudations of gums on the bark of certain trees such as rosin from pines, congo, damar, kauri, and manila gums. Synthetic resins were introduced into the industry during the 1950s and have since become the basis of nearly all paints.

There are numerous types of binders currently available to the paint industry for various applications such as alkyds, polyesters, acrylics, vinyls, natural resins, and oils. The more common resins or polymers used in coatings are described in the following subsections:

Alkyds.   Alkyd resins represent the single largest quantity of solvent-soluble resin produced for use in the surface coating industry. They are relatively low molecular weight oil-modified polyesters prepared by reacting together polyols, dibasic acids, and oil (linseed or soya fatty acids). Two of the most common polyols used are glycerol and pentaerythritol. The most common dibasic acids used are phthalic acid and isophthalic acid. According to the oil or fatty acid content, the alkyds are divided into three broad categories:

- Short oil (to 40 percent)
- Medium oil (40–60 percent)
- Long oil (more than 60 percent) alkyd resins

They are further divided into drying (oxidizing) and nondrying (nonoxidizing) types. Nondrying oil alkyds do not readily form films and, as such, they are mainly used as plasticizers for other binders. Drying oil alkyds can form films (coatings) through oxidative polymerization in a similar manner to that of the natural oils (linseed or soya) from which they are made.

Short drying oil alkyds are typically made of linseed, soya, or dehydrated castor oils. The linseed based alkyds are used in automotive refinishing enamels and in general purpose air drying enamels.

Nondrying, short oil alkyds are generally based on castor or coconut oils. They are used with nitrocellulose for exterior lacquers. Coconut oil alkyds give the best exterior durability; castor oil lacquers have the best film properties.

Medium oil linseed and soya alkyds are used in automotive refinishing and implement enamels. In general, all-round durability of medium oil alkyds is better than their longer or shorter relations.

Long oil length alkyds are almost always prepared from drying and semidrying oils, with pentaerythritol being the preferred polyol. The most common oils used are linseed and the semidrying oils, soya, safflower, sunflower, and tall oil. Their main use is in architecture and maintenance as brushing enamels, undercoats, and primers, and also marine paints. Their slowness to dry and lack of response to forced drying has prevented their use in industrial finishes [4].

Alkyd resins are ideal for pigmented coatings because they are excellent pigment wetters and dispersers. They can be easily modified to meet specific applications. They have good durability, flexibility, solvent resistance, gloss, and color retention. Alkyds are lower in cost relative to other resins. They are good general-purpose coatings for a wide-variety of applications. In architectural coatings, alkyds are used for porch, deck, floor, and trim enamels. In product finishes, they are used for automotive chassis enamels, maintenance primers, and topcoats, and container enamels. However, alkyds have poor resistance to alkaline environments and are susceptible to chemical and sunlight-induced attack. Wrinkling is another cause of failure in alkyd-based coatings. Wrinkling occurs when the surface of the alkyd dries considerably faster than the interior. The surface then contracts, and as the interior is still *mushy*, it is pulled along by the shrinkage of the surface and a wrinkle results. This usually occurs when the coating is very thick. Modifiers such as phenolic and silicone resins are added to overcome the wrinkling problem in alkyd paints [4].

Polyesters. Polyesters are polymers obtained by reacting monomeric polycarboxylic acid and polyalcohols. They are practically free of fatty acids (oils) and have a much simpler structure than that of alkyd. Polyester resins do not undergo oxidative polymerization (curing) and have a different curing mechanism than an alkyd.

Saturated polyesters are produced from a large number of polyfunctional alcohols, for example, 1-6-hexanediol, neopentyl glycol, and polycarboxylic acids (phthalic acid and adipic acid). Most saturated polyester resins have relatively low molecular weights, ranging from 5000 to 10,000 g/mol. These resins do not have the mechanisms for curing, and therefore, coatings prepared from them use cross-linking resins such as melamine-formaldehyde (MF) resin, benzoguanimine-formaldehyde (BF) resin, or epoxy resin.

Polyester resins possess premium performance properties such as exterior durability, gloss, flexibility hardness, color stability, and versatility of cure. Polyesters are used in product finishes for household appliances, food and beverage containers, aircraft and equipment, automotive primers and bake coats, metal furniture, and fixtures. For example, water-soluble saturated polyesters are used in industrial baking paints, and in combination with melamine resin. Polyesters can be formulated in high solids and waterborne formulations to meet the requirements for the low VOC coatings being mandated by the EPA.

Acrylics.   Acrylic resins are the most widely used polymers in the paint and coating industry. The two principal forms of acrylic used in surface coatings are thermoplastic and thermoset. Thermoplastics form a film by the evaporation of the solvent present in the coating formation. Thermosets are cured at ambient or elevated temperatures by reacting them with other polymers. The following monomers are generally used in the synthesis of acrylic polymers (Table 7.6) [10].

Thermoplastic acrylic.   Thermoplastic acrylic resins belong to the two subgroups: solution acrylics and acrylic latex coatings. Solution acrylics (acrylic lacquers) are single component, thermoplastic coatings that dry and cure by solvent evaporation. They are used for wood furniture, automotive topcoats, aerosol paints, and maintenance coatings. They have relatively high molecular weights and are rather low in solid content to achieve workable viscosities. They exhibit good resistance to hydrolysis and ultraviolet degradation, which accounts for their outstanding durability. The demand for acrylic lacquers is, however, declining because of VOC restrictions.

Acrylic latex coatings are a stable, fine dispersion of polymer in water. They are used in a very large amount in the coatings industry. Because of very low VOC, easy application, cleanability with soap, and good service, the acrylic latex make up the bulk of the house-paint and architectural coatings. They are also gaining a significant market share in the exterior paint market because of their resistance to photodegradation.

A latex has basically two parts, a dispersed phase (polymer particles) and the continuous phase (the water the liquid in which the polymer droplets are dispersed), the process is usually referred to as emulsion

TABLE 7.6   Structures of Some Common Acrylic Monomers

Monomer	Structure
Ethyl acrylate	$CH_2=CHCOOCH_2CH_3$
Methyl methacrylate	CH_3 $\|$ $CH_2=CCOOCH_3$
Acrylamide	$CH_2=CHCONH_2$
Hydroxyethyl acrylate	$CH_2=CHCOOCH_2CH_2OH$
Acrylic acid	$CH_2=CHCOOH$
Styrene	$CH_2=CHC_6H_5$

polymerization. Surfactants are used in large amounts to stabilize the latexes. The acrylic monomers are added to a hot concentrated solution of surfactants and the mixtures are agitated leading to the polymer formation. The properties such as viscosity and molecular weight of the polymer are controlled by the selection of monomers and reaction conditions.

The method of film formation on a surface during the application of latex coatings is commonly believed to be a multistep process. First, when a thin film of acrylic latex coating is applied to a surface, evaporation occurs and the polymer molecules (particles) come into physical contact. Second, the particle boundaries merge together (coalescence) and a continuous film is formed. The coalescence of the latex particles is critical in achieving the desired properties of the coating. This step is assisted by incorporation of small amounts of coalescing acids (a high-boiling organic solvent, which is miscible with the continuous phase-water). Finally, some coalescing solvent, left in the voids, evaporates altogether, thus completing the film formation.

Acrylic latex coatings are ideal for house paint and architectural coatings because of their two big advantages to the consumer: low odor and an easy cleanup with water. Acrylic latex coatings can be formulated to meet the extremely low VOC requirements being mandated by the EPA.

Thermoset acrylics. Thermoset acrylics are used in product finishes for metal furniture coatings, automotive topcoats, maintenance coatings, appliance, and other original equipment manufacture finishes. They have major performance advantages for gloss, exterior durability, corrosion resistance, chemical resistance, solvent resistance, and hardness. As they are designed to react chemically after application, they can have lower molecular weights. The coating polymerizes to a permanently solid infusible state upon the application of heat (baking). Once baked, it will not dissolve in the original solvent blend. Thermoset acrylics can be formulated in both high solids and waterborne formulations to meet the requirements for the low VOC coatings. The monomers, such as hydroxylethyl methacrylate, styrene, and n-butylacrylate, are often cross-linked with melamine-formaldehyde (MF) or benzoguanimine-formaldehyde (BF) resins.

Vinyls. Vinyl esters are usually used in waterborne coatings in the form of copolymer dispersions. Typical vinyl esters are vinyl acetate, vinyl propionate, vinyl laurate, and vinyl versatate. Acrylic, maleic, and fumaric acid esters are used as copolymers. Vinyl acetate is lower in cost compared to (meth) acrylic esters. Although vinyl acetate coatings are inferior to acrylics in both photochemical stability and resistance to hydrolysis, this does not prevent them from being used for exterior application. They are primarily used as interior coatings. Most flat interior wall paints are vinyl latexes. They are also used in latex block fillers, which are highly pigmented coatings, essentially used as a primer to fill the imperfections in rough masonry walls prior to the application of a smoother, glossier topcoat.

7.2.5 Solvents

Solvents are volatile liquids added to dissolve or disperse the film-forming constituents of paints and allied products. They evaporate during the drying, and therefore, do not become a part of the dried film. In brief they

1. Regulate application properties

2. Control consistency and character of finish (minimizes defects)

3. Control evaporation rate

4. Adjust solids level that influence film application thickness

5. Adjust and influence coating viscosity (thickness of paint)

6. Are used in resin manufacturing

7. Should also have an acceptable odor, minimal toxicity, and reasonable cost

The solvents generally used in the paint industry may be divided into three classes:

a. Hydrocarbon solvents

b. Oxygenated solvents

c. Water

Hydrocarbon solvents are the most commonly used solvents in paints to carry the pigment and binder. They are divided into three groups: aliphatic, naphthenic, and aromatic. The preferred type of solvent is an odorless aliphatic hydrocarbon (mineral spirits), which can be used in all areas including home. However, mineral spirits do not dissolve all binder resins. Aromatic solvents provide stronger solvency, but with a greater odor. The most common are toluene, xylene, and naphthas.

The principal oxygenated solvents are ketones, esters, glycol esters, and alcohols. They offer much stronger solvency and are widely used as active solvents for synthetic binders. Ketones are characterized by their strong odor, range of water solubility, and evaporation rate. Esters provide solvency nearly equal to ketones but with more pleasing odors.

Glycol ethers possess both alcoholic and ether functional groups and are milder in odor. They display water miscibility, strong solvency, and slow evaporation. N-Butanol and denatured alcohol are the most commonly used oxygenated solvents.

Water is the main ingredient of the continuous phase of most emulsion paints. It is also used alone, or blended with alcohols or etheralcohols, to dissolve water-soluble resins. Any type of resin can be made water-soluble by incorporating sufficient carboxylic groups into the polymer. The advantages of water as a solvent are its availability, cheapness, lack of smell, nontoxicity, and nonflammability. However, it is not an ideal paint solvent because of its limited miscibility with other organic solvents, and because film formers designed to be dissolved or dispersed in water usually remain permanently sensitive to water.

7.2.6 Additives

The major components of paints are

- Binders
- Pigments and extenders
- Solvents
- Additives

The three main components have been discussed earlier in more detail but one class of component remains, namely, additives. Additives are substances that are added in small quantities to a paint to improve or to modify certain properties of the finished paint coatings or of the paint during its manufacture, storage, transport, or application. The amount of additives in a paint can be as little as 0.001 percent and seldom more than 5 percent. The average proportion of a single additive in a formulation is usually around 1.5 percent of the total quantity of the paint formulation.

The additives have a profound influence on the physical and chemical properties of the paint. They are classified according to their function as follows:

Thickening agents. These additives influence the rheological properties of a paint by increasing the viscosity. Organoclays, organically modified laminar silicates, are the most widely used inorganic thickeners in the paint industry. There are a number of organic thickeners, notably hydrogenated castor oil and its derivatives, polyamides, and polyamide-oil, or polyamide-alkyd reaction products, that are successfully used for the optimization of the rheological properties of solvent-borne paints and coating materials.

Surface active agents.   This group consists of three types of additives: wetting and dispersing agents, antifoam agents, and adhesion promoters.

Wetting and dispersing agents are additives that belong to the group of surfactants. They consist of amphiphilic molecules, which facilitate the very important process for pigment and extender dispersion and stabilization in paints and coatings. An unwelcome side effect of adding these surface-active chemicals is the stabilization of entrapped air during the manufacture or application of the paint, in the form of foam bubbles. The bursting of these bubbles during drying of the film of paint lead to surface defects. In order to avoid such problems, suitable defoamers are used. Defoamers on the whole consist of water-insoluble, hydrophobic, organic liquids such as mineral, vegetable, and animal oils as well as polydimethyl siloxanes or mixtures thereof. The necessary dosage or effectiveness of a defoamer depends on the formulation.

Adhesion promoters are the substances that improve adhesive strength of paints in terms of its resistance against mechanical separation from the painted surface. A large number of different chemical adhesion promoters are available. These include silanes, silicones, titanium compounds, zirconates, amides, imines, phosphates, and specially modified polymers. Furthermore, there are binders, plasticizers, and additives, which have the secondary effect of providing good adhesive strength. Adhesion promoters can be used as additives to the paint formulation, or can be employed in the form of a surface pretreatment.

Surface modifiers.   These additives control the mechanical (e.g., surface slip, scratch resistance) and optical properties (e.g., gloss) of a coated surface. Polysiloxane-based and wax-based additives are used to control these surface characteristics. Matt surfaces (low gloss) are often preferred for many different reasons. Matting agents are used to reduce gloss. Natural silicas (sand) are used together with pigments and fillers in wall paints as matting agents. However, there is a current trend toward replacing such products with synthetic silica. Waxes are also used as matting agents but they possess inferior matting efficiency compared to silica. Very often waxes and silica in combination are used if special surface characteristics are required.

Leveling agents and coalescing agents.   These additives are used to control flow and leveling of a paint during and after the application and before the film is formed; it influences, to a large extent, the appearance of the coating. Coalescing refers to the film formation of emulsion paints. Polyacrylates, cellulose acetobutyrate, and other specialty polymers are used as leveling additives. Unmodified low molecular weight, volatile silicones, and modified silicones (dimethylpolysiloxanes) are typically used as surface-flow control agents.

Coalescing agents are, in reality, temporary plasticizers, which promote the coalescence by increasing the amount of plastic flow in latex paints. The types of materials used are ether-alcohol (such as butyl glycol), tributyl phosphate, pine oil, or other strong solvents. All of these have a degree of volatility so that within a period of time they are lost from the film so that the latex hardens to its required properties.

Catalytically active additives. This group includes paint driers and other catalysts that are used to accelerate a chemical reaction occurring during the film-forming process. Driers are organometallic compounds, soluble in organic solvents and binders. They are of two types: primary (or active driers) and secondary (or auxiliary driers). The primary driers are compounds of cobalt and manganese, which have the highest catalytic activity and most pronounced accelerating effect on film formation. Secondary driers are compounds of lead, calcium, zinc, or zirconium; they possess a lower level of catalytic activity. The driers in a paint formulation are very carefully selected as an imbalance in the selection can cause wrinkling if the surface of the coating dries significantly faster than its interior.

Special-effects additives. This group of additives include a number of other substances that are added to paint formulation, for example:

- Antiskinning agents
- Light stabilizers
- Corrosion inhibitors
- Biocides
- Flame retardants

Antiskinning agents are used in coatings that cure by air oxidation, such as alkyds and other oil-based paints. In order to prevent skin-over in the can during storage, additives known as antiskinning agents are used. These agents are volatile mild antioxidants such as methyl ethyl ketoxime and butylaldehyde oxime. The antiskinning effect is based on complexation reaction of the antioxidant with the binder thus blocking the polymerization reaction. Antiskinning agents prevent the formation of a skin or an undesired film during the manufacturing process and storage of paints. They do not influence the drying time, odor, or other film characteristics.

The main application of light stabilizers is in automotive coatings. These coatings have to withstand various environmental influences such as UV light, oxygen, humidity, and air pollutants, leading to color change, loss of gloss, chalking, cracking, and delamination. In order to prevent this, additives known as light stabilizers are used. These are UV absorber organic molecules such as (2-hydroxyphenyl) benzo-triazoles (BTZ), hydroxyphenyl-5-triazines (HPT), 2-hydroxybenzophenone (BP), and oxalic amilides. Recommended use levels are 1 to 3 percent.

Corrosion inhibitors are used to prevent or slow down the corrosion rate of a metal or slow down the individual corrosion reactions. Numerous organic compounds exhibit inhibiting properties. These include acetylene derivatives and cyclic aromatic systems, amines, nitrogen, sulfur and/or oxygen-containing heterocycles, long-chain aldehydes and ketones, carboxylic acids and their derivatives, as well as nitrogen or sulfur compounds such as thiourea derivatives and thiophosphates. The usage levels of corrosion inhibitors vary from less than 1 percent for short-term to 6 percent for long-term corrosion protection.

Biocides are used to kill bacteria, which attack waterborne paints. Waterborne paints can be subjected to the growth of micro-organisms (fungi, algae, and other bacteria) while in the can, resulting in objectionable odors and changes in viscosity and color of the coatings. Certain biocides may have a biocidal action in an environment infested by microorganisms. They act as preservatives and maintain the quality of paints in the can or in storage tanks. Currently used biocides include complex phenols, formaldehyde compounds, and substituted oxazolidins. There are several other additives and some pigments, which are effective in controlling fungal growth. Zinc oxide and barium metaborate are two pigments with mildewcidal properties. Dithiocarbamates and dichlorofluamide have a broad spectrum of effectiveness against many types of fungi.

Flame-retardants are used as additives in the preparation of fire retardant paints. They are decomposed by heat to produce nonflammable components, which are able to blanket the flames. Both inorganic and organic types of flame-retardants are available in the market. The most widely used inorganic flame-retardants are aluminum trihydroxide, magnesium hydroxide, boric acid, and their derivatives. These substances have a flame-retardant action mainly because of their endothermic decomposition reaction and their dilution effect. The disadvantage of these solids is that they are effective in very high filler loads (normally above 60 percent).

Halogen-containing flame-retardants such as chlorinated paraffins, polybromodiphenyl oxides, and polybromodiphenyls are used in conjunction with antimony oxide. On exposure to fire, the halogen gases liberated by decomposition of the resin component of the paint film react with the antimony oxide to produce a vapor of antimony halide that blankets the flame.

The third most commonly used class of flame-retardants is phosphorus-containing compounds such as phosphoric acid esters, diphenyl cresylphosphate, dimethyl methylphosphonate, and dibutyl dihydroxyethyl diphosphate. They have good flame-retarding performance and are effective in small amounts. They are, however, expensive and have low hydrolysis stability.

Intumescent paints or sealants offer the finest line of fire retardant and fire barrier products in the market today.

The three components necessary for fire are fuel, oxygen, and a source of ignition. Intumescent paints produce outstanding results by eliminating two of these three components. The paints automatically react with fire or heat to convert combustible gases and tars to noncombustible carbon char, nitrogen, and carbon dioxide. This chemical reaction causes the formation of carbon char as a dense foam layer. This foam layer acts as an insulating barrier separating the substrate from the flames. The nitrogen and carbon dioxide from this reaction displace available oxygen right at the fuel source, thereby further suppressing the fire. Intumescent paints require the following three basic components.

1. A source of carbon (carbonific): Dipentaerythitol is an example. Carbonific must have hydroxyl groups and an abundance of carbon. The carbonific must also be more thermally stable than the catalyst.

2. A catalyst: This is a source of phosphoric acid and, subsequently, must have a high level of phosphorus. The catalyst decomposes to form the phosphoric acid. This should happen before the carbonific decomposes.

3. A blowing agent: This is used to expand the carbon char into a foam structure. It must be a product that will release a nonflammable gas. Urea and melamine have been used as blowing agents.

7.3   Paint Formulation

The formulation of a paint is a matter of the skill and experience of a paint technologist. It is largely determined by the ratios of the constituents in paints and the nature of the substrate to which the paint is to be applied. For example, a paint for use over concrete pavement will have very different properties from one intended for application to timber floors. Two paints may even be based on the same generic type of vehicle, but the formulation will be quite different in their final composition.

Once the proper constituents of a paint have been selected, these materials are combined together in the proper amounts. The fundamental parameters used in the formulation of a paint are (a) pigment to binder ratio; (b) solid contents; (c) pigment volume concentration; and (e) cost.

The performance capability of a paint depends largely on the capability of a binder in the film to provide a completely continuous matrix for the pigment. If relatively large amounts of pigment are used in a paint formulation then such films would erode rapidly owing to the inability of the relatively small amount of binder in the film. Thus, the weight ratio of the pigment and extender content to that of binder solid content can be usefully correlated with the performance properties of a paint. It is an easily measurable and a helpful concept in paint formulation. In general, paints with a pigment to binder ratio of greater than 4:1 are low gloss, suitable only for some interior applications.

The total solid content of a paint is another simple property that can be readily determined from a percentage weight formula. This is the amount of material that does not evaporate during the formation of a paint film on a surface representing the pigments and binder solids. The solids content of some paints provide useful information to paint formulators.

The concept of pigment volume concentration (PVC) is of far-reaching consequences for the modern paint formulator. It is defined as the percentage of pigment volume in the total volume of solids in the paint.

$$PVC\% = \frac{\text{Volume of Pigment} \times 100}{\text{Volume of Pigment} + \text{Volume of Nonvolatile Binder}}$$

The PVC offers a more scientific approach to the formulation of paints and its effects on important properties such as gloss, opacity, durability, rheology, and washability of paints. In the above formula, because the amounts of the pigment and the binder are on volume basis, it is obvious that two paints can have an identical pigment-binder (wt percent) ratio, but very different PVC values, simply by using pigments of different densities. However, the requirements of high gloss, high opacity, and high durability are found to be of conflicting nature as maximum gloss and durability are achieved at low PVC, and maximum opacity at either moderate or very high PVC. The following tabulation is used by some paint manufacturers as an approximate range of PVC for a given paint:

Flat paints	50–75%	Exterior house paints	28–36%
Semigloss paints	35–45%	Metal primers	25–40%
Gloss paints	25–35%	Wood primers	35–40%

In general, the paints formulated with low PVC show an excess of binder present which results in a well-bound film giving a high gloss level, and good chemical, water, and abrasion resistance. At extremely high PVCs, there may be insufficient binder to firmly bind the pigment particles together and such paints would be flat with a poor degree of wash and abrasion resistance. Paints with intermediate PVCs generally have properties somewhere in between these two performance extremes. However, many properties of film change abruptly at some PVC as PVC is increased in a series of formulations. These changes are not desirable in high performance systems. This level of pigmentation is known as the critical pigment volume concentration (CPVC). CPVC is usually described as the PVC at which there is precisely the right amount of binder to wet the pigment particles and to fill the voids between them. At levels above the CPVC, there is insufficient binder to wet all of the pigment and the air-filled voids will form in dry film. When the PVC is equal to or below the CPVC, the binder forms a continuous film, which is relatively impermeable. The direct calculation of CPVC is not possible, but its value can be estimated by experiment with reasonable accuracy by plotting the change of a specific property such as opacity, durability, or moisture permeability against PVC. An inflection in the curve indicates the CPVC value. A good-quality paint should be formulated at a PVC at least 5 percent above or below the CPVC, depending on the properties required. The relationship between certain film properties and the PVC is summarized in Table 7.7 [5]. The CPVC of a paint system is variable and depends on the nature of vehicles used in the formulation.

The effect of composition, the nature, and the ratios of the constituents on the properties of paints, are factors of utmost importance in paint formulations. An indication of likely service environment, life expectancy of the coatings, method of application, color, surface finish, drying time, and cost are also taken into consideration. All of this information,

TABLE 7.7 Comparative Paint Properties at Low and High Pigment Volume Concentrations (PVC)

Property	Low PVC (<CPVC)	Change at CPVC	High PVC (>CPVC)
Durability	high	marked	low
Permeability	low	marked	high
Blister resistance	low	marked	high
Gloss	high	not normally marked	low
Tensile strength	high	marked	low
Extensibility	high	not normally marked	low
Abrasion resistance	high	marked	low
Opacity	low	not normally marked	low
Cost of raw materials	high	–	low

together with the availability of the constituents dictate the selection of the components of the particular paint. Thus, the technique of paint formulation involves a considerable amount of laboratory development work to achieve optimum results.

7.4   Paint Manufacture

The manufacture of paint is basically a physical process involving weighing, mixing, grinding, tinting, thinning, filtering, and packaging (filling). No chemical reactions are involved. These processes take place in large mixing tanks at approximately room temperature. Figure 7.12 shows the paint manufacture process in proper sequence in the flowchart [6].

The important stages in the large-scale production of paints are discussed in the subsequent sections.

7.4.1   Pigment dispersion

The important stage in the manufacturing process is the initial dispersion operation, which is commonly referred to by an incorrect term, *grinding*. The solid pigments and extenders are usually supplied as a fine powder by the pigment manufacturers. These fine powder particles must be dispersed and evenly distributed throughout in the vehicle or the liquid phase. For this suspension to have a maximum stability in the liquid phase, the surface of each particle should be completely wetted with the liquid vehicle and there should not be any intervening layers of air or adsorbed water. To achieve the fine dispersion, there are a number of types of different dispersion equipment (ball mill, sand mill, roller mill, or other high-energy milling equipment) in common use in the paint industry. In most of these, the principle applied is that of shearing a viscous solution and (sometimes) attrition. Several types of mills are used. The *ball mill* is a steel cylinder mounted horizontally

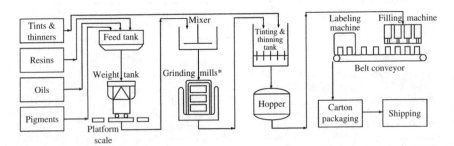

Figure 7.12   Flowchart for paint manufacture.

on its axis equipped with a suitable door for loading and for drawing off the finished product. The mill is partly filled with steel, porcelain, or pebbles. The speed of rotation is such that the balls continuously rise with the motion and then cascade down again, crushing and shearing the pigment. The mill is charged with the vehicle, pigment, and thinners, and run for the time necessary to secure proper dispersion. The paint is then removed [7].

The *sand mill* (Fig. 7.13) consists of a water-cooled cylinder inside of which are a number of rotating discs that can generate rapid movement in the grinding elements (sand grains). The violent agitation of the sand induced by the rotating discs affects shearing of the pigment particles during their dwell-time within the cylinder. Dispersed paint is obtained from the other end through a wire mesh designed to retain the sand grains [7].

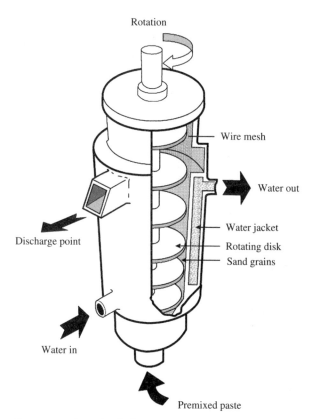

Figure 7.13   A sand grinder.

The *roller mills* are another type of very largely used dispersion equipment. They consist of a number of horizontal steel rolls placed side-by-side and moving in opposite directions, often at different speeds, with very small clearances in between. The triple-roll machine is shown in Fig. 7.14. The three rolls, made either of steel or granite are revolved in the direction indicated in the diagram. E is a flexible steel scraper to remove the finished product. The three rolls are geared to revolve at different speeds to increase shear. Triple roller mills are used for the preparation of paints requiring a low degree of dispersion, such as primers and undercoats. They have largely been superseded in the function by ball mills [7].

Another type of mixer is the *pug mill*, in which roughly two S-shaped, intermeshing blades revolve in opposite directions and at different speeds in an adjacent trough (Fig. 7.15). A high consistency (viscous) pigment-binder paste is subjected to a mechanical breakdown. The dispersion efficiency is, however, rather poor [7].

Attritors are a development from ball mills in which the mill charge is moved about in the mill by the use of a vertical shaft carrying *fingers* at right angles. Rotation of the shaft at about 100 rpm causes movement within the mill charge. The attritor handles mill bases, similar to those employed in ball mills, with a slightly higher pigment loading being tolerated.

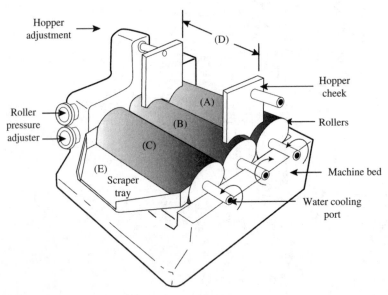

Figure 7.14   A triple roll mill.

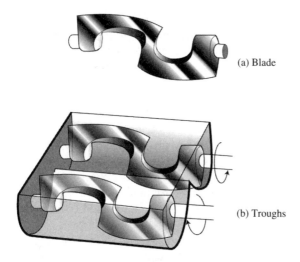

(a) Blade

(b) Troughs

Figure 7.15   (a) Blade for a heavy duty mixer. (b) Troughs of a heavy duty mixer.

Extruders have developed into very efficient pieces of equipment for pigment dispersion of high viscosity liquid dispersion and almost all powder coatings. Two types of extruders are commonly used: single screw and twin screw. A powerful motor turns screws to drive the material through a barrel. The screws and barrel are configured to mix the material thoroughly and apply a high rate of shear. Both types of extruders are capable of excellent dispersion of most pigments and are justified for the high capital and operating cost.

7.4.2 Processing operations

The dispersion of pigments is usually achieved by using a relatively small proportion of the total binder requirement of the paint. The remaining binder and any further liquid additives such as driers and solvents are added to the thoroughly dispersed pigment system and mixed. After mixing, the base paint is transferred to a tinting and thinning tank, where it is thinned and tinted for color. Tinters are basically colored pigments dispersed in a glycol-surfactant blend that are added to base paints to produce color paints. The color of a paint is usually matched to the color of the previous batch of the same paint or to a new shade of an agreed standard. The color matching procedure is a skilled art, the object of which is to adjust the color of the dried film of paint so that it exactly matches that of an agreed shade standard. The use of spectrophotometers and computers has speeded up this process considerably.

The completed product at this stage is tested for viscosity, color, and physical properties pertinent to the formulation being prepared. The paint conforming to the required quality standards is then strained and filled into cans or drums, labeled, packed, and moved to storage.

7.4.3 Classification and types of paints

The paint manufacturing industry produces a variety of products. These products are used to protect, preserve and beautify the objects to which they are applied. In general, paints are classified by their proposed function or service applications such as architectural coatings, industrial coatings, special purpose coatings, varnishes, lacquers, etc. The characteristics of most important classes are summarized below.

Architectural coatings (house paints). This class includes paints and coatings, which are used for the decoration and protection of exterior and interior of buildings. They are divided into (a) *solvent-based* and (b) *water-based paints*. The normal materials used in the painting of buildings include *primers*, *undercoats*, and *finish coats* (*top coats*). In the year 2003 in the United States, the architectural paints were about 40 percent of the value of all paints.

Primers are pigmented coatings that are applied to new surfaces or to old cleaned surfaces, prior to the application of *undercoats* or *top coats*. Its main functions are to achieve adequate adhesion to the substrate and to provide good intercoat adhesion for subsequent coats. They are specifically formulated for particular substrates such as wood, metals, concrete, and other masonry surfaces. Concrete and other masonry surfaces are alkaline and often require special surface treatments. For etching and neutralization of these alkaline surfaces, hydrochloric or phosphoric acid washing is usually done.

The *undercoats* are pigmented paints that are applied to primed surfaces prior to the application of finished coats. The undercoats are high pigment paints with a matte finish and a color to complement that of the ultimate finishing coats.

The *finish coat* or the *top coat* are the final coats for use both over primers or undercoats, and directly on a substrate. They are formulated to provide good adherence to the undercoat, high durability, the desired appearance, and other properties. These properties are invariably controlled by the class of resin used as the principal binder in the top coat. The nature of these various binders is discussed earlier in Sec. 7.2. In this section, the formulation and manufacture of some exterior and interior house paints are discussed.

Exterior building paints. The exterior paints are formulated to meet more hostile atmospheric conditions, such as rain, dew, temperature extremes,

UV radiation, and other pollutants. The paint film must be able to resist mildew growth, cracking, and checking. In addition, the nature and condition of the substrate to which paint is applied is one of the major factors determining the durability of a paint. The cementitious and timber substrates are the two major classes of substrates commonly used in exterior surfacing of the buildings. The majority of the exterior house paint sold in the United States is latex paint. Generally, the paints that are used on building exteriors are 25 percent oil or alkyd and about 75 percent latex.

The cementitious substrates include concrete, masonry, sand-cement, and gypsum plasters. All these substrates retain moisture and are alkaline in nature. The surface alkalinity can result in a chemical attack or saponification of certain types of binders used in paints, notably oils and alkyds, resulting in a marked diminution in the paints' resistance to washing, abrasion, and weathering. Alkyd paints are, therefore, not used on fresh concrete, masonry, and plaster surfaces.

The majority of timber used externally is in the form of solid timber, as opposed to laminates, veneers, and panel products such as plywood. Both softwood and hardwoods are used. The moisture and ultraviolet light (sunlight) are particularly harmful to timber. A coating system having an ability to protect the substrate from water and damaging effects of UV light is used. Pigmented paints, normally based on drying oil alkyd resins, are normally used. The use of clear varnishes is, however, common on many timbers. Latex paints are found to be more resistant than varnish or alkyd paints in performance.

Table 7.8 gives a typical formulation for an exterior latex house paint [4]. The formulation is given in terms of both volume and weight. The total volume is approximately 1000 l. The preferred PVC for an exterior flat latex paint is 45 to 50 percent. Conventional premium titanium dioxide is used along with extenders such as talc, calcium carbonate, and mica. Cellulosic thickeners are added to increase the viscosity during production and also to control the viscosity of the final paint. The propylene glycol controls the drying rate of the paint and also acts as antifreeze to stabilize the paint against coagulation. The function of surfactants, anionic dispersants, and nonionic wetting agents, is to stabilize the pigment dispersion while not interfering with the stability of the acrylic latex dispersion. The antifoam is necessary for controlling foam. Ammonia solution (NH_4OH) is added to adjust the pH in the range of 8.8–9. The high pH assures the stability of the anionic dispersing agents. Finally, water or a cellulosic thickener is added to adjust the viscosity and ratio to the solids at the standard level.

Many compounds are used as fungicides, bactericides, and biocides. Fungicides (ZnO) are used to minimize growth on the paint films after it is applied. Bactericides and biocides such as substituted 1-aza-3,7-dioxabicyclooctanes are added to suppress bacterial growths.

TABLE 7.8   The Formulation for a Typical Exterior Latex House Paint

	Kilograms	Liters
Water	130.0	130.0
Anionic dispersant	4.0	3.03
Nonionic wetting agent	3.0	2.90
Polyphosphate dispersant	2.5	1.00
Preservative/fungicide (10% Hg)	0.6	0.22
Antifoam/defoamer	1.0	1.12
Ammonia solution	1.01	1.11
Calcium carbonate	50.0	18.52
Mica	40.0	14.29
Rutile titanium dioxide	350.0	87.50
Suitable acrylic emulsion (46% NV)	583.0	550.0
Antifoam/defoamer	1.0	1.12
Propylene glycol	30.0	28.85
Coalescent	15.0	15.79
Cellulosic thickener solution (3.5% NV)	100.0	98.04
Water of cellulosic thickener	45.4	45.40
Ammonia solution	1.0	1.11
	1357.5	1000.0

Method of manufacture. Load the vehicle into a clean vessel. Add pigments slowly under high speed dispersion (HSD). Disperse to 40 to 50 μm. Add the remainder in order while stirring. *Adjust* viscosity and pH.

Characteristics
Density	1.357 kg/L
Stormer viscosity	75–80 KU
Mass solids	52.2%
Volume solids	35.7%
PVC	33.7%

The formulation for a typical alkyd oil paint is given in Table 7.9 [4]. The solvent-based alkyd paints make up a very large and important section of the architectural paint market. The choice of alkyd resin depends on whether the resin is based on linseed, soya, or sunflower oils, or blends of these. Straight linseed is not suitable for white finishes as it yellows more than semidrying types when not directly exposed to light. Sunflower and tall oil may be blended with linseed to balance dry, cost, and color retention. During the manufacture, the pigment is added to the vehicle in a clean vessel and dispersed to less than 12 μm in a high-speed disperser, using a solvent as required to maintain a suitable consistency. The viscosity of the finished paint is adjusted with the final solvent.

Alkyd paints are the most foolproof type of coatings and are often preferred for use because of their lower cost and relatively easy-to-make pigment dispersions that do not flocculate. The major limitation of alkyds is their limited exterior durability as compared to latex coatings.

TABLE 7.9   The Formulation for a Typical Alkyd Oil Paint

	Kilograms	Liters
Rutile titanium dioxide	350.0	86.0
Antisag gel (8% NV)	35.0	42.0
Lecithin solution (50%)	4.0	4.4
Long oil alkyd resin (70% NV)	75.0	78.9
White spirit	62.7	79.0
Long oil alkyd resin (50% NV)	620.0	645.8
Cobalt drier (6% Co)	2.5	2.5
Lead drier (24% Pb)	14.0	11.7
Calcium drier (6% Ca)	6.0	6.2
Antiskin solution (25%)	5.0	6.0
White spirit	30.4	37.5
	1204.6	1000.0

Method of manufacture. Premix the vehicle in a clean vessel under HSD. Add the pigment and disperse to less than 12 μm, using the required solvent as required to maintain a suitable consistency. Add the letdown under an efficient stirrer. Adjust the viscosity with the final solvent.

Characteristics
Density   1.26 kg/L
Stormer viscosity   65–58 KU
Mass solids   70.2%
Volume solids   55.9%
PVC   15.6%

Interior flat paints. The interior flat wall paints are the largest volume of trade sales paints. In the retail market, water-based latex paint has almost entirely taken over the market from their oil-based counterparts. The major advantages of latex paints over oil-based paints are (a) fast drying and less sagging; (b) low odor; (c) ease of cleanup; (d) low VOC emission; and (e) less yellowing and embrittlement.

The formulation for a typical PVA-acrylic emulsion-based interior wall paint is given in Table 7.10 [4]. The formula characteristics are as follows:

Density   1.492 kg/L
Stormer viscosity   75–85KU
Mass solids   52.5%
Volume solids   32.8%
PVC   57%

Flat white paints are stocked as white paint and tinting colors are added to make a color chosen by the consumers. Equal white tinting strength is controlled through quality control so that the colors obtained will not differ. The most expensive major component of any white flat paint on a volume basis is the TiO_2. Inert pigment such as talc is added to reduce the cost.

TABLE 7.10   The Formulation of Water-Based Interior Flat Paint

	Kilograms	Liters
Water	280.0	280.0
Cellulosic thickener	1.5	1.0
Anionic dispersant	3.0	2.2
Wetting agent	2.5	2.4
Antifoam	1.0	1.1
Ammonia solution	1.5	1.6
Preservative (mercurial type)	0.5	0.2
Talc	200.0	73.5
Diatomaceous silica	75.0	32.0
Rutile titanium dioxide (special grade for high PVC latex paints)	300.00	81.0
Coalescing agent	10.0	10.5
PVA acrylic emulsion (55% NV)	308.0	280.3
Antifoam	1.5	1.6
Ammonia solution	1.5	1.6
Cellulosic thickener solution (3% NV)	150.0	148.0
Cellulosic thickener solution or water	83.0	83.0
	1419.0	1000.0

Industrial coatings (OEM paints).   Industrial coatings include paints and finishes used in factories on products such as automobiles, magnet wire, aircraft, furniture, appliance finishes, metal cans, chewing gum wrappers, and various other products. Powder coatings and radiation-cured coatings are also included. They are commonly called OEM coatings, that is, *original equipment manufacturer* coatings. In 2003 in the United States, the OEM coatings were about 33 percent of the value of all coatings.

The industrial coatings are custom designed for a particular customers' manufacturing conditions and performance requirements. The number of products in this group is much larger than in the others; a specification is usually received by a paint manufacturer and the formulator has had to add to suit these conditions.

Often the OEM coatings depend on: the nature and condition of the substrate to which paint is applied; application methods and conditions; drying time required; and decorative and protective requirements. The substrate most commonly coated with industrial coatings are iron and steel, but also include other metals such as aluminum and its alloys, zinc-coated steel, brass, bronze, copper, and lead. Nonmetallic substrates include timber and timber products, concrete, cement, glass, ceramics, fabric, paper, leather, and a wide range of different plastic materials. Consequently, industrial coatings are usually formulated for use on either a specific substrate or a group of substrates.

Industrial coatings that are used on cars, trucks, and OEM appliances usually comprise primers or undercoats, and gloss finishing top coats. Primers are used to aid adhesion of the top coat to a surface and to provide a relatively uniform film thickness on all metal surfaces. Primers can also be used to prevent corrosion of a metal surface. A typical formulation of a primer, based on a low viscosity vinyl solution pigmented with a low solubility chromate pigment, is shown as follows [5]:

Component	% by weight
Resin	
Polyvinyl butryal resin	7.2
Zinz tetroxychromate	7.0
Talc	1.0
Isopropyl alcohol	50.0
Toluol	14.8
Etchant	
85% phosphoric acid	3.6
Water	3.2
Isopropyl alcohol	13.2

This primer can be applied by brush, spray, or dipping and it functions by both improving the adhesion of subsequently applied top coatings and by reducing the risk of underfilm corrosion.

Paints used as the final coats are referred to as finishes or top coats. They are based on binders selected to withstand the conditions likely to be experienced in the proposed service environment. Alkyd resins are used extensively for exterior exposure under mild conditions. For exposure to severe conditions, phenolics, acrylics, epoxies, urethanes, and chlorinated rubber are found more effective. A typical formula for a low build, air drying, white gloss finish based on chlorinated rubber is presented here [5]:

Component	% by Weight
Titanium dioxide	17.0
Chlorinated rubber	20.0
Chlorinated paraffin	13.0
Xylene	40.0
White spirit	10.0

Top coats are high gloss and must maintain their appearance for a long time. Until the early 1980s, all top coats were *monocoats*, a single coating composition applied in several coats. Monocoats have been largely supplanted by *base coat-clear coat systems*: a base coat containing the color pigments covered by a transparent coating. Base coat-clear coat systems provide better gloss and gloss retention than monocoats.

Powder coatings. Powder coatings are used by the paint industry usually for metal substrates. The powder is applied to the substrate and fused to a continuous film by baking. The formulation of a powder coating is based on pulverizing solid components, resins, pigments, and a hardener. Thermosetting, thermoplastic, and vitreous enamel powders are available; the major portion of the market is for thermosetting types.

Binders for thermosetting powder coatings are often called a *hardener*. The hardeners are a mixture of a primary resin and a cross-linker. The major types of binders can be limited to polyester, epoxy, hybrid epoxy-polyester, acrylic, and UV cure types. Polyester binders are used for good exterior durability, retention of gloss, and resistance to chalking.

Vinyl chloride copolymer (PVC), polyamides, and thermoplastic polyesters are used as binders for thermoplastic powder coatings. The use of thermoplastic coatings has declined considerably (less than 10 percent of the U.S. market) in recent years because of several disadvantages compared to thermosetting coatings. They are difficult to pulverize to small particle sizes; thus, they can only be applied in relatively thick films. They are more viscous and give poor flow and leveling, even at high baking temperatures.

Ultraviolet-cured powder coatings are used for rapid curing at low temperatures. The curing process is based on both free radical and cationic cure coatings. Free radical cure coatings use acrylated epoxy resins as binders. Cationic UV cure coatings use BP epoxy resins as binders. Photoinitiators such as benzins and acetophenones are used in the formulation. After application, the powders are fused by passing under infrared lamps and then are cured by passing under UV lamps.

Special purpose coatings. Special purpose coatings represent approximately 14 percent of market, which includes specific paint, such as highway marking paint, automotive refinishing, and high performance maintenance paints. The term maintenance paints is generally taken to mean paints for field application, including highway bridges, refineries, factories, power plants, and tank forms. A major requirement of maintenance paints is corrosion protection. Another important requirement is the time interval to be expected between repaintings. Most maintenance paints include at least two types of coatings: a primer and top coat. Primers provide the primary corrosion control, but top coats also have significant effects on corrosion protection by reducing oxygen and water permeability of the combined films. Top coats also provide other properties such as gloss, exterior durability, and abrasion resistance. The pigments used in the formulation of industrial paints are mainly zinc meal, zinc oxide, molybdates, and phosphates. For severe environments, chlorinated rubber, vinyl solutions, epoxies, and cross-linked epoxies are used.

Special paints are used for protecting flammable substrates by retarding flame spread. Such paints contain polyammonium phosphate, which emits a gas at elevated temperatures but lower than charring temperatures. The vehicle softens and is foamed by the gas forming a semirigid foamed char on the surface, which insulates the substrate from further heat.

Many different end uses of special purpose coatings such as marine, aircraft, barrier coatings, and the like are involved and are not included here.

7.4.4 Varnishes

Varnishes are nonpigmented paints, which dry to a hard-gloss, semi-gloss, or flat transparent film by a process comprising evaporation of solvent, followed by oxidation and polymerization of the drying oils and resins.

The varnish is manufactured by cooking the drying oil (usually linseed oil, tung oil, or mixture of the two) and resin together to a high temperature to obtain a homogeneous solution of the proper viscosity. The varnish is then thinned with hydrocarbon solvents to application viscosity. Varnishes were widely used in the 19th and early 20th centuries as spar varnishes for use on the wooden spars of ships, furniture, and floors. The original spar varnish was a phenolic-tung oil varnish; the tung oil provides high cross-linking functionality, and the phenolic resin imparts hardness, increased moisture resistance, and exterior durability. The types of oils and resins and the ratio of oil to resins are the principal factors, which determine the properties of a varnish. The bulk of the market for these traditional types of varnishes have been almost completely replaced by a variety of other products, especially to uralkyds that provide greater abrasion and water resistance. Uralkyds are also called urethane alkyds or urethane oils. They are alkyd resins in which a diisocyanate, usually toluene diisocyanate, has fully or partly replaced the phthalic anhydride usually used in the preparation of alkyds. Uralkyds are superior in performance over alkyds or epoxy esters. These days the term *varnish* refers generally to the transparent coatings, even though few of them today are varnishes in the original meaning of the word.

7.4.5 Lacquers

A lacquer is a solution of a hard linear polymer in an organic solvent. It dries by simple evaporation of the solvent. The film-forming polymers usually used are chlorinated rubber, nitro cellulose, acrylics, vinyl resins, or other high molecular weight linear polymers. The properties of lacquers vary with the main type of film-forming resin used, and their main advantage is rapid drying speed. They are made for application to a wide variety of substrates at all practical temperatures and particularly where oven heating is not available.

Cellulose nitrate is the most widely used film-former for the manufacture of lacquers. They are the fastest drying types and may be conveniently made by dissolving the cellulose nitrate and resins in the solvent mixture in fairly high-speed mixers. Plasticizers (such as vegetable oils, monomeric, and polymeric esters) are added to impart necessary flexibility to nitrocellulose films. Pigments and additives may be added if required. The lacquers may be formulated for application by most of the conventional methods such as cold spraying, hot spraying, dipping, squeeze coatings, and electrostatic application.

The lacquers are largely used in automobile finishes, furniture finishes, metal finishes, and plastic, rubber, paper, and textile finishes.

7.5   Paint Application and Causes for Paint Failure

7.5.1   Techniques of paint application

The most common methods of paint applications are brush and roller, air or airless spray, roll coating, electrostatic spraying, electro deposition, and dip coating. Many factors affect the choice of method to be used for a particular application. These include film thickness, appearance requirements, and operating cost.

Brushes, pads, and hand rollers are the most widely used techniques for the on-site application of architectural paints. Brushes and rollers are available in a number of sizes and designs to suit differing areas. Viscosity characteristics and drying times are critical in using brushes and rollers. The drying time of brush applied paints normally have to be such that optimum flow can occur along with maintaining the film in a state such that the overlapping of adjacent areas of freshly applied paintwork can be accomplished without film disruption along the interface. Paints for roller applications are generally applied at a slightly lower viscosity than when applied by brush. This is achieved by the on-site addition of suitable solvents to a brushing quality paint. Rollers can apply paint considerably faster than brushes.

Pad applications are also used. The most common type of pad consists of a sheet of nylon pile fabric attached to a foam pad that is attached to a flat plastic plate with a handle. Pads hold more paint than a similar width brush and can apply paint up to twice as fast as a brush.

Spray painting is a widely used application technique for most industrial maintenance and commercial architectural jobs. Spray painting is much faster than using brushes, pads, or rollers. A large variety of spray equipment is available, including air, airless, plural spray, and electrostatic.

In conventional air spraying, compressed air and paint are supplied to the spray gun, which atomize and transport the paint to the article being coated where they are deposited forming a uniform film. The air pressure used in this operation is critical and it should be kept at the minimum required amount, to atomize and deposit the paint onto the substrate.

In airless spray, the paint is forced out of an orifice at very high pressure (approximately 1000 to 5000 psi) resulting in atomization of the paint into fine droplets. The very high pressures used in the airless spray techniques permit nearly all paints to be sprayed in their original unthinned state. Airless spraying minimizes over-spraying, enabling high film builds on large surface areas in a relatively short time. The other advantages of airless spraying are ease of painting surfaces in enclosed areas (no spray fog), greater paint economy, very fast application, and less pollution.

Direct roll coating method is used for coating thin-gauge sheets or coil stock. The sheet stock is fed between applicator rollers rotating in the same direction as the moving sheet. The applicator rollers are fed by smaller pick-up rollers that are partially immersed in trays containing the paint. The coat sheet is subsequently fed into an oven for baking.

The electrostatic spraying is a technique designed for the automatic or semiautomatic coating of articles on a conveyer system. The atomized paint is attracted to the conductive object to be painted by an electrostatic potential between the two. This process now finds widespread use in the automobile industry. Several car producers have installed fully automated electrostatic multiple gun systems capable of applying high solid and water-based coatings.

Electro deposition consists of depositing paint on a conductive surface from a water bath containing the paint. During operation, current is passed through the cell causing the negatively charged paint particles to diffuse to the anodically polarized object. At the anode, the paint is deposited onto the surface and this effectively insulates those areas of the article from further deposition. The system is limited to one-coat application owing to the shielding and insulting effect of the deposited paint films. This process is used in the application of primers to the chassis and bodywork of vehicles.

In dip coatings the article is completely immersed in a large tank containing a certain quantity of paint to be applied. The article is then pulled out; excess coating drains back into the dip tank. This technique is used for metal primer application such as motorcar bodies.

7.5.2   Causes for paint failure

There are a number of reasons why a paint system may fail. These failures may be ascribed to any one of a number of causes and may therefore have a corresponding number of remedies.

a. Defects in the liquid paint

b. Defects during application

c. Defects during drying or curing

d. Defects in the dry film

The following alphabetical listing of defects within the above four groups covers most of the causes for paint failure. There are still more reasons that can cause coatings to fail which are not included here [4].

Aeration (bubbling). "Incorporation of bubbles of air in paint during stirring, shaking or application." This can lead to *foaming* during application of the coating, or *cissing*, or *cratering*, during drying. Aeration can be controlled by the addition of proprietary defoamers in the case of latex paints, or bubble release agents in the case of solvent paints. Aerated paints will exhibit subnormal density values, which provide an easy test for this defect at the manufacturing stage.

Aging. "Degeneration occurring in a coating during the passage of time and/or heating."

Bleeding. "Discoloration caused by migration of components from the underlying film." Substrates that can cause problems are those coated with tar- or bitumen-based materials, paints made on certain red and yellow organic pigments (which are partially soluble in solvent), some wallpapers, and timber stains that contain soluble dyes. The remedy is to use a specially formulated sealer or an aluminum paint.

Blooming. "The formation of a thin film on the surface of a paint film thereby causing the reduction of the luster or veiling its depth of color." This defect occurs mostly in stoving enamels (particularly blacks) in gas ovens. Lacquers also exhibit this defect, especially when used with low-quality thinners under certain ambient conditions.

Blistering. "Isolated convex deformation of paint film in the form of blisters arising from the detachment of one or more of the coats." This is often the consequence of faulty surface preparation, leading to poor primer-substrate adhesion. Dark coatings are more prone to this defect than light coatings. The only effective remedy is removal of the surface coating, careful preparation of the substrate, and repainting with the correct materials. Painting under very hot ambient conditions should be avoided.

Blushing. "The formation of milky opalescence in clear finishes caused by the deposition of moisture from the atmosphere and/or precipitation of one or more of the solid constituents of the finish." This defect is generally associated with quick-drying lacquers. The rapid evaporation of solvent causes the cooling of the substrate and the consequent condensation of moisture. The remedy is to adjust the evaporation rate of the solvents used, or preheat the article being coated.

Bridging. "The separation of a paint film from the substrate at internal corners or other depressions due to shrinkage of the film or the formation of paint film over a depression or crack." Undercoats or primers that do not have adequate filling properties will give rise to this defect. Poor surface preparation is another cause. The remedy is to provide adequate surface preparation, and apply an undercoat with good filling properties. A lower application viscosity may also be helpful.

Brush marks. "Lines of unevenness that remain in the dried paint film after brush application." Brush marks and ropiness are associated with poor flow and sticky application. These defects are more often encountered in highly pigmented products and in certain latex paint formulations. Too rapid recovery of consistency in a thixotropic system will also cause these defects. The remedy may be the addition of a flow promoter, reduction in consistency, or modification of the rheological properties.

Can corrosion. This may be caused by incorrect pH of latex paints, or incorrect choice of ingredients leading to acidic by-products on storage. The remedy is careful selection of can lining, or perhaps the addition of anti-corrosive agents to the paint, or improved formulation and adjustment.

Chalking. "Change involving the release of one or more of the constituents of the film, in the form of loosely adherent fine powder." This is generally a result of the gradual breakdown of the binder because of the action of the weather. Careful selection of pigment types and levels and the use of more durable binder types retard the process. In flat white paints, chalking will enable the finish to be self-cleaning. A chalked surface requires washing down, or sealing with a penetrating sealer, before painting.

Checking. "Breaks in the surface of a paint film that do not render the underlying surface visible when the film is viewed at a magnification of 10X." Slight checking is not a serious defect, as it indicates a relieving of shrinkage stresses in a paint film.

Cheesy film. "The rather soft and mechanically weak condition of a dry-to-touch film but not a fully cured film."

Coagulation. This refers to the premature coalescing of emulsion resin particles in the paint. This is also termed *breaking* of the emulsion. Excessive stirring, solvent addition, or addition of coalescing agents may be the cause. Because universal colorants contain solvents (typically glycols), they may have the same effect if added too quickly or without poststirring. There is no truly effective remedy for coagulated paint. Straining (followed by addition of further latex) may partially recover a batch, and permit blending off.

Cobwebbing. "The formation of the filaments of partly dried paint during the spray application of a fast-drying paint." This can be caused by an incorrect solvent blend in the coating, or by spraying too far from the article. The remedy in each case is obvious.

Coverage. "The spreading rate, expressed in square meters per liter." Poor coverage is a defect related to either *sticky application* because the *viscosity* of the paint is too high, resulting in too much paint being applied, or to an absorbent substrate. In the latter case, reduction with the appropriate thinner for the first coat only will provide the remedy.

Cracking. "Formation of breaks in the paint film that expose the underlying surface." This is the most severe class of defects, which include *checking, crocodiling*, and *embrittlement*. These phenomena do not necessarily indicate that anything was or is wrong, if they are the natural consequences of normal *aging* of the film. This process of breakdown leading to cracking in the paint film is essentially shrinkage of the film. Much of the cracking noted over a timber substrate is caused by splitting and grain opening of the substrate and not defective paint. The only effective remedy for cracked paint is total removal and repainting.

Crissing. "The recession of a wet paint film from a surface leaving small areas uncoated." This is a consequence of improper wetting of the substrate by the paint. Frequently it is an aggravated form of pinholing. Where crissing is because of high surface tension inherent in the coating, specific proprietary additives can be used to remedy the situation. Examples are cellulose acetate butyrate for polyurethane lacquers, and anionic or nonionic surfactants for latex paints.

Crocodiling (alligatoring). "The formation of wide crisscross cracks in a paint film." Here the cracks are pronounced and expose the underlying paint films.

Embrittlement. This can occur where the curing process continues throughout the life of the coating—for instance, alkyd enamel drying by oxidative cross-linking.

Erosion. "Attrition of the film by natural weathering which may expose the substrate." This is normal in any paint system. It becomes a defect only if it occurs within the expected lifetime of the coating. In this case, the cause may be the incorrect choice of binder and pigment types, or poor quality control.

Fading. This can be caused by poor light-fastness of the pigments used, or by *chalking*. The use of the cheaper, low-fastness red, and yellow organic pigments can represent a serious problem in exterior quality surface coatings.

Fat edge. "Accumulation of paint at the edge of a painted surface." (See "sagging").

Floating. "Separation of pigment which occurs during drying, curing or storage which results in streaks or patchiness in the surface of the film and produces a variegated effect." Close examination will reveal Bénard (hexagonal) cells. This is because of differences in pigment concentration between the edges and centers of the cells, caused by convection currents in the drying film. Thixotropic paints will minimize this defect and the use of proprietary materials may also assist.

Flooding. "An extreme form of floating in which pigment floats to produce a uniform color over the whole surface which is markedly different from that of a newly applied wet film." Again, thixotropic systems or specialist additives will provide a remedy.

Flow. "The ability of a paint to spread to a uniform thickness after application." See "brush marks" as a special case of poor flow. The remedies suggested there apply here also.

Foaming. "This is the formation of a stable gas-in-liquid dispersion, in which the bubbles do not coalesce with each other or with the continuous gas phase. It is a defect commonly encountered in application by a roller, particularly with latex paints. The remedy is the addition of an antifoam agent in the manufacture of the paint, and/or a reduction in the speed of the roller."

Gassing. This is *aeration* as a result of a chemical reaction within the liquid paint during storage. It can result in explosion of cans, with resultant hazards to health and property. The action of water on aluminum or zinc-based paints, or acid on calcium carbonate will give this defect. In the case of aluminum paints, an air vent in the lid (covered with a paper sticker) is a wise precaution. Such paints may also be held in bulk until gassing testing is completed. The formulator should also consider including a small addition of water or acid scavenger in the paint.

Gelling (livering).   "Deterioration of a paint or varnish by the partial or complete changing of the medium into a jelly-like condition." The cause of this condition may be a chemical reaction between certain pigments and vehicles (such as zinc oxide and an acidic vehicle) or between atmospheric oxygen and oxidisable or polymerisable oils in the vehicle. A paint that has gelled to a livery mass that will not disperse on stirring, even with added solvent, is unrecoverable.

Hazing.   Loss of gloss after drying. It is usually caused by application of a gloss paint on a ground coat that has not hardened sufficiently; or excess driers in the final coat; or partial solution of organic pigments in the paint.

Lifting.   "The softening, and wrinkling, of a dry coat by solvents in a subsequent coat being applied." Usually, it is the action of strong solvents that cause this effect. Coatings that dry by oxidation are particularly prone to lifting. It is important to observe the recommended recoating times nominated by the supplier, as the rectification involves sanding and recoating.

Fly-off.   "The throwing-off of particles of paint from a paint roller." This is a particular instance of poor rheological control.

Mold.   The growth of mold is associated with dampness, either of the substrate or of the surrounding atmosphere. It is recognized by black or variable-colored spots or colonies which may be on, in, or beneath the paint film and can occur on almost any type of building material. The growth may penetrate the underlying plaster or brickwork and become difficult to eradicate.

Opacity (hiding power).   "The ability of a paint to obliterate the color difference of a substrate". Insufficient opacity, or failure to cover adequately, may be a consequence of insufficient covering pigment in the formulation. Where poor opacity is claimed on products of known good quality, the following causes may apply:

a. Overreduction or overspreading

b. Pigment settlement not redispersed

c. Poor application technique

Peeling (flaking). "Loss of adhesion resulting in detachment and curling out of the paint film." Peeling is essentially a manifestation of poor adhesion, either between the paint and the substrate or between successive coats of paint.

The effect of poor adhesion may not be apparent until something occurs to disturb the film, such as the action of heat or light or ageing, repeated wetting and drying, the exudation of resins, or the crystallization of salts beneath the film. The only remedy is complete removal of all peeling and flaking paint, and repainting.

Pinholes. "Minute holes in a dry film which form during application and drying of paint."

Sagging. "Excessive flow of paint on vertical surfaces causing imperfections with thick lower edges in the paint film." Poor application technique and the condition of the substrate are major causes of the fault; however, the rheology of the paint can influence results obtained. Many paints, currently sold, have thixotropy deliberately built in, to facilitate the application of heavier coats with lesser tendency to sag. The rate of recovery of viscosity after application is the key to reducing sagging. In production, paints can be checked for poor flow or sagging tendency by various types of sag index blades or combs. These deposit tracks of paint of varying film thickness, and the resultant tracks, if left in a vertical position, give an indication of the flow behavior to be expected in use.

Tackiness. "The degree of stickiness of a paint film after a given drying time."

Settling. "Separation of paint in a container in which the pigments and other dense insoluble materials accumulate and aggregate at the bottom." The law of gravity applies to paint, as does Stokes' law. An increase in consistency will help, as will a thixotropic rheology. Various additives are available to do this.

Skinning. "The formation of a tough, skin-like covering on liquid paints and varnishes when exposed to air." A skin sometimes forms across the surface of a paint during storage in sealed or unsealed, full or partly filled containers. If the skin is continuous and easily removed, it is not as troublesome as a slight, discontinuous skin, which may easily become mixed with the remaining paint.

The formation of skin is because of oxidation and polymerization of the medium at the air-liquid interface. Antiskinning agents, usually volatile antioxidants, are generally added to paint to prevent skinning. A proportion may be lost, by evaporation, if the batch is left for an excessive time before filling. Because air (oxygen) is generally necessary, the best way of preventing skinning is to keep the air away as much as possible. When skin is encountered in a full container, the seal of the lid may be the cause.

Viscosity increase (thickening). A slight increase in viscosity during storage of a paint is not uncommon, but rapid or excessive thickening is either because of instability of the medium or because of a reaction between the pigment and the medium known as "feeding." This can sometimes be corrected by adjusting the drier content, or by the use of antiskinning agents, stronger solvents, or certain chemical additives.

Wrinkling. "The development of wrinkles in a paint film during drying." This defect is closely associated with *drying problems*. Its cause is the surface of the film drying too rapidly before the underlying layer has firmed up. Correct balance of metal driers and solvents will cure this defect. Excessive film thickness may also be a factor.

7.6   Testing and Quality Control

The paint is applied to a substrate to provide a proper appearance, ample protection, adequate functionality, and sufficient durability. It is the influence of the paint on the value of the end product that makes the properties so important that they must be characterized, tested, and controlled at each step from raw material manufacture through paint manufacture, storage, and final application.

There are a range of miscellaneous tests for evaluating the properties of paints. The aim of testing may be to determine one of the following:

1. Package properties: Package tests include viscosity, skinning, settling weight per gallon, flash point, freeze-thaw stability, and fineness of grind.

2. Application properties: Application properties include ease of brushing, spraying, or rolling, leveling, sag, spatter, and drying time.

3. Film appearance: Film appearance tests include gloss, color, opacity, color acceptance, and color development.

4. Film performance: Film resistance tests are for hardness, abrasion, adhesion, flexibility, impact, scrubbability, and chemical and water resistance.

5. Durability: Durability tests include weatherometer, salt, spray, humidity, and exterior exposure.

ASTM (American Society for Testing and Materials) annually publishes books describing tests. There are other standard testing methods published by ISO (International Standards Organization) and U.S. federal standards tests. The use of standard methods enables direct comparisons of results and quick recognition of differences in properties among different producers and consumers. The largest single collection of the tests mentioned in this paragraph has been compiled by the ASTM in volumes 06.01, 06.02, and 06.03: *Paint-Tests for Formulated Products and Applied Coatings.* More details on these tests may be found in ASTM or similar standards.

7.7   Environmental Impacts and Risks

By far the most important environmental impact from paints and coatings is the release of volatile organic compounds (VOC) from drying. The second largest source of man-made VOC emissions comes from the paint and coating industry. Three end effects of VOC emissions into the atmosphere are important: formation of eye irritants, particulates, and toxic oxidants, especially ozone. Ozone, a high reactive form of oxygen, is a health risk at very low concentrations, and is the ultimate risk factor associated with VOC emissions. With the rapid growth of VOC emissions from man-made sources since 1900, ozone levels on many days of the year in many parts of the world, especially in and around cities, have exceeded the levels that cause respiratory problems, vegetation damage, and material degradation. A program of voluntary monitoring was proposed by the industry to ensure that emissions of VOCs from paints and coatings fall within prescribed limits. To reduce VOC emissions from the manufacture of paints and coatings, control techniques include condensers, or absorbers, or both on solvent handling operations, and scrubbers and afterburners on cooking operations. Afterburners can reduce VOCs by 99 percent.

Current US Environmental Protection Agency (EPA) regulations treat almost all solvents used in paint manufacture (except water, acetone, carbon dioxide, silicone fluids, and fluorinated solvents) as equally undesirable. However, it was recognized by the regulators that some paints require a higher VOC than others for adequate performance. Based on a study, the EPA established different maximum VOC guidelines for major applications. During most of the 1990s, the EPA guidelines ranged from 0.23 to 0.52 kg/L (1.9 to 4.3 lb/gal) for most major industrial coating operations. Further, much tighter EPA regulations are expected in the new millennium.

There has been a significant shift during the past 20 years in the use of formulations based on petroleum solvents to formulations based on water as a primary solvent. In addition to reducing VOC emissions, water-based formulations offer advantages such as easier cleanup, and less odor. Consequently, their market acceptance is much greater than that of other low-emission paints and coatings.

The use of supercritical carbon dioxide (CO_2) as a component in a solvent mixture is another ingenious technique to reduce VOC emission by 50 percent or more. This technique takes the advantage of the fact that CO_2 is a supercritical fluid below its critical temperature (31.3°C) and critical pressure (7.4 MPa). Solid coating and supercritical CO_2 are metered into a proportioning spray gun in such a ratio so as to reduce the viscosity to the level needed for proper atomization. Airless spray guns are used.

There are, however, still several applications where the necessary performance can be achieved only by using solvent-based systems. Research is continuing to further reduce solvent content while retaining its beneficial properties.

Other environmental impacts arise from the presence of toxic solid materials in the paint formulations and the handling of postconsumer paints. In contrast to the immediate effects of VOCs, solids persist, and can create problems long after coating is applied. For example, lead was phased out of most paints in the late 1970s. However, many surfaces painted prior to the phase-out, such as walls and window frames, are typically painted over rather than removed, and can persist, carrying their toxic burden for many generations. The problem of children in older houses ingesting paint chips will remain for some time. Similarly, some specialized coatings still contain problem materials, such as use of chromium and cadmium for tough protective coatings of steel. The efforts are being made by both manufacturers and regulators to deal with such problems of international consequences.

The landfilling of paint containers with leftover contents is another environmental issue. In most jurisdictions these are not accepted in landfill sites because of their potential for contamination of the soil, so waste paint is normally collected at a special depot, along with other household hazardous waste. The paint industry has developed techniques for collecting paint from these waste depots, testing for contamination, and reformulating the paint into a usable product.

Paint and coatings industries affect water quality in a variety of ways. Most contamination of waterways occur either from solvents contained in process wastewater discharge, or runoff from vehicles, ships, and aircraft bearing protective coatings with toxic metals. There are many federal and local regulations that are applied to control storm water runoff and wastewater discharge from paint and coatings manufacturing units.

References

1. Reisch, Marc, *Chemical & Engineering News*, American Chemical Society, 79(45), p. 23–30, November 5, 2001.
2. Reisch, Marc, *Chemical & Engineering News*, American Chemical Society, 81(44), p. 23–24, November 3, 2003.

3. *Riegel's Handbook of Industrial Chemistry,* 8th ed., Kent J. A. (ed.), Van Nostrand Reinhold, New York, 1983.
4. *Surface Coatings,* Volumes 1 and 2, prepared by the Oil and Color Chemists' Association, Australia, Tafe Educational Books, Randwick, Australia, 1983 and 1984.
5. Boxall, J., and Von Fraunhofer J. A., *Concise Paint Technology,* Paul Elek (Scientific Books) Limited, London, 1977.
6. *Shreve's Chemical Process Industries,* 5th ed., Austin G. T. (ed.), McGraw-Hill, New York, 1984.
7. Turner, G. P. A., *Introduction to Paint Chemistry,* 2d ed., Chapman and Hall, London, 1980.
8. *Kirth-Othmer Encyclopedia of Chemical Technology,* 3d ed., Vol. 16, John Wiley and Sons, New York, 1983.
9. Wicks, Jr. Z. W., Jones F. N., and Pappas S. P., *Organic Coatings, Science and Technology,* 2d ed., Wiley-Interscience, New York, 1999.
10. Weldon, D. G., *Failure Analysis of Paints and Coatings,* John Wiley & Sons Ltd., Chichester, U.K., 2002.

8

Dyes: Chemistry and Applications

Mohammad Farhat Ali

8.1 Introduction

Dyes are colored organic compounds that are used to impart color to various substrates, including paper, leather, fur, hair, drugs, cosmetics, waxes, greases, plastics, and textile materials. The history of dyes goes back to prehistoric times. *Indigo*, the oldest known dye, was used by the ancient Egyptians to dye mummy clothes. *Tyrian purple*, obtained from *Murex* snails found near the city of Tyre, was used by the Romans to dye the togas of the emperors. From prehistoric times until the mid-19th century, natural dyes, such as indigo, Tyrian purple, alizarin, and logwood, were used exclusively until the discovery of *Mauve* by W. H. Perkin in 1856. Perkin's discovery marked the beginning of the dye manufacturing industry in Europe. The Germans and Swiss rapidly became leaders in the production of synthetic dyes and the rest of the world imported most of their chemical dyes from Germany. The outbreak of World War I cut off the supply of German dyes and this led the United States to establish a domestic industry. American companies, primarily DuPont, moved quickly to fill the vacuum left by the cutting off of German supplies. The dye industry has always been highly competitive; the industry has lately experienced major setbacks in terms of profitability and overall attractiveness particularly in Europe and the United States. Several major American companies have dropped dye manufacture since 1978. Major changes have taken place during the last 20 years, and today Asia (India, Japan, Korea, and China) has become the largest dyestuff market, accounting for about 42 percent of the value of the global dyestuff market [1, 2, 3].

World demand for dyes and organic pigments is forecasted to increase 5.1 percent per year to more than US$14 billion in 2004. Gains will result from improvement in the textile industry, which accounts for about 70 percent of the global dye demand [4]. These demand figures include the forecast for plastic application, where more brilliant and striking colors are generating consumer appeal.

8.2 Colorants

The two major types of colorants produced today are dyes and pigments. Pigments, both inorganic and organic types, are almost always applied in an aggregated or crystalline insoluble form that requires a binder to form a coating on the surface of a substrate. Pigments do not interact with the substrate and hence do not destroy the crystal structure of the substrate. Dyes are normally water-soluble or water-dispersible organic compounds that are capable of being absorbed into the substrate destroying the crystal structure of the substance. The dye molecules are usually chemically bonded to the surface and become a part of the material on which they are applied. The primary use of dyes is in the textile industry, although substantial quantities are consumed for coloring such diverse materials as leather, paper, plastics, petroleum products, and food [5]. To be of commercial interest, dyes must have high color intensity and produce dyeing of some permanence. The color intensity of the dye molecule depends on how strongly it absorbs radiation in the visible region, which extends from 400 to 700 nm. It was observed earlier that only some types of organic structures give rise to color. The partial structures necessary for color (unsaturated groups that can undergo π-π^* and n-π^* transitions) were called *chromophores*. Examples of some chromophores are:

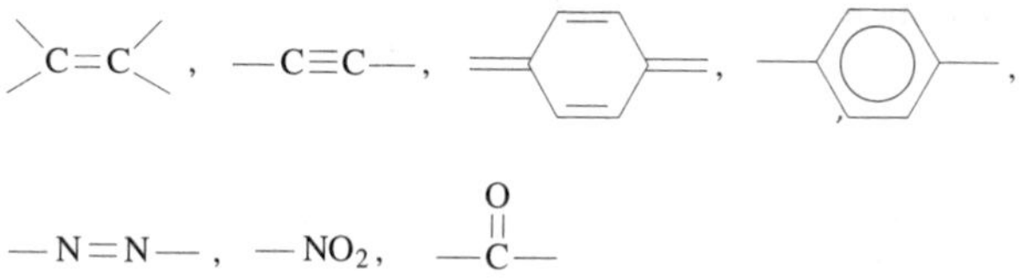

$$\ce{>C=C<} \;,\;\; \ce{-C#C-} \;,\;\; \text{(benzene ring)} \;,\;\; \text{(benzene ring)} \;,$$

$$\ce{-N=N-} \;,\;\; \ce{-NO_2} \;,\;\; \underset{\displaystyle \|}{\overset{\displaystyle O}{\ce{-C-}}}$$

It was also observed that the presence of some other groups caused an intensification of color. These groups are called *auxochromes*. Auxochromes are groups that cannot undergo π-π^* transitions, but can undergo transition of n electrons. Examples of some auxochromes are:

$$\ce{-OH}, \ce{-OR}, \ce{-NH_2}, \ce{-NHR}, \ce{-NR_2}, \ce{-X}$$

As the coloration with dyes is based on physico-chemical equilibrium processes, namely diffusion and sorption of dye molecules or ions followed by chemical reactions in the substrate, the shade and fastness of a given dye may vary depending on the substrate. The development of new substrates present new problems, give rise to new dyes and dyeing methods and thus influence dyeing technology to a wider scale.

8.3 Classification of Dyes

Dyes may be classified according to their chemical structure or the method by which they are applied to the substrate. The dye manufacturers and dye chemists prefer the former approach of classifying dyes according to the chemical type. The dye users, however, prefer the latter approach of classification according to the application method.

Classification by application or usage is the principal system adopted by the color index (CI) [6]. The classification of dyes according to their usage is summarized in Table 8.1, which is arranged according to the CI application classification. It shows the principal substrates, the methods of application, and the representative chemical types for each application class [7, 8].

The definition, chemical nature, the type of substrate best suited for each class, and the mechanism by which the dye is retained by the substrate of the various application classes are discussed in the following sections.

Acid dyes.   Acid dyes are water-soluble anionic dyes, containing one or more sulfonic acid substituent or other acidic groups. An example of the class is acid yellow 36.

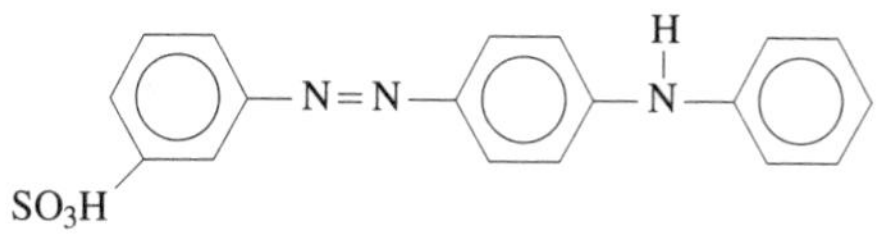

TABLE 8.1   Application Classification of Dyes [7]

Class	Principal substrates	Method of application	Chemical types
Acid	Nylon, wool, silk, paper, inks, and leather	Usually from neutral to acidic dyebaths	Azo (including premetalized), anthraquinone, triphenylmethane, azine, xanthene, nitro, and nitroso
Azoic components and compositions	Cotton, rayon, cellulose acetate, and polyester	Fiber impregnated with coupling component and treated with a solution of stabilized diazonium salt	Azo
Basic	Paper, polyacrylonitrile, modified nylon, polyester, and inks	Applied from acidic dyebaths	Cyanine, hemicyanine, diazahemicyanine, diphenylmethane, triarylmethane, azo, azine, xanthene, acridine, oxazine, and anthraquinone
Direct	Cotton, rayon, paper, leather, and nylon	Applied from neutral or slightly alkaline baths containing additional electrolyte	Azo, phthalocyanine, stilbene, and oxazine
Disperse	Polyester, polyamide, acetate, acrylic, and plastics	Fine aqueous dispersions often applied by high temperature or pressure or lower temperature carrier methods; dye may be padded on cloth and baked or thermofixed	Azo, anthraquinone, styryl, nitro, and benzodifuranone
Fluorescent brighteners	Soaps and detergents, and all fibers, oils, paints, and plastics	From solution, dispersion or suspension in a mass	Stilbene, pyrazoles, coumarin, and naphthalimides
Food, drug, and cosmetic	Foods, drugs, and cosmetics		Azo, anthraquinone, carotenoid, and triarylmethane
Mordant	Wool, leather, and anodized aluminum	Applied in conjunction with Cr salts	Azo and anthraquinone
Oxidation bases	Hair, fur, and cotton	Aromatic amines and phenols oxidized on the substrate	Aniline black and indeterminate structures

(Continued)

TABLE 8.1   Application Classification of Dyes [7] (*Continued*)

Class	Principal substrates	Method of application	Chemical types
Reactive	Cotton, wool, silk, and nylon	Reactive site on dye reacts with functional group on fiber to bind dye covalently under influence of heat and pH (alkaline)	Azo, anthraquinone, phthalocyanine, formazan, oxazine, and basic
Solvent	Plastics, gasoline, varnishes, lacquers, stains, inks, fats, oils, and waxes	Dissolution in the substrate	Azo, triphenyl-methane, anthra-quinone, and phthalocyanine
Sulfur	Cotton and rayon	Aromatic substrate vatted with sodium sulfide and reoxidized to insoluble sulfur-containing products on fiber	Indeterminate structures
Vat	Cotton, rayon, and wool	Water-insoluble dyes solubilized by reducing with sodium hydrogensulfite, then exhausted on fiber and reoxidized	Anthraquinone (including polycyclic quinines) and indigoids

Acid dyes are applied from acidic dye baths to nylon, silk, wool, and modified acrylics. They are also used to some extent for paper, leather, ink-jet printing, food, and cosmetics [7]. The dyeing process is reversible and may be described as follows:

$$Dye^- + H^+ + Fiber \rightleftharpoons Dye^-H^+ - Fiber$$

The ionic bonding between the dye and the fiber is the result of the reaction of the amino groups on the fiber with acid groups on the dye. Generally the fastness of this dye depends on the rate with which the dye can diffuse through the fiber under the conditions of washing. Metal complex (cobalt or chromium) acid dyes are used mainly on wool for improved fastness.

Azoic dyes. Azo dyes contain at least one azo group ($-N{=}N-$) attached to one or often two aromatic rings. They are produced in textile fibers (usually cotton, rayon, and polyester), by diazotization of a primary aromatic amine followed by coupling of the resulting diazonium salt with an electron-rich nucleophile (azo coupling). A variety of hues can be obtained by proper choice of diazo and coupling components. The production of bluish red azoic dye from the following two components is an example [5].

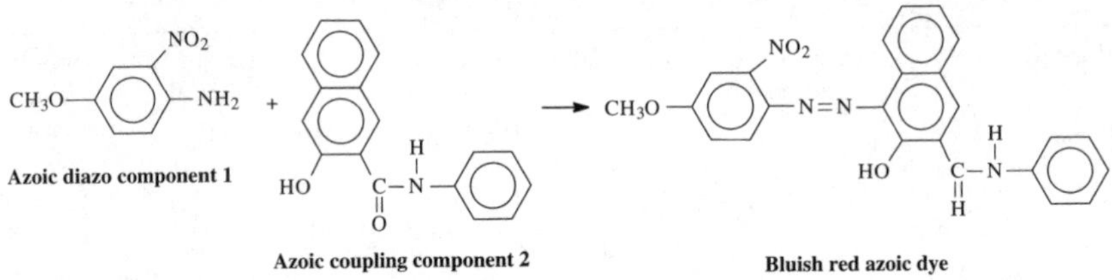

Azoic diazo component 1   Azoic coupling component 2   Bluish red azoic dye

Basic (cationic) dyes.   Basic dyes are water-soluble and produce colored cations in solution. They are mostly amino and substituted amino compounds soluble in acid and made insoluble by the solution being made basic. They become attached to the fibers by formation of salt linkages (ionic bonds) with anionic groups in the fiber. They are used to dye paper, polyacrylonitrile, modified nylons, and modified polyesters. In solvents other than water, they form writing and printing inks. The principal chemical classes are triaryl methane or xanthenes. Basic brown 1 is an example of a cationic dye that is readily protonated under the pH 2 to 5 conditions of dyeing [5].

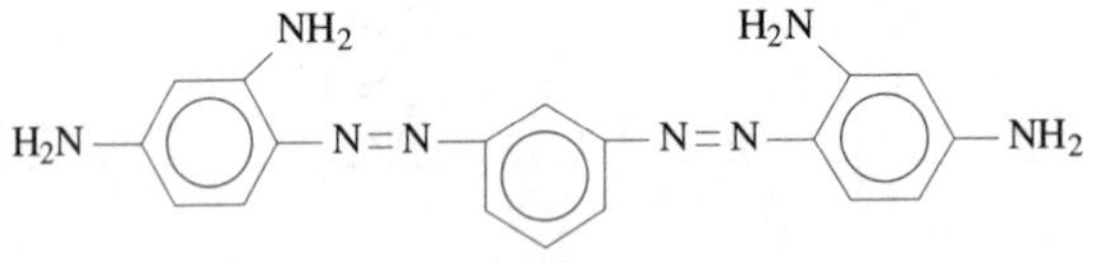

Basic brown 1

Direct dyes.   Direct dyes are water-soluble anionic dyes, but are not classified as acid dyes because the acid groups are not the means of attachment to the fiber. They are used for the direct dyeing of cotton and regenerated cellulose, paper, and leather. They are also used to dye union goods (mixed cotton, and wool or silk) and to a lesser extent nylon fiber. Most of the dyes in this class are polyazo compounds, along with some stilbenes, phthalocyanines, and oxazines. The solubility of the dye in the dye bath is often reduced by adding common salt or Glauber's salt. The presence of excess sodium ions favors establishment of equilibrium with a minimum of dye remaining in the dye bath. Direct orange 26 is a typical direct dye [5].

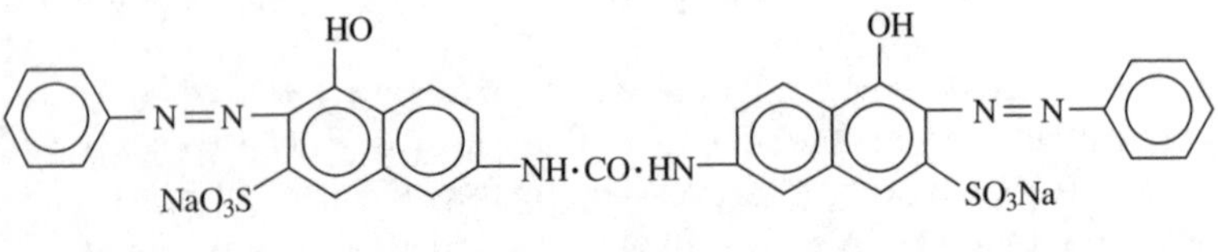

Direct orange 26

Disperse dyes.   Disperse dyes are substantially water-insoluble nonionic dyes for application to synthetic hydrophobic fibers from aqueous dispersions. Disperse dyes are applied as very finely divided materials that are adsorbed onto the fibers with which they then form a solid solution. Dispersed dyes are primarily used for polyester and acetate fibers. Simple soluble azo, styryl benzodi furanone, and insoluble anthraquinone are the most common disperse dyes. Disperse yellow 3, disperse red 4, and disperse blue 27 are good examples of disperse dyes [5]:

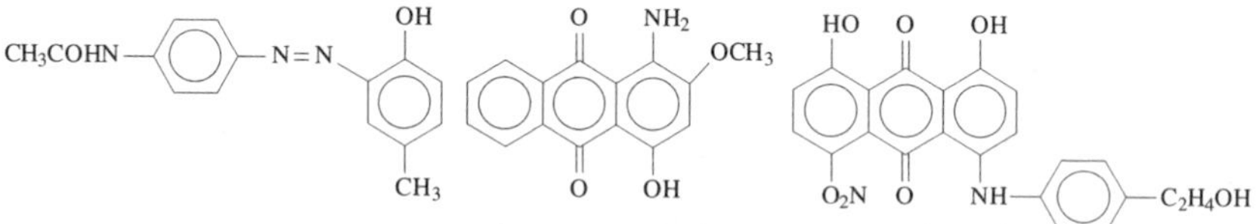

Disperse yellow 3   Disperse red 4   Disperse blue 27

Fiber-reactive dyes.   These dyes react with the cellulosic fiber to form a covalent bond. This produces dyed fiber with extremely high wash fastness properties. Cotton, rayon, and some nylons are dyed by this relatively simple dye. The principal chemical classes of reactive dyes are azo, triphendioxazine, phthalocyanine, formazan, and anthraquinone. An example of this type is the reactive blue 5 dye shown below [5]:

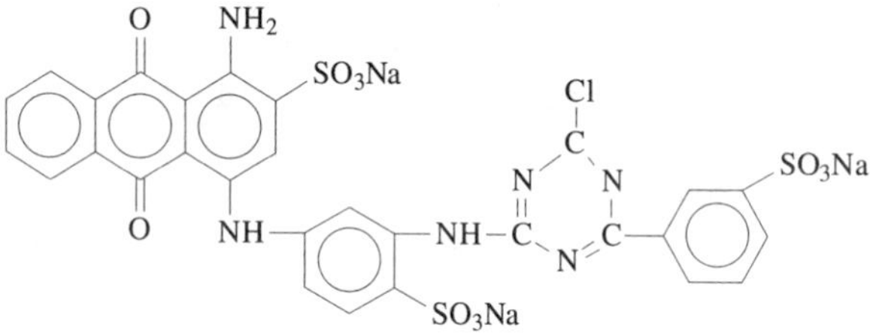

Reactive blue 5

Mordant dyes.   Some dyes combine with metal salts (mordanting) to form insoluble colored complexes (lakes). These materials are usually used for the dyeing of cotton, wool, or other protein fiber. The fiber is first treated with an aluminum, chromium, and iron salt and then contacted with a lake-forming dye (azo and anthraquinone derivatives). The metallic precipitate is formed in the fiber producing very fast colors highly resistant to both light and washing. Alizarin is the best-known anthraquinone derivative as an example of a mordant dye as shown in the following reaction. The hydroxyl groups attached to anthraquinone nuclei are capable of reaction with metals in the mordant material (aluminum hydroxide) to form mordant dyes (aluminum *lake*).

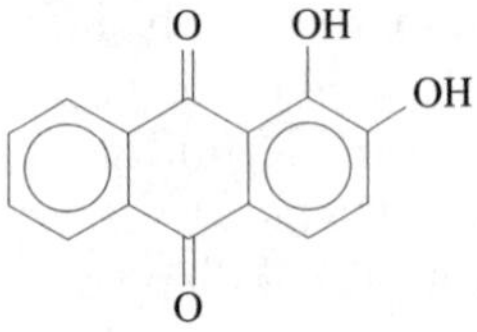

Mordant red 11

Mordant dyes have declined in importance because easier methods using developing dyes and vat dyes have replaced this process [5].

Sulfur dyes.   Sulfur dyes are applied to cotton from an alkaline reducing bath with sodium sulfide as the reducing agent. These dyes are water-insoluble but they are soluble in their reduced form and exhibit affinity for cellulose. They dye by adsorption; but on exposure to air, they are oxidized to reform the original insoluble dye inside the fiber. The cost is low and they have good fastness to light, washings, and acids. The actual structures of sulfur dyes are largely unknown although it is considered that they possess sulfur-containing heterocyclic rings [5].

Vat dyes.   The vat dyes are insoluble complex polycyclic molecules based on the quinone structure (keto-forms). They are reduced with sodium hydrosulfite in a strongly alkaline medium to give soluble leuco forms that have a great affinity for cellulose. After the reduced dye has been absorbed on the fiber, the leuco forms are reoxidized to the insoluble keto forms. The dyeings produced in this way have high wash and light-fastness. An example of a vat dye is vat blue 4 (indanthrene) [5].

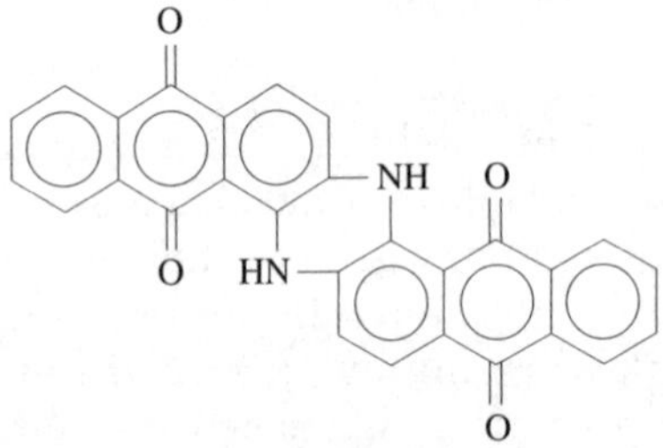

Vat blue 4

Solvent dyes.   These dyes are water-insoluble but soluble in alcohols, chlorinated hydrocarbons, or liquid ammonia. They are used for coloring synthetics, plastics, gasoline, oils, and waxes. The dyes are predominantly azo and anthraquinone, but phthalocyanine and tri-arylmethane dyes are also used [7]. Perchloroethylene is the preferred solvent.

Oxidation dyes. These dyes are produced on the substrate by oxidation of colorless compounds such as aromatic amines and phenols. For example, aniline may be oxidized in cotton with sodium bichromate to produce aniline black. The color produced by aniline black is characterized by a full bluish black shade and excellent fastness. However, oxidation dyes are rapidly decreasing in importance because aniline, other aromatic amines, and bichromate that are used for oxidation, are toxic [7].

Fluorescent brighteners. Fluorescent brighteners or fluorescent whitening agents (FWAs) are colorless to weakly colored organic compounds that, in solution or applied to a substrate, absorb ultraviolet light (e.g., from daylight at approximately 300 to 430 nm) and remit most of the absorbed energy as blue fluorescent light approximately between 400 and 500 nm. FWAs are used to make yellowish laundry appear dazzling white. These compounds are added to soaps and detergents to produce greater brilliance in laundry washings. FWAs are stilbene, pyrazoles, coumarin, and naphthalimides. Like dyes, FWAs are available in classes analogues to acid, basic, direct, and disperse dyes for application to all substrates, including paper. FWAs are used for improving the appearance of recycled paper. An example of an FWA is 4,4′-bis (ethoxy-carbonylvinyl) stilbene, which can be obtained by the reaction stilbene-4,4′-dicarbaldehyde with triethyl phosphonoacetate in the presence of sodium methoxide [7].

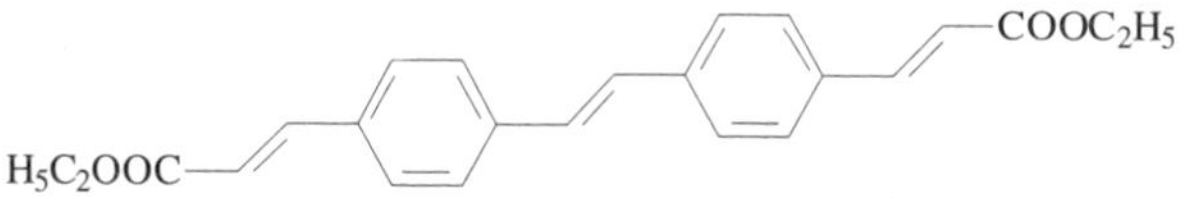

Food, drug, and cosmetic dyes. Most synthetic and natural dyes commonly used in food, drugs, and cosmetics are carefully controlled materials, regulated by the government agencies in the EEC, the United States, and Japan. These currently consist of very few (under 100) dyes and are listed in the approved list. Regulations list the approved color additives and conditions under which they may be safely used, including the amounts that may be used.

The most frequently used synthetic dyes for food, drugs, and cosmetics belong to azo, anthraquinone, carotenoid, and triarylmethane chemical types. The following two azo series food dyes are good examples [7]:

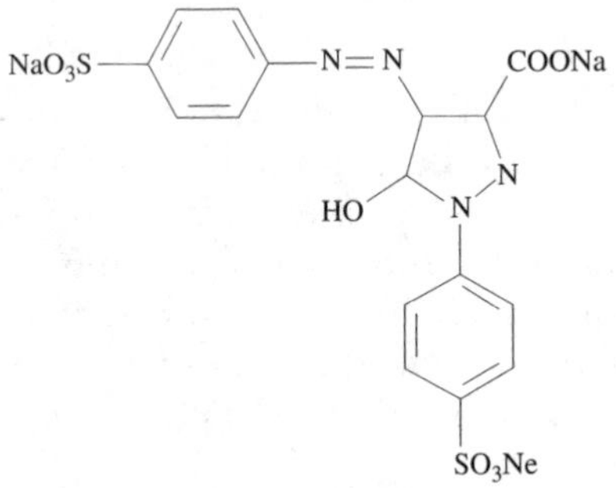

C.I. Food yellow 4, 19140 [1934-21-0],

E 102, Tartrazine

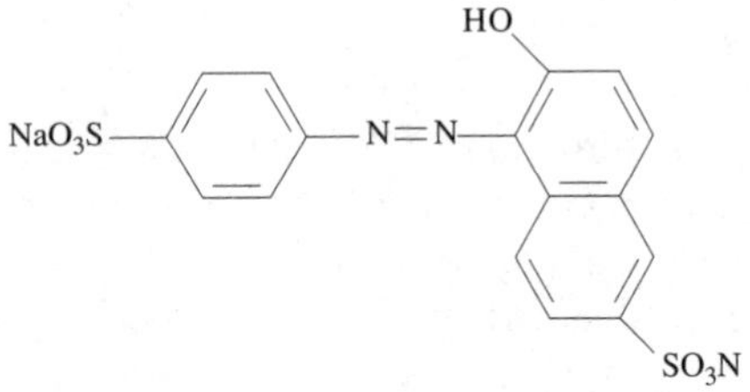

C.I. Food yellow 3, 15985, [2783-94-0],

Yellow E110, Orange S

Ingrain dyes. These dyes are produced on textile fibers by applying precursors. The examples are azoic and oxidation dyes. Other dyes in this group are phthalocyanine compounds. The very bright blue dye, copper phthalocyanine, is used in textile printing.

8.4   Textile Fibers

The world textile industry is one of the largest consumers of dyestuffs. An understanding of the chemistry of textile fibers is necessary to select an appropriate dye from each of the several dye classes so that the textile product requirements for proper shade, fastness, and economics are achieved. The properties of some of the more commercially important natural and synthetic fibers are briefly discussed in this section. The natural fibers may be from plant sources (such as cotton and flax), animal sources (such as wool and silk), or chemically modified natural materials (such as rayon and acetate fibers). The synthetic fibers include nylon, polyester, acrylics, polyolefins, and spindex. The various types of fiber along with the type of dye needed are summarized in Table 8.2.

Cotton.   The cotton fiber is essentially cellulosic in nature and may be chemically described as poly (1,4-B-D-anhyrdoglucopyranose), with the following repeat unit (about 3000 units):

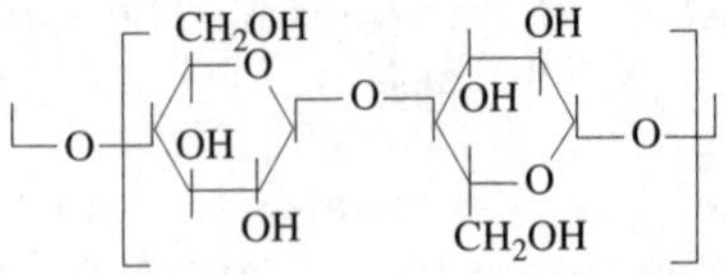

TABLE 8.2   Fiber-Dye Property Requirements

Fiber name	Type/general classification	Chemical constitution	Ionic nature in dyebath
Cotton, linen, and other vegetable fibers	Natural hydrophilic	Cellulose	Anionic
Viscose rayon	Synthetic, hydrophilic	Regenerated cellulose	Anionic
Wool, silk, hair	Natural, hydrophilic	Complex proteins	Cationic
Nylon	Synthetic, somewhat hydrophobic	Polyamide	Usually cationic
Acrylics	Synthetic, hydrophobic	Modified poly-acrylonitriles	Anionic
Acetate	Synthetic, hydrophobic	Acetylated cellulose	Nonionic
Triacetate	Synthetic, hydrophobic	Acetylated cellulose	Nonionic
Polyester	Synthetic, hydrophobic	Polyester	Usually nonionic
Polypropylene	Synthetic, hydrophobic	Polyolefin	Nonionic

The structure has primary and secondary alcohol groups uniformly interspersed throughout the length of the polymer chain. These hydroxy units impart high water absorption characteristics to the fiber and can act as reactive sites. The cotton fibers are hydrophilic and swell in water. It is hydrolyzed by hot acid and swollen by concentrated alkali. The cotton is treated with caustic soda solution (12 to 25 percent) under tension to develop a silk-luster and stop longitudinal shrinkage. This process is called *mercerization*. Mercerized cotton exhibits increased moisture and dye absorption.

The dyeing of cotton fiber is accomplished by three principal processes. Cotton may be chemically reacted with fiber-reactive dyes in solution. The dyeing takes place by reaction with hydroxyl groups in cotton. A second method is the use of substantive dyes that diffuse directly into fiber from a dye solution. The dyeing rate is increased by the addition of electrolytes. The third method is referred to as mordant dyeing in which the dye in solution reacts with metals previously applied to the fiber to form insoluble colored compounds on the cotton. Vat dyes are another important class of dyes for cotton. These are applied in a soluble reduced form and after application they are oxidized, forming an insoluble molecule [8].

Rayon.   Rayon, the first commercial manmade fiber, is composed of cellulose in a quite pure form. It is produced by the treatment of wood pulp with alkali and carbon disulfide to form a viscous solution of cellulose xanthate. This viscous solution, called viscose, is extruded through spinnerets into an aqueous acid bath that coagulates the cellulose xanthate, decomposes it, and regenerates the cellulose.

Rayon fibers are easily wetted by water and provide easy access to dye molecules. Dyeing may take place by absorption or by reaction with the hydroxyl groups. Rayon fibers may also be dyed with mordant and vat dyes.

Cellulose acetates.   Cellulose acetate is a well-known derivative of cellulose and has found many uses as a fiber. Acetate, diacetate, and triacetate are similar in chemical structure with acetate having about 83 percent of the hydroxyl groups acetylated, and not less than 92 percent hydroxyl groups are acetylated in triacetate. Triacetate and diacetate fibers are manufactured by the acetylation of refined wood pulp or purified cotton linters. The acetylation reaction is chemically quite simple and may be visualized in the following manner:

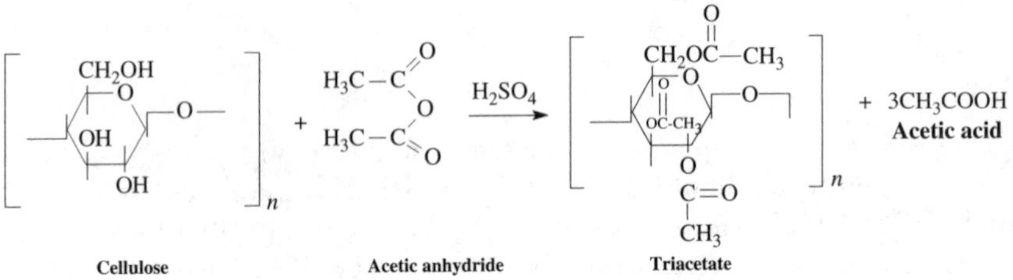

Cellulose   Acetic anhydride   Triacetate

Acetate or diacetate is made by the saponification of one of the acetylated groups, thus restoring some of the hydroxyl in cellulose units. The conversion of the hydroxyl group causes these fibers to be hydrophobic and changes their dyeing characteristics drastically from those of cotton and rayon. The higher the degree of acetylation the more hydrophobic the acetate becomes. Triacetate is the most hydrophobic. The dyeing of acetate fiber cannot be done with acid, basic, and reactive dyes because of lack of sites for attachment. Dyeing is usually affected using dyes of low water solubility that become dissolved in the fiber.

Wool and silk.   Wool is animal hair from the body of sheep. Silk is a lustrous, tough elastic fiber produced by silkworms. Both wool and silk fibers are protein substances with both acidic and basic properties. The building blocks for these fibers are amino acids. The α amino acids (about 20) in wool are arranged in a polypeptide chain in the following manner:

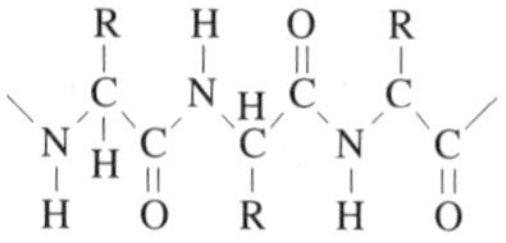

R   H   O   R
| | || |
C   C   C
N   C   N   N   C
| H   C   C | | ||
H || | H   O
H   O   R

Silk, like wool, is a protein fiber, but of much simpler structure. It comprises six α amino acids, and is the only continuous filament fiber.

Owing to the presence of many amine, carboxylic acid, amide, and other polar groups, wool and silk are hydrophilic in nature, wetted by water, and are dyed with either acid or basic dyes through the formation of ionic bonds (salt linkages). They may also be dyed with reactive dyes that form covalent bonds with available amino groups.

Polyamides. Nylon 6,6 was the first synthetic polyamide developed that reached technical importance as a fiber. Later, nylon 6 also became a commercial product and assumed an important position among synthetic fibers. In both nylon polymers one chain end consists of an amino group, which can be presented in the free state or in the acetylated form, as shown in the following structures:

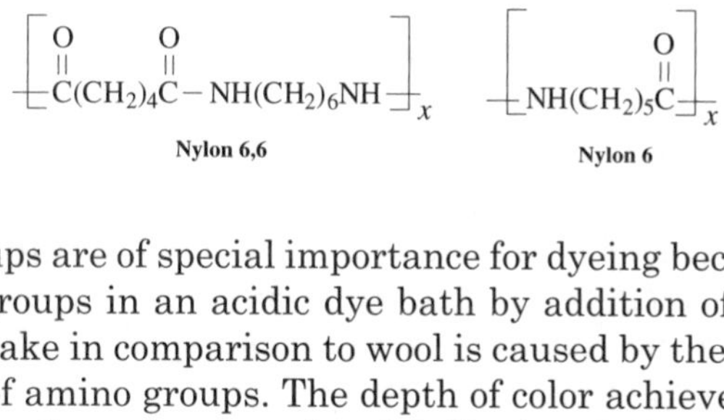

$$\left[\begin{matrix} O & O \\ || & || \\ C(CH_2)_4C-NH(CH_2)_6NH \end{matrix}\right]_x \qquad \left[\begin{matrix} O \\ || \\ NH(CH_2)_5C \end{matrix}\right]_x$$

Nylon 6,6   Nylon 6

Amino groups are of special importance for dyeing because they form ammonium groups in an acidic dye bath by addition of a proton. The lower dye uptake in comparison to wool is caused by the comparatively low number of amino groups. The depth of color achieved in nylon 6 is somewhat less than that in nylon 6,6. During dyeing a pH below 2.4 should be avoided, because fiber damage occurs at low pH.

Reactive dyes that bond to available amino groups may also be used.

Acrylics. Acrylics are produced by the polymerization of acrylonitrile. They have a chemical structure essentially comprising the repeating unit, $[-CH_2-CH(CN)-]_n$, with up to 15 percent of the polymer comprising one or two other monomeric units. As comonomers, vinyl acetate and an acrylate or methacrylate ester is used to vary the properties of the polymer for both ease of processing into a fiber and for improved fiber properties [8].

Acrylic fibers are hydrophobic with excellent chemical stability. As there are no functional groups present in the acrylics, the fiber producers found ways to modify the basic polymer by incorporating acidic groups in the polymer. The acidic group frequently used is the sulfonic acid or its salts that is carried into the polymer chain as a substituent of vinyl benzene, alkoxy benzene, or diamino stilbene monomer. Although the monomer is in the gel state, it can be treated with sulfonic acid derivatives and upon drying and scouring it will retain sufficient acid groups for dyeing with basic dyes [8].

Polyolefins. The polyolefin fibers of importance are polyethylene and polypropylene. They are linear polymers, produced by the polymerization of ethane and propane, respectively. These polymers are completely inert and are hydrophobic. The dyeing is done using special disperse dyes that are usually introduced into the polymer before the fibers are spun. The polypropylene fibers can be modified by cospinning with other polymers for the principal objective of improving and modifying dyeability. The sites for dyeing in polypropylene can be incorporated by cospinning with polyamides, polyamines, polytriazoles, and polypyrrolidones. Likewise, the addition of metal salts or complexes to polypropylene can produce similar modifications and improve dyeability. The metals most frequently used are aluminum, zinc, nickel, cobalt, cadmium, calcium, barium, and magnesium.

8.5 The Application of Dyes

The nature of substrate selects the type of dye needed, and methods of dye application. The development of new substrates led to new dyes and dyeing methods and influenced the dyeing technology in a fundamental manner. The dyeing of any substance (e.g., textile fiber) is based on a physico-chemical equilibrium process, namely diffusion and sorption of dye molecules or ions. These processes may be followed by chemical reactions in the substrate, for example, in the application of vat, reactive, azoic, and chrome dyes [9].

Modern dyeing technology consists of dozens of processes that have to be selected according to the substrate, to the available dyes and dyeing equipments, fastness properties necessary for the respective use of the dyed goods, economic considerations, and many other factors. The mass, chemical constitution, and geometry of dye molecules play an important role, both in the kinetics of dyeing and in imparting substantivity to textile fibers [10].

The process of dye application involves the transfer of dye from a solution in a dye bath to the fiber, the dye preferentially adsorbs onto and diffuses into the fiber. The driving force for this adsorption process is the difference in chemical potential between the dye in the solution phase and the dye in the fiber phase. In order for a dye to move from the aqueous dye bath to the fiber phase the combination of dye and fiber must be at a lower energy level than dye and water. This may be achieved by the proper selection of dye for the particular fiber type.

The methods of application of dyes in textile dyeing and printing have undergone several modifications to meet the requirements of the new synthetic fibers and their blends with the natural fibers and new classes of dyes. However, the basic operations of dyeing remain the same and include the following:

a. Preparation of the fiber

b. Preparation of the dye bath

c. Application of the dye

d. Finishing

Fiber preparation ordinarily involves scouring to remove foreign material and ensure even access to dye liquor from the dye bath. The textile material generally needs a pretreatment before dyeing. Wool must be washed to remove wax and dirt and sometimes bleached. Cotton must be boiled and bleached to remove pectins and cotton seeds and is mercerized. Sizes and spinning oils must be eliminated [7].

The dyeing of fiber from an aqueous dye bath depends on the dye-fiber interaction. Depending on the nature of dye and the nature of fiber, the dye is fixed onto the fiber chemically or physically. Table 8.1 shows the methods of application for various dye classes and principal substrates [7].

Additives such as wetting agents, salts, carriers, retarders, and others may be added to the dye bath along with the dye if required to facilitate the dyeing process.

Dyes are applied to textile fibers by batch as well as continuous methods. Batch dyeing may be a hand operation by the simple process of moving the fibrous material in an open bath containing the dye liquor. This process is still practiced and has some advantages, but machine dyeing is increasingly used. In machine dyeing (e.g., the Lancashire jigger machine) [10], the yarn or cloth is moved in the dye bath, which is kept stationary except for the agitation of the liquor because of the movement of the yarn or cloth. Batch methods for machine dyeing depend on the physical structure of fibers in the textile material to be dyed. They include jig dyeing, beck dyeing, jet dyeing, package dyeing, skein dyeing, pad-batch, beam dyeing, and others. Each of these batch methods employs a different type of machine.

Continuous dyeing is designed for long runs using a padding machine. The machine consists of a dye beck of relatively small volume provided with guide rollers together with two or more squeezing rollers for removal of excess liquor. More than one padding machine in series may be used, and they may be followed by working units in which the fabric is given the various rinsing, soaping, and other after treatments (finishing steps) that may be necessary. The drying equipment normally works with infrared heat or with a hot air stream or a combination of both. Drying should start contact-free to avoid smearing of the fabric and soiling of the equipment [10].

8.6   Intermediates

Intermediates in the dye industry are referred to the organic cyclic type compounds used for the manufacture of synthetic dyes. These compounds are considered as chemical stepping stones between the parent organic raw materials and the final products. The starting raw materials are mainly aromatic hydrocarbons such as benzene, toluene, naphthalene, and the like, which are derived from petroleum. The products are various dyes.

The manufacture of a dye from primary raw materials involves a number of prior synthetic stages and transformations, commonly referred to as *unit processes*. Such processes include nitration, sulfonation, diazotization, oxidation, reduction, chlorination, and others. The products, precursors of the dyes themselves, are collectively known as *intermediates*. Intermediates are produced by a variety of reactions. Many dye intermediates are manufactured by repeated, and often difficult, chemical reactions to obtain the desired product. Such conversion may be exemplified by the manufacture of a relatively simple intermediate, for example, N,-N-diabenzylaniline disulfonic acid. This conversion requires a number of unit processes, namely the nitration of benzene, the reduction of nitrobenzene, to give aniline, the alkylation of aniline leading to N,N-dibenzylaniline the sulfonation of which gives, finally, the disulfonic acid [11].

Relatively few reactions are available for direct introduction of substituent groups into the aromatic nuclei that are sufficiently general to be applicable to benzene, naphthalene, and higher polynuclear hydrocarbons. Although each of these reactions has specific characteristics, they can be classified into two somewhat differentiated types:

Type 1   } Nitration   Type 2   } Sulfonation
(normal)   } Halogenation   (abnormal) } Friedal-crafts acylation
               (Br_2, Cl_2) Friedal-crafts alkylation

Different substitution reactions are influenced by substituent groups already present in the benzene ring. Rules of substitution are covered in detail in several books on organic chemistry. Substituents are important beyond their contribution to molecular geometry in that they can alter the solubility of the molecule in acid or basic solvents and possess directional effects on ensuing nucleophilic substitution [1]. Some of the more common unit processes employed in the manufacture of dye intermediates are discussed with emphasis on the general character of these processes below:

Sulfonation. Most aromatic compounds can be sulfonated by the action of concentrated or fuming sulfuric acid to form sulfonic acids. The reaction may be represented as:

$$Ar\text{-}H + H_2SO_4 \rightarrow Ar\text{-}SO_3H + H_2O$$

The readiness with which sulfonation takes place, however, varies considerably. Benzene and toluene can be sulfonated in the cold, whereas anthraquinone requires fuming sulfuric acid and high temperatures [10]. Unlike the other electrophilic substitution reactions of benzene, sulfonation is a readily reversible reaction. If water is not removed continuously during the reaction then the hydrolysis of sulfonic acid will lead to the inverse reaction to the starting product containing no sulfonate grouping. The overall process of sulfonation, if water is present in the reaction mixture, is represented by the following equation:

$$Ar\text{-}H + H_2SO_4 \rightleftharpoons ArSO_3H + H_2O$$

Sulfonation of aromatic compounds is very important in the manufacture of dyes. As sulfonic acids are generally water-soluble, the sulfonation of aromatic compounds provides a means to make aromatic compounds water-soluble. Also, the sulfonic acid group is easily displaced by a variety of other groups. Therefore, aryl-sulfonic acids are useful synthetic intermediates.

Benzene is usually sulfonated by means of fuming sulfuric acid ($H_2SO_4 + SO_3$); a temperature of 80 to 100°C is used at the end for completing the reaction.

Toluene reacts more readily than benzene under similar conditions forming a mixture of the o- and p-acids. At low temperatures the o- is the major product, and at high temperatures the p-acids.

Naphthalene sulfonates to give a mixture of 1- and 2- naphthalenesulfonic acids:

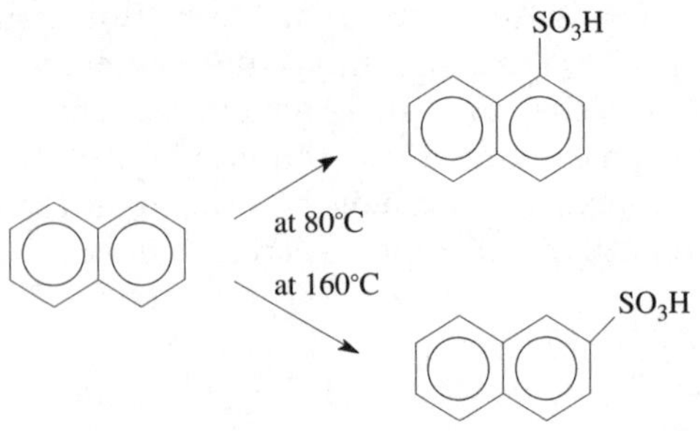

At low temperatures naphthalene-1-sulfonic acid predominates whereas at higher temperatures (up to 160°C) the 2-product is formed in higher yields (85 percent). Further sulfonation at 150 to 160°C gives rise to di-, tri-, and tetra-sulfonic acids. The important naphthalene-1, 3,6-trisulfonic acid can be obtained directly from naphthalene by progressive sulfonation with fuming sulfuric acid first at 40°C, then at 60°C, and finally at 150 to 155°C [11].

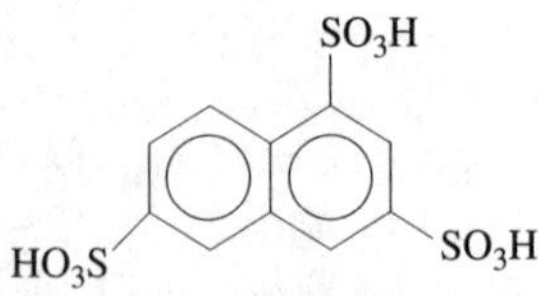

The intermediate derived from naphthalene-1,3,6-trisulfonic acid along with the class of derived dyes are shown in Table 8.3 [11].

2-naphthol undergoes sulfonation to produce two important monosulfonic acids: schaeffer's acid and crocein acid. At 100°C schaeffer's acid predominates; at lower temperatures more crocein acid is formed.

TABLE 8.3   Intermediates Derived from Naphthalene-1,3,6-Trisulfonic Acid

Systematic name	Other name	Class(es) of derived dye(s)
1-Amino-8-naphthol-3,6-disulfonic acid	H acid	Azo, Hydroxyketone
1,8-Dihydroxynaphthalene-3,6-disulfonic acid	Chromotropic acid	Azo
1-Naphthol-3,6,8-tri-sulfonic acid	—	Azo, Hydroxyketone
1-Aminonaphthalene-3,6,8-trisulfonic acid	Koch acid	—

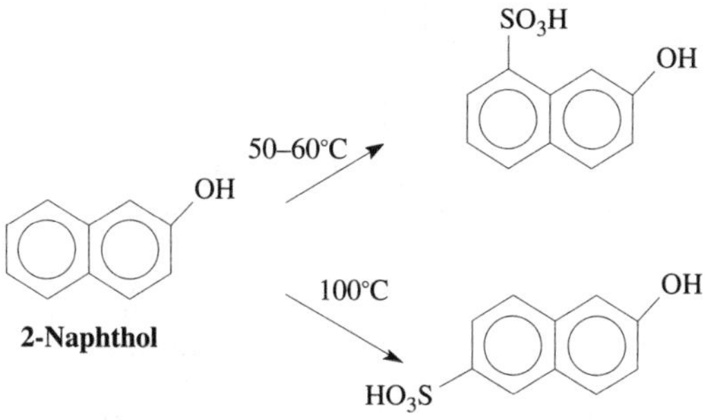

2-Naphthol

Further sulfonation of 2-naphthol leads to the formation of two isomeric disulfonic acids. At low temperatures, G acid is preponderant, whereas at higher temperatures R acid predominates. Both R and G acids are important intermediates for azo dyes.

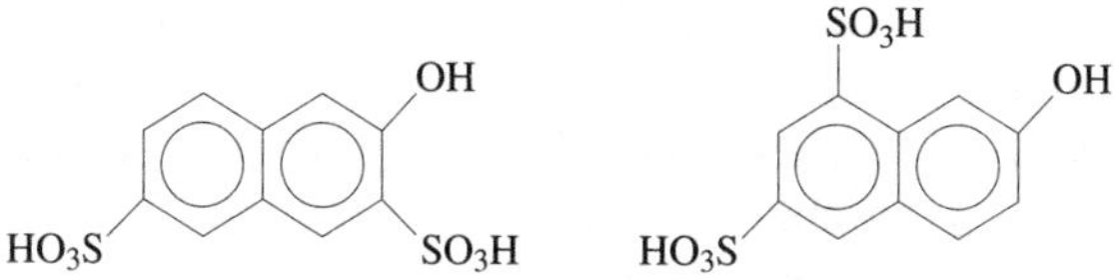

R acid

G acid

Anthraquinone is sulfonated by reacting with fuming sulfuric acid at 120 to 140°C. The product is mostly the 2-sulfonic acid and small quantities of the 2,6 and 2,7-disulfonic acids.

The presence of mercuric salts in the sulfonation mixtures favors the formation of anthraquinone-1-sulfonic acid; further sulfonation leads to a complex mixture of disulfonic acids, chiefly 1,5 and 1,8 with lesser quantities of 1,6 and 1,7-isomers [11].

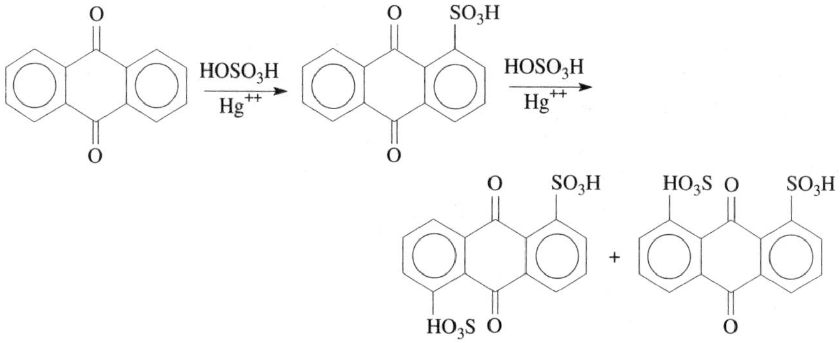

The compounds listed in Table 8.4 are intermediates for the anthraquinone series of dyes [11].

TABLE 8.4   Anthraquinone Intermediates

Systematic name	Other name
1-Aminoanthraquinone	α-Aminoanthraquinone
2-Aminoanthraquinone	β-Aminoanthraquinone
1-Amino-4-bromoanthaquinone-2-sulphonic acid	Bromamine acid
1-Amino-4-bromo-2-methylanthraquinone	—
1-Amino-2,4-dibromoanthraquinone	—
1,2-dihydroxyanthraquinone	Alizarin
1,4-dihydroxyanthraquinone	Quunizarin
1,5-dyhydroxyanthraquinone	Anthrarufin
1,8-dihydroxyanthraquinone	Chrysazin

Halogenation. The benzene nucleus is readily substituted by chlorine or bromine atoms on interaction with the halogens in the presence of ferric chloride or bromide, aluminum chloride, or iodine. The replacement of nuclear hydrogen by chlorine or bromine in the benzene series is mostly done directly using gaseous chlorine or liquid bromine. Chlorination may also be achieved by reacting with thionyl chloride, phosphorus oxychlorides, phosphorus pentachloride, or sulfuryl chloride. For example, the dye disperse violet 28 is prepared by the chlorination of lenco-1,4-diaminoanthraquinone with sulfyryl chloride by the following reaction [5]:

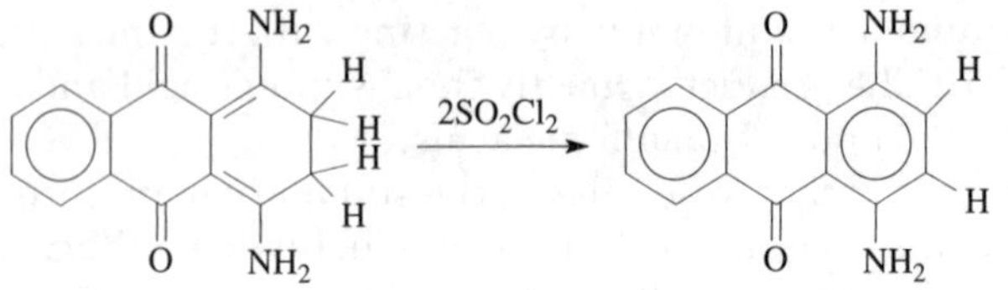

Disperse violet 28

Toluene undergoes catalyzed chlorination or bromination in the same manner as benzene, with equal quantities of o- and p-chlorotoluenes being produced. When, however, the catalyst is omitted and the hydrocarbon is treated with either halogen at the reflux temperature, preferably with exposure to light, the chlorine or bromine atoms enter the methyl side chain rather than the nucleus, with formation in succession of the mono-, di-, and trihalo derivatives.

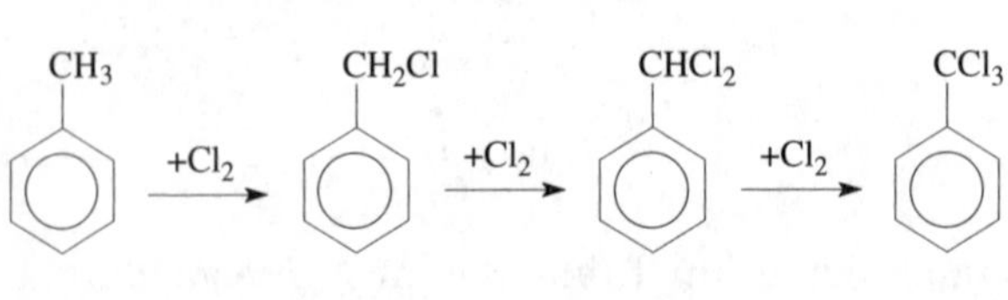

Toluene   Benzyl chloride   Benzal chloride   Benzotrichloride

Benzyl chloride is used in the manufacture of ethylbenzylaniline and dibenzylaniline for triarylmethane dyes.

The direct chlorination of naphthalenes results in the formation of numerous isomers. This reaction is seldom used.

In the anthraquinone series an indirect chlorination method is used. The advantage is taken from the fact that the —SO_3H group may be replaced by —Cl through the action of hot concentrated hydrochloric acid and 10 to 15 percent aqueous sodium chlorate. Such a case is the preparation of 1,5-dichloroanthraquinone [11].

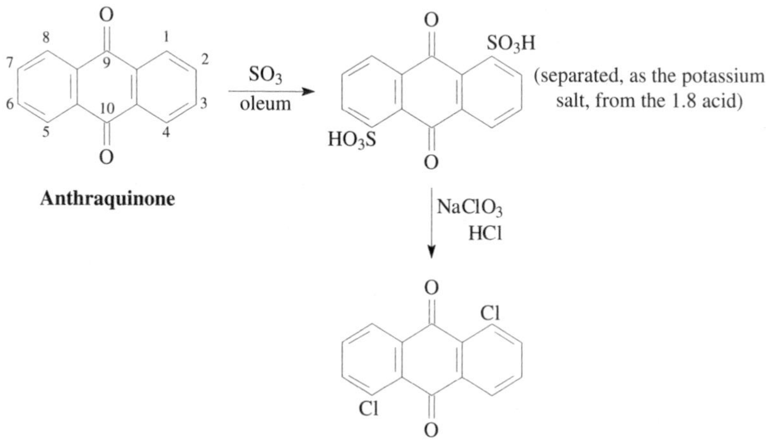

Anthraquinone

(separated, as the potassium salt, from the 1.8 acid)

1,5-dichloroanthraquinone

Nitration. Nitration conducted with nitric acid alone or in combination with sulfuric acid, provides an efficient method for preparation of mono-, di-, and trinitro derivatives of many aromatic hydrocarbons and of hydroxy, halo, and other substitution products. The polynitro compounds accessible by the direct route have the m-orientation, and o- and p- dinitro compounds, though obtainable by indirect methods, are rare. Products of nitration are useful dye intermediates.

Nitrobenzene can be produced on a technical scale in yields up to 98 percent by nitration of benzene with mixed acid. The sulfuric acid serves as a solvent and generates the nitronium ion, which is the attacking electrophile [5]. Nitrobenzene itself may be further nitrated with a mixed acid giving m-dinitrobenzene. The o- and p-dinitrobenzenes are removed as water-soluble products by treating the nitration product with aqueous sodium sulfite, sodium o- and p-nitrobenzenesulfonates being formed. The residual m-nitrobenzene is separated from the aqueous layer, washed with water and finally dried [11].

The nitration of toluene is carried out as in the case of benzene, but the temperature is maintained below 20°C to avoid oxidation of the methyl group. A mixture of o-, p-, and m-nitrotoluene is obtained in the approximate percentage of 63, 35, and 2. On cooling to −10°, part of the p-nitrotoluene crystallizes and is separated in a centrifuge. The residual oil is separated by a combination of fractional distillation and a process of crystallization known as *sweating*.

When benzene derivatives (such as phenols and chlorobenzene) are nitrated, isomers of the desired product are obtained. For example, nitration of phenol by nitric acid gives o- and p-nitrophenol:

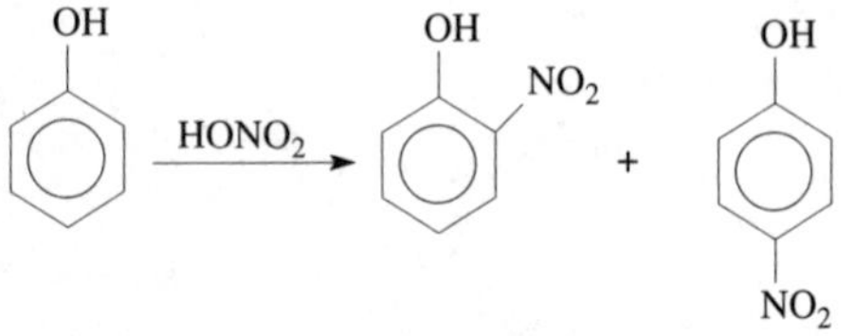

The nitration of chlorobenzene, dichlorobenzene, and chlorotoluene leads to substances of considerable value in dye chemistry in that nitro groups in position 2- and 4- relative to chlorine having an activating effect [11].

Mononitration of naphthalene with mixed acid at a temperature of 35 to 50°C gives a good yield of 1-nitronaphthalene; this is the main source of 1-naphthylamine and its derivatives.

Anthraquinone on nitration at about 50°C gives mainly 1-nitro-anthraquinone. At 80 to 95°C, dinitration occurs to give a mixture of 1,5- and 1,8- isomers. These isomers are important as starting materials for the preparation of other intermediates.

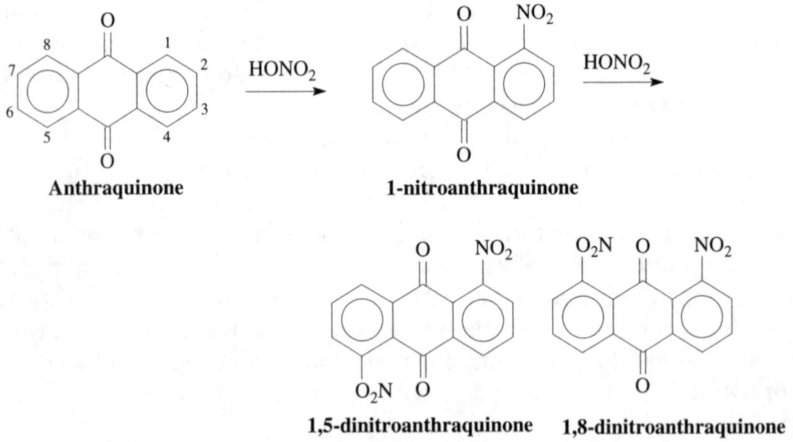

Anthraquinone   1-nitroanthraquinone

1,5-dinitroanthraquinone   1,8-dinitroanthraquinone

Amination. The process of making an amine (RNH_2) is generally referred to as amination. The methods commonly used are (a) reduction of a nitro compound and (b) action of ammonia on a chloro-, hydroxy-, or sulfonic acid compound.

The main method, both in the laboratory and in technical practice, for the introduction of an amino group into an aromatic compound, is nitration and reduction. Reduction of nitro compounds is accomplished by: (1) catalytic hydrogenation, (2) iron reduction (Béchamp method), (3) sulfide reduction, or (4) zinc reduction in an alkaline medium.

The most widely used method of converting nitro compounds into an amine is the Béchamp method, which uses iron turnings in water containing a small quantity of hydrochloric acid. The reaction may be represented by the Eq. [5].

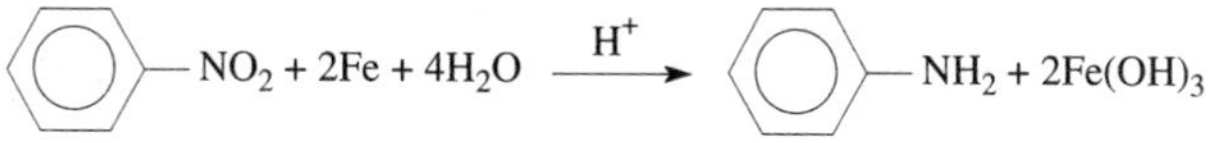

$$\text{C}_6\text{H}_5-NO_2 + 2Fe + 4H_2O \xrightarrow{\ H^+\ } \text{C}_6\text{H}_5-NH_2 + 2Fe(OH)_3$$

The Béchamp method can also be used in reducing the nitro derivatives of sulfonic acids. A typical example being the manufacture of metanilic acid and related substances [11].

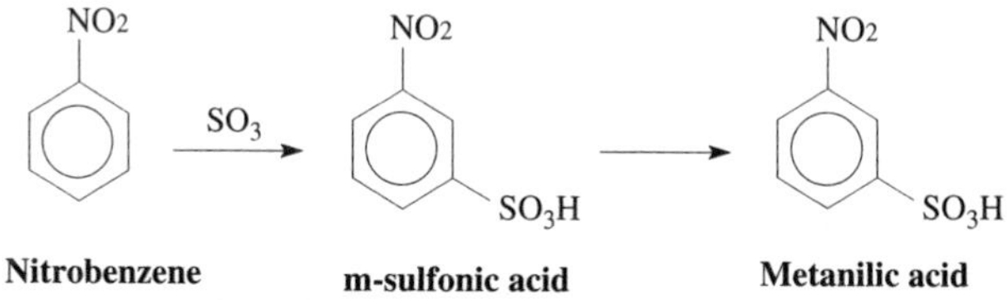

Nitrobenzene   m-sulfonic acid   Metanilic acid

An example of direct action of ammonia on substitute aromatic compounds is the conversion of p- nitrochlorobenzene to p-nitroaniline. This reaction may be carried out continuously with 40 percent aqueous ammonia under 200 atmosphere at 235 to 240°C [5].

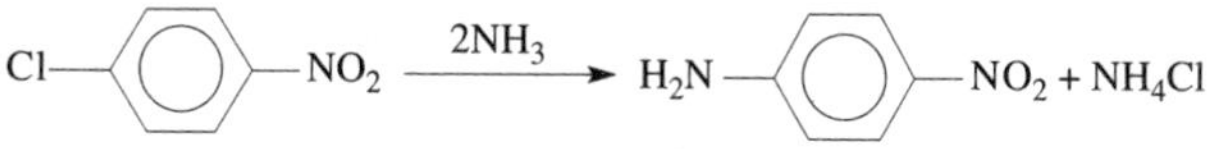

$$Cl-\text{C}_6\text{H}_4-NO_2 \xrightarrow{\ 2NH_3\ } H_2N-\text{C}_6\text{H}_4-NO_2 + NH_4Cl$$

Aniline can also be prepared by the amination of phenols under a pressure of 15 to 16 atm at a temperature of 380 to 385°C in the presence of a silica- alumina catalyst (Halcon process).

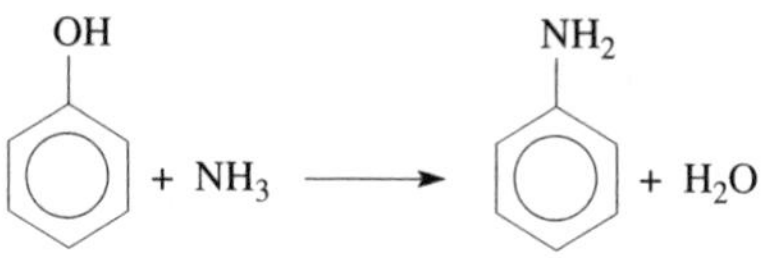

$$\text{C}_6\text{H}_5OH + NH_3 \longrightarrow \text{C}_6\text{H}_5NH_2 + H_2O$$

The Bucherer reversible reaction of conversion of naphthols to anilines by aminolysis is of outstanding importance in naphthalene chemistry. The reaction is used for the replacement of aromatic amino groups by hydroxy groups, in the naphthalene series. The reaction is valuable in the synthesis of naphthalene dye intermediates and may be represented as follows:

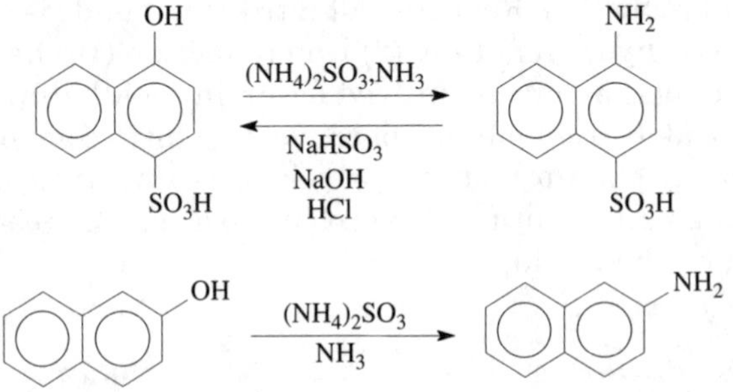

Two important azo dye intermediates, Amino J acid and Amino G acid, are also produced by the Bucherer reaction represented as follows:

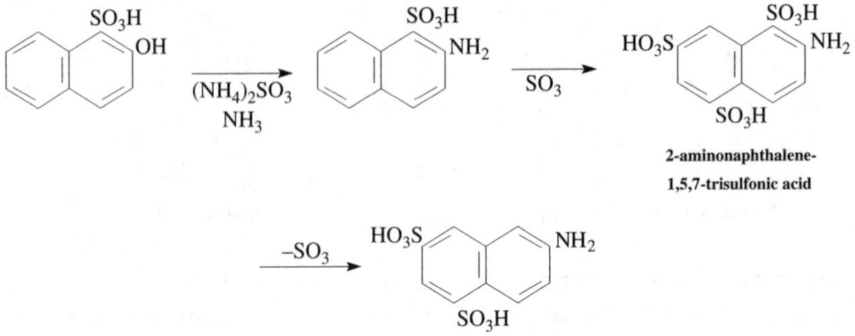

2-aminonaphthalene-
1,5,7-trisulfonic acid

Amino J acid

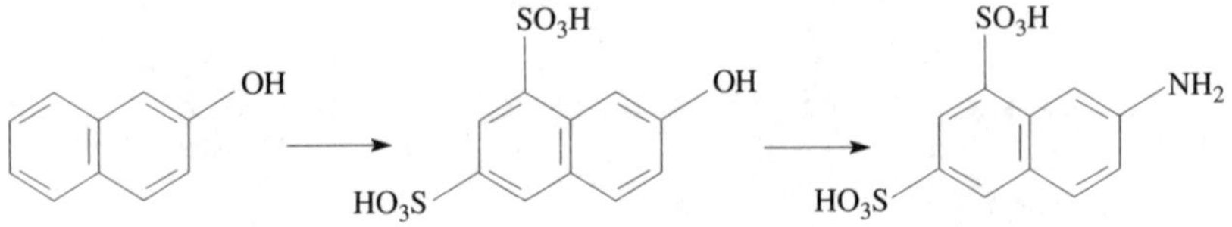

2-naphthol-6,8-disulfonic acid   Amino G acid

Hydroxylation.   The main general method for the hydroxylation of aromatic compounds is the alkali fusion of sulfonic acids:

$$Ar—SO_3Na + 2NaOH \longrightarrow Ar—ONa + Na_2SO_3 + H_2O$$

A second general reaction is hydrolysis of chloro compounds:

$$Ar-Cl + 2NaOH \longrightarrow Ar-ONa + NaCl$$

The phenates thus obtained can be converted to phenol by reaction with an acid.

In a similar way resorcinol is made by fusion of m-benzene disulfonic acid:

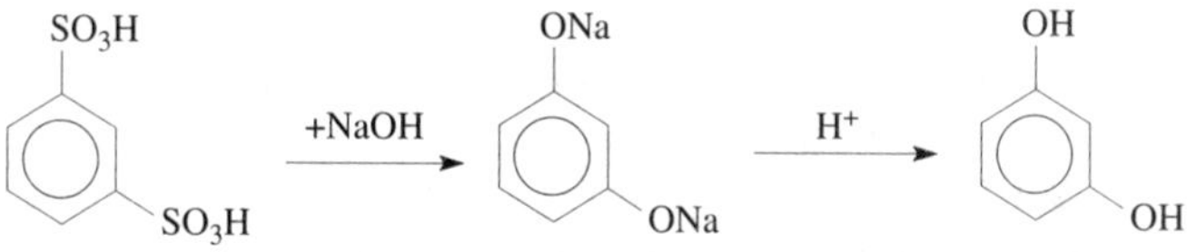

Resorcinol

2-naphthalene sulfonic acid can be converted to 2-naphthol by the hydrolysis with sodium hydroxide:

2-anthraquinosulfonic acid can be converted to the hydroxy compound by heating with calcium hydroxide in water. Alkaline fusion of the same sulfonic acid gives alizarin [5].

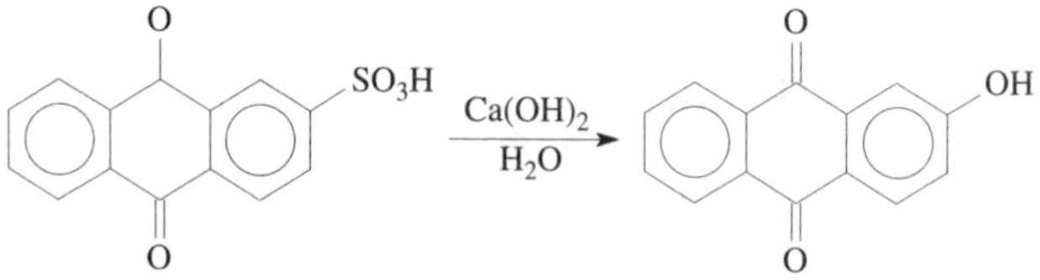

2-hydroxyanthraquinone

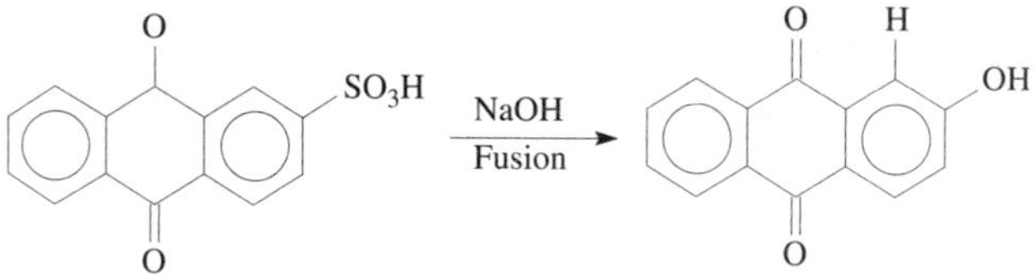

Alizarin

Oxidation.   Controlled oxidation of aromatic compounds is an important process. The most common reagent is air or oxygen gas in the presence of a catalyst. One of the classical examples is the catalytic vapor phase oxidation of naphthalene to phthalic anhydride. This reaction is carried out over a vanadium pentoxide catalyst at 450°C.

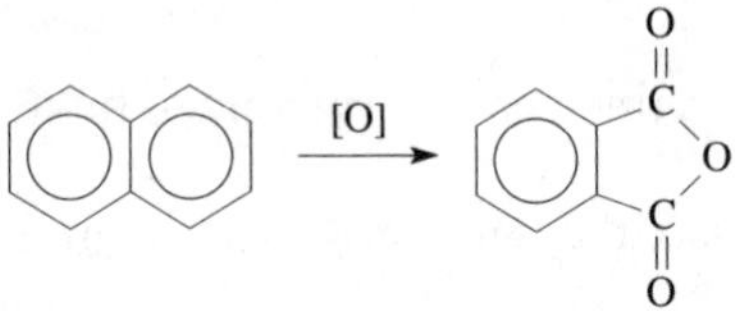

Phthalic anhydride is used in very large quantities in the plastic industry and plays a significant role in dye chemistry.

Currently, phthalic acid is mainly produced by the catalyzed oxidation of o-xylene obtained by a cracking process from petroleum.

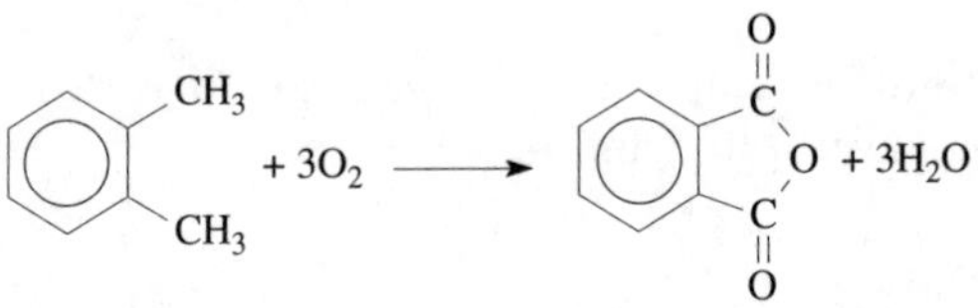

The oxidation of the methyl group in toluene derivatives to the aldehyde is an important stage in triarylmethane dye manufacture. p- toluene sulfonyl chloride is first hydrolyzed in concentrated sulfuric acid and then subjected to the oxidizing action of a manganese compound to produce benzaldehyde-4-sulfonic acid [11].

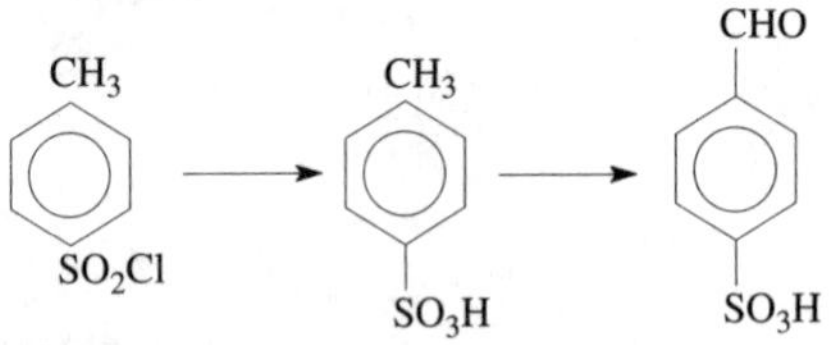

Similarly, the oxidation of the methyl group in an anthraquinone derivative leads to the formation of the corresponding carboxylic acid [5].

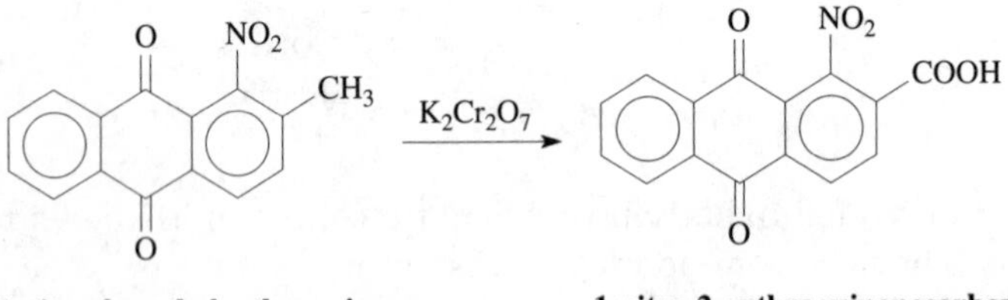

1-nitro-2-methylanthraquinone   1-nitro-2-anthraquinonecarboxylic acid

8.6.1 Miscellaneous reactions

Alkylation. Akyylation on nitrogen and alkylation on oxygen are of some importance in dye chemistry. Large quantities of mono- and di-alkylamines are manufactured annually for use in the organic chemical industry. The usual alkylating agents are methanol, ethanol, methyl and ethyl chlorides, and dimethyl and diethyl sulfates. N,N-Dimethylaniline is probably the most important of the alkylanilines and is made by heating a mixture of aniline, methanol, and concentrated sulfuric acid at 230 to 235°C and a pressure of 25 to 30 atm.

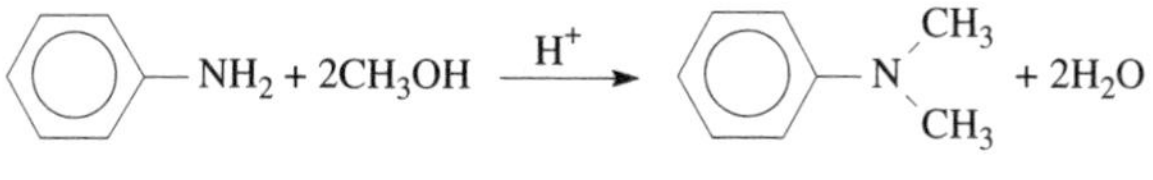

$$\text{Ph}-NH_2 + 2CH_3OH \xrightarrow{H^+} \text{Ph}-N(CH_3)_2 + 2H_2O$$

Phenol may be methylated with methyl sulfate in cool alkaline medium to give anisole:

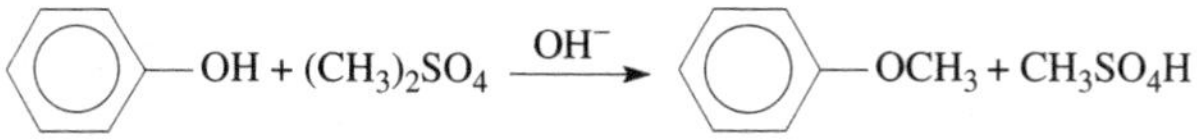

$$\text{Ph}-OH + (CH_3)_2SO_4 \xrightarrow{OH^-} \text{Ph}-OCH_3 + CH_3SO_4H$$

Arylation. The introduction of arylamino groups is often needed in the synthesis of intermediates and dyes. An important intermediate, 3-ethoxy-4-methyldiphenylamine is prepared from resorcinol and p-methylaniline as follows:

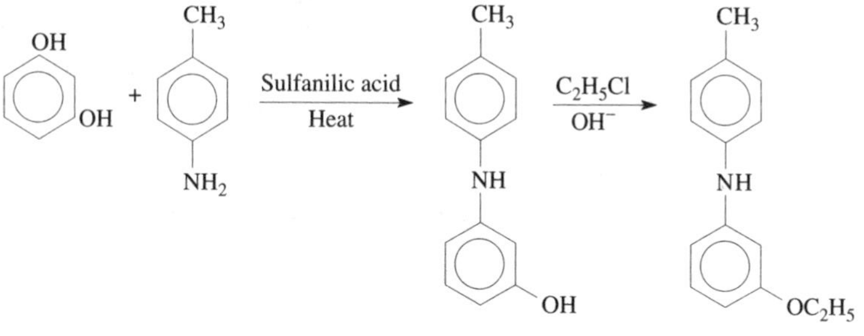

Condensation and addition. In both condensation and addition reactions, two or more molecules combine by the elimination of a simple molecule (condensation), or the reaction is stopped after the molecules are joined (addition). There are only a few intermediates manufactured in any considerable quantity by these reactions. An example of a condensation reaction is the formation of the diphenylamine derivative, commonly called nitro delta acid [5].

$$O_2N\!-\!\langle\bigcirc\rangle\!-\!Cl + H_2N\langle\bigcirc\rangle\!-\!NH_2 \longrightarrow O_2N\langle\bigcirc\rangle\!-\!NH\!-\!\langle\bigcirc\rangle\!-\!NH_2$$

Nitro delta acid

The formation of cyanuric acid from cyanogen chloride is a good example of an addition reaction.

$$3ClCN \longrightarrow$$

Cyanogen chloride   Cyanuric chloride

8.7 Manufacture of Dyes

The plant required for the manufacture of dyes usually depends on unit processes. The unit processes such as nitration, sulfonation, reduction, and diazotization are treated extensively in the textbooks of organic chemistry and will not be included here. The general operation sequences in dye and intermediate manufacture are shown in Fig. 8.1.

The raw materials comprising the reactants and the reaction medium (water or solvent) are delivered to the reactor. The bomb-shaped reactors, made from cast-iron, stainless steel, or steel lined with rubber, glass (enamel), brick, or carbon blocks, have generally replaced older wooden vats. These reactors have capacities of 2 to 40 m^3 (Ca 500 to 10,000 gal) and are equipped with mechanical agitators, thermometers, heaters, condensers, pH-meters, and so on, depending on the nature of the operation. Crystallization of the final product is done under careful control to obtain the most desirable physical form. Solid products are separated

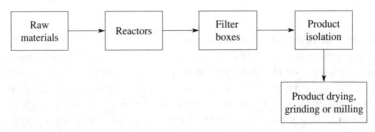

Figure 8.1 Flow sheet for the manufacture of dyes.

from liquids using an efficient filtration system such as centrifuges, filter boxes, belt filters, or filter presses, with a view to both speed and maximum liquid removal.

Drying of the solid product is carried out using ovens, agitated vacuum dryers, spray dryers, or any other types of dryers.

The final stage in dye manufacture is grinding or milling. Dry grinding is usually carried out in impact mills equipped to control dust. More modern methods such as continuous drying allow the production of materials that do not require a final comminution stage [7].

8.8 Environmental and Health Aspects

The protection of environment, technological, and ecological considerations are of increasing importance for manufacturers and consumers of dyes. Primary considerations in dye manufacture and use include personal safety, atmospheric emissions, wastewater quality, and appropriate waste disposal. These considerations are of major importance in choosing major plant sites and are sometimes economically restrictive to existing operations [7, 12].

The toxic nature of some dyes and intermediates has long been recognized. Dermatologists have reported cases of skin reactions suspected to be caused by textile dyes. The positive link between benzidine derivatives and 2-naphthylamine with bladder cancer is also well documented. In fact, these studies prompted the introduction of stringent government regulations to control the exposure of workers to dye dust and the discharge of chemicals, colors, and other effluents.

The Ecological and Toxicological Association of the Dyestuffs Manufacturing Industry (ETAD) have published brochures on safe handling of dyes as well as product stewardship of dyes. ETAD cooperates with government departments and agencies and other public institutions in dealing with ecological and toxicological requirements concerning dyes. Currently, the three principal regulatory agencies worldwide are European Core Inventory (ECOIN) and European Inventory of Existing Commercial Substances (EINECS) in Europe, Toxic Substances Central Act (TOSCA) in the United States, and Ministry of Technology and Industry (MITI) in Japan. Each of these requires the preparation of an inventory and the notification of new chemicals. An ETAD online database on publication of toxicological and ecological effects concerning dyes (and pigments), as well as on appropriate legislation, is available for ETAD member companies [13].

The discharge of intensely colored dyes in wastewater is aesthetically displeasing and of more concern owing to the presence of toxic heavy metals and other carcinogens along with it. Wastewaters from both dye manufacturing plants and dye houses are treated before leaving the plant. Various treatment methods such as neutralization of acidic and alkaline effluents, removal of heavy metals, aerobic and anaerobic biological treatments, and other newer methods such as chemical oxidation are used [7]. Color is removed by adsorption on activated carbon, or by chemical treatment of the effluent with a flocculating agent such as ferric (Fe^{3+}) or aluminum (Al^{3+}) ions.

References

1. Austin, G. T., (Ed.), *Shreve's Chemical Process Industries*, 5th ed., McGraw-Hill, New York, N.Y., 1975.
2. Kirk, R. E. and Othmer, D. F., *Encyclopedia of Chemical Technology*, 4th ed., Wiley-Interscience, New York, N.Y., 1999.
3. Will, R., Ishakwa Y., and Leder, A., *Synthetic Dyes, Abstract of SCUP report*, SRI International, December 2000.
4. Market Report, *World Dyes and Organic Pigments, Fredonia Group*. Available at www.mindbranch.com, June 2000.
5. Kent, J. A. (Ed.), *Riegel's Handbook of Industrial Chemistry*, 9th ed., Chapman & Hall, New York, N.Y., 1992.
6. *Color Index*, 3rd ed., 5 vols., Society of Dyers and Colorist, Bradford, England, and American Association of Textile Chemists & Colorists, Lowell, MA, 1971.
7. Hunger, K. (Ed.), *Industrial Dyes, Chemistry, Properties, Applications*, Wiley-VCH, Darmstadt, Germany, 2003.
8. Carter, M. E., *Essential Fiber Chemistry*, Marcel Dekker Inc., New York, N.Y., 1971.
9. Zollinger, H., *Color Chemistry*, VCH, Weinheim, Germany, 1991.
10. Venkataraman, K., *The Chemistry of Synthetic Dyes*, Vol. 1, Academic Press Inc., New York, N.Y., 1952.
11. Abrahart, E. N., *Dyes and Their Intermediates*, Pergamon Press Ltd., London, 1977.
12. *Ullmann's Encyclopedia of Industrial Chemistry*, 6th ed., Vol. 11, Wiley-VCH, Darmstadt, Germany, 2003.
13. ETAD, *Ecological and Toxicological Association of Dyes and Organic Pigment Manufacturers*, Bassel. Available at www.etad.com., August 2004.

Industrial Fermentation

Manfred J. Mirbach and Bassam El Ali

9.1   Introduction and History

Fermentation can be defined as the alteration or production of products with the help of microorganisms. Fermentation has been used to conserve and alter food and feed since ancient times. Actually, it was the method of choice to convert fresh agricultural products into durable food items for many thousand years. In everyday life, we also know the reverse process, namely the uncontrolled decay of food or organic matter in general. Under controlled conditions fermentation is a useful process. Yogurt, salami, sauerkraut, soy sauce, vinegar, and kefir are just a few examples of fermented food products that we still know of today.

Fermentation can be spontaneous or be induced by specifically added microorganisms. An everyday example of such an induced fermentation is the addition of baking yeast to flour to make bread or cakes. As with bread, fermentation can be done in a normal environment where many different microorganisms are present. A more sophisticated way is to exclude unwanted microorganisms by sterilization of the materials before adding a starter culture.

Since around 1800, the mechanism of fermentation has been studied in a scientific way. It started when German scientist Erxleben discovered that yeast induces fermentation. Louis Pasteur, a French scientist, made many contributions to microbiology. He explained that bacteria produce lactic acid, which then conserves the food. Pasteur also noticed that unwanted fermentation can be stopped by heat treatment of the substrate (*pasteurization*). This technique is still widely applied today to treat milk or fruit juices. Actually, the production of neat lactic acid was also the first nonfood industrial application of fermentation.

The first aseptic fermentation (exclusion of unwanted microorganisms) on an industrial scale was the production of acetone, butanol, and butandiol for rubber production. After World War I the production shifted to organic acids, when acetone and butanol became available from other sources.

An important milestone was the introduction of biological wastewater treatment by fermentation. Traditionally, wastewater containing human or animal excrement was sprayed on the fields as fertilizer or simply discharged into rivers and lakes. This caused microbial pollution and was the cause of many infectious diseases, like typhus and cholera. During the 19th century, modern industrialization started and many people migrated from the agricultural area to the big cities. Public hygiene became a major task. Therefore, it was a big step forward when public sewage systems and biological wastewater treatment plants were introduced. Life in the big cities would be unbearable without wastewater treatment, which is perhaps the most widely used fermentation process, even today.

Another breakthrough in fermentation and human welfare was the discovery of penicillin. It was the first antibiotic and the first really effective medication against bacterial infections. It was also the first high-cost product of fermentation and it started the development of high-tech fermentation reactors.

Amino acid production by fermentation started around 1960 in Japan. Initially glutamic acid was the main product. It was sold as sodium salt, monosodium glutamate (MSG), a flavor enhancer on oriental cuisine. Other amino acids soon followed. They are used in food and feed to increase the efficiency of low protein substrates. Microbiologically produced enzymes were introduced around 1970. They are used in grain processing, sugar production, fruit juice clarification, and as detergent additives (Table 9.1).

Since around 1980 the development of genetic engineering made it possible to tailor microorganisms to perform specific tasks. Today it is quite common to alter the DNA of bacteria and to introduce selective genes from other species. This allows the production of products with high selectivity and rates that were previously not believed possible. Insulin was the first commercial product using genetically engineered bacteria for fermentation.

Today, many different fermentation processes are applied in industry. They range from large-scale low-tech processes, like wastewater treatment to very sophisticated biotechnology processes to produce

TABLE 9.1   History of Fermentation

Time	Event
Since >5000 years	Spontaneous fermentation to produce bread, vinegar, soy sauce
Since 2500 years	Fermentation of sugar containing crops to produce wine and beer
Since 1500 years	Commercial use of fermentation processes in Asia (except beer and wine)
Since 500 years	Commercial use of fermentation in Europe (except beer and wine)
1818	Discovery of yeast as the origin of fermentation (Erxleben)
1857	Scientific explanation of lactic acid formation (Pasteur)
1881	First production of neat lactic acid by industrial fermentation
Since around 1900	Public wastewater treatment plants
1910	Industrial production of butanol and acetone by aseptic fermentation
1925	Industrial production of organic acids
1928/1929	Discovery of penicillin; commercial production since 1941
Since 1960	Industrial production of amino acids
Since 1970	Industrial production of enzymes
Since 1980	Introduction of genetically modified microorganisms, production of insulin

TABLE 9.2   Overview of Industrial Fermentation Products

Category	Examples	Uses or Remarks
Food	Sour dough, soy sauce, yogurt, kefir, cheese, pickles, salami, anchovy, sauerkraut, vinegar, beer, wine, cocoa, coffee, tea	Conservation of perishable food by the formation of lactic acid and ethanol
Feed	Silage	Conservation of green plants by organic acids
Cell mass	Yeast, lactic acid bacteria, single cell protein	Used as starter cultures, animal feed
Organic solvents	Ethanol, glycerol, acetone, butanediol	Cosmetics, pharmaceuticals
Organic acids	Lactic, citric, acetic, acrylic, formic acid	Food, textiles, chemical intermediates
Amino acids	L-lysine, L-tryptophan, L-phenylalanine, glutamic acid	Food and feed additives
Antibiotics	Penicillin, streptomycin, tetramycin, tetracycline	Human and veterinary medicines
Vitamins	B12, biotin, riboflavin	Food and feed supplements
Enzymes	Amylase, cellulase, protease, lipase, lab	Food processing, tanning, detergents additives
Biopolymers	Lanthan, dextran, polyhydroxybutyrate	Food additives, medical devices, packaging
Speciality pharmaceuticals	Insulin, interferon, erythropoietin (EPO)	Human medicines
Environmental	Waste and wastewater treatment	Public hygiene
Energy	Ethanol from carbohydrates and methane from organic waste	Fuel additives or heat generation

expensive pharmaceuticals with genetically modified microorganisms. Examples are listed in Table 9.2.

9.2   Biochemical and Processing Aspects

9.2.1   Overview

Nearly all fermentation processes follow the same principle. The central unit is the fermenter in which the microorganisms grow and where they produce the desired products. The substrate is the feed of the microorganisms; it also contains any other starting materials that are required for the process. The fermentation is started by adding the seed microorganisms, which are present in the starter culture. The starter culture is also called *inoculum*. The starter microorganisms are produced in small inoculum fermenters before being added to the main large-scale

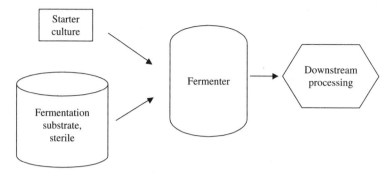

Figure 9.1   Schematic flow chart of a fermentation process.

production fermenters. At the end of the fermentation process a complex broth is obtained containing bacteria, products, unconverted substrate, side products, water, and so on. It needs further work-up steps for separation and purification before the product is pure enough to be marketed. The name *downstream processing* is used for all the steps that follow the actual biochemical reaction.

The four parts of a fermentation process are discussed in more detail below. (Fig. 9.1)

9.2.2   Microorganisms

Microorganisms used in fermentation are usually single cells or cell aggregates—often bacteria, sometimes fungi, algae, or cells of plant or animal origin. A bacterial cell comprises an outer cell wall lined with a cell membrane that keeps the cell content from leaking out, but allows the transport of nutrients in, and of metabolites out. The cell liquid contains everything that the cell needs to live and to proliferate, for instance proteins, enzymes, and vitamins. The DNA is the carrier of most of the genetic information. Plasmids are DNA units that are independent of the chromosomal DNA. They are important for the transfers of genetic information into other cells. Chemically, a cell mainly comprises water, protein, and a large number of minor compounds. Breaking of the cell wall (lyses) kills the organism and releases the content of the cell into the surrounding medium.

The energy to keep the cell alive comes from absorption of light or from oxidation of organic or inorganic compounds. If the oxidizing agent is oxygen, the microorganisms are called aerobic. Anaerobic bacteria survive in an oxygen free environment, because they use chemically bound oxygen from nitrate, sulphate, or carbon dioxide. The biomass of the cell mainly comprises the elements carbon, hydrogen, oxygen, sulfur, and nitrogen. Therefore, these substrates must be added to enable the cells to grow and multiply. Organic substrates are usually the source of carbon, but it can also be carbon dioxide for phototropic species. Phototropic microorganisms use the energy of the (sun) light to convert carbon dioxide to organic matter. Examples are green algae and bacteria. The hydrogen comes from the organic substrates or from water and sometimes from other inorganic hydrogen compounds. Sulfur and nitrogen come from organic sources or from inorganic ions, such as sulfate, sulfide, nitrate, or ammonium. In addition, a number of minor elements (minerals) are required to support growth.

Many fermentation processes use sugars as the substrate. The principle of the microbial metabolization of glucose is described in Fig. 9.2. The first step is the cleavage of the glucose (glucolysis); it is in reality a multistep reaction, which results in the formation of glyceraldehyde-3-phosphate. A series of complex enzyme-induced reactions leads to pyruvate. Depending on the predominating enzymes, pyruvate reacts to L-lactic acid (with lactic dehydrogenase) or acetaldehyde and ethanol (with pyruvic decarboxylase and alcohol dehydrogenase).

Primary metabolites.   During cell growth the nutrients of the substrate are converted to cell mass. The chemical compounds produced in this process are called *primary metabolites*. The cell mass itself mainly comprises proteins, but a number of primary waste products are also formed, for instance carbon dioxide, lactic acid, ethanol, and so on. Primary metabolites are produced in parallel with the cell mass. There are exceptions, however, in which the metabolites are still formed after the cell growth has ceased. The most important example of such an exception is the production of citric acid.

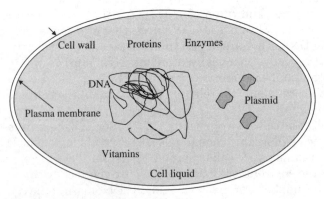

Figure 9.2   Scheme of a bacterial cell.

Secondary metabolites. The formation of secondary metabolites is not directly related to cell growth; rather they are formed because of some other, often unknown, reason. They are the side products of bacterial life. In nature, they are produced in low concentration, but through laboratory mutation and selection, cells can be optimized to overproduce these metabolites. Many antibiotics and vitamins are secondary metabolites. The formation of secondary metabolites is not directly proportional to primary metabolism and cell growth. Therefore, optimum medium composition and process conditions to maximize the product yield may be different from those which are optimal for cell growth.

Primary metabolites are often released into the surrounding medium, whereas secondary metabolites tend to remain inside the cell and can be recovered only after lyses of the cell walls. Some metabolites are toxic; therefore any fermentation must be monitored for toxins. Two types are distinguished: exotoxins are released into the fermentation broth, endotoxins remain inside the cell and are sometimes difficult to detect (Fig. 9.3).

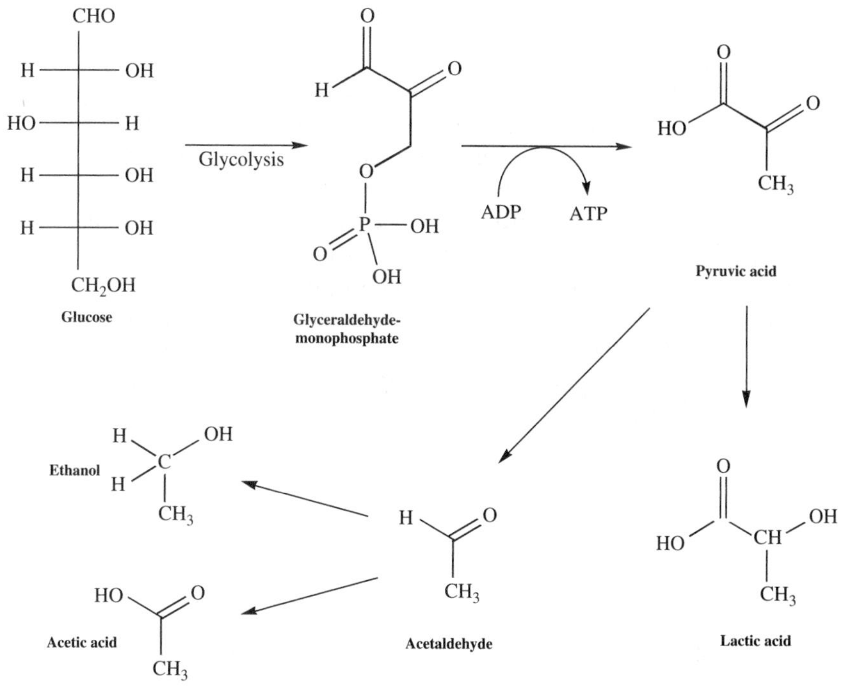

Figure 9.3 Microbial metabolism (primary) of glucose to lactate or ethanol. ADP = adenosine diphosphate; ATP = adenosine triphosphate.

9.2.3   Culture development

Naturally occurring mixed populations of microorganisms (wild type) do not give a satisfactory yield of the target product. Improvements are necessary to make a fermentation process economically feasible. The first step is the selection of the best culture with respect to selectivity and growth characteristics, such as pH, mechanical stress, and temperature sensitivity. This selection is a tedious process based on trial and error screening of a large number of strains. Mass screening techniques have been developed for this purpose, for example, agar plates that are doped with specific inhibitors or indicators. The primary screening results in several potentially useful isolates, which go into secondary screening. Here, false positives are eliminated and the best strains are selected by using a small-scale fermentation technique with shake flasks.

Although primary and secondary screening yields, hopefully, the best candidate, the best natural (wild type) strain is still not good enough for industrial production. Further development is necessary to improve the technical properties of the culture, its stability, and yield.

The *genetic improvement* technique induces deliberate mutations in the DNA of the cells. Such mutations can be induced chemically, by ultraviolet light, or by ionizing radiation. This change is random, that means positive or negative with respect to the intended purpose. Therefore, a new selection process is needed to find the improved strains. The mutated cells are again screened; the best candidates are selected, again mutated, screened and so on, until a satisfactory strain is obtained. Chemical substances induce mutations by reaction with amino acids of the DNA chain. Nitrous acid (HNO_2), for example, reacts with guanine under deamination leading to xanthene (Eq. 9.1). Methylation of the amino groups is also possible, for example, with N-methyl-N'-nitro-N-nitrosoguanidine, a strong mutagen, but without lethal effects. A third type of mutation is the insertion of alien molecules between two amino acids, thereby altering the macroscopic structure of the DNA.

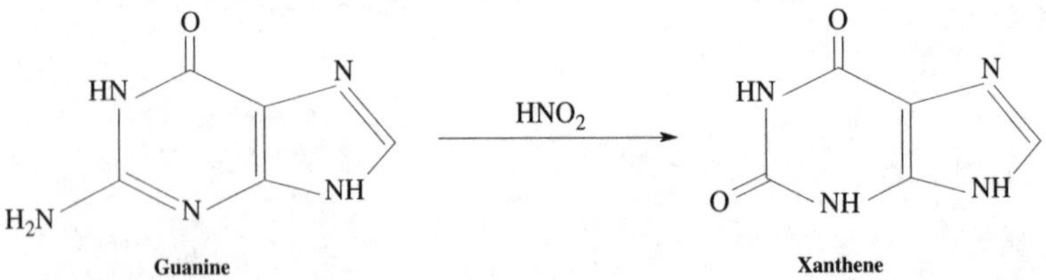

$$ (9.1) $$

DNA absorbs UV light with a wavelength of <260 nm, leading to photochemical reactions, for instance, the dimerisation of pyrimidine (Eq. 9.2).

Ionizing radiation (x-rays, electron beams, gamma radiation, and so on) is less selective. It leads to a random cleavage of the DNA chains.

$$ \text{(9.2)} $$

The most advanced method to improve the microorganisms is by changing the cells in a controlled way through genetic engineering. The exchange of genetic information is normally limited to cells of the same type and species. Membranes and other mechanisms inhibit the transfer of genes or DNA between different cell types. Genetic engineering has changed this. Today, it is possible to transfer genetic properties between completely different species, for instance from plants to bacteria or from bacteria to plants. The principle is not difficult to understand (Fig. 9.4). The DNA of a cell is cut into fragments with specific enzymes. The DNA fragment that is responsible for the production of the target product (e.g., insulin) is selected and transferred into the plasmid (DNA)

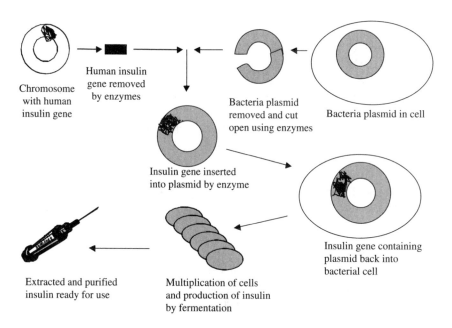

Figure 9.4   Principle of genetic engineering; example: insulin from transgenic bacteria. (*Source*: www.chadevans.co.uk.)

of a microorganism (e.g., *Escherichia coli* = *E. coli*). The genetically engineered cells can be cultivated in fermenters like a *normal* cell. The only difference is that the genetically modified microorganisms (GMO) produce a substance (e.g., insulin) that it would never generate without modification.

Genetic engineering of microorganisms is a key step in modern biotechnology. It makes it possible that bacteria produce valuable substances which are difficult to obtain otherwise. Nevertheless, the production technology behind these advancements in science is nearly always fermentation, an old technology with very modern applications.

Elicitors are microorganisms or chemicals that help the bacteria to produce the target product. For example, the production of the important pharmaceuticals morphine and codeine by *Papaver somniferum* was increased 18-fold by the addition of *Verticillium dahliae.*

Once the right culture is obtained, it must be stored under conditions that retain their genetic stability and viability. One method is to keep the microorganisms on an agar plate in an incubator. Agar is a substrate containing all the nutrients necessary to microorganisms. It is usually sold in ready-to-use shallow round dishes. A cell culture can stay alive on agar plate for some time until the nutrients are used up. Part of it is then transferred to a new plate, and so on. Maintenance of microorganisms on agar plates requires continuous attention by skilled personnel. Another common method is lyophilization (freeze drying). The cell suspension is shock frozen and the water is removed by evaporation at low temperature under reduced pressure. Freeze-dried microorganisms can be stored for a long time with minimum maintenance, but only robust cell types survive the procedure. A third method is cryopreservation of the cells at very low temperature. Cell suspensions in aqueous glycerol or DMSO are shock frozen and stored in liquid nitrogen or dry ice. As the cold chain must not be interrupted, a supply of liquid nitrogen must always be available; otherwise this method cannot be used.

Independent of the method of storage, the revitalization of the microorganisms requires several steps. Starting from the test-tube or agar plate scale, the scale up proceeds stepwise via laboratory inoculums to seed fermenters, which produce enough cells to feed production fermenters.

9.2.4   Process development

Screening for the best possible reaction conditions, the optimal media, and the most appropriate microorganisms is the first step in process development. This screening process is performed on a mL scale in shake flasks. These are 50 or 100 mL Erlenmeyer flasks that are gently agitated under controlled temperature. This is a simple, inexpensive way to get basic qualitative information about the reaction parameters.

In the next step, more sophisticated laboratory fermenters with a volume in the liter range are used to simulate the technical process better and to measure pH change, oxygen concentration, growth rate, mass transfer and so on during the fermentation. The obtained quantitative data are used to design the next step of the development, the pilot plant fermenter. In the pilot stage, the volume of the fermenter can be several hundred liters and the set-up is as close as possible to the intended large-scale process.

An old rule during development is that the scale should not be increased by more than 100-fold from step to step; it is particularly true for the fermentation processes. Fermentation uses living organisms, which lead to a number of problems that are not important in conventional (abiotic) chemical processes. Mixing, especially oxygen transfer, is such an example. It takes only a few seconds in the laboratory reactor, but it can last several minutes in large-scale reactors, leaving a large part of the microorganisms under suboptimal conditions. In the worst case they die. It is not possible to enhance mixing by a more vigorous agitation or stirring, because the microorganisms are sensitive to mechanical stress. Heat transfer is another critical factor. An efficient cooling system is needed in a large-scale reactor to keep the temperature in the optimal range. Already small temperature changes can stop the fermentation altogether or favor different types of microorganisms that lead the reaction in a wrong direction.

Another parameter unique to fermentation is the stability of the biological system. In this connection, stability means how many generations the microorganisms can be used for production before they degenerate. Often, nonproducing strains start to dominate after some time, leading to a reduced yield or quality of the fermentation product. Bacteria replicate on a time scale of hours. That means that they have several generations per day. As we have discussed in the previous chapter, commercial microorganisms are mutants or genetically modified cells, optimized for yield and production rate. Such strains are often less robust and may mutate or grow more slowly than wild-type strains, which have a competitive advantage and may dominate the fermentation process after several generations. An unstable culture can lead to serious disruption of the production in a large-scale plant. There were examples where commercial plants had to close soon after start-up, because it was impossible to obtain stable conditions.

The cost of the fermentation medium becomes a critical factor upon scale-up. The best growth medium may not always be the most economical and the cheapest not always optimal for production. The constituents of the medium must reflect the composition of the biomass, which mainly comprises carbon, oxygen, hydrogen, sulphur, magnesium, potassium, and trace elements. The raw materials for the fermentation are often side products of food processing. Sugars, like glucose, sucrose, or lactose, corn syrup, sugar beet, or sugar cane molasses serve as carbon sources. Urea, ammonia, soy flour, cotton seed meal, corn steep liquor or brewer's yeast are also nitrogen sources.

9.2.5 Bioreactors

Bioreactors must fulfill a number of requirements to be suited for large-scale production. They must allow efficient mixing without exerting too much mechanical stress on the microorganisms. They must allow temperature and pH control and effective introduction of oxygen (for aerobic processes). They must allow on-line measurement of the process parameters and must be easy to clean and sterilize between operations (Fig. 9.5).

Bioreactors come in many different designs and shapes. The stirred tank reactors are common. They comprise a cylindrical tank, a mechanical stirrer, and a guidance system for the liquid to reduce stress and enhance mixing. Pneumatic reactors use air or oxygen to mix the fermentation broth. The gas is introduced near the bottom of the reactor and induces circulation of the liquid.

Reactors can operate in batch, fed batch, and continuous mode. The batch mode is the traditional way. All ingredients are filled into the reactor prior to inoculation. After the starter culture is added, the fermentation starts slowly and proceeds until the nutrients are consumed.

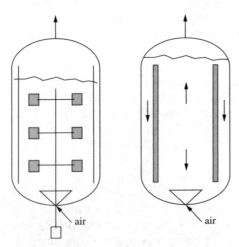

Figure 9.5 Principle of fermentation reactors: stirred tank reactor (left); air lift reactor (right). (*Source*: Schugerl, K., 1982)

The fermentation broth is removed from the reactor and the cells or other products are harvested. The reactor is cleaned and sterilized before a new batch is started. Advantages of the batch process are simple operation and low risk for contamination. The main disadvantage is the low productivity with respect to time and reactor volume.

Fed batch processes have a higher production per volume and often also a higher product concentration in the final fermentation broth. They need, however, continuous sterilization and an automatic feeding system. A fed batch process starts with the inoculum and a small part of the medium. More medium is added after the fermentation has reached a certain rate. The cells continue to grow until the reactor is full. The broth is removed from the reactor and the products are harvested. A small part of the fermentation broth is left in the reactor as the inoculum for the next batch.

Continuous processes have an even better productivity, especially for slow fermentations. Their disadvantages are their sensitivity to contamination by unwanted microorganisms, and to accumulation of side products, which can interfere with the fermentation. In the continuous mode, the starting culture and medium are filled into the reactor and more nutrients are added continuously as the cells are growing. Part of the fermentation broth is removed at a suitable rate to keep the volume constant. All media must be sterilized before they enter the reactor, which can lead to problems during routine operation.

Monitoring of the fermentation progress is important for process control. Dissolved oxygen, pH value, composition of the off-gas, number of cells per volume, and concentration of the target products are measured routinely. A variety of analytical methods are used for this purpose. A special requirement is that the sampling must take place under sterile conditions and that all in reactor sensors must be sterilizable.

To maintain sterile conditions is a major task in modern fermentation. Not only the reactor, but also all nutrients, oxygen, water, and auxiliary equipment must be sterilized before use and their sterility must be monitored by suitable techniques. Thermal sterilization is usually carried out by treatment with superheated steam at 121°C for 15 min. Sometimes higher temperatures are necessary, but they can be used only when the substrate is heat stable. Sterilization by filtration is an alternative that is used for gases or nonviscous liquids. The pore size of the filters must not be larger than 0.2 μm to retain small bacteria and spores. Other means for sterilization are irradiation with UV light or ionizing radiation or treatment with chemical antimicrobials. Sterile conditions must be maintained during the whole fermentation process. This requires special seals, special pumps, and in particular, intensive training and discipline of the operating staff.

Monitoring for contamination is a daily activity. Equipment, water, air, and nutrients are analyzed by cell counters, microscopy, or plating on agar plates. However, the sampling and sample handling is difficult. Secondary contamination of the samples must be avoided, because that would lead to false positive findings and would disturb the production unnecessarily. Therefore, high quality standards are also necessary in the testing laboratory. Examples of bioreactors are shown in Fig. 9.6.

(a)

(b)

Figure 9.6   Examples of fermenters. (a) Large-scale fermenter for the production of ethanol (Copyright Lurgi AG, Germany). (b) High-tech industrial scale fermenter; only the top part with the connections for loading and downstream processing is shown, the lower part extends to the floor below (right). (*Source*: Lonza Biologics; Portsmouth USA)

9.2.6 Downstream processing

Another important problem in the production of chemicals by fermentation is that the products are obtained in diluted form in an aqueous soup that contains many components. Concentrating the solutions and separating the products from the other products of the fermentation broth is tedious and often the main cost factor. About 60 to 95 percent of the total cost is for product recovery.

The first step of downstream processing is separation of the cells. A simple way to achieve this is by sedimentation. If the cells are heavier than the liquid, they accumulate at the bottom of the tank when agitation is stopped. Addition of flocculants can accelerate the sedimentation by helping the formation of larger particles. If sedimentation by gravitation alone is too slow, the centrifugation can help. The opposite effect is flotation. It is applied when the cells float on the surface of the liquid and can be accelerated by flotation aids. Filtration through a filter medium is essentially interdependent of gravity and the density of the cells. Its practical application, however, can be difficult because of the widely differing particle sizes of the components of the fermentation broth. A modern variant of the classical deep-bed filters are membrane filters. These are semipermeable polymer tubes or plates that separate liquids and particles.

In many processes the cells themselves are the product (e.g., single cell protein). In this case, the separated cells are dried and packed. If the desired product was released into the broth, the filtrate is the actual medium for further processing (e.g., ethanol). The third possibility is that the product is produced, but remains in the cell (e.g., enzymes). In this case, the cell walls must be disrupted to release the product. High-speed ball mills or high-speed stirrers can do this mechanically. Cell walls are also destroyed by high pressure; to be more correct, by the release of high pressure. Under high pressure a gas, for example nitrogen, dissolves in the cell liquid. Upon sudden relief of the pressure, the dissolved gas is released in bubbles that burst the cell. Other methods for cell lyses are ultrasound or enzymes. In any case, the disruption of the cell wall (lyses) releases the cell liquid and leaves the cell wall fragments behind.

The liquid phase is a complex mixture of large and small molecules from which the product must be isolated. If the desired product is a small molecule, the next step is precipitation of the proteins by addition of acid, base, or organic solvents. The denatured solid proteins can then be removed by filtration or centrifugation. The small molecule is either extracted with a suitable solvent and crystallized or distilled. If the proteins themselves are the target product, denaturation is not allowed. Therefore, mild conditions must be used to isolate the protein. Filtration through fine membranes (ultrafiltration) and chromatography over modified silica are suitable techniques. The high molecular weight fraction is further purified by chromatographic techniques, such as preparative HPLC (high performance liquid chromatography, separating according to polarity), GPC (gel permeation chromatography, separating according to molecular weight), or electrophoresis (separating according to electric charge).

The last step in downstream processing is the final purification and conditioning of the product. Chemicals and proteins are often recrystallized and heat or freeze dried. They are stored and sold in bags or drums. Very sensitive products are not fully isolated but sold as concentrates. Liquid products like ethanol or acetic acid are distilled and sold in tanks, drums, or bottles.

9.2.7 Animal and plant cell cultures

In most fermentation processes, microorganisms like bacteria or yeast are grown as the target product or as the producer of the target product. Microorganisms are relatively robust and grow rapidly, provided the conditions are right. Some applications, however, need the cultivation of animal or plant cells. Animal and plant cells are very sensitive to mechanical stress and temperature variations. They are grown in suspension, or on micro-carrier support, or on both. The cultivation volume range is laboratory scale (3 to 20 L), pilot plant scale (5 to 75 L), and industrial scale (20 to 1000 L). They need gently stirred reactors, a well-defined and controlled environment, and an aseptic design for safe operation over long cultivation periods. All reactor components must be sterilizable and easy to clean as animal cell cultures may be pathogenic.

Animal or plant cell cultures are used to produce vaccines, monoclonal antibodies, blood components, and other important medicinal products. The involved mechanisms are biological in nature and are therefore not discussed in detail here.

9.3   Food and Feed Treatment by Fermentation

9.3.1   Food conservation

As mentioned in the introduction, fermentation has been used since ancient times to conserve and alter food. Also today, it is still applied on a very large scale for this purpose. A few typical examples are described in this chapter. The principle is similar in most cases. Lactic acid produced by bacteria protects the food from deterioration by inhibiting the growth of mold and other microorganisms. Most vitamins and nutrients of the food are preserved during fermentation. Three examples are discussed in more detail below: The production of sauerkraut, soy sauce, and milk products (Table 9.3).

TABLE 9.3   Examples of Food Items Produced by Fermentation

Food products	Ingredients	Organisms	Region/country
Sauerkraut, Kimchi	Cabbage	*Lactobacillus*	Worldwide
Soy sauce	Wheat, soy bean meal	*Lactobacillus*, yeast	East Asia
Cheese, yogurt	Milk	*Lactobacillus*	Worldwide
Kefir	Milk	*Streptococcus lactis, Lactobacillus bulgaricus*	South-western Asia
Tarhana	Wheat meal, yogurt	*Lactis*	Turkey
Salami	Beef	*Pediococcus cerevisiae*	Europe, world
Katsuobushi	Tuna	*Aspergillus glaucus*	Japan
Cocoa beans	Cacoa fruit	*Candida krusei*	Africa, South America
Coffee beans	Coffee cherries	*Erwinia dissolvens, Saccharomyces spp.*	Brazil, Congo, India
Beer (with or without alcohol)	Barley, hops	Yeast	Worldwide
Olives	Green olives	*Lactobacillus*	Worldwide
Pickles	Cucumbers	*Lactobacillus*	Worldwide
Sake	Rice	*Sacharomyces saki*	Japan
Wine	Grapes	*Saccharomyces*	Worldwide
Vinegar	Cider, wine	*Acetobacter* species	Worldwide

The early sailors used sauerkraut to fight scurvy, a disease that is caused by vitamin C deficiency. Sauerkraut is the German name for fermented white cabbage produced in a batch process following a traditional recipe. The cabbage heads are cut into 1-to 3-mm wide strips and placed in large concrete tanks in intermittent layers with salt. The liquor of the previous batch is added as the starter culture. The tank is sealed and remains undisturbed for 4 to 6 weeks. The reaction sequence is illustrated in Fig. 9.7. At the very beginning of the fermentation aerobic bacteria start to grow (Fig. 9.7a). They would lead to deterioration of the cabbage, if there would be enough oxygen available. This is not the case and the aerobic bacteria die, after all oxygen is consumed. Now the process becomes anaerobic and the lactobacillus cells from the starter culture grow rapidly (Fig. 9.7b) and produce lactic acid. The mixture becomes more and more acidic and the acid inhibits the further growth of the lactobacillus cells and of all other microorganisms. After a few weeks, the whole process comes to an end and the product is removed and packed in drums or plastic bags. The fermentation reaction also starts without an added starter culture, but with some delay. The reason for this delay is that natural lactobacillus bacteria are present in relatively small numbers

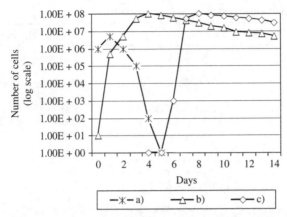

Figure 9.7 Development of the number of cells during different phases of the fermentation of cabbage: (a) aerobic bacteria decrease because of lack of oxygen; (b) fermentation starts immediately after a starter culture is added (hetero fermentation); (c) fermentation starts delayed, when no starter culture is added (homo fermentation) (*Source*: Praeve et al., 1982)

in fresh cabbage. They need time to multiply to a level that leads to a significant lactic acid production (Fig. 9.7c).

Soy sauce is a dark brown salty liquid with a peculiar aroma and a meaty taste. It is the chief seasoning agent in oriental cuisine, but it is becoming increasingly popular in many other regions of the world. It is produced from salt, water, wheat, and soybeans, originally in the batch mode. Today's processes are continuous and much faster than the traditional batch fermentation. They allow the production of 100 million L/year in one factory. The heart of the manufacturing process is a complex sequence of fermentation steps in which the carbohydrates are converted to ethanol and lactic acid and the proteins are broken down to peptides and amino acids.

Figure 9.8 describes a batch process for the production of Japanese soy sauce (koikuchi). Soybeans and defatted soybean meals are cooked in continuous pressure-cookers and mixed with roasted and coarsely broken wheat. The mass is inoculated with *Aspergillus* spores and incubated in shallow vats with perforated bottoms that allow air to be forced through the mass. After three days of incubation at around 30°C, mould growth covers the entire mass. This mass is called *koji*. Koji is the essential ingredient of most fermented products of East Asia. It is a concentrated source of enzymes necessary for breaking up the large molecules of the carbohydrates and proteins.

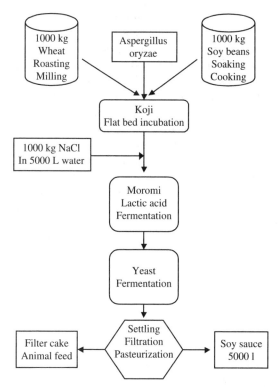

Figure 9.8   Schematic flow chart for the production of soy sauce. (*Source*: Praeve et al., 1982)

The addition of salt and the exclusion of air change the conditions and favor the growth of lactic acid producing organisms. The koji is mixed with brine containing 22 to 25 percent salt (weight by volume) and transferred to deep fermentation tanks. Lactic acid bacteria and yeast cultures are added and the slurry (*moromi*) is allowed to ferment at a controlled temperature. The high salt concentration effectively inhibits growth of undesirable *wild* microorganisms. During this stage the starch is transformed to sugars, which are fermented to lactic acid and ethanol. The pH drops from nearly neutral to 4.7 to 4.8. The moromi is held in the fermentation tanks for 6 to 8 months.

After the termination of the fermentation, the solid particles settle at the bottom and the majority of the liquid is recovered by decanting. The remaining concentrate is filtered and the liquid soy sauce is pasteurized to inactivate the remaining microorganisms. The filter cake contains the cell mass and can be used as animal feed.

Without processing, milk turns sour rather rapidly. To avoid deterioration of such a valuable food item, milk has been fermented since ancient times. The products vary from region to region, dependent on the available microorganisms and climate conditions. The principle, however, is the same in all processes, namely that lactic acid is produced by fermentation. At acidic pH the casein cells break up and precipitate. Depending on the target, they are either separated (e.g., to produce cheese) or rehomogenized to stay in the product (e.g., yogurt) (Fig. 9.9).

Although today food fermentation is well understood and controlled, there are still a few risks left. The most common problem is contamination by unwanted microorganisms. They can spoil the food by misguided fermentation and the production of substances with annoying odor or bad taste, such as butyric acid, hydrogen sulfate, or aromatic amines. The growth of pathogenic microorganisms must also be avoided, because they would lead to acute illness (food poisoning). A different

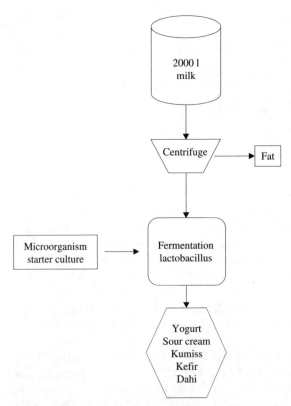

Figure 9.9   Schematic flow chart for the production of milk products. (*Source*: Praeve et al., 1982)

problem is residues of antibiotics in animal products, because they inhibit the fermentation process.

9.3.2 Feed and agriculture

Not only human food, but also green animal feed must be protected from deterioration to provide supplies during the nongrowing season. The same fermentation principles are applied as described for food. The largest volume product is *silage*. It is produced from fresh plants or plant parts, such as green maize plants and foliage of sunflower plants, field beans, or sugar beets. The green plants are compressed and shielded from air (oxygen) by strong plastic bags or concrete silos. As described for sauerkraut, first the aerobic bacteria start to grow until all oxygen is consumed. Then, the anaerobic lactic acid–producing bacteria takes over and produces acid, which protects the silage from rotting. The fermentation starts spontaneously, but it can be supported by adding minerals, sugar, and inhibitors for aerobic microorganisms. The temperature must be in a range that is tolerable for the lactobacillus microorganism. In moderate cool climates, silage can be produced outdoors without special control. This is not possible in very hot or cold climates.

Natural (e.g., *Bacillus thuringiensis*) or genetically engineered microorganisms are used to fight insects and plant diseases in agriculture. These microbial pesticides are produced on an industrial scale by fermentation. The bacteria are multiplied in the fermenter, then removed from the reactor, partially dried and applied to the infested crop as living bacteria. The bacteria produce toxins that reduce the insect population. As an alternative method it is possible to isolate the toxins that are produced by the microorganisms and apply only the toxins to the field (See also Chap. 11, Agrochemicals).

9.3.3 Single cell protein (SCP)

After the process of fermentation is over, the exhausted bacteria can be separated from the broth by filtration. This cell mass has a number of names, such as *microbial biomass* or *single cell protein* (SCP). Microbial biomass is a side product of all fermentation processes but in some cases it is actually the sole target product. Bacterial cells have a high content of protein, but are low in fat and cholesterol. This explains the names *single cell protein* (SCP) or *microbial protein*. SCP is mainly used as an additive in animal feed to enhance protein content. In principle, it is also safe for human food use, but the acceptance has been low until now.

For SCP production it is desired to optimize the fermentation conditions for maximum cell growth. This means that sufficient carbon, nitrogen, water, and other nutrients must be supplied to keep the cells growing at the highest possible rate. An efficient addition of oxygen is particularly important for aerobic processes, because oxygen transfer is often the rate-limiting factor. Starting materials for the production of a single-cell protein can be anything from agricultural commodities to petrochemicals. Large quantities of organic material are available from the production of pulp and paper, sugar, canned food, and so on. In some countries agricultural crops like sugar cane, maize, or sorghum are used as feedstock for SCP production. Many of the natural feedstocks must be pretreated to convert starch and cellulose to sugars that can be digested by the microorganisms. The fermentation conditions for SCP are very similar to those for ethanol production, which are described in the next chapter.

This chapter concentrates on the possibility of producing SCP from petrochemical feed stocks, such as n-paraffin, methane, or methanol. Between 1960 and 1980, the idea to produce SCP from crude oil sources found a lot of attention and several large-scale plants with capacities of several 100,00 tons/year were built, for instance in southern Italy. Hopes were high at the time, but the development made only slow progress. One of the reasons could be the choice of the location, which is far away from both, the source of the feedstock and the consumers of the product. Another is that oil and natural gas are expensive feed stocks, because they also have other uses. Anyway, the technology for protein production from chemicals exists and may be applied with more success in other areas of the world, where more favorable starting conditions exist.

The fermentation is usually continuous; it proceeds under sterile conditions, at constant temperature, and is started with a defined starter culture to avoid side products as far as possible. Several processes were developed: Shell had originally introduced a process that used methane (natural gas) as the feedstock for SCP production. The microorganisms are cultured in an aqueous medium at temperatures of 42 to 45°C and at a pH value of 6.8 under semisterile conditions. The final fermentation broth contains protein at a concentration of 25 g/L. The biomass is concentrated in large sedimentation tanks and then spray-dried. The mass balance equation (Eq. 9.3) shows that large volumes of oxygen are needed and that carbon dioxide and heat must be removed from the reactor.

$$3 \text{ kg O}_2 + 1.2 \text{ kg CH}_4 \rightarrow 1 \text{ kg cells} + 1.2 \text{ kg CO}_2$$
$$+ 2 \text{ kg water} + 13.2 \text{ kcal} \qquad (9.3)$$

Several types of microorganisms are needed for an optimized continuous process. *Methylococcus* species metabolize the methane; *Pseudomonas, Nordica,* and *Moraxella* species are present to convert other hydrocarbons and side-products.

Methylomas microorganisms are used to convert methanol to single-cell protein. The process conditions are similar to the methane process, but the cells are harvested by electrochemical aggregation and filtration. Yeast cells (*Candida lipolytica*) can convert n-paraffins to SCP. The process developed by BP uses a continuous stirred tank reactor under sterile conditions. The SCP is harvested by centrifugation and then spray-dried. The mass balance equation (Eq. 9.4) shows that less heat is generated and that a little less oxygen is needed than for the methane process.

$$1.12 \text{ kg paraffin} + 2.56 \text{ kg O}_2 \rightarrow 0.13 \text{ kg CO}_2$$
$$+ 1.08 \text{ kg H}_2\text{O} + 8 \text{ kcal} \qquad (9.4)$$

9.4   Industrial Chemicals by Fermentation

9.4.1   Ethanol

Ethanol is a primary alcohol with many industrial uses. It can be produced from sugar containing feedstock by fermentation. Alcoholic fermentation is one of the oldest and most important examples of industrial fermentation. Traditionally, this process has been used to produce alcoholic beverages, but today it also plays an outstanding role in the chemical and automotive industry. The largest potential use of ethanol is as car fuel either neat or as an octane booster and oxygenate in normal gasoline. In the United States, it is heavily promoted as a replacement of MTBE (methyl-t-butylether). Ethanol is also an important solvent and starting material for cosmetics and pharmaceuticals and is also widely used as a disinfectant in medicine.

Ethanol is produced from carbohydrate materials by yeasts in an extracellular process. The overall biochemical reaction is represented by (Eq. 9.5).

$$C_6H_{12}O_6 \rightarrow 2 \text{ C}_2\text{H}_5\text{OH} + 2 \text{ CO}_2 + \text{energy} \qquad (9.5)$$

Sugar containing plant material can be used without chemical pretreatment either directly as mash or after extraction with water. Examples are fruits, sugar beets, sugar cane, wheat sorghum, and so on. Starch containing agricultural commodities or waste products is pretreated with enzymes. Cellulose materials, such as wood, are cooked with acid to break up the polymeric carbohydrate bonds and to produce monomeric or dimeric sugars.

1. Feedstock preparation: Sugarcane or sorghum must be crushed to extract their simple sugars. Starches are converted to sugars in two stages, liquefaction and saccharification, by adding water, enzymes, and heat (enzymatic hydrolysis).

2. Fermentation: The mash is transferred to the fermentation tank and cooled to the optimum temperature (around 30°C). Care has to be taken to assure that no infection (other organisms that compete with the yeast for the glucose) occurs. Then the appropriate proportion of yeast is added. The yeast will begin producing alcohol up to a concentration of 8 to 12 percent and then become inactive as the alcohol content becomes too high.

3. Separation: The mash is now ready for distillation. A simple one step *stripper* distillation separates the liquid from the solids. The residue of this distillation is a slurry comprising microbial biomass and water, called stillage. It is removed to prevent clogging problems during the next step, fractionated distillation. It is often used to produce secondary products, such as animal feed additives or seasonings or it is converted to methane and burned as an energy source.

4. Distillation: Distillation separates the ethanol from the water in a rectifying column. The product is 96 percent ethanol. It cannot be further enriched by distillation because of azeotrope formation, but must be dehydrated by other means.

5. Dehydration: Anhydrous ethanol is required for blending gasoline. It can be obtained by additional dehydration, for example, with molecular sieves or carrier-assisted distillation.

9.4.2 Other industrial alcohols

By changing the reaction conditions to aerobic or by using different microorganisms, it is possible to produce other alcohols and acetone. Today, these products are, however, available in large quantities from petrochemical sources and the fermentation route is mainly of historical interest.

Fermentation by aerobic bacteria, such as *Aerobacter* and *Erwinia*, produces butane-2,3-diol with concentrations up to 10 percent. In the early 20th century, diol was an important product, as it could be converted to butane-1,3-diene, which could be polymerized to give synthetic rubber. At that time, natural rubber supplies were limited and the synthesis of butadiene from petrochemicals not yet developed.

ABE (acetone, butanol, and ethanol) fermentation has a long history of commercial use and perhaps the greatest potential for an industrial comeback. Acetone, butanol, and ethanol can all be isolated from this remarkable metabolic system; carbon dioxide and hydrogen are additional products. The solvents were used as paint solvents in the expanding automobile industry. Ultimately these processes proved uncompetitive because of poor yields, low product concentrations, and problems with viruses attacking the fermenting bacteria.

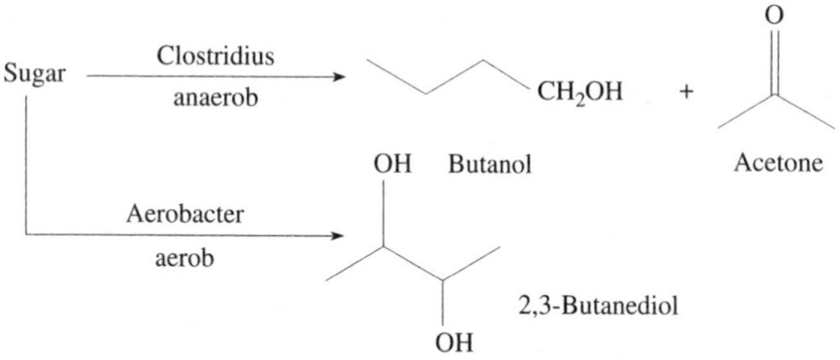

$$(9.6)$$

Recently genetic engineering was applied to transfer relevant genes to more hardy and solvent-tolerant clostridium microorganisms. This led to a 30 percent increase in product concentration that now makes the process commercially viable.

Glycerol is also no longer produced industrially by fermentation, but is an important example of how microbial metabolism can be manipulated. The conditions and the microorganisms are very similar to ethanol fermentation. However, when sodium hydrogen sulfite is added, glycerol is produced instead of ethanol, because the hydrogen sulfite blocks the primary metabolic pathway to ethanol. The glycerol was used for explosive production during World Wars I and II, as well as for applications in the cosmetics and pharmaceuticals industries.

9.4.3 Organic acids

The formation of lactic acid and its role as a food preservative were already discussed in connection with food fermentations, where it is produced in small concentrations. It is also possible to isolate it as a neat acid to convert the acid to the corresponding esters. Ethyl and butyl esters are good solvents for polymers and resins. Ethyl lactate, for instance, is used in the electronics industry to remove salts and fat from circuit boards; it is also a component in paint strippers. Ethyl and butyl esters are approved food additives. This illustrates their low toxicity.

Acetic acid is produced by oxidation of ethanol by *Acetobacter* organisms. It is either used in diluted form as vinegar or distilled to give neat (100 percent pure) acetic acid. For many centuries, acetic acid was produced only via the fermentation route. Since the advancement of the petrochemical industry, it is also produced synthetically, at least for industrial use.

By changing the fermentation conditions to aerobic, using *Aspergillus niger* microorganisms, it is possible to produce citric acid from sugar-containing feedstock. These three examples show how versatile fermentation is and how minor modifications lead to different products (Eq. 9.7).

Modern industrial scale processes produce many thousand tons of citric acid per year. The substrate comprises the glucose or saccharose solution and salts. The sugar substrate is fed into a cation exchanger to remove interfering ions and subsequently sterilized in a continuous sterilizer. The citric acid is produced batchwise in high yield by submerged fermentation with *Aspergillus niger*. Bubble columns are used as reactors. After fermentation, the broth is stored in harvest tanks so that the fermenters can be prepared immediately for fermentation of the next batch. The cells are removed by a vacuum filter separation. Proteins and other organic ingredients are precipitated by adding a precipitation agent approved for use in the food industry. All insoluble particles such as cells, coagulated proteins, and others are removed by continuous membrane filtration. Impurities are removed with anion and cation exchange resins and activated carbon. The clear, colorless citric acid solution is concentrated in a high-efficiency evaporation unit. The concentrated citric acid is crystallized; the crystals are dried, sifted, and packed.

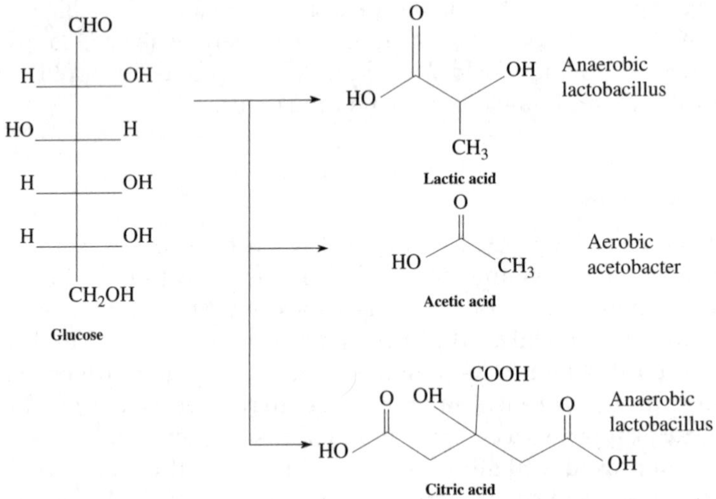

$$(9.7)$$

9.4.4 Amino acids

L-glutamic acid or its salt, monosodium glutamate (MSG), is used as an additive to human food to enhance the taste. Although seaweed had been used in Asia to enhance food flavor for over 1000 years, it was not until 1908 that the essential component responsible for the flavor phenomenon was identified as glutamic acid. From 1910 until 1956, monosodium glutamate was extracted from sea weed, a slow and costly method. In 1956, Ajinomoto, a Japanese company, succeeded in producing glutamic acid by means of fermentation. Today, L-glutamic acid or MSG is generally made by microbial fermentation using genetically modified bacteria.

The fermentation uses glucose-containing organic feedstock; it is aerobic and the L-glutamic acid is excreted by the cell into the surrounding liquid medium. The glutamic acid is separated from the fermentation broth by filtration; the filtrate is concentrated and the acid is allowed to crystallize. MSG is manufactured on a large scale in many countries and is an additive in many food items. The worldwide production is estimated to be 800,000 t.

The industrial application of fermentation for the production of amino acids as feed additives has almost a 40-year history. Production of L-lysine by fermentation was started in Japan during the 1960s. In addition to DL-methionine and L-lysine, L-threonine and L-tryptophan were introduced in the late 1980s. With the progress in biotechnology, the cost of production of each amino acid has been significantly reduced. Amino acids for feed now play very important roles in improving protein use in animal feeding (Table 9.4).

TABLE 9.4   Examples of Industrial Amino Acids

Amino acid	Starting material	Microorganism	Remarks
L-alanine	L-aspartic acid	*Pseudomonas dacunhae*	
L-aspartic acid	Fumaric acid	*Escherichia coli* (aspartase)	
L-Dopa L-tyrosine	o-Catechol or phenol, ammonia, pyruvate	*Erwinia herbicola*	
Arginine Isoleucine Lysine Threonine Tryptophane Valine	Glucose, ammonium sulfate	Genetically modified bacteria	Production: several 100,000 t/year, each
L-glutamic acid = monosodium glutamate (MSG)	Sugar containing materials	*Coryne bacterium glutamicum* or genetically modified bacteria	Production: 800,000 t/year

Amino acids can be produced as mixtures or as single compounds. Special microbial strains are responsible for the production of single amino acids. Figure 9.11 shows a schematic flow chart of the L-lysine production. The medium contains glucose as the carbon source, ammonium sulphate, urea or ammonia as nitrogen sources, and other nutrients, such as minerals and vitamins. The product is a complex, concentrated broth containing nutrients, cells, products, and side products. In the first work-up step the cells are removed by filtration or centrifugation. Part of the cells are recycled as starter culture for the next batch; the other part are spray-dried and used as animal feed or other purposes. The solution is clarified with charcoal to remove large organic molecules and colorants. The amino acid is extracted from the fermentation broth with ion exchange resin treatment, etc. The recovered amino acid is concentrated and cooled down in crystallization vessels, from which the product is removed by filtration. The yield of L-lysine from glucose or sugar is over 50 percent. This high yield is the main reason for the economic success of the process. Crystalline amino acids are added to the feed to balance the nutritional value. Processes for the production of the remaining limiting amino acids, isoleucine, valine, and arginine, are being developed. The production of L-lysine alone is today 330,000 tons/year.

Consumer concerns regarding BSE (*mad cow*) disease essentially stopped the usage of animal protein in feed. Therefore, supplementing diets with amino acids produced by fermentation is a noncritical alternative.

Until now, examples were discussed in which amino acids are produced from mixed organic matter substrates. It is also possible to start with defined chemical compounds. An example is the synthesis of L-alanine from fumaric acid in a two-step reaction. Other examples for a highly selective fermentation are the synthesis of L-Dopa from ortho-catechol and of L-tyrosine from phenol.

9.4.5   Vitamins

Vitamins are produced by fermentation of sugar containing starting materials and special additives by bacteria or yeast. They are produced inside the cell and not released into the fermentation broth. The process parameters are similar to those described for the other examples; the difference being the additives, which are essential components of the vitamins.

Vitamin A1 (retinal) is produced from β-carotene, which can be obtained by fermentation of corn, soybean meal, kerosene, thiamin, and α-ionone. The dry-mass after fermentation contains 120 to 150 g product/kg.

Vitamin B2 (riboflavin) is produced by yeast from glucose, urea, and mineral salts in an aerobic fermentation.

Vitamin B12 (cyanocobalamine) is produced by bacteria from glucose, corn, and cobalt salts in anaerobic (3 days) and then an aerobic fermentation (also 3 days). The starting point for synthesis of vitamin C is the selective of oxidation of the sugar compound D-sorbit to L-sorbose using *Acetobacter suboxidans* bacteria. L-sorbose is then converted to L-ascorbic acid, better known as vitamin C.

Vitamin D2 is formed by photochemical cleavage of ergosterin, which is a side-product of many fermentation processes. Microorganisms usually contain up to 3 percent of ergosterin.

Usually the vitamins are added to animal feed as bacterial dry mass without isolation. They are also isolated in crystalline form and used to enrich food for human use.

9.4.6   Industrial enzymes

Enzymes are the active components in the cells, where they induce the chemical transformations. They can be removed from the cells without loss of activity and sold as separate products. These isolated enzymes are used in many industrial processes, especially in food production. They are more stable and easier to handle than the original microorganisms from which they were isolated. The enzymes are often obtained from the *waste* bacterial biomass that remains after food fermentation processes.

The names of enzymes comprise two parts, the first part describes their action and the second part, *–ase*, stands for enzyme. Alkaline protease, for instance, is an enzyme that cleaves proteins. It is present in many bacteria and fungi. Proteases are produced industrially on a large scale and are added to detergents to enhance the hydrolysis of protein-containing stains. About 80 percent of all household detergents contain proteases as microencapsulated solids (about 0.02 percent).

Acidic proteases are isolated from yeast and are used in the food industry to help produce cheese, soy sauce, and baking products. α-Amylases cleave starch into amylase and amylopectine, which is applied in paper manufacturing and in the food industry. Glucomylases help to hydrolyze oligosaccharides to glucose in fruit juice, thereby enhancing the taste and removing turbidity.

Enzymes play an important role in clinical and biochemical analysis. They are a key component in many immunoassays that are used as diagnostic tests in clinical chemistry. The best-known example is probably the indicator strips that are used to diagnose diabetes by measuring sugar levels in urine or blood. The principle is that a drop of body fluid comes in contact with the enzyme peroxidase, which generates hydrogen peroxide on contact with sugar. The hydrogen peroxide reacts with an added suitable organic compound and forms a dye, which can be detected visually by a color change of the strip.

9.5 Pharmaceutical Products by Fermentation

9.5.1 Pharmaceuticals by direct fermentation

The pharmaceutical industry is the driving force behind the development of modern biotechnology. Numerous compounds and processes have been introduced and many more are under development. Although most research is devoted to the biological and pharmacological problems, the key step in the actual production of biotech pharmaceuticals is fermentation. This is demonstrated by the examples penicillin, insulin, interferon, and erythropoietin (EPO)—to name just a few. The history of penicillin illustrates a typical development of a new product from a scientific curiosity to one of the most important drugs in modern times. Penicillin changed the world! It was the first highly efficient antibiotic pharmaceutical that allowed an effective treatment of bacterial infections. At the time (around 1940 to 1950) it was such an improvement that it was called the *miracle drug*.

Penicillin was discovered by chance in 1928 by Alexander Fleming. He observed that the growth of a bacteria culture was inhibited by a fungus *Penicillum notatum*. He published his results but did not pursue its industrial development actively. Ten years later, H. Florey and coworkers had produced enough purified penicillin to treat just one patient. This test, however, was sufficient to prove that it was a viable drug. From then on many people and companies participated in the development of new fermentation technologies, new microorganisms, new downstream processing, and so on to make a large-scale production possible. Penicillin did not only change the medical world, but also the fermentation technology. The naturally growing (wild type) *Penicillum notatum* produced penicillin with a yield of 10 mg/L. Therefore, the first task was the search for a more productive species. Eventually, *Penicillium chrysogenum* was identified as the most productive species. To enhance penicillin production further, the old method of growing *Penicillum* mold on the surface of the medium in liter-sized flasks was replaced by fermentation in large aerated tanks. This allowed the mold to grow throughout the entire tank and not just on the surface of the medium. Today, penicillin and other antibiotics are produced in large-scale fermenters holding several hundred cubic meters of medium and the yield has increased 5000 fold to 50 g/L.

Equation 9.8 shows a simplified scheme of the biosynthesis of penicillin. It starts with the amino acids L-α-aminoadipic acid and L-cysteine from penicillin N in a complex reaction sequence. When phenyl acetic acid is added to the fermentation medium, the side chain of the molecule is modified and the resulting product is called penicillin G. Today, several hundred antibiotics are on the market, most of them have at least one fermentation step in the production process. The production of antibiotics is in the order of 50,000 t/year.

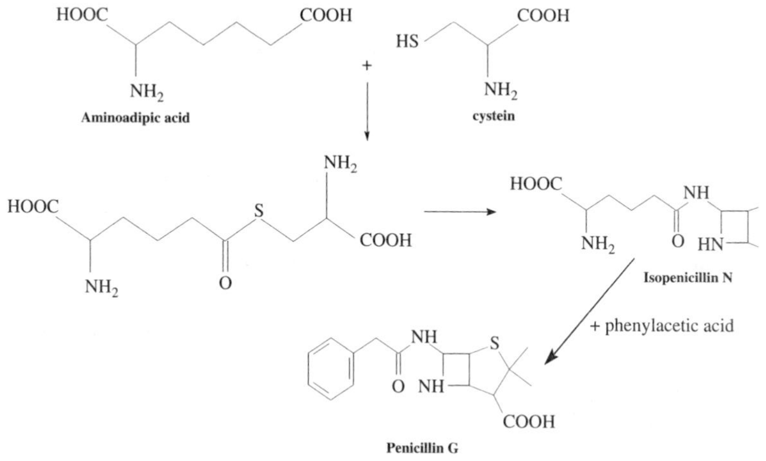

Unfortunately, bacteria develop a resistance against penicillin. Therefore, it is necessary to continuously develop new antibiotics. This can be done by modification of penicillin either by adding new functional groups or by using other microorganisms that produce different classes of antibiotics.

9.5.2 Pharmaceuticals via biotransformation

Biotransformations are chemical reactions that are induced by enzymes in the cells. Sometimes it is possible to isolate the enzymes and to carry out the chemical reaction in a separate reactor in the absence of living cells. Starting materials are single chemical compounds or mixtures of related compounds, which are converted to the product with high selectivity. Specificity has several levels: Conversion of one compound in a mixture of similar compounds or conversion of only one functional group on a complex molecule with many functional groups.

Many biotransformations are difficult to achieve by conventional synthesis. A classical example is the synthesis of chiral molecules.

A compound is chiral when it can occur in two forms that are mirror images of each other. The two forms (enantiomers) are very similar, but not identical, for instance, like the right and the left hand of the same person. Classical synthesis produces both enantiomers in a 1 to 1 ratio. They cannot be separated by normal physical means. Nature is, however, more selective. Here, only single enantiomers are formed. This can be used to separate D,L enantiomers of amino acids. The enzyme L-amylase produces selectively the L-amino acid from a mixture of the DL-acylamino acids. The D-acylamino acid remains unchanged and can be separated easily by extraction or crystallization.

Separation of enantiomers is important in the pharmaceutical industry, because often only one enantiomer has the desired efficacy, whereas the other causes unwanted side effects.

$$
\begin{array}{c}
\underset{\underset{\textrm{NH}-\textrm{CO}-\textrm{CH}_3}{|}}{\textrm{R}-\textrm{CH}-\textrm{COOH}} \longrightarrow \underset{\underset{\textrm{NH}_2}{|}}{\textrm{R}-\textrm{CH}-\textrm{COOH}} + \textrm{CH}_3\textrm{COOH}
\end{array}
\tag{9.9}
$$

D,L acylamino acid   L-amino acid   acetic acid

It is also possible to convert nonchiral readily available industrial organic chemicals into valuable chiral natural-analogue products. This is demonstrated by the conversion of achiral fumaric acid to L(–)-malic acid with fumarase as the active enzyme. The same compound is converted to the amino acid L(+)-aspartic acid by *Escherichia* bacteria that contain the enzyme aspartase. If pseudomonas bacteria are added, another amino acid L-alanine is formed (Eq. 9.10).

Fermentation of the inexpensive industrial chemicals benzaldehyde and acetaldehyde with *Sacchromyces cervisiae* microorganisms leads to (R)-phenylacetylcarbinol, which is converted to the important drug substance (1R, 2S)-ephedrine.

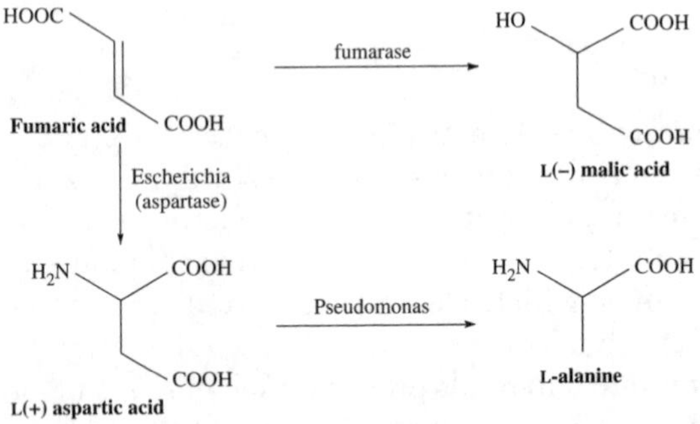

(9.10)

Steroids is the name of a class of chemical compounds that are of great importance in nature, for instance, as hormones. In the pharmaceutical industry they are active ingredients in many drugs. Their synthesis or conversion by chemical means is difficult because of their complicated chemical structure. Therefore, conversions with microorganisms are welcome alternatives. The first commercial biotechnological steroid conversion started with a steroid named 4-androsten-3,17-dione; it was converted to testosterone, the male sexual hormone, by fermentation with yeast (Schering, 1937). Progesterone, itself, can be fermented with *Rhizopus nigricans* to 11-α-hydroxiprogesterone, which is used to synthesize cortisone, another very important pharmaceutical drug substance.

There are different ways to add the educt to the fermenter. The most common is to add it during the growth phase of the cells. The process is similar to a normal fermentation; the only difference is that an additional compound is added. Another method is to grow the cells in a separate fermenter until a large amount of microorganisms is produced. The mixture is filtered and the solid cells are transferred to the actual reaction vessel, which contains the chemical to be transformed. This *stationary method* allows better control and is less prone to infections by unwanted microorganisms. A third alternative is to immobilize the cells by fixation on an inert carrier material, for example, a porous polymer. Here, the advantage is that the cells can be more easily separated from the reaction solution. The disadvantage is the often low activity (Table 9.5).

Biotransformations have a number of advantages over normal chemical reactions. They are very specific. They allow conversion of otherwise unreactive groups in a molecule and they can be carried out under mild conditions in an aqueous solution. The main disadvantage is that it is often difficult and expensive to isolate (*harvest*) the products from the reaction mixture. Therefore, biotransformations are applied when high

TABLE 9.5   Examples of Pharmaceuticals Produced by Fermentation and Biotransfermation

Pharmaceutical	Use
Penicillin	Antibiotic
Tetracycline	Antibiotic
Streptomycin	Antibiotic
Cephalosporin	Antibiotic
Insulin	Antidiabetic
Cortisone	Antiinflammatory
Cyclosporine	Immunosuppressant
Testosterone	Hormone
Prostaglandins	Stimulant, antihypertension
Ephedrine	Antiasthmatic
Interferones	Antiviral, e.g., HIV

specificity is required that is difficult to achieve by conventional means and when the added value is large. This means that a valuable, expensive product is produced from inexpensive, readily available starting materials.

9.5.3 Biopolymers

Many membranes, proteins, and nucleotides that are present in living organisms are polymers. However, in this chapter the term *biopolymers* refers to polymers that are used as materials in industry. Industrial biopolymers are still niche products, but they are gaining rapidly in importance, as they have advantages in special applications. Here are a few examples: Water-soluble carbohydrate (= polysaccharide) polymers modify the properties of aqueous systems. They can thicken, emulsify, stabilize, flocculate, swell, and suspend, or form gels, films, and membranes. Other important aspects are that polysaccharides come from natural, renewable sources, that they are biocompatible and biodegradable. For example, xanthan gum is a water-soluble heteropolysaccharide with a very high molecular weight (>1 million) produced by the bacterium *xanthomonas campestris*. It is used in food processing as a stabilizer for sauces and dressings. Another example is Scleroglucan, a water-soluble nonionic natural polymer produced by the fungi *sclerotium rofsii*. Scleroglucan has technical applications in the oil-drilling industry for thickening drilling muds and enhancing recovery. Biopolymers are also used in adhesives, water color, printing inks, cosmetics, and in the pharmaceutical industry.

Polylactides are made from lactic acid and are used for orthopedic repair materials. They can be absorbed by the body and are used for the treatment of porous bone fractures and joint reconstruction. Dextran is a substitute for blood plasma in medicine. It is produced by fermentation of saccharose by *Leuconostoc mesenteroides* microorganisms. After the fermentation is completed (about 24 h), the cell mass is separated and the dextran is precipitated by addition of ethanol to the liquid phase.

The butyrate or octanoate copolymer and butyrate or hexanoate or decanoate terpolymer have properties similar to those of higher-grade LLDPE (linear low-density polyethylene) and higher-grade PET (polyethylene terephthalate). They can be molded or converted into films, fibers, and nonwoven fabrics. The biopolymer is produced by low-cost fermentation or from wastestream substrates.

Polyhydroxyalkanoic acids (PHAs) have been extensively researched since the 1970s because of the potential applications of these compounds as biodegradable substitutes for synthetic polymers. The most successful PHA products are the polyhydroxybutyrates (PHBs). The bacterium

Alicagenes eutropha produces a copolymer of hydroxyvalerate and hydroxybutyrate when deprived of key nutrients, such as amino acids and minerals. The product, biopol, represents up to 90 percent of the dry weight of the bacterium. It is comparable to polypropene in physical properties, has better flexibility at low temperatures, and is biodegradable to CO_2 and water within months. However, the polymer (trade name Biopol) is not currently cost-competitive with synthetic polymers because of the high costs of the fermentation substrates and the fermentation plants.

Most biopolymers are produced as extracellular metabolites by fermentation in bioreactors leading to special technical problems caused by the very viscous solutions that make mass transfer and mixing in the fermentation fluids difficult. Large volumes of water and solvents are needed for dilution and extraction, respectively.

9.6   Environmental Biotechnology

When modern industrialization started in the 19th century, many people migrated from the agricultural area to the big cities. Public hygiene became a major problem. Human excrement and waste was discharged into open channels, rivers, and lakes. The pollution was disastrous and hygiene-related epidemic diseases, like cholera and typhus, occurred frequently. Therefore, it was an important step forward when public water collection systems and treatment plants were introduced at the end of the 19th century.

In an industrialized society every person produces about 200 to 400 L of wastewater; factories and other commercial enterprises release varying volumes of water. The degree of pollution of the wastewater is measured as biological oxygen demand (BOD_5) or chemical oxygen demand (COD). The BOD_5 is the amount of oxygen that is consumed during the microbial conversion of organic matter in 5 days. The COD is the amount of potassium permanganate solution needed to titrate a defined volume of the wastewater. Public wastewater in industrialized countries has a BOD_5 of approximately 60 mg/L.

Modern biological wastewater treatment plants use a combination of aerobic and anaerobic fermentation reactors to remove organic matter from the wastewater. In the aerobic part the microorganisms feed on the organic matter in the wastewater and convert it to microbial biomass and carbon dioxide. In the anaerobic part the microbial biomass of the aerobic part is digested by a second type of microorganism that produces methane as it grows. The anaerobic microorganisms die immediately when they come into contact with air. That means that they are not infectious and do not present a risk to humans and the environment when they are released from the reactor.

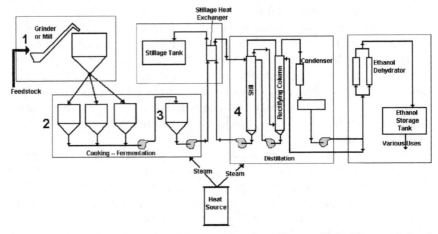

Figure 9.10 Flow chart of an ethanol fermentation plant. (*Source*: United States. National Agricultural Library; Office of Alcohol Fuels; Solar Energy Research Institute. Fuel from Farms: A Guide to Small-Scale Ethanol Production. Golden, Colo.: Technical Information Office, Solar Energy Research Institute, 1982; published at www://dnr.state.la.us.)

A schematic flow diagram of a wastewater treatment plant is shown in Figs. 9.9 to 9.12. In primary physical treatment, solid material is separated from the liquid by screens, settling tanks, and skimming devices. This removes about 50 percent of the pollutants. The remaining organic material is subjected to biological treatment.

In smaller plants, the water is treated in open basin-type reactors (aerated basin). They are inexpensive to build and easy to maintain. The oxygen is supplied by bubbling air through the water or by uptake from the ambient air with vigorous agitation of the water. The bacteria in the reactor feed on the organic matter, consume oxygen, and generate carbon dioxide. The bacteria are macroscopically seen as sludge. This sludge is heavier than the water and can be separated by sedimentation in a clarification basin. Part of the sludge is recycled as inoculums to the aerated basin. The rest is subjected to anaerobic treatment. In the large treatment plants of big cities, the open basins are replaced by more sophisticated reactors. For instance, bubble columns, which can be 30 m high, or deep-shaft reactors with a height of up to 100 m, are partly buried in the ground. At this point 90 to 95 percent of the biodegradable matter is removed from the wastewater. The remaining 5 to 10 percent is treated in clarifier basins. The water is then filtered and sometimes disinfected with sodium hypochlorite. The treated water is essentially free of pathogenic microorganisms and can be used for irrigation or discharged into rivers or lakes without any risk to the environment.

Most of the solid collected in the primary and secondary treatment steps are transferred to the *digester*. This is an anaerobic fermentation

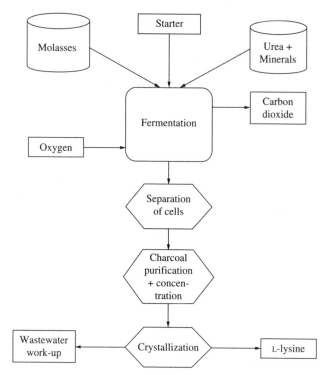

Figure 9.11 Production of L-lysine as an example for industrial amino acids.

Figure 9.12 A large-scale sludge fermenter for the biogas production and sludge treatment in a public sewage treatment plant. The scaffolding illustrates the size of the fermenter, which is about 30 m high. (*Source*: Fischer fixing systems, Germany)

reactor, often egg-shaped, in which anaerobic microorganisms convert organic matter to methane. The mass of the solid waste is reduced by some 70 percent, most pathogenic organisms are killed, and the odor potential is largely eliminated. The produced methane can be used to generate electricity or heat; the remaining solid can be incinerated or discharged.

Biological wastewater treatment is very efficient in removing organic matter and biodegradable chemicals. It is rather inefficient in removing inorganic ions, especially nitrate and phosphate. Nonbiodegradable organic compounds, such as polychlorinated hydrocarbons (PCB), highly branched hydrocarbons, or some pharmaceuticals (e.g., steroids) also pass through treatment without change. Another problem arises when antibacterial compounds reach the treatment facility. They kill the bacteria in the bioreactors and can severely disturb plant operation. Therefore, the discharge of disinfectants and antibiotics—and actually all pharmaceuticals—to the public sewer system must be avoided (Fig. 9.13).

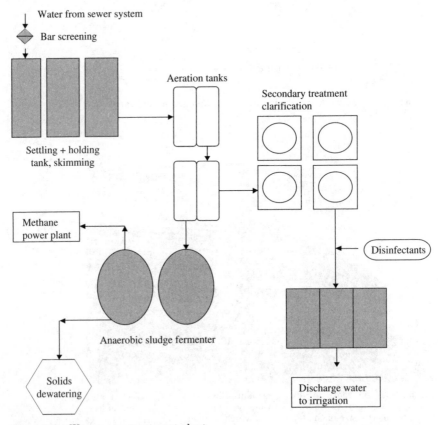

Figure 9.13 Wastewater treatment plant.

Fermentation is also used to treat industrial chemical or organic waste. The principle is very similar to the described anaerobic sludge treatment. That means that the organic material is converted to methane. Examples include waste containing cotton, rubber, plastics, fats, explosives, and detergents. The waste can be transferred to special treatment plants or be treated in situ in the open field where the waste was buried. Open-field microbiological treatment of spills or deposits of hazardous chemicals is a potentially attractive and inexpensive remediation method and has attracted a lot of research attention. So far, however, only a few examples have been successful.

Another example of the application of fermentation is the removal of organic compounds from exhaust air. Such *biofilters* are often trickle-bed reactors, in which the microorganisms grow on a solid support, such as wood chips or porous stones. Water is trickled through the reactor, whereas the exhaust air flows in the opposite direction. The bacteria digest the organic components and destroy odor-causing chemicals. Biofilters are applied in municipal wastewater treatment, food production, paint, paper, and timber industries or soil remediation. They provide an attractive alternative to thermal, chemical, and adsorptive processes for cost-effective treatment of air pollutants.

9.7   Social and Economic Aspects

Biotechnology is a synonym for modern technology. The term is frequently used, but it seems that different people understand it differently. The OECD defines biotechnology as "The application of Science and Technology to living organisms as well as parts, products and models thereof, to alter living or non-living materials for the production of knowledge, goods and services."

The actual production process in most industrial biotech applications is fermentation. *Genetic engineering* is a method to genetically modify microorganisms or cells of plants, and animals that are used as starters for the production of products by industrial fermentation. As described in this chapter, fermentation has many uses and is of vast social and economic importance. It spans a wide range of products, from soy sauce to interferon, from antibiotics to biogas. Some products, like food or vitamins, are mature and will see a stable market, but with decreasing prices. Other products, especially speciality pharmaceuticals and biopolymers, are expected to gain economic importance in the future.

The economic value of food, feed, and biotech pharmaceuticals is enormous. Although fermentation is a key step in the production of these products, it contributes only a small part to the total cost. This is illustrated by antibiotics. The market value of the finished drug is certainly much higher than US$20 billion per year. A toll manufacturer carrying out only the fermentation would get a fraction of this sum, probably <5 percent. Therefore, the value of the fermentation itself is difficult to estimate, but could be in the order of US$10 billion worldwide, in 2000.

The social aspects are also interesting and the consequences are difficult to predict. There are a number of undisputed benefits connected with the production of food, feed, vitamins, and pharmaceuticals by fermentation. The starting materials are from renewable resources; the products are useful and low risk; the production takes place under mild conditions and the by-products are biodegradable and harmless. Some of the most important pharmaceuticals are produced by fermentation; insulin, penicillin, and tetracycline are just a few examples. They have changed the quality of life—at least for those people who have access to them.

The production of ethanol by fermentation and its use as a car fuel may serve as an example to demonstrate the social benefits and risks of fermentation. Ethanol is an efficient fuel, it can be produced from carbohydrates (sugar cane, maize, and so on), that means from renewable resources. This seems to be a plus, and it is one, as long as agricultural by-products are used. However, there is a different point of view: When farmers can make a better profit with raw materials for fuel, they will produce it. But who will produce our food, when the arable land is used to make car fuel? With a growing world population we can hardly afford this.

Generally, experts expect that fermentation processes have economic advantages only for the production of expensive chemicals, not for mass products. They expect more applications in the pharmaceutical field, where the active substances are valuable and difficult to produce by conventional chemical means. Therefore, we may not have to worry about the ethanol example discussed in the previous paragraph, as the economics are not favorable.

Bibliography

Anderson, T. M., Industrial fermentation processes, *Encyclopedia of Microbiology, Vol. 2*, Academic Press, N.Y. 2000.

Hirose, Y., Production and isolation of enzymes, *Enzyme Catalysis in Organic Synthesis*, Vol. 1, pp. 41–66, Drauz, K. and Waldmann, H. (Eds.), Wiley-VCH, Weinheim, 2002.

Maisch, W. F., Fermentation processes and products, *Corn Chemistry and Technology*, 2nd ed., 695–721, White, P.J. and Johnson, L.A. (Eds.), AACC Monograph 2003.

Notermans, S. and Rombouts, F., (Eds.), *Frontiers in Microbial Preservation and Fermentation*, Elsevier Science Ltd., Oxford, 2002.

Praeve, P., Faust, U., Sittig, W., Sukatsch, D.A. (Eds.), *Handbuch der Biotechnologie*, Akademische Verlagsgesellschaft, Wiesbaden 1982.

Saha, B. C., Commodity chemicals production by fermentation: An overview, *ACS Symposium Series*, 862, 3–17, 2003.

Sanchez, S., and Demain, A.L., Metabolic regulation of fermentation processes, *Enzyme and Microbial Technology*, 31, 895–906, 2002.

Schugerl, K., New bioreactors for aerobic processes, *International Chemical Engineering*, 22, 591, 1982.

Waites, M. J., Morgan, N.L., Rockey, J.S., and Higton, G., *Industrial Microbiology*, Blackwell Publishing, 2001.

Vert, M., Polymers from fermentation. Poly(lactic acid)s and their precursors, the lactic acids, *Actualité Chimique*, 11, 12, 79–82, 2002, (in French).

Encyclopedia of Amino Acids. Available at www.ajinomoto.com/amino/index.html.

Information Note on the Use and Potential of Biomass Energy in OECD Countries, Directorate for Food, Agriculture and Fisheries, OECD, COM/ENV/AGR/CA()()147/ FINAL, 28 Nov 2000. Available at www.oecd.org.

Microbiology, What it's all about. Available at www.microbeworld.org.

Economic Aspects of Biotechnologies related to Human Health; Part II: Biotechnology, *Medical Innovation and the Economy: The Key Relationships*, by Working Party on Biotechnology, OECD, DSTI/STP/BIO(98)8/FINAL 09 Nov 1998. Available at www.oecd.org.

The Pharmaceutical Industry

Bassam El Ali and Manfred J. Mirbach

10.1   Introduction

The pharmaceutical industry is one of the most important sectors of health care worldwide. Pharmaceutical materials are all manufactured in very small quantities relative to other types of compounds, but their dollar value is exceedingly high.

The pharmaceutical industry must invest more in research and development than any other industry to discover innovative drugs and therapies to fulfill medical needs. Because of the high R&D cost and to meet growth expectations and goals for new product launches, many companies are constantly reorganizing their research operations to be more effective [1].

Competition between companies arises not only from new products but also from generic drugs. Generic products are copies of the original products. They are produced by specialized manufacturers, after the patent of the original product is expired or when it is not patented in certain countries. In the United States, sales of generic drugs have grown to as much as 40 percent of total pharmaceutical sales. Changes in health care systems are having a profound impact on drug purchasing. Nevertheless, the worldwide market for pharmaceuticals is expected to grow about 6 to 12 percent per year through 2005. This growth is fueled by an increasing world population, by higher life expectancy, and by changes in lifestyle.

10.2   Use and Economic Aspects

World pharmaceutical sales grew 18 percent between 2001 and 2002 to reach US$400.5 billion. North America was the largest market and one of the most dynamic regions, with 17 percent growth to US$203.6 billion (Table 10.1) [2a]. The numbers are based on the IMS world review that tracks actual sales of approximately 90 percent of all prescription drugs and certain over-the-counter (OTC) products in more than 70 countries.

TABLE 10.1   2002 World Drug Market by Region

World market	2002 Sales (US$ billion)	% of global sales
North America	203.6	51
European Union	90.6	22
Japan	46.9	12
Asia, Africa, and Australia	31.6	8
Latin America	16.5	4
Rest of Europe	11.3	3
Total	400.5	100

SOURCE: *IMS World Review 2003 and IMS Consulting.*

TABLE 10.2   Leading Therapy Classes in 2002 Global Pharmaceutical Sales

World therapy classes	2002 Sales (US$ billion)	% of global sales
Antiulcerants	21.9	6
Cholesterol and triglyceride reducers	21.7	5
Antidepressants	17.1	5
Antirheumatic nonsteroidals	11.3	3
Calcium antagonists, plain	9.9	3
Antipsychotics	9.5	2
Erythropoietins	8.1	2
Oral antidiabetics	8.0	2
ACE inhibitors, plain	7.8	2
Cephalosporins and combinations	7.6	2
Total	122.9	32

SOURCE: *IMS World Review 2003.*

North America, Europe, and Japan accounted for 90 percent of the worldwide pharmaceutical spending in 2002.

If the products are grouped according to the therapy classes, the top ten pharmaceuticals accounted for 31 percent of the total world market in 2002. Cholesterol and triglyceride reducers, antipsychotics, oral antidiabetics, and erythropoietins are growing fastest, with antipsychotics and oral antidiabetics sales each up to 30 percent (Table 10.2) [2a].

The top ten best-selling drugs worldwide accounted for sales of US$44.7 billion in 2002. Within the total world market, Lipitor was the top-selling drug in 2002, with US$8.6 billion in sales, compared with US$5.4 billion in 2000 (Table 10.3) [2].

TABLE 10.3   Leading Products in 2002 Global Pharmaceutical Sales

Trade name	Common name	2002 sales (US$ billion)
Lipitor	Atorvastatin calcium	8.6
Zocor	Simvastatin	6.2
Losec or Prilosec	Omeprazol	5.2
Zyprexa	Olanzapine	4.0
Norvase	Amlodipine	4.0
Ogastro or Prevacid	Lansoprazole	3.6
Celebrex	Celecoxib	3.1
Erypo	Epoetin alfa	3.8
Seroxat or Paxil	Paroxetine	3.3
Zoloft	Sertraline	2.9
Total ten leading products		44.7

Another fast-growing worldwide drug was Zyprexa with sales of US$4.0 billion in 2002. Pharmacia and Pfizer's competing COX-2 inhibitor, Celebrex, generated US$3.1 billion in sales with a 32 percent increase compared to the previous year. Zoloft replaced Vioxx, which dropped out completely in the top ten drugs [2]. Vioxx was withdrawn from the market in 2004 because of severe side effects.

The chemical structures, names, and therapeutic areas of the top ten best-selling drugs based on the IMS worldwide review 2003 are given hereafter [11].

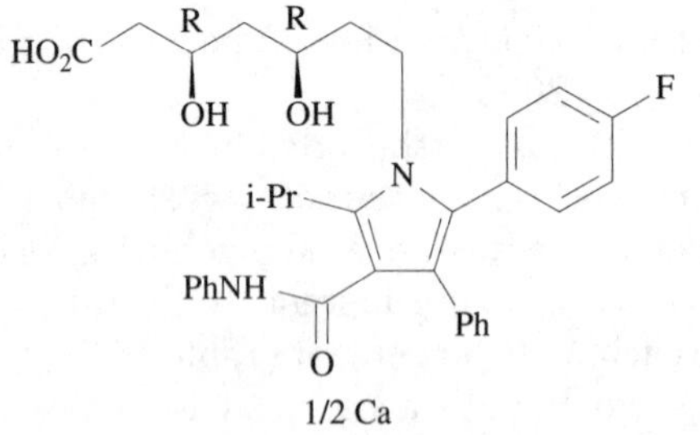

Lipitor; cholesterol reducer

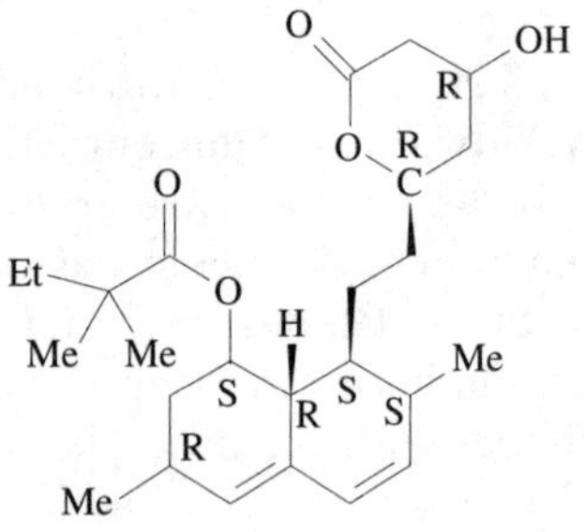

Zocor; cholesterol reducer

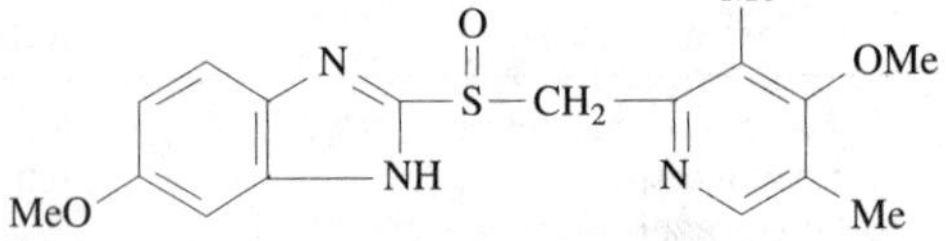

Losec/Prilosec; antiulcerants

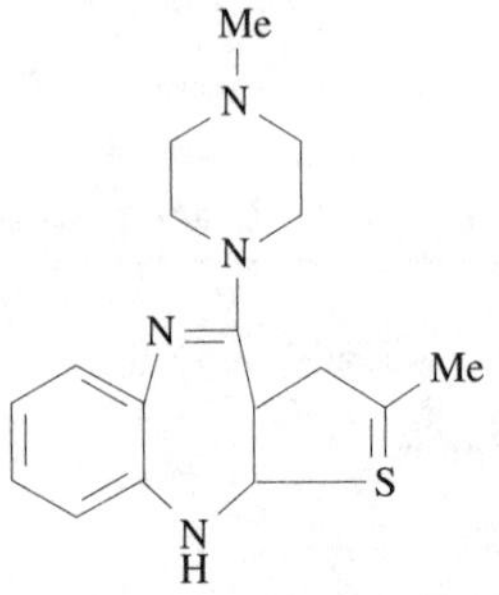

Zyprexa; antipsychotic

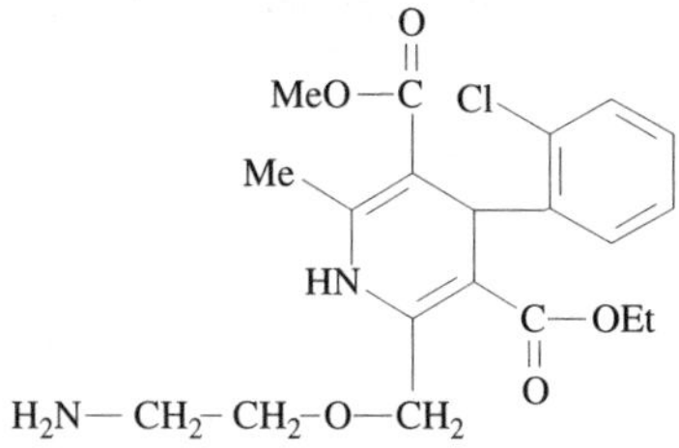

Norvase; calcium antagonists

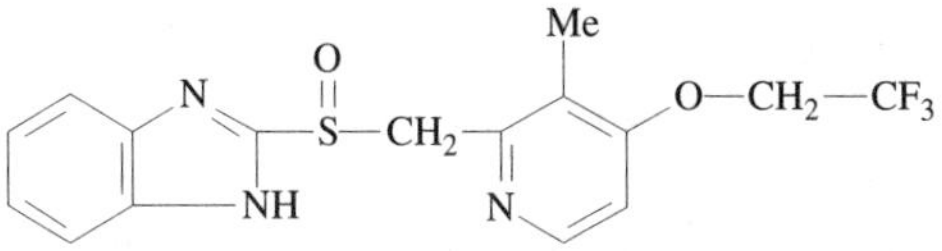

Ogastro/Prevacid; antiulcerants

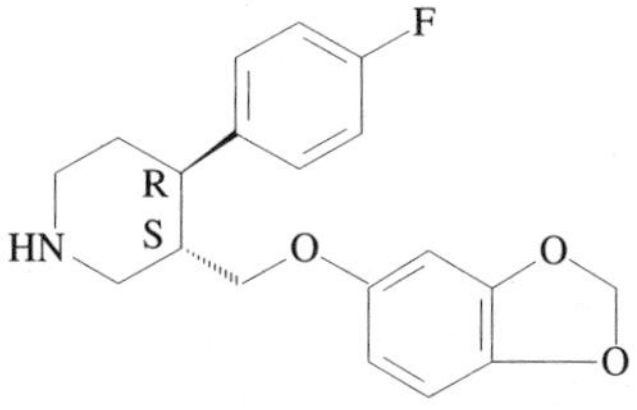

HCl
Seroxat/Paxil; antidepressant

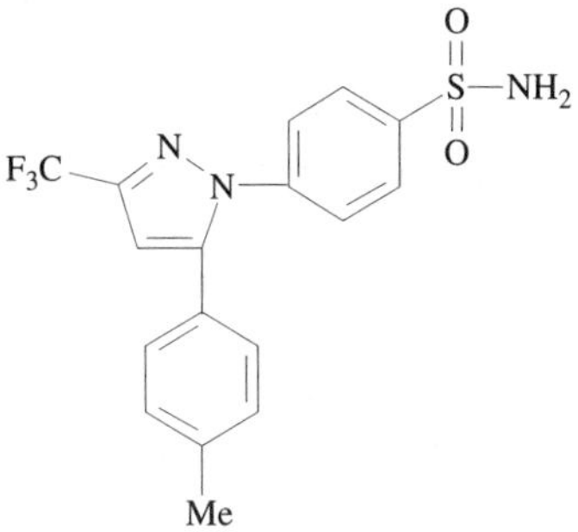

Celebrex; antirheumatic

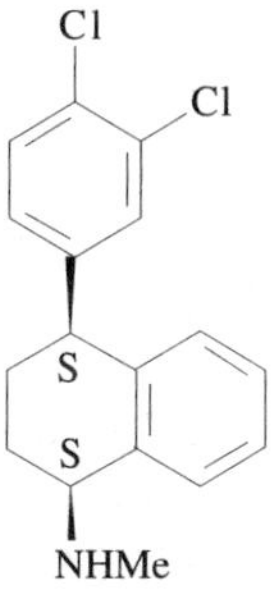

HCl
Zoloft; antidepressant

Despite decades of prosperity, drug firms are reshaping R&D to drive growth, boost productivity, and exploit new technologies. Simply to maintain about 10 percent annual growth, or roughly US$1 billion in new revenues, a company must bring two or three new products to the market each year.

The discovery of drugs is a process in continual evolution using new technologies to enhance the quality and the quantity of products.

TABLE 10.4   Cost and Resources in Pharmaceutical in R&D

R&D state	% of spending	% of employees
Synthesis or extraction	11.0	12.9
Biological screening or pharmacological testing	16.0	18.3
Toxicology or safety	6.0	7.8
Dosage formulation	9.6	11.6
Clinical evaluation	34.5	24.3
Regulatory	3.8	4.8

SOURCE: *Pharmaceutical Research & Manufacturers of America, Industry Profile, 1997.*

The cost of developing a new drug is increasing steadily, mainly regulatory requirements that must be met to get marketing authorizations in the various countries. It is also important to note that most of the R&D cost is associated with the clinical evaluation; next is biological screening/ pharmacology testing. Chemical synthesis or extraction is in the third place. Although doctors and staffs in clinical trial function account for only 24.3 percent of the total R&D staff, they spend 34.5 percent of the total available funds (Table 10.4).

The investment, from start to finish, is on the order of US$500 million per new product. However, for areas where therapy already exists, the companies have to make sure that they offer a significant improvement over the existing products to get a marketing approval.

Improvements include increased effectiveness, fewer side effects, a new mode of action, or a new drug form, for example, pills versus injections, that will improve patient compliance and quality of life.

The focus of drug discovery research has been shifting from acute to chronic illnesses, such as Alzheimer's, cancer, obesity, AIDS, asthma, arthritis, and neurologic conditions.

Companies are always looking for several measures of success: increased productivity, decreasing costs and time for drug development, innovative products, and continued sales growth.

With the US$1.17 billion pharmaceutical market, over 2400 pharmacies, and more than 4600 registered drugs, both generic and patent, Saudi Arabia is the largest consumer of pharmaceuticals in the Gulf region. The market for drug products in Saudi Arabia is growing by an estimated 15 percent annually [2b]. Despite recent government efforts to bolster the domestic pharmaceutical industry, the Saudi market is still heavily reliant on imports. According to the National Commercial Bank, the Kingdom imported US$973 million worth of pharmaceuticals in 2000, representing a 57 percent increase since 1995. Imports come from Europe (35 percent),

Egypt, Jordan, the United Arab Emirates (combined 12 percent), the United States (7 percent), Japan, Canada, India, China, and other Asian countries. According to the U.S. Foreign & Commercial Service, the Saudi market for imported drugs and pharmaceuticals is divided among the following categories: Penicillin (33.4 percent), Hormones (14.3 percent), Vitamins (7.2 percent), and others (45.1 percent). Antirheumatics, antagonists, broad-spectrum oral penicillin, cephalosporin, nonnarcotic analgesics, macrolides, antihistamines, and topical corticosteroids are the most important pharmaceutical products imported to Saudi Arabia [2b].

To promote domestic production, at the beginning of 2001 the Consultative Council approved a plan to set up a regulatory body similar to the U.S. Food and Drug Administration. The authority is responsible for licensing of pharmaceutical products and manufacturing facilities. The Ministry of Health has also introduced a 10 percent subsidy to local manufacturers to enable them to sell their products at a competitive price. Finally, the government has been purchasing local pharmaceuticals in bulk in recent years. Despite the fact that these measures have increased local production, only 17 percent of Saudi Arabia's pharmaceutical market was manufactured domestically in 2000. Banaja Saudi Import Company is the market leader in both manufacture and distribution of pharmaceuticals. Regarding distribution, Banaja has established partnerships with 12 leading international companies, such as GlaxoSmithKline and Hoffmann La Roche. Banaja distributes three of the top ten products in the Saudi market: Zantac, Dermovate, and Supradyn. Banaja is also a local manufacturer of pharmaceuticals. It has a 51 percent ownership of Glaxo Saudi Arabia Ltd., which is the first joint venture between a Saudi investor and a foreign pharmaceutical company in Saudi Arabia. Saudi Pharmaceutical Industry & Medical Appliances Corporation (SPIMACO) is the second largest producer of pharmaceuticals in Saudi Arabia with total sales of over US$126 million in 2000. SPIMACO, a joint-stock company with fully paid-up capital of US$160 million, currently produces antirheumatics, antibiotics, antimicrobials, antituberculins, antihypertensives, antidiabetics, antihistamines, antiallergics, cough sedatives, expectorants, and mild topical corticosteroids. Other major local manufacturers in Saudi Arabia include the Saudi Arabian Japanese Pharmaceutical Company, Ltd. (SAJA), a joint venture between Tamer Industries and a consortium of Japanese partners [2b].

10.3   Discovery and Development of Drugs

10.3.1   Introduction

The development of a new drug is a long, tedious, and expensive process. It takes about 8 to 12 years from the time a new drug substance is discovered until a marketing authorization is obtained. The associated cost easily reaches US$ several hundred million and there is no guarantee that a promising new substance will become a commercial success.

The starting point is always connected to the health problem that needs a cure. This is the main target. The next step is to find a chemical substance that shows promising activity against the target disease. New leads are sometimes discovered incidentally and more often they are the results of long-term observation of nature. In modern times, they are derived from random screening or random synthesis or computer-based expert systems. If a good lead is found, it is chemically modified until the efficacy is optimal and the unwanted side effects are minimal.

The drug discovery and early development phase is followed by the preclinical phase, during which the route of synthesis, the safety, and the general properties of a drug substance are established. This is a prerequisite before it can enter the clinical phase, which is testing on humans. The tentative conclusion of the development process is the granting of marketing authorizations by the competent authorities of the countries in which the drug is to be sold. However, even after a drug is in the market, research continues to improve its dosage forms or to discover new applications. It is tested, for instance, whether the drug substance is also suited to treat other diseases. Sometimes, adverse side effects are observed that need investigation, or the galenic form is changed to make the product easier to use for both patient and doctor.

The well-known drug Aspirin^{TM} can demonstrate the various stages in the life of a drug substance. It was known since ancient times that the bark and leaves of the willow tree can provide relief from pain and fever. In 1832, the German chemist Piria isolated salicylic acid as the active substance in the plant material. Salicylic acid became the lead substance for further development. Soon a synthetic route was developed that replaced the tedious extraction process. Salicylic acid was used as a pain remedy, but because of its high acidity it had severe side effects when applied orally. In 1887, Felix Hoffmann, a chemist at Bayer AG, realized that the ester form of salicylic acid, acetylsalicylic acid, had fewer side effects and was even more efficacious than the free acid. The Bayer Company marketed the ester under the trade name Aspirin, which soon became the number one selling drug worldwide and is still a synonym for pain relief today.

However, the story was not over. About 80 years after the introduction of Aspirin as a pain reliever, it was established that acetylsalicylic acid prevents heart attacks and strokes and reduces the risk of death during a suspected heart attack. In 1982, John R. Vane was awarded the Nobel Prize for medicine for discovering the basic mechanism of the action of acetylsalicylic acid.

Only one out of 10,000 substances that enter the development pipeline pass all the tests and become new drugs in the market. Because of the huge financial risk of drug development, the survival of a pharmaceutical company depends on a successful pipeline. Therefore, companies invest heavily in research and new technologies. Consequently, the process of drug discovery has changed dramatically over the last 20 years, especially owing to the introduction of computerized robotic systems.

10.3.2   Classical drug discovery and early development

As shown in the Aspirin example, the first step of drug discovery is to define a target disease, for which a remedy is to be developed. The next step is the design of a suitable screening model, since it is not possible to use human patients for screening of unknown substances. The model can be a cell culture, for instance bacteria, or a laboratory animal, usually a mouse, in which the disease is artificially induced. In classical drug discovery, two sources of possible new drugs are used: natural products and chemical compounds that were synthesized for this purpose. These compounds are tested on the models and those that show a positive effect are selected as hits and the others are shelved for possible later use. Further research work is required to confirm the result in a second test and also to investigate that the substance has no deleterious toxic effects on mammals. A confirmed hit is called a *lead*.

The lead compound serves as a model for further development. Similar substances are synthesized and compared to the lead. The chemical structure is elucidated and carefully analyzed to determine which part of the molecule is responsible for the pharmacological effect. It is called the *pharmacophore*. The original lead compound can be modified to change its physical and physiological properties, but the pharmacophore remains the same. This principle is illustrated by the β-lactam pharmacophore in structures **1** to **4**.

The modified substances must again be tested, modified, tested—and so on, until an optimal drug substance is obtained (Table 10.5). The synthesis of a new substance by classical means takes about 1 to 2 weeks, its biological testing another 2 to 4 weeks in each optimization step. This shows that classical drug discovery is a time-consuming process. Many chemists are needed for working on the synthesis of new compounds and it takes many years to develop a drug by this method.

Sometimes luck helped the scientists to discover a new drug. The most famous example of such an incidental discovery of a new lead drug is the story of penicillin by Alexander Fleming. Penicillin is so important because it opened the route to a whole new class of highly active antibiotics. The molecule contains the characteristic β-lactic group that is the lead structure for other substances. The simplest modification is changing the side chains of the molecule leading to other penicillins, **1**. The replacement of the 5-membered thiazolidine ring to an analogous 6-membered ring leads

TABLE 10.5   Tasks in Drug Discovery and Early Development

Discovery and early development	Classical approach	High throughput approach
Target identification	Based on medical symptoms	Based on receptor binding
Screening	Using laboratory animals or cell cultures	Using receptor binding assays with fast detection techniques
Hit confirmation => lead	Use of second animal species	Use of second assay type
Lead optimization	Classical synthesis of 10 to 20 similar compounds, testing on laboratory animals	High throughput synthesis of 1000 to 100,000 compounds tested with HTS receptor binding assays

to cephalosporin, **2**, structure. The structure can be simplified even further to give clavulanic acid, **3**, and norcardicin, **4**, which are both powerful antibiotics although they are no longer used as penicillins.

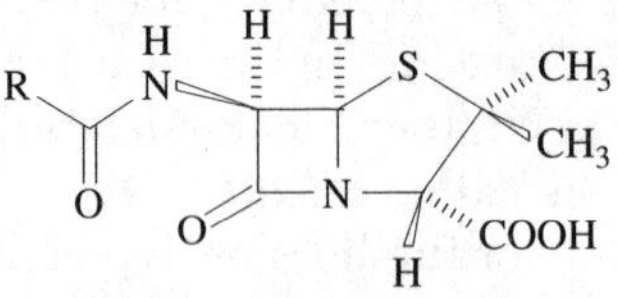

1

R=PhOCH$_2$, penicillin V;
R=CH$_2$Ph, penicillin G

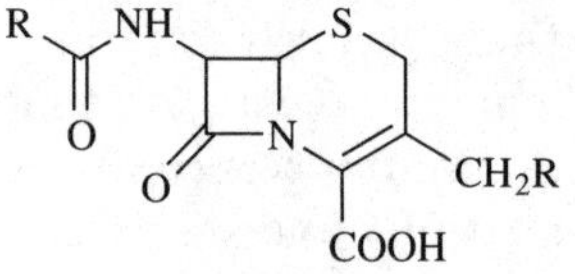

2, Cephalosporin

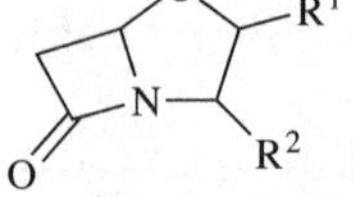

3, Clavulanic acid

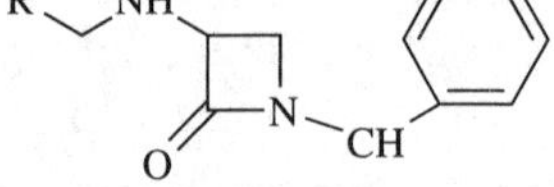

4, Norcaradcine

The history of penicillin has been described many times; the story of the discovery of benzodiazipine tranquilizer drugs is less known. Leo Sternbach was a chemist in the research laboratories of Hoffmann-La Roche (Roche). The target was to find a better drug against sleeping disorders and anxiety than the then available barbiturates that had severe side effects. Sternbach had started with a class of compounds, called benzheptoxdiazines, familiar to him from his earlier work and that he thought can be promising. However, after synthesizing more than 40 new compounds that failed the screening tests, he was about to give up. In one of his last experiments in the project, he isolated a white powder with the laboratory code Ro5-0690. He put it on the shelf and forgot it.

Eighteen months later his technician was cleaning the laboratory when he found the bottle and asked what to do with it. First, he considered discarding it but then Sternberg sent it for screening to the biological unit. Some time later he received the result that his compound was more active in mice than any other reference compound.

It turned out that the nearly discarded substance had a novel structure, **5**, a chlordiazepoxide. It was developed further and became known under the trade name *Librium*^{TM}, which was the first antianxiety agent. Librium became the lead compound for the development of many other compounds with related structures. The best known is the diazepam *Valium*^{TM}, **6**, a tranquilizer used to treat anxiety, muscle spasm, and symptoms of epilepsy. Today, about 40 diazepines are in the market. Nearly forgotten on the laboratory shelf, Librium and Valium became blockbuster drugs that fueled the growth of the Roche company. (Drugs are called blockbusters when sales are much higher than US$ 1 billion per year.)

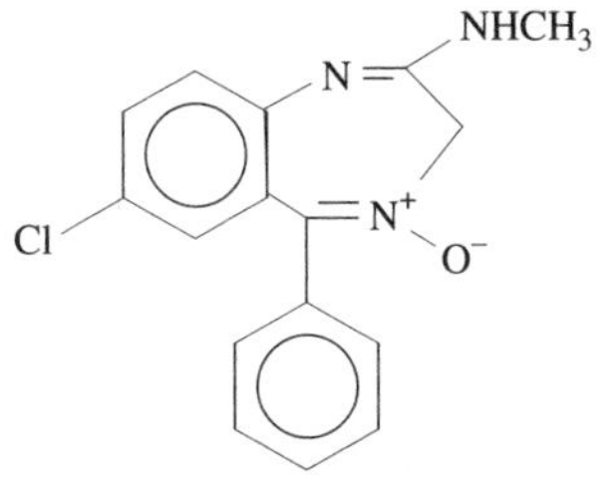

5, Librium 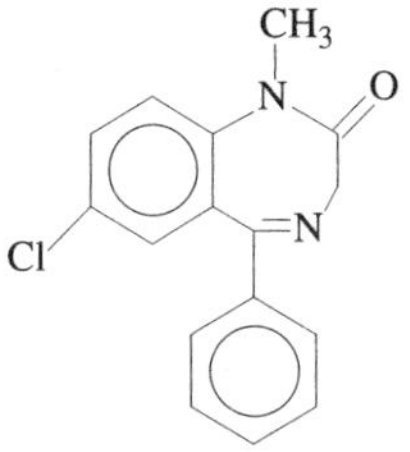   **6, Valium**

10.3.3 Modern drug discovery

It was very time-consuming to screen substances in animal tests for efficacy and to prepare lead analogues by classical synthesis. Therefore, it is not surprising that the researchers looked for alternatives. Today pharmaceutical companies apply computerized robotic systems in drug discovery. The starting point is the chemical library. This library does not contain books, but chemicals. Many companies keep small samples of all chemical compounds that they ever synthesized or extracted from the plant material. The amount of the samples is usually small and they are kept in microplates in dedicated temperature-controlled storage facilities. Chemical libraries of large pharmaceutical companies contain several million different compounds.

The size of the library is increased continuously using high throughput synthesis. The concept of *combinatorial chemistry* is applied, meaning that available analogues of starting material A are combined with analogues of B. This leads to a very large number of reactions requiring handling by robotic systems. Let us assume there are three analogues of A (A1, A2, and A3) and three analogues of B (B1, B2, and B3); then the following combinations are possible:

A1 + B1 => X1,   A1 + B2 => X2,   A1 + B3 => X3

A2 + B1 => X4,   A2 + B2 => X5,   A2 + B3 => X6

A3 + B1 => X7,   A3 + B2 => X8,   A3 + B3 => X9

The reaction between amino acid esters, **7**, and aldehydes, **8**, may serve as an example of a real combinatorial reaction that was used to produce 126 N-alkylated-α-amino methyl esters **10** via the intermediate **9**. The identity and yield of the products were determined by LC-MS analysis [4].

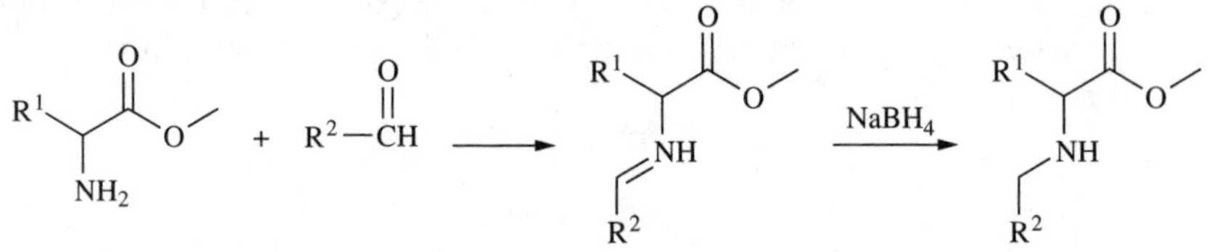

7   8   9   10

This example illustrates that the combinatorial approach yields a large number of different products. The combinatorial library can be reagent based; that means all available reagents of types A and B are combined with each other, or based on structure-activity relationship (SAR). SAR can be used to select starting compounds with chemical structures that are known to have pharmacological activity. On the other side, it can also help to avoid structures that have negative effects, such as high toxicity, low bioavailability, or slow metabolism. If SAR is applied, the library is product based, that means it is more focused than a random reagent-based library.

The relationship between chemical structure and activity has been recognized from early on, as the examples in Sec. 10.3.2 illustrate. Today, computerized expert systems allow the virtual screening of millions of possible structures with the objective to find the best candidates for development. However, the accuracy of such *in-silico* predictions is still low and it will take time before they can replace experimental discovery methods.

To have a large library is only a plus, when a high throughput technique is used for screening (high throughput screening [HTS]). Conventional techniques are precluded, because they are too slow and insufficient substance is available for testing on animals. Binding assays and microplate robotic techniques are used for HTS. When the physiological target is identified, a receptor binding assay is developed, which should be a characteristic indicator of the intended biological

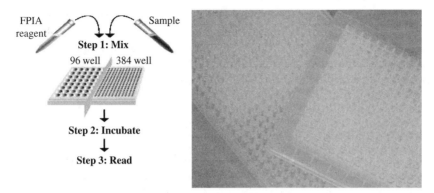

Figure 10.1 Basic setup of a high throughput binding assay using 96 or 384 well microplates (left). Photo of real microplates (dimensions of a plate: 12 cm × 8 cm). (*Source: Adapted from www.caymanchem.com*)

effect. Then all or selected substances from the library are screened with the assay. Very small amounts of the substances are transferred to plastic plates that have tiny holes, keeping the reagents for the assay. A plate can have 96, 384, or several thousand wells. This means screening takes place on a microliter scale with unbelievable high speed, many thousand substances per day (Figs. 10.1 and 10.2).

Figure 10.2 Modern high throughput screening (HTS) laboratory. (*Source: Courtesy TECAN Deutschland GmbH, Germany.*)

To find a good assay for the desired pharmaceutical target is the critical step for successful HTS. Let us take a protein as an example for a target. Only substances that bind to protein can have physiological activity; they provide a *hit*. Substances that do not bind fail the test and are no longer considered for this application.

Screening a million compounds usually results in a few thousand hits. These hits are investigated further to either confirm their activity or eliminate false positives. At the end of the screening part a few promising candidates are selected as leads for further development.

So far, the screening was random and selected the best candidates from the existing library. Perhaps a slight modification of the structure would increase the efficacy? This question is tested by high throughput synthesis of new compounds with structures similar to the lead compound. This new set of compounds is carried through a similar screening and confirmation process as described above. At the end of the exploratory phase 3 to 10 drug candidates are selected for further testing in the preclinical phase.

10.3.4 Preclinical testing

The objective of the preclinical phase is to assure that new substances are safe for humans before they are tested in patients. The tasks of the preclinical phase are summarized in Table 10.6.

First, a few kilograms of the test substance must be synthesized to have enough material for further testing. The purity and the impurities are determined by spectroscopic and chromatographic techniques and basic

TABLE 10.6   Tasks in Preclinical Development

Preclinical phase	Typical tasks
Route of synthesis	Development of an industrial synthesis method
Properties and stability	Identification of impurities, physicochemical properties, stability upon storage
Toxicity	Acute toxicity in rats and mice (single-dose application), 4 weeks to 2 years; repeated dose toxicity in rats and dogs Genetic toxicity in vitro and in vivo, for example, mutagenicity Reproductive toxicity in rats and rabbits
Pharmacokinetics and metabolism (ADME)	Absorption into blood circulation Distribution in body Metabolism (= chemical reactions) of drug in the body Elimination of drug from body
Development of drug formulation	Development of tablets, syrups, injectables, and so on, containing the drug substance

physicochemical properties are established, especially water solubility and stability.

The toxicity testing starts simultaneously with analytical studies. Single application of different doses of the test substance to rats or mice gives an indication of the acute oral toxicity. If the results are satisfactory, chronic toxicity studies are conducted. Here, doses of the test substance are given repeatedly to animals for a long time to investigate the long-term effects.

Another type of study is described by the term ADME (absorption, distribution, metabolism, and elimination). The rate of absorption describes how fast the drug substance enters the bloodstream and is a measure of bioavailability. The distribution in the body is measured to confirm that the substance reaches the target organ and that it does not accumulate in other organs. Metabolism is the chemical conversion of the substance in the body to other compounds and elimination describes how fast and by which route the substance and its metabolites are eliminated from the body.

The distribution of the substance in the body is often studied with substances that have carbon-14 as a radioactive label in the molecule. The labeled substance must be specifically synthesized, but it allows a direct visualization of the distribution in the animal body. In a process called *whole body radiography*, thin slices of the dead deep-frozen animal are prepared and scanned for distribution of radioactivity. Images of the various two-dimensional slices are then combined to give a three-dimensional picture. In a parallel experiment, the target organs are isolated and extracted with solvents. The extracts are used to isolate and identify the metabolites.

Although there is currently a lot of research to replace the preclinical safety tests in animals with in vitro high throughput methods, so far success is limited to some metabolite and genetic toxicity screens. These methods have not been accepted yet by the regulatory authorities in the various countries as valid alternatives to animal tests. The reason is that in these screening assays too many false results are obtained to accept them as the only basis for safety assessments.

10.3.5 Clinical testing

Finally, after 3 to 5 years of research and development, the new drug substance can be applied to humans. Commonly, three clinical phases and the postmarking monitoring are distinguished. Clinical studies conducted with humans are directed by physicians. Nevertheless, clinical and analytical chemistry plays an important role in monitoring the results of the tests (Table 10.7).

In phase I, the substance is applied to healthy human volunteers, starting with a very low dose and then slowly increasing until the planned

TABLE 10.7   Tasks in Clinical Development and Postmarketing Research

Clinical testing	Description of work
Phase 1	Testing of pharmacokinetics and side effects in healthy human volunteers
Phase 2	Proof of concept and efficacy in small groups of human patients
Phase 3	Testing of efficacy and long-term effects in large groups of patients in different countries
Postmarketing research	Development of new indications and formulations

pharmaceutical dose is reached. The purpose of the test is twofold. The clinical parameters are monitored to detect any side effect that may not have been observed in the animal tests. The second is to establish the pharmacokinetics in humans. For this purpose blood samples are taken at predetermined intervals and analyzed for drug substance concentration usually by LC-MS-MS (liquid chromatography with tandem mass spectrometry). The kinetics of absorption and elimination and the bioavailability parameters are calculated on the basis of analytical results.

When a drug is administered orally as a tablet, it takes some time before it dissolves in the stomach and transfers to the blood circulation. During this absorption phase, the level of the drug increases with time and reaches a maximum after about 1 to 2 h. As soon as the drug reaches the circulation, it is also metabolized, mainly in the liver. The drug is slowly eliminated from the body and as a consequence the blood level and also the effect of the drug decrease (Fig. 10.3).

After the basic pharmacokinetic parameters are known, the tests are repeated with volunteers that take other medication or eat different diets to see whether these parameters influence the pharmacology of the new drug.

In phases II and III, the drug is administered to patients who have the adverse health condition which the drug is intended to treat. The efficacy is monitored by comparing the patients that receive the new drug with others that receive an existing reference drug or a placebo. A placebo is a preparation that looks exactly like the real drug, but does not contain the active substance. The condition and the clinical analysis results of all patients are carefully recorded and statistically analyzed. At the end all findings are summarized in a report.

Clinical trials are subject to approval by ethics committees or governmental evaluation boards. All involved test persons must be informed about the potential risks and benefits and must confirm this in writing. The ethical principals of clinical trials are internationally accepted and applied ("Declaration of Helsinki") [5].

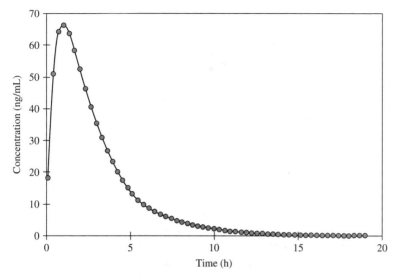

Figure 10.3 Typical curve of the blood levels of a drug after a single oral application. (*Source*: *Adapted from PK Tutorial, available from www.pksummits.com*)

10.4 Classification and the Chemistry of Pharmaceutical Products

Pharmaceutical products can be readily classified and arranged by their therapeutic functions. Classification according to chemical criteria is more difficult, because the pharmaceutical industry employs more complicated steps in its manufacturing processes than most other chemical industries. In this section, we will briefly describe the various therapeutic classes of drugs and we will also include the reactions involved in the synthesis of some representative examples of drug substances.

10.4.1 The analgesics

Analgesics represent an important class of drugs that is used primarily for the relief of severe pain. They are classified as narcotic and nonnarcotic.

The opiates are perhaps the oldest drugs known to man. Opium contains a complex mixture of almost 25 alkaloids. The principal alkaloid in the mixture, and the one responsible for analgesic activity, is morphine, **11**. Pure morphine is especially good for treating dull, constant pain, and periodic pain. Unfortunately, it has a large number of side effects, which include the depression of the respiratory center, excitation, nausea, euphoria, dependence, and others [6].

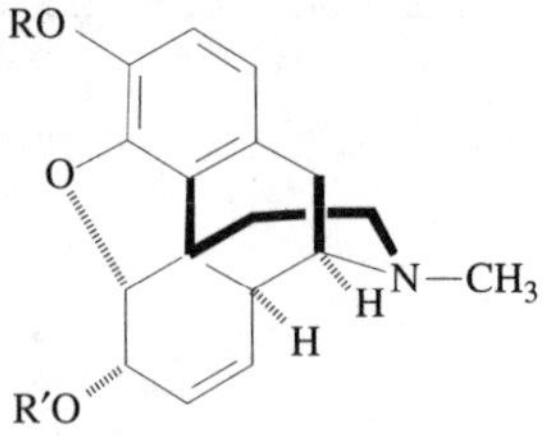

11; R = R′= H   Morphine

12; R = CH$_3$, R′ = H   Codeine

$$\overset{O}{\overset{||}{}}$$

13; R = R′ = CH$_3$C   Heroin

The first and the easiest morphine-based analgesics could be prepared by peripherical modifications of the molecule. Codeine, **12**, is the methyl ether of morphine and also present in opium. Codeine is used for treating moderate pain and coughs. The analgesic activity drops drastically by methylation of morphine; for example, codeine is only 0.1 percent as active as morphine. It has, however, significantly less side effects than morphine itself. Clearly, a free phenolic group is crucial for analgesic activity [12]. The diamorphine or heroin, **13**, a morphine analogue, is more active than morphine by a factor of two. The codeine is produced from morphine by the alkylation reaction in presence of a quaternary nitrogen alkylating agent such as phenyl trimethyl ammonium hydroxide. The morphine is dissolved in absolute alcohol in the presence of potassium ethoxide. The process leads to the formation of dimethyl aniline and ethanol that are recovered and recirculated [7, 8].

$$\text{Morphine} + C_6H_5N(CH_3)_3OH + C_2H_5OK \xrightarrow[\text{High pressure}]{130°C} \text{Codeine}$$

$$+C_6H_5N(CH_3)_2 + C_2H_5OH + KCl$$

A newer compound (**14**, bremazocine) has a longer duration, has 200 times the activity of morphine, appears to have no addictive properties, and does not depress breathing.

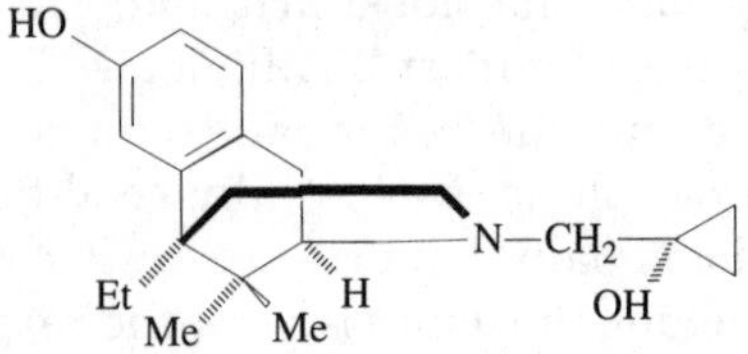

14, Bremazocine

The nonnarcotic analgesics are widely used as nonprescription drugs. They are also efficient for the relief of mild pain and symptoms of rheumatoid arthritis [9]. For example, acetaminophen, **15**, sold under the trade name Tylenol, is a widely used nonprescriptive drug. Acetaminophen is usually prepared by reacting p-aminophenol with a mixture of glacial acetic acid and acetic anhydride [8].

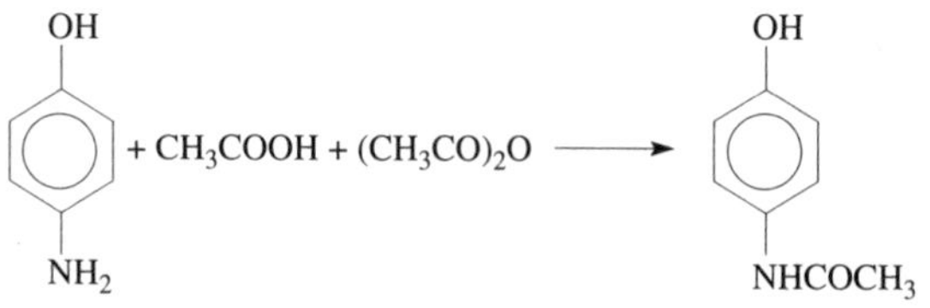

OH

$\bigcirc$ + CH_3COOH + $(CH_3CO)_2O$ $\longrightarrow$ $\bigcirc$

NH$_2$

OH

NHCOCH$_3$

15, Acetaminophene

Phenacetin, **16**, is also a nonprescription drug that represents the ethyl ether of acetaminophen and is prepared from p-ethoxaniline [8].

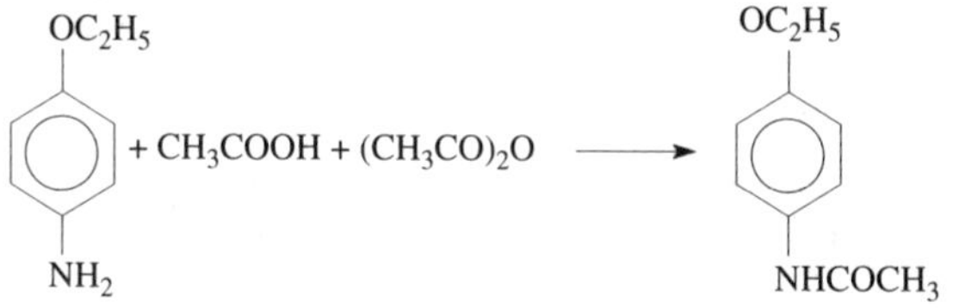

OC$_2$H$_5$

$\bigcirc$ + CH_3COOH + $(CH_3CO)_2O$ $\longrightarrow$ $\bigcirc$

NH$_2$

OC$_2$H$_5$

NHCOCH$_3$

16, Phenacetin

The acetyl ester derivative of salicylic acid is an old analgesic known as Aspirin, **17** [8].

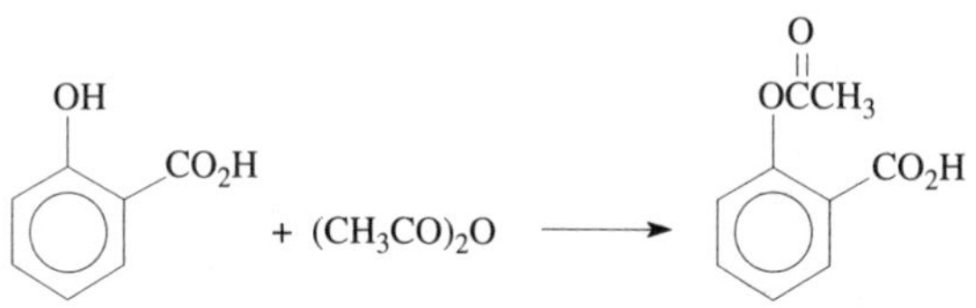

OH

CO$_2$H

$\bigcirc$ + $(CH_3CO)_2O$ $\longrightarrow$

Salicyclic acid

O
‖
OCCH$_3$

CO$_2$H

$\bigcirc$

17, Aspirin

(S)-Naproxen, **18**, and S-ibuprofen, **19**, are important and widely used analgesics with annual sales of about US$1.4 billion and a production volume of about 8000 tons. A technical feasible use of R,R-tartaric acid as chiral auxiliary was demonstrated in the Zambon Process for S-naproxen manufacture [3]. The diastereoselective bromination is followed by bromine hydrogenolysis and hydrolysis to produce S-naproxen in 75 percent overall yield.

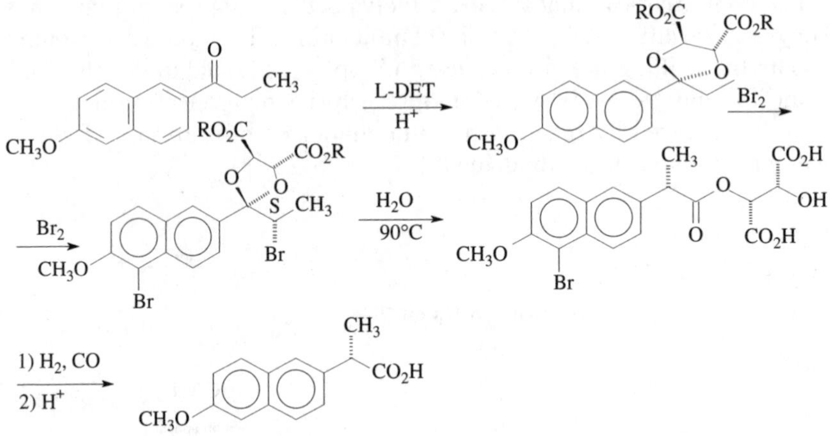

(S)-Naproxen, **18**

An elegant method for the production of ibuprofen, **19**, was developed by Hoechst-Celanese. This method involves two steps: catalytic hydrogenation and catalytic carbonylation [10].

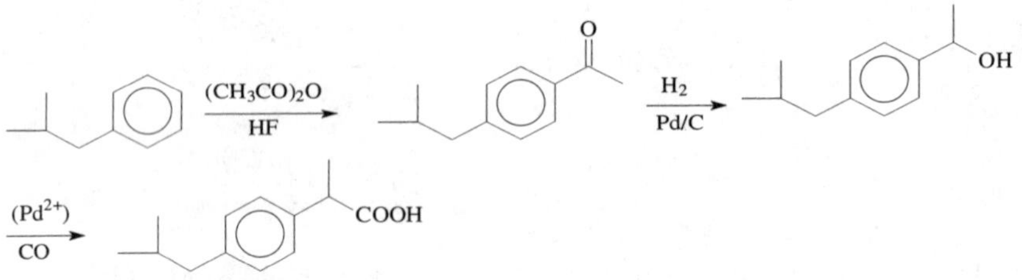

Ibuprofen, **19**

(S)-ketoprofen, **20**, is also an important analgesic used to relieve strong-to-mild pain. It is produced by asymmetric hydrogenation followed by oxidation.

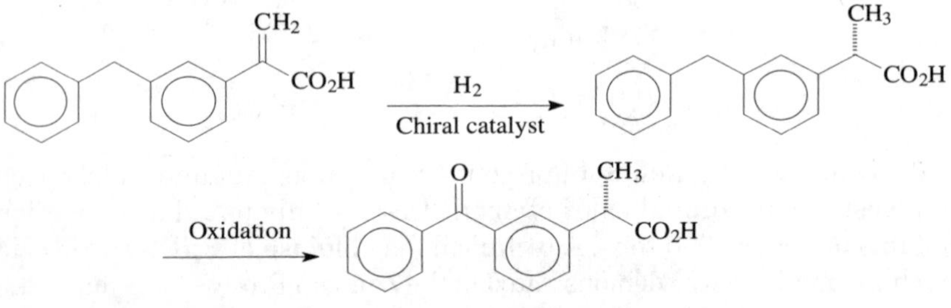

(S)-Ketoprofen, **20**

10.4.2   Antiallergy and antiasthmatic drugs

Allergic reactions occur when the body's immune system reacts to a foreign substance that is typically not toxic. Antihistamines such as Claritin, Hismanal, and Zyrtec are used to relieve allergies, such as seasonal hay fever and other forms of allergies, by counteracting the effect of histamines in the body that are the transmitters of the allergic symptoms. The first generation of antihistamines, known as piperadines, include azatadine, **21**, hydroxyzine HCl, **22**, and others [3, 9, 11].

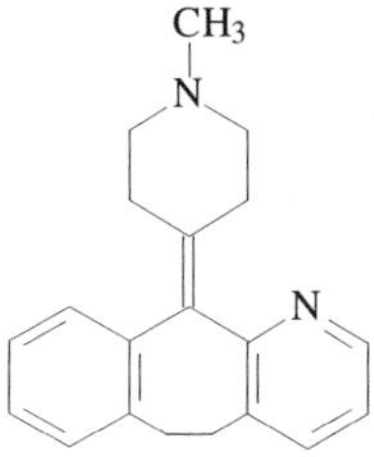

Azatadine, **21** 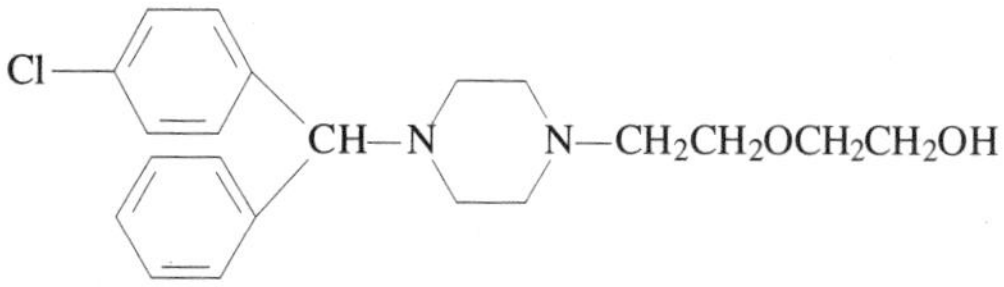   Hydroxyzine HCl, **22**

The second generation of antihistamines includes stronger drugs that inhibit the release of histamine and other inflammatory mediators from several cell types. Examples of this group include terfenidine, **23** and loratadine, **24** [11].

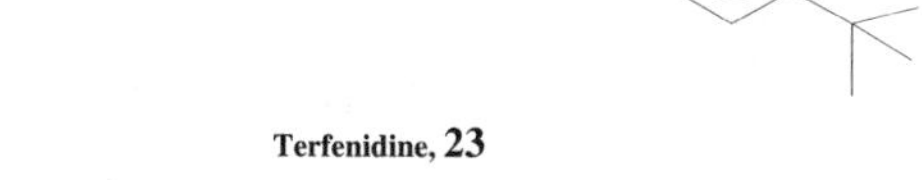

Terfenidine, **23**

Loratadine (Claritin), **24**, is prepared in good yield by the following reaction [12]:

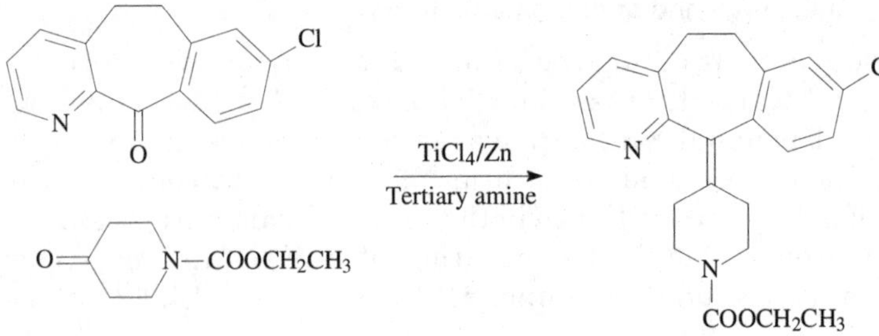

Loratadine, **24**
(Claritin)

Cetrizine, **25**, is sold as Zyrtec and is widely used as an antiallergic and is most effective against rash/hives. Cetrizine is an orally active racemic compound [11].

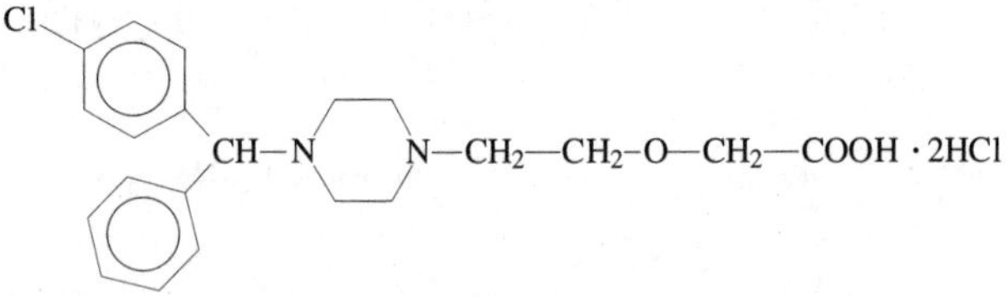

Cetrizine (Zyrtec), **25**

Antiasthmatic and bronchodilators (e.g., Singulair, Flovent, Ventrolin, and Spiropent) prevent the effects of asthma by treating the underlying causes of airway inflammation. They also provide symptomatic relief by expanding the bronchial airways by relaxing the bronchial muscles [3, 11]. Presently, the prevailing medical therapy includes the usage of selective B_2-receptor stimulants for the treatment of bronchial asthma. Xanthine derivatives, such as theofylline, **26**, and enprofylline, **27**, were found to possess the property of effecting smooth muscle relaxation [13].

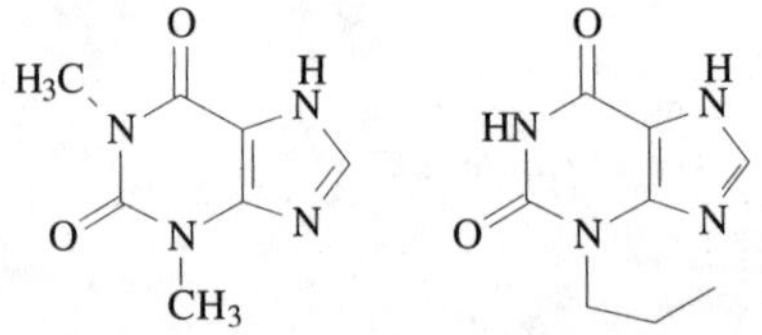

Theofylline, **26**   Enprofylline, **27**

Enprofylline, **28**, is produced in a multistep process of conversion of aminopropyluracil, **29** [3].

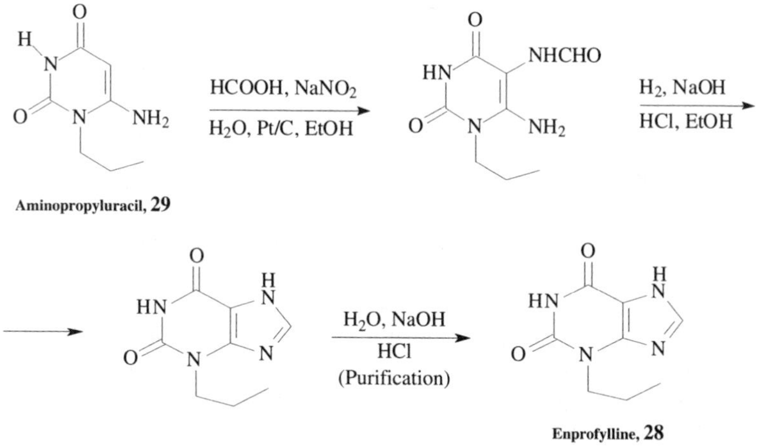

Aminopropyluracil, **29**

Enprofylline, **28**
(Pure)

10.4.3 Antibacterials and antibiotics

The fight against bacterial infection is one of the greatest success stories of medicinal chemistry. Bacteria were first identified in the 1670s by Van Leeuwenhoek, following his invention of the miscroscope [6].

Sulfonamides (known as sulfa drugs) represent the best example of antibacterial agents before the discovery of penicillin.

In 1935, a red dye called Prontosil, **30**, was discovered to have antibacterial properties *in vivo* (i.e., when given to laboratory animals). No antibacterial effect was observed *in vitro* (i.e., Prontosil could not kill bacteria grown in a test tube). This result remained a mystery until it was discovered that Prontosil was not in fact the antibacterial agent. Instead, it was found that the Prontosil was metabolized by bacteria present in the small intestine of the test animal, and broken down to sulfanilamide **31** [6].

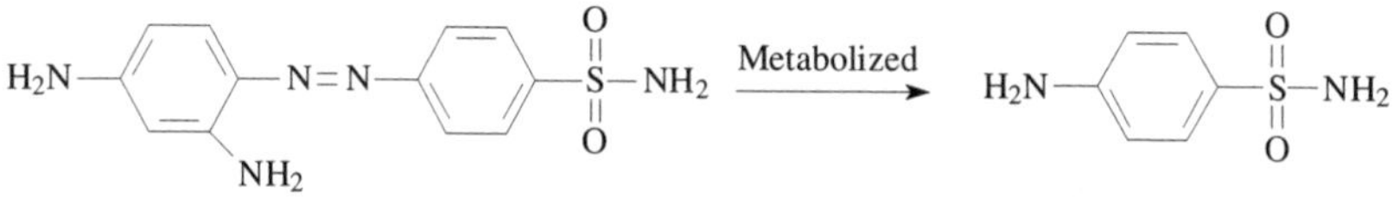

Prontosil, **30**   Sulfanilamide, **31**

The synthesis of a large number of sulfonamide analogues, **32**, led to the following conclusions [6]:

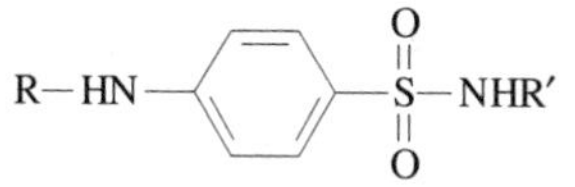

32

1. The p-amino group is essential for the activity and must be unsubstituted (R═H).
2. The aromatic ring and the sulfonamide group are both needed.
3. The aromatic ring must be *para*-substituted only.
4. The sulfonamide nitrogen must be secondary.

It is interesting to observe that certain human genotypes are more susceptible to one type of sulfonamides than others. For example, the Japanese and Chinese metabolize sulfathiazole, **33**, more quickly than Caucasians (e.g., Europeans) thereby reducing its efficacy [6]. It was discovered that this specificity problem of sulfathiazole could be overcome by replacing the thiazole ring in sulfathiazole with a pyrimidine ring to give sulfadiazine, **34**.

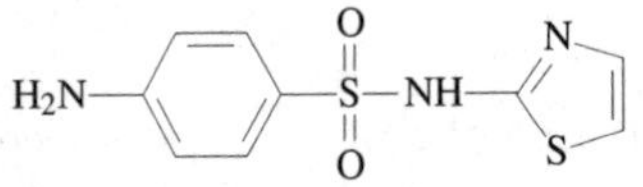

Sulfathiazole, **33** 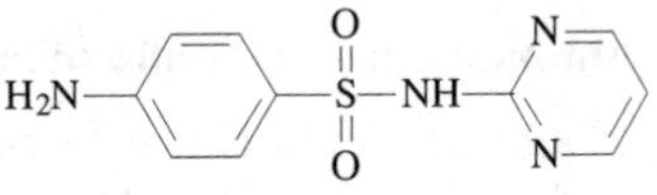   Sulfadiazine, **34**

Sulfonamides maintain a significant role even today, because they are often used in combination with other antibacterials. For example, trimethyoprim, **35**, has a diaminopyrimidine structure that has proved to be a highly selective, orally active, antibacterial, and antimalarial agent [6]. Trimethyoprim is combined to sulfamethoxazole, **36**, for the treatment of bacterial respiratory tract infections and gastrointestinal infections [9].

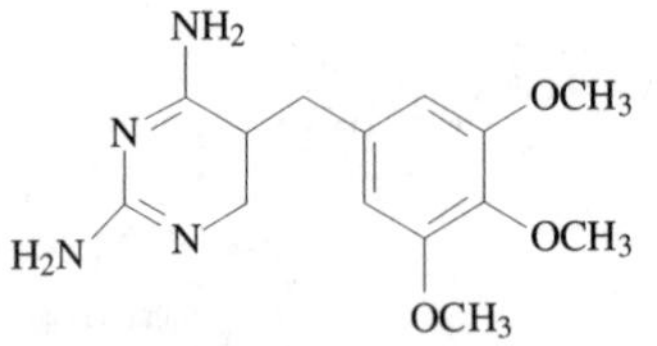

Trimethyoprim, **35** 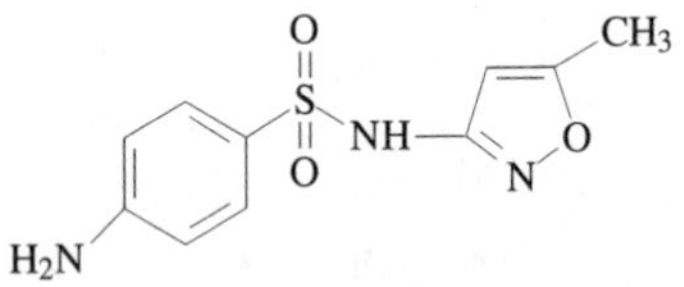   Sulfamethoxazole, **36**

For most indications, sulfa drugs have been largely replaced by antibiotics because they have a relatively narrow antibacterial spectra, low potency, and cause rapid development of resistance in the bacteria.

There are two major classes of chemicals that act as antibiotics—penicillins and cephalosporins. The penicillins are still the most active agents for the treatment of a variety of major infectious diseases. Penicillin contains a highly unstable-looking bicyclic system consisting of a four-membered β-lactam ring fused to a five-membered thiazolidine ring. The skeleton of the molecule suggests that it is derived from the amino acids cytosine and valine.

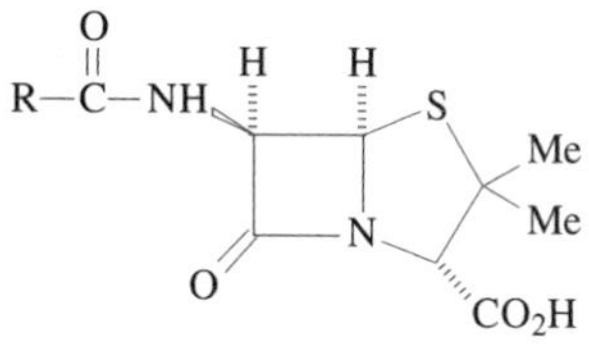

General structure of penicillin

The penicillin derivatives include:

Benzyl penicillin (PEN G); R = ⟨◯⟩—CH₂; **37**

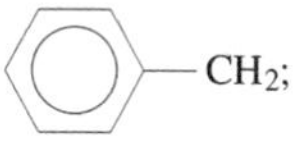

Penicillin G is a nontoxic drug and active against gram-positive bacilli (e.g., staphylococci, meningitis, and gonorrhoea) and many gram-negative cocci. It is not active over a wide range of bacteria and ineffective when taken orally, since it is hydrolyzed by the stomach acid. Penicillin G can only be administered by injection.

Phenoxy-methyl penicillin (PEN V); R = ⟨◯⟩—O—CH₂; **38**

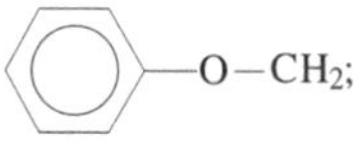

Penicillin V has an electronegative oxygen on the acyl side-chain with the electron withdrawing effect required. The molecule has better stability than penicillin G and is stable enough to survive the acid in the stomach. Thus, it can be given orally [6].

Further work has been done to solve the problem of acid sensitivity of the penicillin derivatives by incorporating into the side-chain a five-membered heterocycle which was designed to act as a steric shield and also to be electron withdrawing. Oxacillin **39**, cloxacillin **40**, and flucloxacillin **41**, are acid-resistant and penicillinase-resistant [6]. The latter is an enzyme produced by bacteria that causes resistance, because it induces the decomposition of the penicillin molecule.

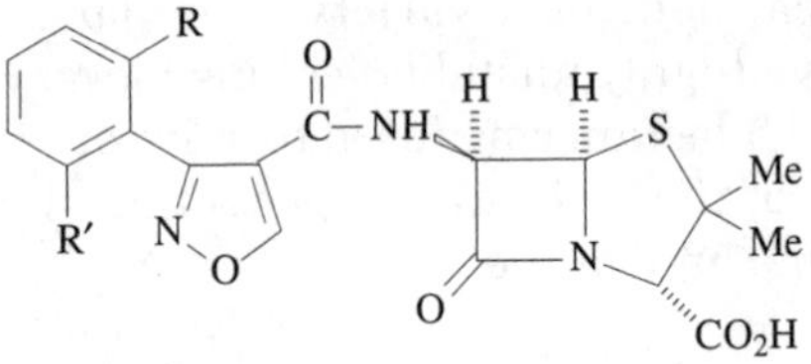

Oxacillin (R = R'= H) (39)
Cloxacillin (R = Cl; R'= H) (40)
Flucloxacillin (R = Cl, R'= F) (41)

Some penicillin derivatives have a useful activity against both gram-positive and gram-negative bacteria. There are two classes of such broad-spectrum antibiotics and both have an alpha-hydrophilic group. However, in one class the hydrophilic group is an amino function as in ampicillin, 42, or amoxicillin, 43, whereas in the other the hydrophilic group is an acid group as in carbenicillin, 44.

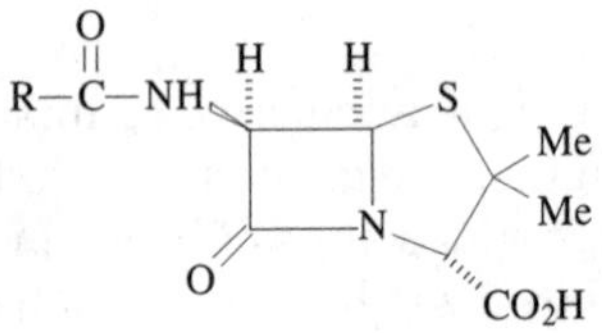

Ampicillin (42) R =

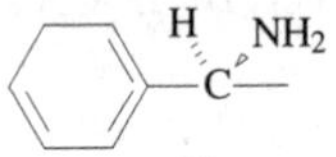

Amoxycillin (43) R =

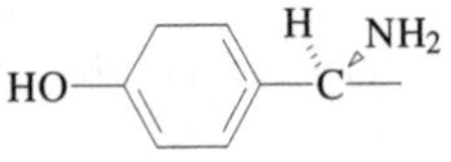

Carbenicillin (44); R =

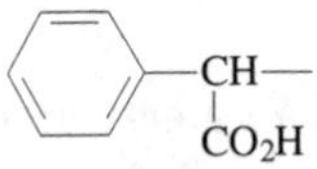

The second major group of β-lactam antibiotics includes cephalosporins and cephamycins. The analog X is H are cephalosporins, and those X is OCH$_3$ are called cephamycins.

Cefazoline, 45, is a first-generation cephalosporin, ceroxitim and cefaclor, 46, are second-generation, ceftizoxime, 47 and cefoxitin, 48, are third-generation cephalosporins [14, 15].

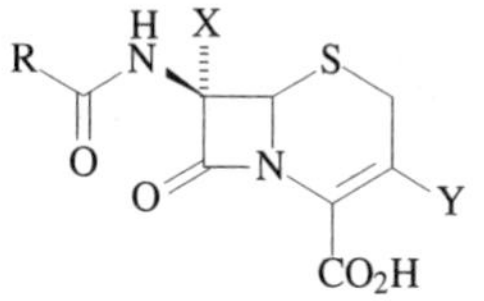

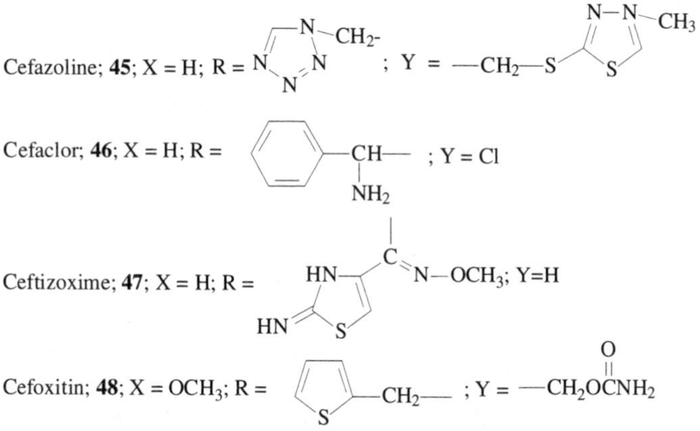

Cefazoline; **45**; X = H; R = (tetrazole-N-CH₂-) ; Y = —CH₂—S—(thiadiazole-CH₃)

Cefaclor; **46**; X = H; R = (phenyl)—CH— ; Y = Cl
                                        |
                                       NH₂

Ceftizoxime; **47**; X = H; R = (thiazole HN, HN-S) C=N—OCH₃; Y=H

Cefoxitin; **48**; X = OCH₃; R = (thiophene-S)—CH₂— ; Y = —CH₂OCNH₂ (O double bond)

10.4.4 Antidepressants

The new era of therapeutics for the treatment of depression began in the late 1950s with the introduction of both the monoamine oxidase (MAO) inhibitors and the tricyclic antidepressants. The first MAO inhibitor was iproniazid, **49**, which initially was used as an antituberculosis drug until it was observed that patients taking it exhibited excitement and euphoria [9].

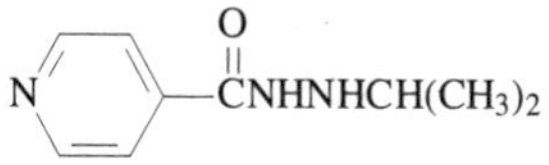

$$N \text{(pyridine)}—\overset{O}{\overset{\|}{C}}NHNHCH(CH_3)_2$$

Iproniazid, 49

Some of the major drugs in the category of MAOs include phenelzine, **50**, moclobemide, **51**, toloxatone, **52**, and others. Tranylcypromine (trans-phenyl-cyclopropylamine), **53**, was one of the first MAO inactivators approved for clinical use [6].

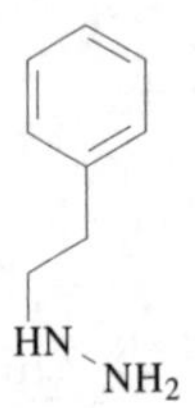

Phenelzine, **50** 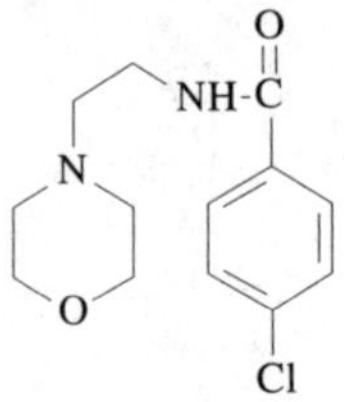   Moclobemide, **51** 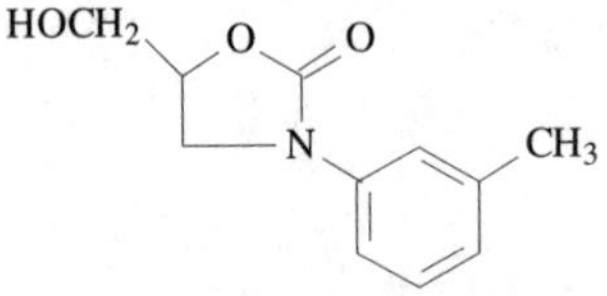   Toloxatone, **52**

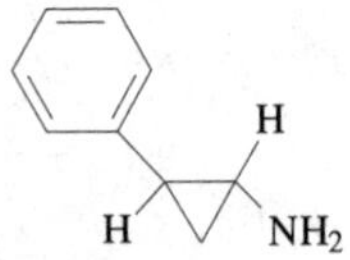

Tranylcypromine, **53**

Among the tricylic antidepressants, imipramine, **54**, was the first commercially available drug, which was introduced in 1960 [9].

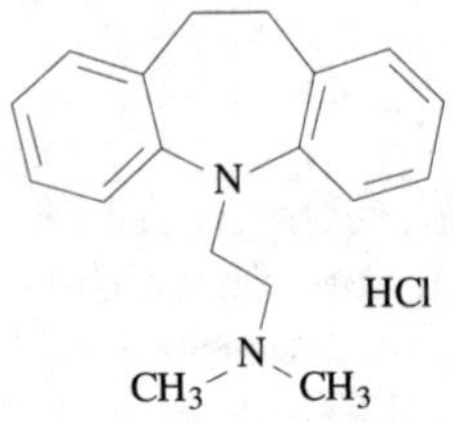

Imipramine HCl, **54**

Other tricyclic antidepressants such as amitriphyline, **55**, nortriptyline, **56**, and clomipramine, **57**, are used for the treatment of obsessive-compulsive disorders.

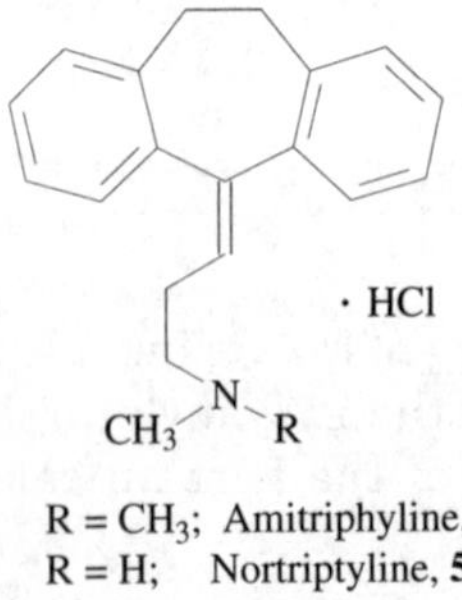

R = CH$_3$; Amitriphyline, **55**
R = H;   Nortriptyline, **56**

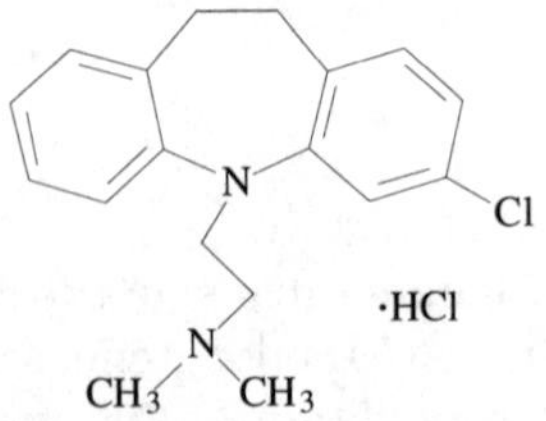

Clomipramine, **57**

10.4.5 Antiepileptics

Epilepsy is a disease that is characterized by recurring convulsive seizures. Phenobarbital, **58**, possess specific usefulness in epilepsy. In general, barbituric acid derivatives are synthesized from phenylethylmalonic diethyl ester [8].

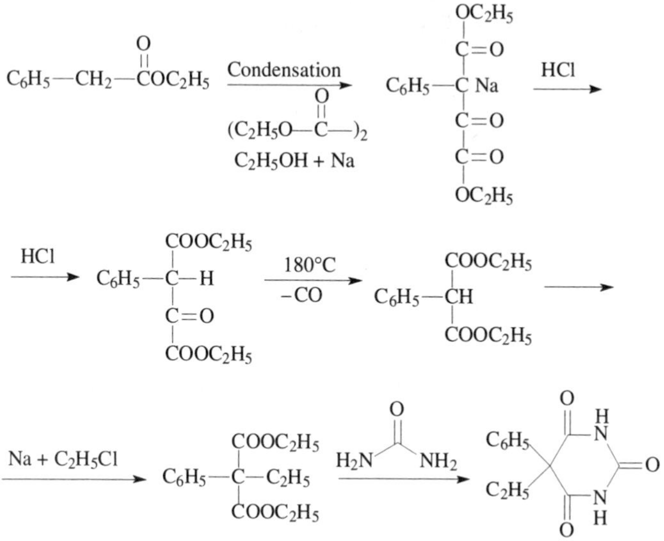

Phenobarbital, **58**

Other drugs are also used for the treatment of epilepsy, phenytoin, **59**, carbomazepine, **60**, and valproic acid, **61** are well-known anticonvulsants.

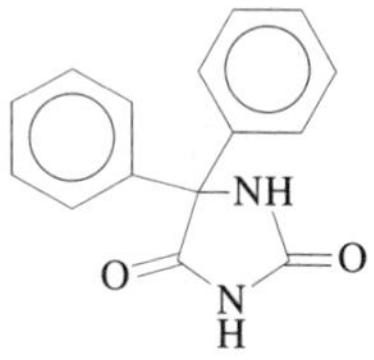

Phenytoin, **59** 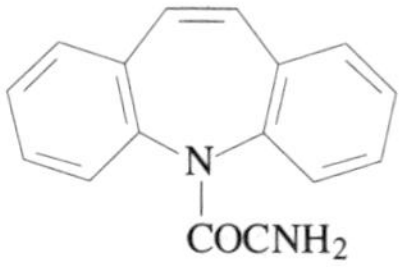   Carbomazepine, **60** 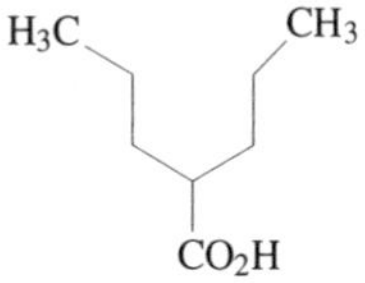   Valproic acid, **61**

10.4.6 Antihypertensives

Antihypertensive drugs are grouped into four main categories that include adrenergic blockers (e.g., cardura, **62**, minizide, **63**), adrenergic stimulants (e.g., aldoclor, clorpres), [α] and [β] adrenergic blockers (e.g., normodyne, **64**), angiotensin converting enzyme inhibitors (e.g., aceon, **65**, captopril, **66**) and angiotensin converting enzyme inhibitors with calcium channel blockers (e.g., nifedipine, **67** tarka). Most of these agents have a very low incidence of side effects and are used with patients with moderate hypertension [11]. The most important factors that must be taken into consideration are the patient's age, race, concomitant diseases and therapies, and life style. Drugs in all categories have application in the treatment of congestive heart failure.

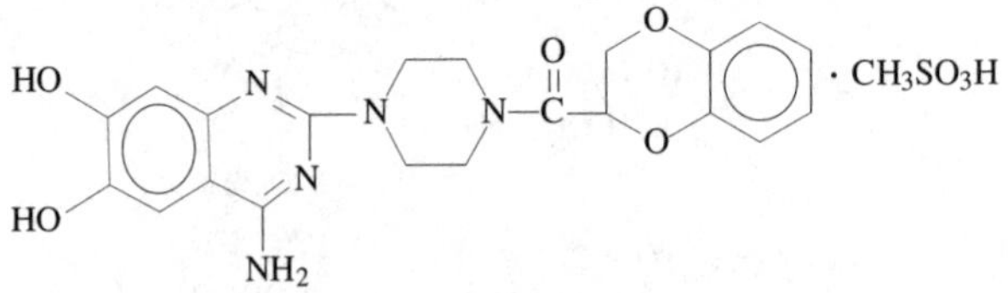

Cardura, **62**

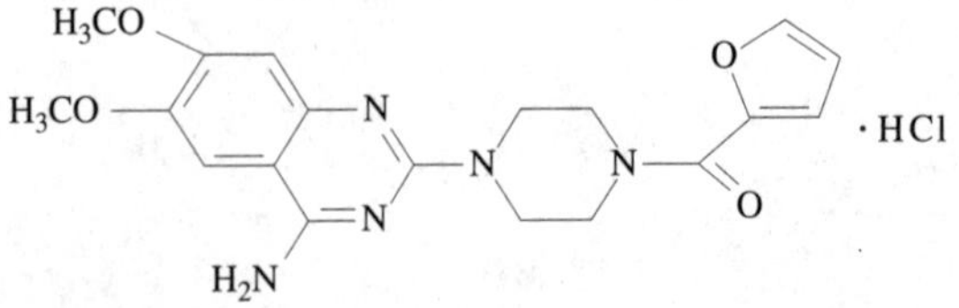

Minizide, **63**

Aldocolor combines methyldopa, **68**, and chlorthiozide, **69**. Clorpes is a combination of clonidine hydrochloride, **70** (a centrally acting antihypertensive agent), and chlortholidone, **71** (a diuretic).

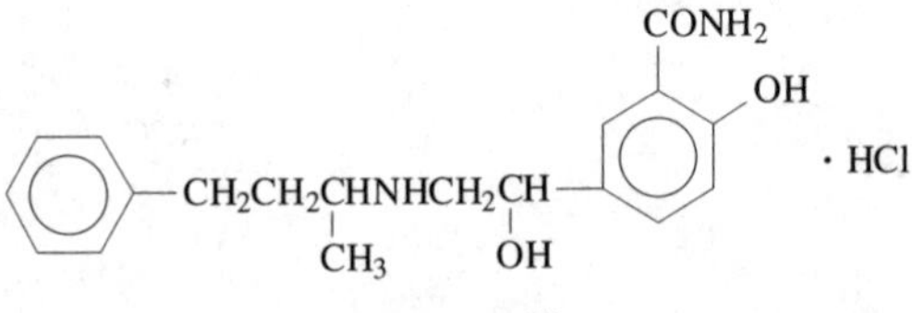

Normodyne, **64**

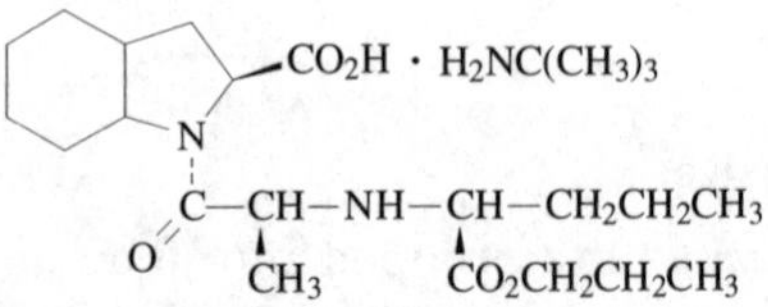

Aceon, **65**

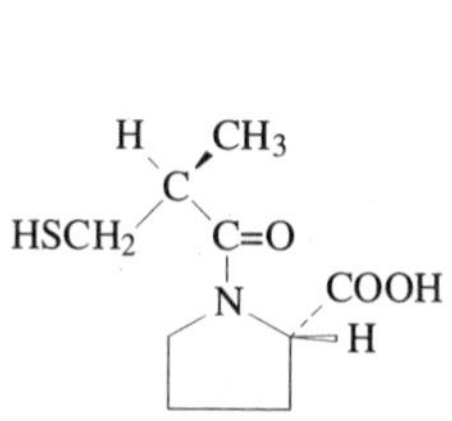

H₃C, N, CH₃ structure

Captopril, 66

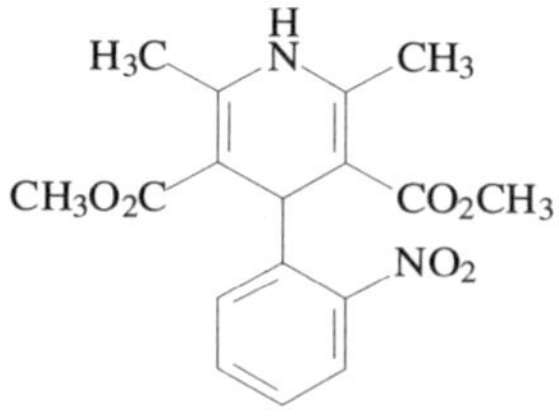

Nifedipine, 67

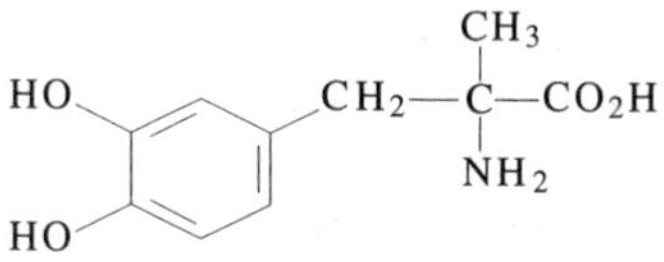

Methyldopa, 68
(antihypertensive)

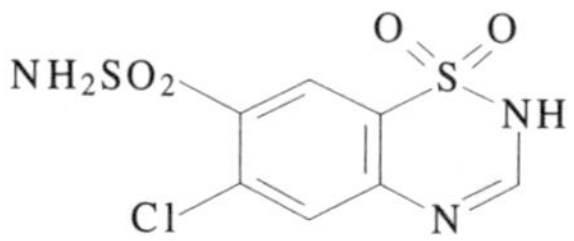

Chlorothiozide, 69
(diuretic and antihypertensive)

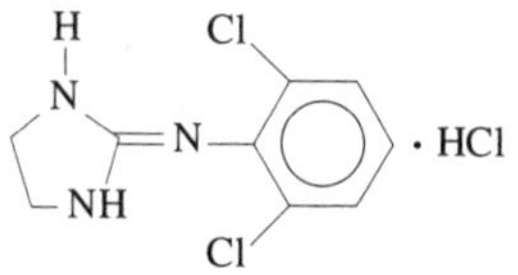

Clonidine, 70

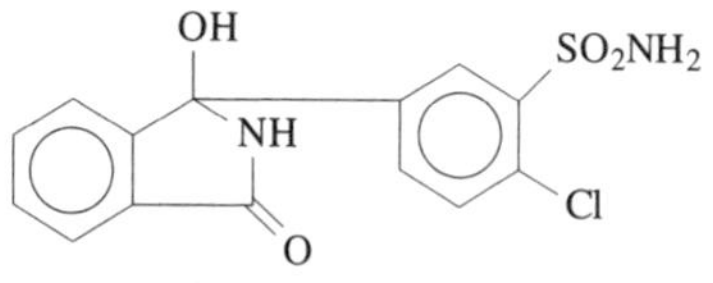

Chlorthalidone, 71

10.4.7 Antiulcers

The methods available for treating ulcers were limited and unsatisfactory before the intoduction of the cimetidine program in 1964. Ulcers are localized erosions of the mucous membrane of the stomach. For a long time it was not known how these ulcers arise, but the presence of gastric acid (HCl), released by cells known as parietal cells in the stomach, aggravates the problem and delays recovery. In the 1960s, sodium bicarbonate or calcium carbonate were used to neutralize gastric acid. Later, cimetidine, **72**, was used as an antiulcer medication. Cimetidine, **72**, acts by blocking the histamine molecules in the stomach from signaling. Cimetidine is prepared by reacting a substituted guanidine on an aminothio compound [8].

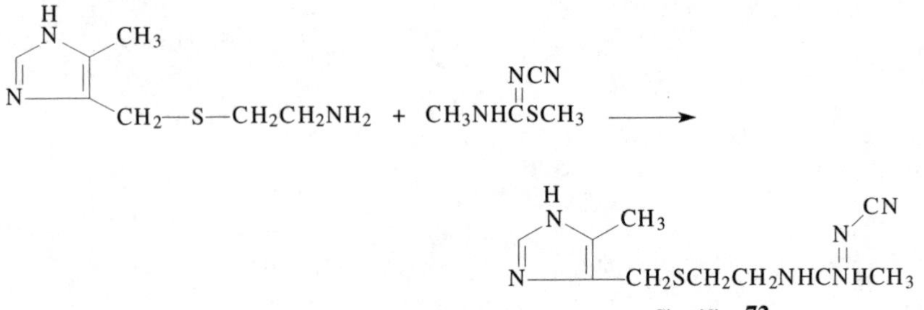

$$\text{imidazole-CH}_2\text{-S-CH}_2\text{CH}_2\text{NH}_2 \;+\; CH_3NHCSCH_3 \;\longrightarrow$$

Cimetidine, **72**

Further studies on cimetidine analogues showed that the imidazole ring could be replaced with other oxygen—or nitrogen—containing heterocyclic rings [6]. Ranitidine, **73**, with a furan ring, has fewer side effects than cimetidine, lasts longer, and is 10 times more active. Famotidine, **74**, and nizatidine, **75**, are more active antiulcers than cimetidine in vitro. Omeprazole, **76**, is also an active antiulcer agent.

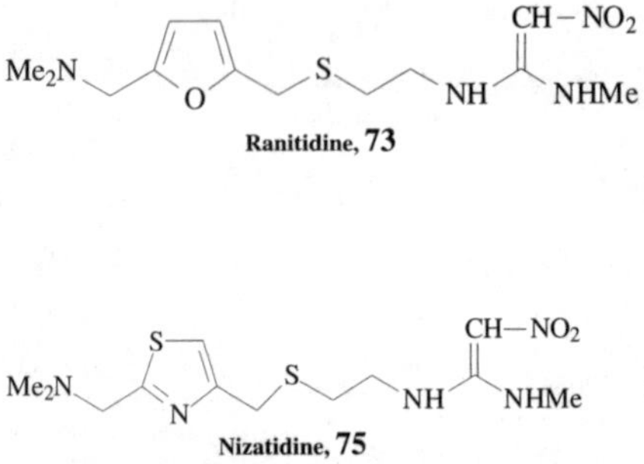

Ranitidine, **73**

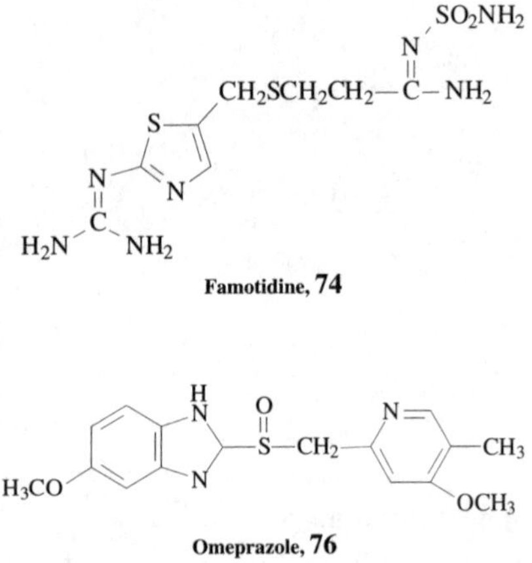

Famotidine, **74**

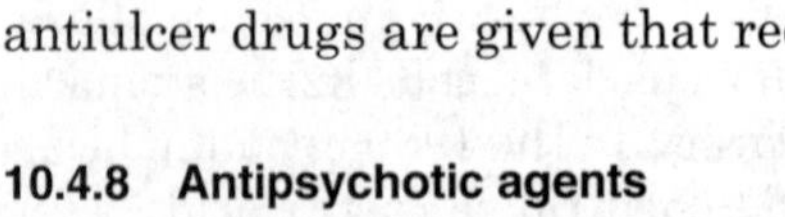

Nizatidine, **75**

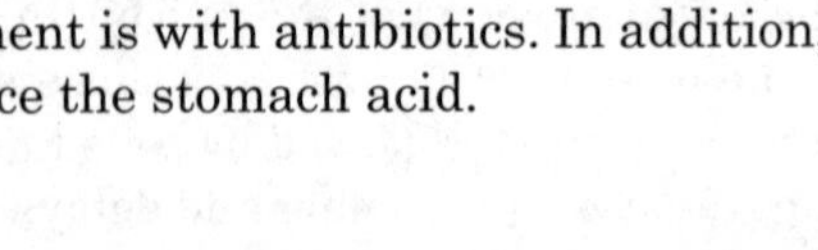

Omeprazole, **76**

Today, it is known that the bacteria *Helicobacter pylori* causes most ulcers. Therefore, the proper treatment is with antibiotics. In addition, antiulcer drugs are given that reduce the stomach acid.

10.4.8   Antipsychotic agents

These drugs are usually used to treat patients with psychotic or other serious related illnesses. They have common side effects such as muscle spasms, restlessness, and Parkinsonism. However, these drugs are considered safe, because the side effects are transient. Clozaril, **77**, is the most active drug with almost 1 percent incidence of agranulocytosis. Other antipsychotics agents such as geodon, **78**, thiothixene, **79**, and zyprexa, **80**, are also widely used as antipsychotic drugs [9, 11].

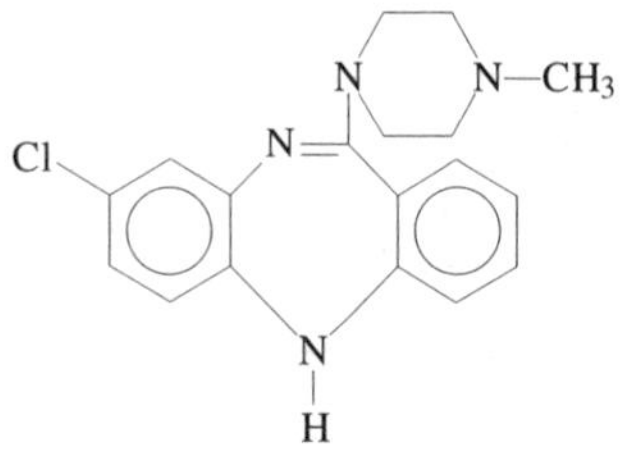

Clozaril, **77**

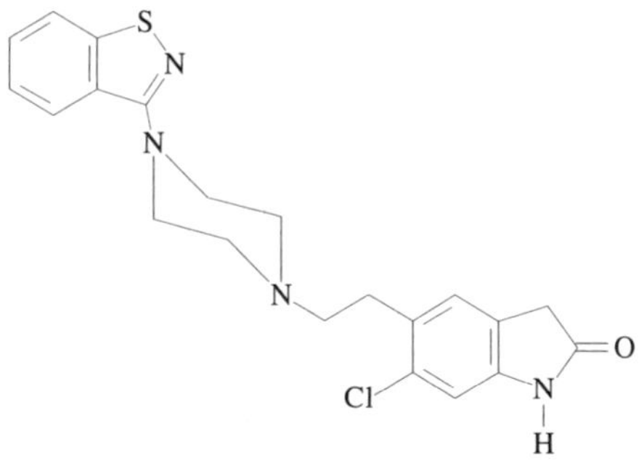

Geodon, **78**

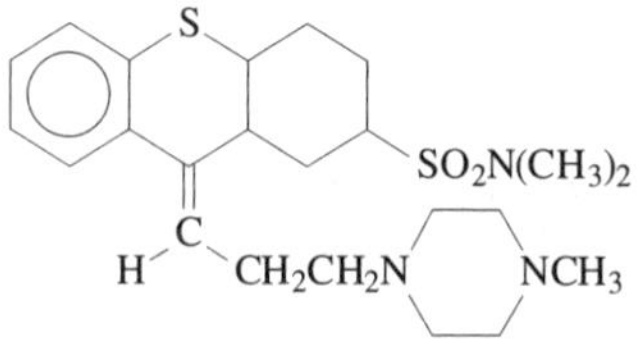

Thiothixene, **79**

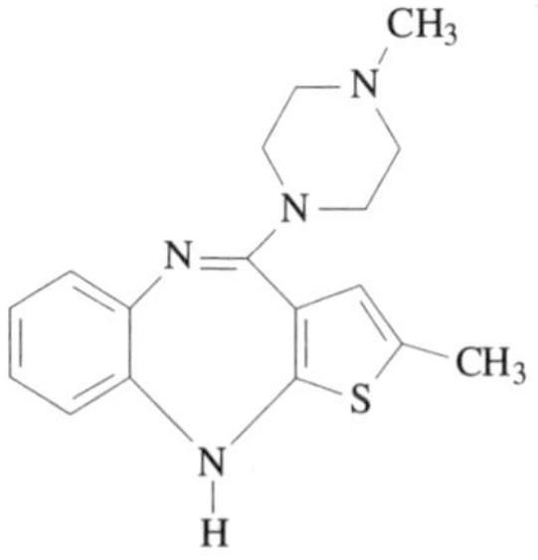

Zyprexa, **80**

10.4.9 Diuretics

Diuretics are highly efficient drugs for the treatment of edema associated with congestive heart failure. They are also used to increase the volume of urine excreted by the kidneys [9]. For example, duranide, **81**, a dichlorinated benzene disulfonamide, is an oral carbonic anhydrase inhibitor. Duranide reduces intraocular pressure by partially suppressing the secretion of aqueous humor [11]. Diuril, **82**, has an anti-hypertensive activity and is issued to control blood pressure [9]. Edecrin, **83**, is an unsaturated ketone derivative of an aryloxyacetic acid. Edecrin is used in the treatment of the edema associated with congestive heart failure, renal disease, and cirrhosis of the liver [11]. Amiloride, **84**, is also used as an adjunctive treatment with thiazide diuretics in congestive heart failure hypertension.

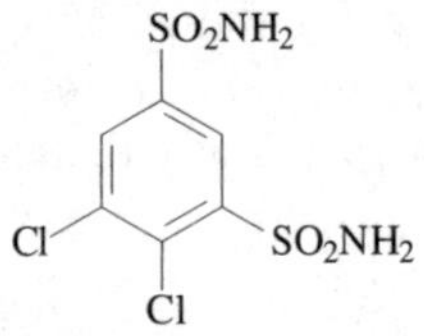

Duranide, **81**

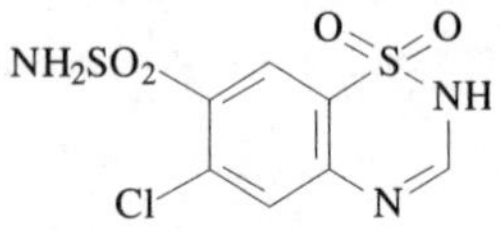

Diuril, **82**

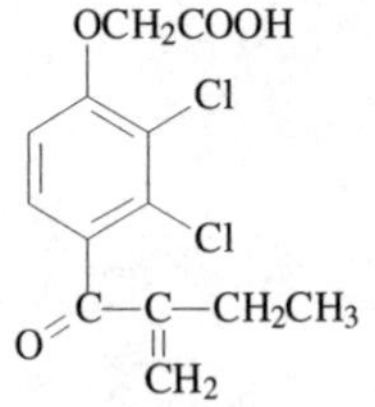

Edecrin, **83**

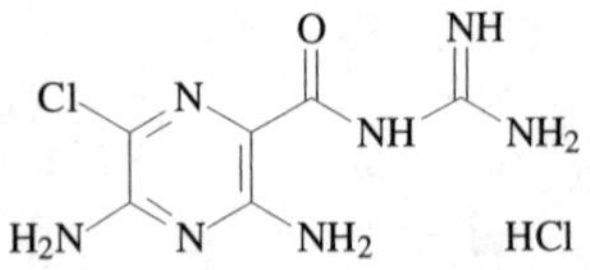

Amiloride hydrochloride, **84**

10.4.10   Contraceptives

Millions of women throughout the world take oral contraceptives (*the pill*). The most common type of oral contraceptive is a combination of a synthetic estrogen such as mestranol, **85**, or ethinylestradiol, **86**, and a progestin such as levonorgestrel, **87**, or norethynodrel, **88**. The contraceptives are sold under different trade names such as Levlen, Brevicon, Modicon, Necon, Ovcon, and others.

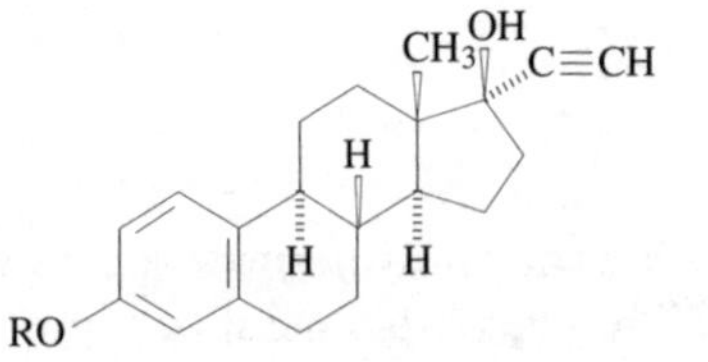

R = CH$_3$; Mestranol, **85**
R = H; Ethinyl estradiol, **86**

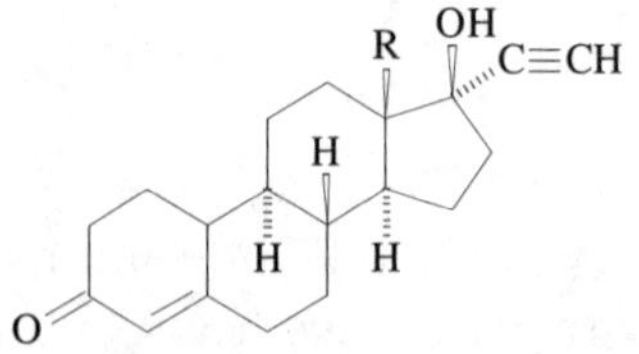

R = CH$_2$CH$_3$; Levonorgestrel, **87**
R = CH$_3$; Norethynodrel, **88**

10.4.11   Vitamins

Vitamin B$_1$, **89**, is essential for daily growth and the prevention of beriberi [8]. The commercial vitamin B$_1$ is obtained by the condensation of 6-amino-5-bromomethyl-2-methylpyrimidine hydrobromide with 5-(hydroxyethyl)-4-methylthiozole [8].

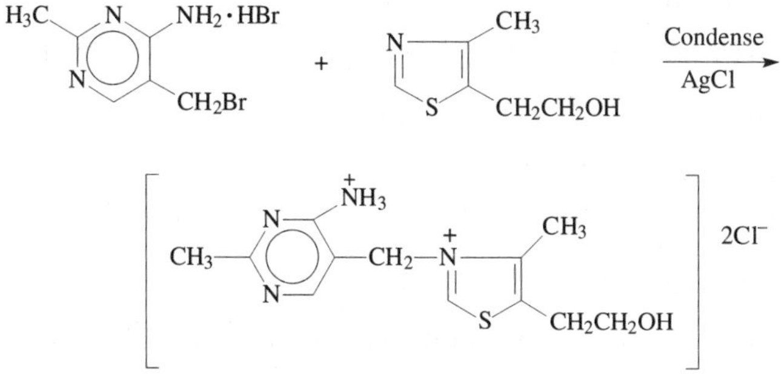

Thiamine chloride
vitamin B₁, **89**

Riboflavin, **90**, is an important and necessary element of most living cells. It is a very stable compound and it is considered to be the growth factor of vitamin B_2. Riboflavin is added on a large scale to bread, flour, and other dietary and pharmaceutical preparations [8].

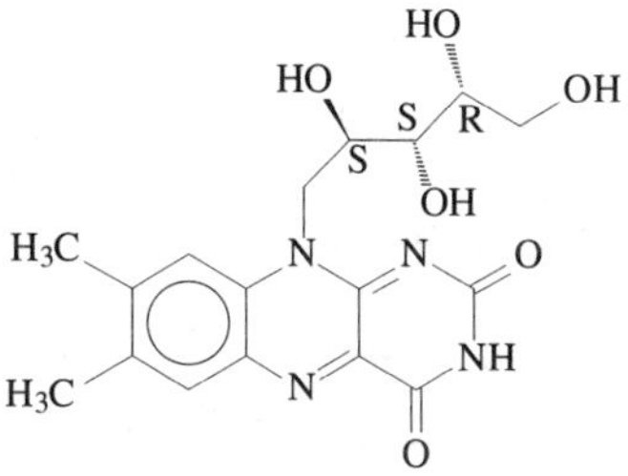

Riboflavin, **90**
(Vitamin B₂)

10.5 Industrial Processes in Pharmaceutical Industry

The production of pharmaceutical products can be divided into three main stages:

1. Research and development

2. The conversion of organic chemicals or natural products into bulk pharmaceutical substances or ingredients

3. The formulation of the final pharmaceutical products

10.5.1 Research and development [16]

The research and development of new drugs includes four phases:

1. Preclinical research and development
2. Clinical research and development
3. New drug application (NDA)
4. Postmarketing surveillance

The preclinical research and development starts after the discovery and the isolation of a promising compound in the laboratory. The biological activity and safety is extensively studied in the laboratory and tested on animals. This phase is usually completed in 6 years.

The clinical research and development is typically conducted in three phases, with each phase involving progressively more people. The aim of the first phase is to establish the safety of the drug. It involves a small number of healthy volunteers and lasts about 1 year. The effectiveness of the drug is determined in the second phase, which lasts about 2 years. In the third phase, the drug is used in clinics and hospitals to confirm the results of the earlier tests. The clinical research and development phase takes about 6 years.

Finally, the pharmaceutical company files a new drug application (NDA) for approval. The NDA should contain comprehensive data including the safety of the drug and its effectiveness. The regulations worldwide require specific and detailed information about the NDA:

The components and composition of the drug

The methods and controls used in the manufacturing

Processing and packaging of the drug

Data from all preclinical and clinical investigations

10.5.2 Chemical manufacturing [16]

Bulk pharmaceutical substances are usually complex organic chemical compounds, which are used to manufacture the final dosage form of drugs. These substances are usually manufactured either by (Table 10.8):

1. Chemical synthesis or
2. Fermentation or
3. Isolation or recovery from natural sources

Chemical synthesis [16]. Synthetic active ingredients and other compounds used in the production of pharmaceutical products are prepared by chemical synthesis using a batch processing. A complex series of reactions

TABLE 10.8   Pharmaceuticals by Bulk Manufacturing Process [16]

Chemical synthesis	Natural product extraction	Fermentation
Antibiotics	Antineoplastic agents	Antibiotics
Antihistamines	Enzymes and digestive aids	Antineoplastic agents
Cardiovascular agents	CNS depressants	Therapeutic nutrients
CNS (central nervous system)	Hematological agents	Vitamins
stimulants	Insulin	Steroids
CNS depressant	Vaccines	
Hormones		
Vitamins		

includes isolation and characterization of the intermediate products obtained in the various stages. Coproducts, such as salts and spent acids, metals, and catalysts may be recovered and reused or sold for reuse.

The intermediates and the final product are sometimes prepared at different manufacturing sites. This means that the intermediate must be isolated, packed and transferred to the other site for further processing.

The pharmaceutical industry is characterized by the diversity in the process design. Therefore, it is impossible to provide a single process flow diagram. However, most of the processes involve common steps such as reaction vessel, separation units, purification, wastewater treatment, air emission control, and others. A simplified process flow diagram for drug synthesis is given in Fig. 10.4 [16].

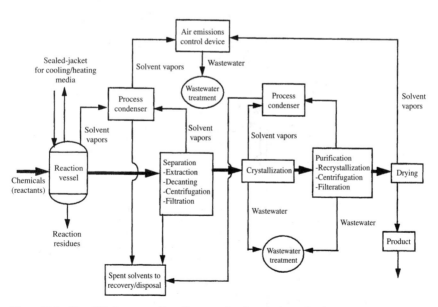

Figure 10.4   Simplified process flow diagram for chemical synthesis.

A kettle-type reactor is the most common type of reactor vessel in the pharmaceutical industry. The capacity of these reactors typically ranges from 50 to several thousand gallons. The reactors, made of either stainless steel or glass-lined carbon steel, are equipped to provide a range of monitoring capabilities usually required during batch processing (Fig. 10.5) [16]. Solvents and starting materials are charged into the reactor vessel, liquid ingredients by pumping or through vacuum and solid starting materials by manual or mechanical means. The reactor can be operated at atmospheric, high, or reduced pressures and at elevated temperature. In addition, these reactors can act as mixers, heaters, holding tanks, crystallizers, and evaporators [16]. Reactors are usually combined with condensers to recover solvents from process operations. Often, they are also attached to various air pollution control devices to remove volatile organics or other compounds from vented gases.

Natural and biological extraction [16].   The active ingredients for some important drugs are sometimes isolated from natural sources such as plants, roots, parasitic fungi, or animal glands. Medicines such as insulin, morphine, anticancer drugs, and others with unique properties are produced by natural product extraction [16].

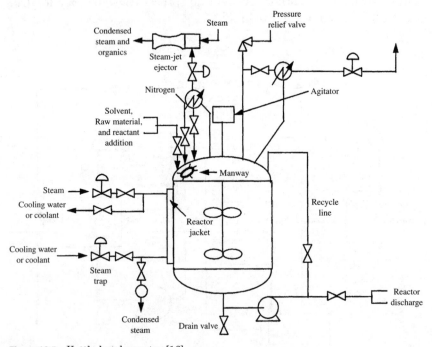

Figure 10.5   Kettle-batch reactor [16].

The raw materials usually contain very low concentration of the desired active ingredient. Therefore, the process is designed to be able to isolate a volume of the final product, which is very small compared to the volume of the raw materials. The final active ingredient is isolated by precipitation, purification, and solvent extraction methods. During precipitation, the solubility of the materials depends on the pH adjustment, salt formation, or addition of an antisolvent [16]. The active ingredient is removed by extraction with one type of solvent, and the fats and oils are removed by another type of solvent without damaging the active ingredients. For example, ammonia is used as a method of controlling the pH when extracting from animal and plant sources. Ammonium salts are used as buffering chemicals. Ammonium salts often prevent unwanted precipitation and have the advantage of not reacting with animal and plant tissues [16, 22].

Fermentation [16].   Some important pharmaceuticals, including steroids, antibiotics, and certain food additives (such as vitamins) are produced by fermentation. In fermentation, microorganisms (e.g., bacteria, yeast, or fungi) are inoculated in a liquid broth supplemented with nutrients (e.g., temperature, pH, oxygen). These microorganisms produce the desired product (e.g., antibiotic, steroid, vitamin, and so on) as a by-product of normal metabolism. The process of fermentation includes three steps:

1. *Seed preparation*: The primary seed fermentation is performed using shaking-flask culture techniques. Once grown, the suspension is then transferred to further seed stages. The purpose of the seed preparation is to generate enough inoculums for the production of a fermenter.

2. *Fermentation*: The fermenter is usually agitated and aerated. The pH, temperature, and dissolved oxygen content of the fermentation broth may be monitored during fermentation. A fermenter *broth* is produced, filtered or centrifuged, and separated.

3. *Product recovery*: Product recovery is achieved by three different methods: solvent extraction, direct precipitation, and ion exchange or adsorption. In solvent extraction the active ingredient is removed from the aqueous broth by an organic solvent followed by crystallization. The direct precipitation method of product recovery involves precipitation of the active ingredient, as a metal salt from the broth using, for example, copper (Cu) and Zinc (Zn) as precipitating agents. Ion exchange or adsorption may be used for product recovery. Ion exchange resin or activated carbon is contacted with the broth and the product is adsorbed on the surface. The product is subsequently recovered from resin by using an organic solvent or basic aqueous solution. Crystallization is always the last step of purification of the active ingredient.

10.6   Manufacturing of Pharmaceutical Products

10.6.1   The manufacturing of aspirin

The commercial process for the production of aspirin (or acetylsalicylic acid) involves a one-pot acylation reaction. Acetic anhydride reacts with salicylic acid in the presence of a small amount of sulfuric acid to produce acetylsalicylic acid and acetic acid.

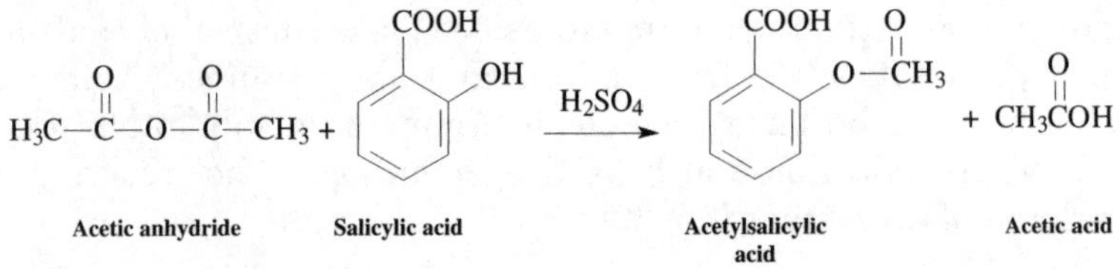

| Acetic anhydride | Salicylic acid | Acetylsalicylic acid | Acetic acid |

Acetic anhydride is produced by the direct oxidation of ethylene in the presence of air. The synthesis of salicylic acid involves the combination of several reactants. Sodium hydroxide (NaOH) reacts with phenol (C_6H_5OH) to give sodium phenolate and water. Sodium phenolate reacts with carbon dioxide (CO_2) to obtain sodium salicylate. The subsequent acidification with H_2SO_4 leads to pure salicylic acid and sodium sulfate.

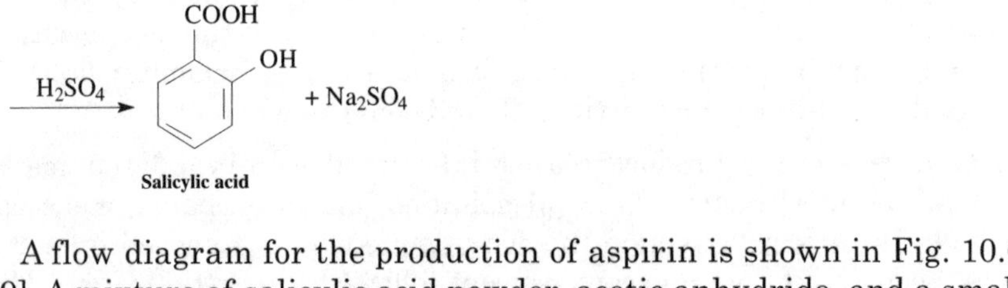

| Phenol | Sodium phenolate | Sodium salicylate |

Salicylic acid

A flow diagram for the production of aspirin is shown in Fig. 10.6 [9]. A mixture of salicylic acid powder, acetic anhydride, and a small amount of sulfuric acid is charged into a batch reactor. The mixture is stirred for 2 to 3 hours and then transferred to a crystallizing kettle. A portion of the mother liquor is recycled in the next run, and the other part of the slurry is centrifuged. The resulting crystalline material is dried in a rotary drier to yield acetylsalicylic acid (aspirin). Excess mother liquor is distilled and the excess of acetic anhydride is recycled.

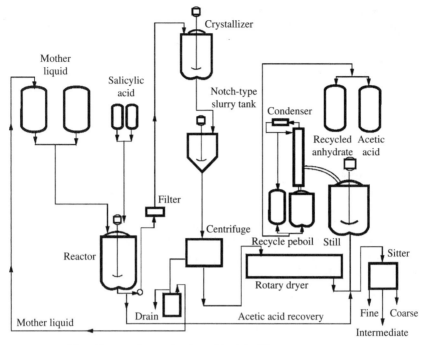

Figure 10.6   Flow diagram for production of Aspirin [9].

10.6.2   The manufacture of pyribenzamine [9]

The production of pyribenzamine (or tripelennamine) takes place in four steps.

1. Dimethylaminoethanol reacts with thionyl chloride in toluene as a solvent. The mixture is heated in the reactor at the boiling point to complete the reaction. Sulfur dioxide gas is evolved and absorbed in the scrubber.

$$\underset{\text{Dimethylaminoethanol}}{HO\overset{CH_3}{\underset{CH_3}{N}}} \quad \xrightarrow{SOCl_2} \quad \underset{\text{2-chloro-N,N-dimethylethylamine}}{Cl\overset{CH_3}{\underset{CH_3}{N}} \cdot HCl} \; + \; SO_2$$

The mixture is cooled to promote crystallization of 2-chloro-N,N-dimethylethylamine as the intermediate product.

2. The second step involves the reaction of 2-amino-pyridine with benzaldehyde and formic acid. The reaction is heated at the boiling

point until completion. After cooling, the mixture is acidified by the addition of aqueous HCl. The second intermediate product 2-benzyl-aminopyridine is separated by centrifuging after a normal workup of the reaction.

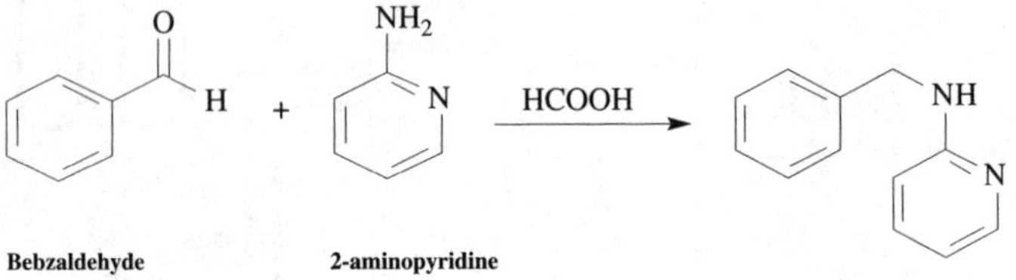

Bebzaldehyde   2-aminopyridine

2-benzylaminopyridine

3. The 2-chloro-N,N-dimethylethylamine hydrochloride formed in the first step is added to 30% NaOH. Toluene is added to the reactor, and the mixture is stirred. The lower aqueous phase is eliminated, and the organic phase is dried.

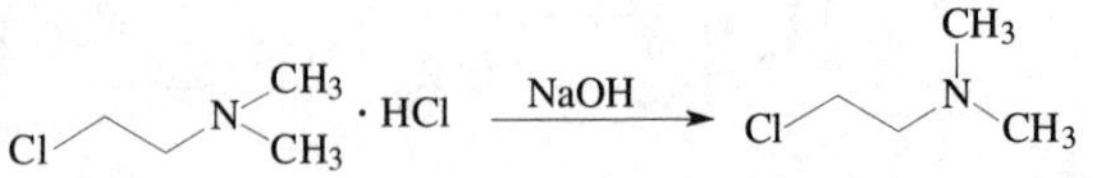

2-chloro-N,N-dimethylethylamine

4. The last step of the process includes the reaction of 2-chloro-N,N-dimethylethylamine with 2-benzylaminopyridine in toluene as a solvent. After completion, the mixture is passed through a filter to remove the sodium chloride by-product. Tripelennamine is separated by concentrating the filtrate under reduced pressure. (Fig. 10.7).

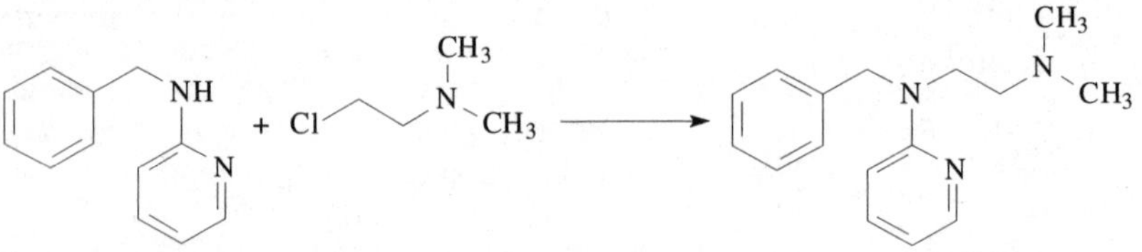

Tripelennamine

10.6.3   Formulation, mixing, and compounding

The manufactured bulk drug substance is converted into its final form by mixing, compounding, or formulating.

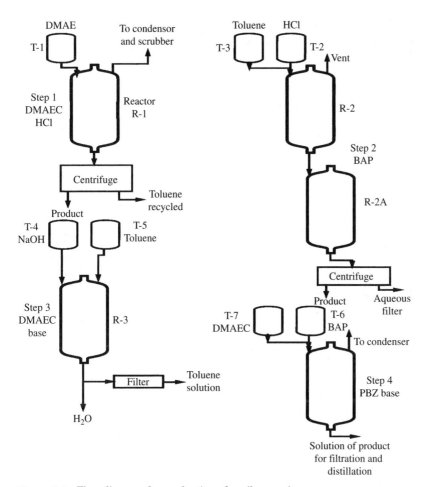

Figure 10.7   Flow diagram for production of pyribenzamine.

The most common dosage forms of pharmaceutical products include tablets, capsules, liquids, creams, ointments, as well as aerosols, patches, and injectable dosages [16, 17].

Table 10.9 contains a list of the most common pharmaceutical dosage forms and their uses [17, 18].

Tablets [16, 17].   Tablets are the most popular type of dosage form, since they offer convenience, stability, accuracy and precision, and bioavailability of the active ingredients [17]. Tablets are prepared by combining the active ingredient with a filler, such as sugar or starch, and a binder, such as corn syrup or starch. The filler is added to ensure that

TABLE 10.9   Pharmaceutical Dosage Forms

Dosage form	Constituents, properties	Uses
	Solids	
Powders, bulk	Comminuted or blended, dissolved, or mixed with water	External, internal
Effervescent	CO_2-releasing base ingredients	Oral
Insufflation	Insufflator propels medicated powder into body cavity	Body cavities
Lyophilized	Reconstitution by pharmacists of unstable products	Various uses including parenteral and oral
Capsules	Small-dose bulk powder enclosed in gelatin shell, active ingredient plus diluent	Internal
Troches, lozenges	Prepared by piping and cutting or disk candy technology; compounded with glycerogelatin	Slow dissolution in mouth
Compressed tablets	Dissolved or mixed with water; great variety of shapes and formulations	Oral and external
Pellets	For prolonged action	Implantation
Coated tablets	Coating protective, slow release	Oral
	Liquid solutions	
Syrups	Sweetener, solvent, medicinal agent	Flavoring agent, medicinal
Spirits	Alcohol, water, volatile substances	Flavor or medicinal
Collodions	Pyroxylin in ether, medicinal agent (caster oil camphor)	External for corns and bunions
Parental solutions	Sterile, pyrogen-free, isotonic, pH close to that of blood; oily or aqueous solution	Intravenous, intramuscular subcutaneous injection
Ophthalmic	Sterile, isotonic, pH close to that of tears; viscosity builder	Eye treatment
Nasal	Aqueous, isotonic, pH close to that of nasal fluids; sprays or drops	Nose treatment
Mouthwash, gargles	Aqueous, antiseptic	Refreshment, short term bacterial control
Inhalations	Administered with mechanical devices	Medication of trachea or bronchioles
	Liquid dispersions	
Suspensions	Powder suspended in water, alcohol, glycol, or an oil	Oral dosing, skin application
Emulsions, lotions	Oil-in-water or water-in-oil	Oral, external or injection

(*Continued*)

TABLE 10.9   Pharmaceutical Dosage Forms (*Continued*)

Dosage form	Constituents, properties	Uses
	Semisolid and plastic dispersions	
Ointments	Hydrocarbon (oily), adsorptive water-washable, or water-soluble bases; emulsifying agents, glycols, medicating agent	External
Pastes and cerates	Ointment with high dispersed solids and waxes, respectively	External
Suppositories	Theobroma oil, glycinerated gelatin, or polyethylene glycol base plus medicinal agent	Insertion into body cavity

SOURCE: Adapted from Zanowaik, P., 1995, Pharmaceuticals in Kirk-Othmer, *Encyclopedia of Chemical Technology*, Vol. 18, 4th ed. [18].

the active pharmaceutical ingredient is diluted to the correct concentration. A binder is needed to bind tablet particles together. In addition, a lubricant, such as magnesium stearate or polyethylene glycol, may be added to facilitate equipment operation, or to slow disintegration or dissolution of the tablet in the stomach.

Tablets are produced directly by compression of powder blends or granulations, which include a small percentage of fine-particle powders. Wet granulation is used for drugs that are not sensitive to moisture and heat. The powdered drug is mixed with the filler. The mixture is then wetted and blended with the binder, forming a solution [16, 17]. The coarse granules are dried on trays in hot air ovens or fluid-bed dryers. Granules are mixed with lubricants such as magnesium stearate and then compressed into tablets [16, 17].

Slugging or dry granulation is used when the drug is not stable under the conditions of wet granulation or when the tablet ingredients have sufficient inherent binding or cohesive properties [16, 17]. In slugging, all the ingredients are blended and compressed on heavy-duty tablet presses to obtain larger tablets (20 to 30 grams in weight). The large tablets are then grounded and screened to a desired mesh size then compressed into final tablets. Dry granulation includes weighing, mixing, slugging, dry screening, lubrication, and compression [16, 17].

Direct compression is a simple and time-saving process. The ingredients are blended and then compressed into the final tablet. The problem is that not all substances can be compressed directly, necessitating an intermediate granulation step.

Compressed tablet formulations contain different types of ingredients necessary for proper preparation and therapeutic performance. The ingredients needed include diluents, disintegrating or binding (adhesive) additives, and lubricants. Tablets are designed to be dissolved slowly in the mouth and should not disintegrate quickly. Lactose or dicalcium phosphates are common diluents, whereas starch and cellulose derivatives are used as disintegrating agents. Lubricants are used to facilitate the flow of granulation. Combinations of silica, corn starch, magnesium stearate, and high molecular weight polyethylene glycols are used. Most lubricants are hydrophobic and slow down disintegration and drug dissolution. Colors and flavors are added to increase the elegance and acceptability of the product. Sometimes colors are used for identification [16, 17].

Coating is used to protect the ingredients from moisture, oxygen, or light, or to mask unpleasant taste or appearance. Coating may also be used to impart distinctive colors to facilitate patient recognition [16, 17]. Enteric coating may delay the release of active ingredients in the stomach and prolong therapeutic activity. The latter are used for drugs that are unstable to gastric pH or enzymes, cause nausea and vomiting or irritation to the stomach, or should be present in high concentrations in the intestines [17]. Their effectiveness depends on the varying pH patterns of the gastrointestinal tract, the enzymes present for dissolution, and aqueous solubility [17].

Sugar coating is applied in rotating, pear-shaped pans. The coating solution is poured onto the tablets. In many operations, aqueous solutions are now used instead of solvent-based (usually methylene chloride) solutions. As the drums rotate, the tablets become coated. After coating, tablets are dried in the drum and may be sent to another rotary drum for polishing. Sugar coating increases the tablet weight, is time-consuming, and requires skilled operators [16, 17]. Film coating in pans is a much quicker procedure than sugar coating. The coating is much thinner and the process can be easily automated or programmed. Polymer solutions are used (e.g., cellulose derivatives) that form films upon drying. Plasticizers improve flexibility [17].

Air-suspension coating, known as the Wurster process, uses a cylindrical chamber in which the cores are suspended in a controlled stream of air. The process is much quicker than pan-film coating; however, care must be taken to avoid destruction of the cores by attrition in the air steam.

Capsules.   The most common solid oral dosage form after tablets is the capsule. There are two forms of capsules:

- Soft-gelatin capsules that contain glycerol as well as gelatin maintain plasticity even when dried
- Hard capsules are formed by dipping metal pins into a solution of gelatin at a specific temperature

Hard capsules are made in two sections, cap and body, which are then filled. Soft-gelatin capsules have their shell formed and filled in succession in one manufacturing procedure. Soft-gelatin capsules are generally filled with nonaqueous solutions. Soft-shelled capsules are formed by placing two continuous gelatin films between rotary die plates. Commercially filled soft gelatin capsules come in a wide choice of sizes and shapes: they may be round, oval, oblong, tube, or suppository-shaped [16, 17].

Liquid dosage forms.   The liquid products are prepared by dissolving the ingredients in the appropriate solvent systems. Dyes, flavors, sweeteners, and antimicrobial preservatives are added to mask unpleasant taste or appearance, and to prevent mold and bacterial growth. The final products are stored in large tanks before final packaging. If the liquid is used for injection or ophthalmic use, the liquid must be sterilized. Solutions for external or oral use do not require sterilization but generally contain antimicrobial preservatives [16, 17].

Creams, ointments, and pastes [17].   Creams are semisolid emulsions and are either oil-in-water or water-in-oil. Generally, the ingredients of the two phases are heated separately to 70–80°C while they are mixed and stirred vigorously to achieve emulsification. A solid ingredient can be added to the appropriate phase before emulsification or may be dispersed at some point after the emulsification step.

Ointments are prepared by melting together the active ingredient with a base, such as petroleum derivative or wax. The powdered drug components are added while stirring and the mixture is cooled. The product then is passed through a roller mill to achieve the particle size range desired for the dispersed solid.

Suppositories [17].   Suppositories are semirigid and plastic dosage forms designed to be delivered to body cavities such as the rectum, vagina, or urethra. They either melt at body temperature (cocoa butter) or are dissolved in the fluid of the cavity (polyethylene glycols or glycerogelatin). Suppositories can be used for systemic therapy (rectal suppositories) or for local treatment.

Cocoa butter–based suppositories can be prepared manually by pharmacists by mixing the ingredients to a pliable consistency in a mortar.

Parenteral dosage forms [17].   Parenteral dosage forms are administered by injection into body fluid systems. They are of special utility for unconscious patients. Parental dosage forms are available in the following types:

1. Solutions (ready to use)—preferable aqueous solutions are used but oily vehicles may be used for solubility reasons

2. Dry solids are dissolved in a suitable vehicle system just prior to injection. These products are prepared by lyophilization, for example, when the component drug is unstable in the desired aqueous vehicle

3. Dry solids used to prepare suspensions in a suitable vehicle system just before injection

4. Suspensions (ready-to-use)

5. Emulsions (ready-to-use)

The therapeutic response of a drug is more readily controlled by parental administration than by oral dosing, because the irregularities of intestinal absorption are avoided [9].

The vehicle or solvent, which is of greatest importance for parental products, is specially distilled, pyrogen-free water. The container of choice is glass. Sterile water is used for the solution or suspension of drugs that are prepared just before injection.

Nonaqueous vehicles, such as cottonseed oil or peanut oil, and esters (e.g., isopropyl myristate) may be used as solvent systems for parenteral drugs. Mineral oil and paraffins should not be used, as they are not metabolized and may irritate tissue.

Various additives are needed for stability, sterility, and isotonicity, including antimicrobial preservatives, antioxidants, chelating agents, and buffers.

10.7   Quality Control [9, 17]

A separate and independent group of scientists within the company takes the responsibility of providing the assurance of purity, potency, and stability for component ingredients used in the preparation of dosage forms. Various chemical, physical, and biological tests are carried out to ensure the required high standards of safety, purity, and effectiveness of drugs. This includes control of all raw materials, including packaging components and labels, control of the manufacturing process, and analysis of the intermediates and the final product.

The official governmental drug agencies assure adequate levels of quality control through its "Good Manufacturing Practice" (GMP) regulations and regular factory inspections. For example, tablets and capsules are examined for the following properties:

a. Identity

b. Content of active drug

c. Size

d. Physical appearance

e. Disintegration and/or dissolution time

f. Friability (mechanical stability of tablets)

g. Variability in weight

All injectable solutions are tested for sterility and absence of pyrogens. Various chromatographic techniques, such as high performance liquid chromatography (HPLC), and other analytical techniques are used in the quality control laboratory to determine the purity and the amount of active drug present in the different dosage forms.

References

1. Thayer, A. M., Pharmaceuticals: Redesigning R&D, *Chemical & Engineering News*, 76(8), 25–37, 1998.
2. *IMS World Review*, IMS HEALTH, 28 February, 2003.
3. Gadamasetti, K. G. *Process Chemistry in the Pharmaceutical Industry*, Marcel Dekker, New York, N.Y., 3–17, 1999.
4. Chamoin, S., H.-J. Roth, C. Borek, A. Jary, and H. Maillot, *Chimia* 51, 241–247, 2003.
5. World Medical Association Declaration of Helsinki, *Ethical Principals of Medical Research Involving Human Subjects*, accepted on June 1963, latest amendment 2002. Available at www.wma.net/e/policy/b3.htm.
6. Graham, P. L., *An Introduction to Medicinal Chemistry*, Oxford University Press, New York, N.Y., 246, 1995.
7. Groggins, G., *Unit Processes in Organic Synthesis*, 5th ed., McGraw-Hill, New York, N.Y., 946, 1985.
8. Austin, G. T. *Shreve's Chemical Process*, 5th ed., McGraw Hill Book Company, New York, N.Y., 705, 1985.
9. Kent, J. A., *Riegel's Handbook of Industrial Chemistry*, Van Nostrand Reinhold, New York, N.Y., 987, 1992.
10. Elango, N., M. A. Murphy, B. L. Smith, K. G. Davenport, G. N. Motto, and G. L. Moss, *European Patent Application 0284310*, (to Hoechst-Celanese), 1988.
11. *Physician's Desk Reference*, 56th ed., "Medical Economics," 2002.
12. Stampa, A., Process for the preparation of Loratadine, *MedChem*, GP 6, 084,100, 2002.
13. Person, C. G. A., *Trends Pharmacol. Science*, 3, 312, 1982.
14. Mandell, G. L., *Principles and Practice of Infectious Diseases* (Eds.); G. L. Mandell, R. G. Douglas, J. E. Bennett, 2nd ed., Wiley, New York, 180, 1985.
15. Sammes, P. G. *Topics in Antibiotics Chemistry*, Vol. 4, 1980; C. E. Horwood, A. G. Brown, and S. M. Roberts, *Recent Advances in the Chemistry of β-Lactam Antibiotics*, Royal Society of Chemistry, London, 1985.
16. Environmental Protection Agency (EPA), *Profile of the Pharmaceutical Manufacturing Industry*, EPA/310-R-97-005; September 1997.
17. Zanowiak, P., Pharmaceuticals, *Encyclopedia of Chemical Technology*, 3rd ed., Wiley, New York, Vol. 17, 272, 1982.
18. Zanowiak, P., Pharmaceuticals, in *Encyclopedia of Chemical Technology*, 4th ed., Vol. 18, 1995.

Agrochemicals

Manfred J. Mirbach and Bassam El Ali

11.1 Introduction and History

Global food consumption will double in the next 25 years. The reasons are the increase of the world population and a higher consumption of every single person. The area of arable land, however, will remain constant. Consequently, the available agricultural land area per person

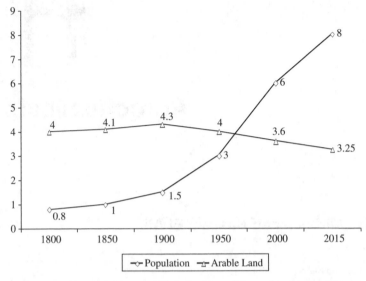

Figure 11.1   Change of the world population and area of arable land (population in billion, arable land in relative numbers). (*Source: FAO statistics, available from www.fao.org and www.syngenta.com*)

has decreased, for example, from 0.5 ha per person in 1950 to 0.25 ha per person today. In 2050, only 0.15 ha of arable land will be available for each person. This illustrates the challenge that agriculture faces. To feed the growing world population, farms must produce more food on less land (Fig. 11.1).

This challenge is not new. About 200 years ago the world's population was about one-quarter of that today and a much larger fraction of these people worked on farms. Nevertheless, famine, malnutrition, waves of pests, and diseases were very common in most parts of the world even though people were practicing agriculture in a similar way as we do today—seeding, growing, harvesting, and storing. The major difference was that they used traditional field management.

The oldest system of food gathering is that of the nomads. When their basic food supply gets exhausted in one place, they migrate to a different area. In other areas, nature helped to keep the land fertile. For instance, in the Nile valley the river flooded the land nearly every year and regenerated the agricultural soil to make it ready for a new harvest. The rotational crop system also has a long tradition. One crop is planted in the first year, a different one in the second year, and then the land is left fallow for a few years to regenerate, before it is used again for crop production.

Mineral fertilizers.   It is estimated that traditional farming without chemical and mechanical support can feed approximately 1.5 billion people. Today Earth has a population of 6 billion. That means something has happened in between that led to an enhanced productivity in agriculture. The first step to modern field management was the introduction of mineral fertilizers. Beginning with the 19th century, the world population grew rapidly and a global famine was looming. Therefore, scientists worked hard to find solutions to improve crop yield. The first important discovery was that the organic carbon in plants comes from photosynthesis using the carbon dioxide in air and not from organic matter contained in the soil.

$$n\,CO_2 + n\,H_2O \quad \xrightarrow{\text{light}} \quad \left(\begin{array}{c} OH \\ | \\ -C- \\ | \\ H \end{array} \right)_n + O_2 \qquad (11.1)$$

The soil supplies the inorganic minerals and water. Justus Liebig, an eminent German chemist and teacher, discovered that four elements are essential for plant growth: nitrogen, phosphorous, calcium, and potassium. Later, additional essential elements (*trace elements*) were discovered. The growth rate of plants is limited by the component that is minimum in the accessible soil layer. When one element is in short supply, adding large quantities of other elements does not increase the yield. It is like a barrel in which the planks are of uneven height. When it is filled with water the lowest plank always determines the capacity, no matter how high the others are (Fig. 11.2).

In Liebig's day, nitrogen was the limiting element in soil. Therefore, he proposed to use natural nitrogen sources, like Chile saltpeter, guano, and manure to fertilize the fields and indeed the productivity increased rapidly.

A new food crisis loomed at the end of the 19th century, when the supplies of Chile saltpeter were nearly exhausted and the little that was left was reserved for military purposes. The supply of food for the world population was again at risk. Desperately, new sources of nitrogen fertilizers were sought. There is an abundance of nitrogen in the air. However, plants are unable to use it, because they cannot split N_2 molecules. Discovery of a catalytic reaction, called the *Haber Bosch Process*, solved this problem by conversion of nitrogen and hydrogen to ammonia. Ammonia is the starting material of all other nitrogen compounds. Nitric acid, for instance, is produced by catalytic oxidation of ammonia. Today, nitrogen fertilizers, like ammonium nitrate, ammonium sulfate, or urea are available in unlimited quantities and nitrogen is no longer the limiting factor in agriculture.

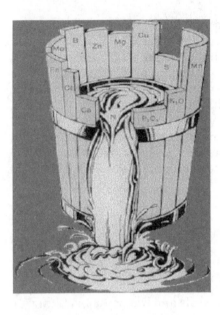

Figure 11.2 The barrel picture used by Liebig to illustrate the principle of the minimum needed essential elements. The capacity of the barrel is limited by the length of the shortest stave (in case of crop, nitrogen), and can only be increased by lengthening that stave. It would not help to increase the height of the others.

Crop protection.   Increasing the yield solves only half the problem. The crop must also be protected from diseases and pests. Without protection, 50 to 90 percent of the harvest is destroyed by pests. This is illustrated by numerous disastrous crop losses in the past. Following are a few examples that occurred before modern pest management was available. The coffee industry of Sri Lanka collapsed completely because of an infestation by coffee rust (*Hemileia vastatrix*). About 50 percent of the annual cacao harvest was destroyed by cacao bugs (*Miridae*). Furthermore, potato blight pest in Ireland caused a disastrous famine that forced millions of Irish to emigrate. The famine in India at the beginning of the 20th century was caused by a rice fungus that destroyed the rice harvest.

Farmers have tried to fight these and other pests throughout history, but their means were limited. Sulfur is the first documented material used in the war against infection. It was used 3000 years ago by the Chinese as a somewhat effective fumigant. Some 2500 years later, arsenic was introduced as an insecticide and nicotine and strychnine in extracts from tobacco leaves and strychnos seeds, respectively, were used as rodenticides. Later, extracts from plants (chrysanthemum, tobacco) or inorganic compounds, for example, Bordeaux mixture (copper sulfate, calcium hydroxide, and water) and Paris green (copper arsenite), were used as insecticides or fungicides. With the rise of synthetic organic chemistry after 1850, many new substances were discovered and tested for biological efficacy. Today about 1000 chemical substances are produced as active ingredients that are used in some 10,000 different products.

Genetic engineering. The latest step of agricultural management is the development of genetically engineered plants that produce food crops of high quality and yield and are resistant to pests and adverse climatic conditions. Although this technology is still in its infancy, we can be optimistic that it will help to feed the world population in the future if the general public accepts it.

11.2   Chemical Pest Control

The basis of all plant growth is the photochemical conversion of carbon dioxide and water to carbohydrates with the assistance of chlorophyll as photocatalyst. Therefore, H_2O and CO_2 are the most important agrochemicals. However, as air and water are provided by nature, we do not see them as chemical products. Agrochemicals in the common sense are fertilizers, pesticides, and other chemicals that help to protect the quality of agricultural commodities. Feed additives, such as vitamins, or veterinary medicines, such as antibiotics, are usually not considered agrochemicals, although they are also synthetic chemicals and are used in agriculture. Therefore, this chapter concentrates on pesticides as the main topic.

Pesticides are applied to control pests and plant diseases. Pesticides used in agriculture to protect living plants and freshly harvested crops are called *plant protection chemicals.* When pesticides are used to protect stored food, processed goods, public hygiene, and dead objects, they are called *biocides.* Pesticides are divided into subgroups named after the pest they fight (Table 11.1). Examples discussed in this book are

TABLE 11.1   Classification of Pesticides According to Target Pest or Function

Class	Target pest	Remarks
Insecticide	Insects	Kills insects or larvae
Fungicide	Fungi, mold	Controls plant diseases
Herbicide	Weeds, plants	Total herbicide kills all plants Selective herbicide controls weeds
Rodenticide	Rats, mice	Controls rodents
Plant growth regulator	None	Controls the size of plants, e.g., keep stems of cereals short
Acaricide	Mites	Controls mites, aphids, and so on
Pheromone	Insects	Attracts insects into traps, controls mating
Repellent	Insects	Repels insects without killing them
Nematicide	Nematodes, worms	Kills worms and similar parasites

TABLE 11.2 Classification of Pesticides
According to Chemical Structure

Name	Structural element
Carbamate	$\begin{array}{c} O \\ \| \\ R\text{-}NH\text{-}C\text{-}O\text{-}R' \end{array}$
Dithiocarbamate	$\begin{array}{c} S \\ \| \\ R\text{-}NH\text{-}C\text{-}S\text{-}R' \end{array}$
Organophosphate	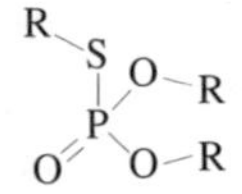
Organochlorine	See DDT
Pyrethroids	See pyrethrum
Sulfonylurea	$\begin{array}{c} O \\ \| \\ R\text{-}SO_2\text{-}NH\text{-}C\text{-}NH\text{-}R' \end{array}$
Triazole	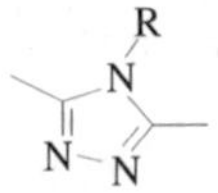

insecticides, herbicides, fungicides, and miscellaneous agrochemicals. The latter group includes rodenticides, plant growth regulators, harvest aids, and postharvest preservatives. Nonagricultural uses of pesticides, such as application of insecticides for disease vector control, in public areas and private homes, are also included in this chapter.

Another method to classify pesticides is related to their main chemical structural elements or their mode of action. Examples are organochlorines, organophosphates, carbamates, pyrethroids, and so on (Table 11.2). In commercial products, the active ingredients are formulated (mixed) with other compounds, such as solvents, surfactants, stabilizers, and so on, that make the pesticides ready for use on farms and for private pest control. The formulations for crop protection are usually concentrates that are diluted with water before being applied on a field.

11.2.1 Herbicides

Herbicides are the most important pesticide class in terms of production volume and market value. In agriculture they are used to control weeds. Weeds are unwanted plants, such as grasses, sedges, and broadleaf plants, that compete with the desired crop plants for nutrition, water, and land, thereby reducing the yield. Herbicides also have nonagricultural uses as they help in erasing vegetation on streets, railroad tracks, sports fields, and other public areas.

Herbicides can be active through leaves (foliage active) or by uptake from the soil (soil active) through roots. They can be applied pre- or postemergence of the target crop. Nonselective or total herbicides kill all plants that are present during application. They are used on fields before emergence of the target crop to remove the competing weeds. Other applications of total herbicides are selective spraying of the ground under the trees in fruit orchards and plantations (cacao, banana, and so on).

Selective herbicides are active against certain species only, for example, broadleaf weeds or perennial grasses. They can also be applied postemergence of the crop plant. For example, some sulfonylurea (e.g., nicosulfuron) can be applied to maize fields during full growth to remove competing weeds without doing any harm to the target crop. Modern herbicides are very potent and need only very low application rates (e.g., 50 g/ha) to be effective. Broadleaf selective herbicides are applied to turf or grassland to control leafy weeds and bushes. Quizalafop-P is an example of a herbicide that controls annual and perennial weeds in potatoes, sugar beet, oilseed rape, vegetables, and so on.

In industrialized countries, many crops are harvested with automatic machines. Their use is impaired by leaves, weeds, or broken plants. An example is the use of herbicides as defoliants in cotton. The leaves are removed by foliar application of paraquat or triazophos prior to mechanical picking of the cotton balls. Another example is the application of a growth inhibitor to keep the plants short. This prevents breaking of the stems during hail storms or heavy rain. A similar result is obtained when a total herbicide is applied to wheat or rye plants 1 to 2 weeks before the regular harvest date. The plants turn yellow and dry and are then ready for automatic combine harvesting. This reduces the risk of the crop being destroyed by bad weather shortly before harvest, when the plants are most vulnerable to breaking. A third example is the harvest of olives or nuts, during which the trees are shaken mechanically until the fruits fall off. They are then collected by hand or dedicated machines. Any weeds or bushes under the trees would interfere with this process and are removed by herbicides.

The most important herbicide on the market is glyphosate. It was originally developed as a total herbicide with many uses in crop and non-crop areas. It has many advantages, that is, it is very effective and the treated plants die nearly immediately. On the other hand, it has no long-term effect. A field treated with glyphosate can be used for planting a few days after the application. In addition, glyphosate has a very low toxicity for mammals and is rather benign to the environment. It is very popular in many countries and sold under different names, such as *rodeo, roundup,* or *touchdown.* The only disadvantage is the high application rate of 2 to 5 kg/ha.

Examples of herbicides are listed in Tables 11.3 and 11.4.

TABLE 11.3   Foliage Active Herbicides (Examples)

Common name or chemical name[*]	Chemical class	Mode of action	Typical uses	Structure
Paraquat or 1,1'-dimethyl-4,4'-bipyridinium dichloride	Bipyridilium	Interrupts photosynthesis, nonselective	Orchards, plantations, defoliant for cotton, aquatic weeds	11-1
Bromoxinil or 3,5-dibromo-4-hydroxybenzonitrile	Benzonitrile	Inhibits photosynthetic electron transport, selective for certain annual broad leave weeds	Cereals, maize, sorghum, turf	11-2
2,4-D or (2,4-dichlorophenoxy) acetic acid	Chlorophenoxy acids	Affect cell membrane and RNA synthesis, selective for broad leave weeds; esters active through leaves	Cereals, maize, sorghum, rice	11-3
Dicamba or 3,6-dichloro-2-methoxybenzoic acid	Chlorobenzoic acids	Affect cell membrane and RNA synthesis, absorbed through leaves and soil; for broad leaves, brushes	Cereals, pastures, range land	11-4
Quizalofop or 2-[4-[(6-chloro-2-quinoxaliny)loxyphenoxy] propanoic acid	Aryloxyphenoxy acid	Inhibition of fatty acid synthesis; selective for grass weeds	Potatoes, soy beans, cotton, flax	11-5
Glyphosate or N-(phosphono-methyl) glycine	Glycine derivative	Inhibits amino acid synthesis; nonselective, fast-acting herbicides	Used as general weed control and weed control in transgenic maize	11-6
Nicosulfuron or 2-[[[[4,6-dimeth-oxy-2-pyrimidinyl) amino]carbonyl]amino] sulfenyl]-N,N-dimethyl-3-pyridinecarboxamide	Sulfenylureas	Absorbed through leaves and roots; selective control of some annual grass weeds and broad leaves	Selective control of broadleaf weeds in maize	11-7
Imazapyr / 2-[4,5-dihydro-4-methyl-4-(1-methyl ethyl)-5-oxo-1H-imidazol-2-yl]-3-pyridi necarboxylic acid	Imidazolinones	Amino acid synthesis inhibitor; nonselective	Noncrop areas, railroad tracks, plantations	11-8

[*] According to *Chemical Abstracts* or IUPAC.

TABLE 11.4 Soil Active Herbicides (Examples)

Common name or chemical name[*]	Chemical class	Mode of action	Typical uses	Structure
Diuron or N´-(3,4-dichlorophenyl)-1,1-dimethyurea	Phenylurea	Inhibits photosynthesis, absorbed mainly by the roots	Total control of weeds and mosses on noncrop land and under fruit trees	11–9
Isoproturon or 3-(4-isopropylphenyl)-1,1-dimethylurea	Phenylurea	Inhibits photosynthesis	Control of annual weeds in winter wheat and barley, rye	11–10
Atrazine or 6-chloro-N-ethyl-N´-(1-methyethyl)-1,3,5-triazine-2,4-diamine	s-Triazine	Inhibits photosynthesis	Control of annual weeds in maize, sugar cane, pineapples, nuts, and noncrop areas	11–11
Pendimethalin or N-(1-ethylpropyl) –3,4-dimethyl-2,6-dinitrobenzenamine	Dinitro aniline	Inhibits root growth, must be applied before emergence	Control of annual weeds in cereals, onions, soy beans, potatoes, cotton	11–12
Aclonifen or 2-chloro-6-nitro-3-phenoxybenzenamine	Diphenyl ether	Inhibits carotenoid biosynthesis	Preemergence control of weeds in winter wheat, potatoes, etc.	11–13
Alachlor or 2-chloro-2´,6´-diethyl-N-methoxymethyl-acetanilide	Substituted amide	Inhibits protein synthesis and root elongation	Preemergence-control of annual weeds in cotton, brassicas, peanuts, soy beans, etc.	11–14

[*] According to *Chemical Abstracts* or IUPAC.

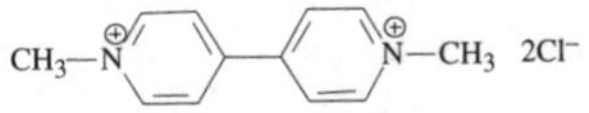

$$CH_3-N^{\oplus}\!\!\!\!\bigcirc\!\!\!\!\bigcirc\!\!\!\!N^{\oplus}-CH_3 \quad 2Cl^-$$

11-1: Paraquat

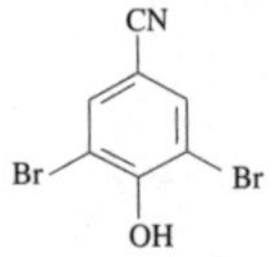

(structure with CN, Br, Br, OH)

11-2: Bromoxynil

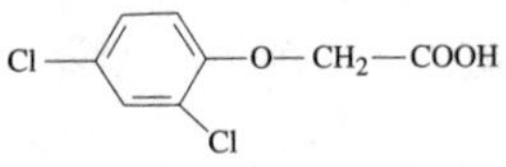

$$Cl-\bigcirc\!\!\!\!-O-CH_2-COOH$$
(with Cl)

11-3: 2,4 - D

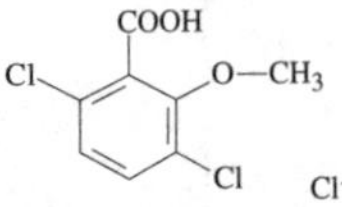

(structure with COOH, O—CH$_3$, Cl, Cl)

11-4: Dicamba

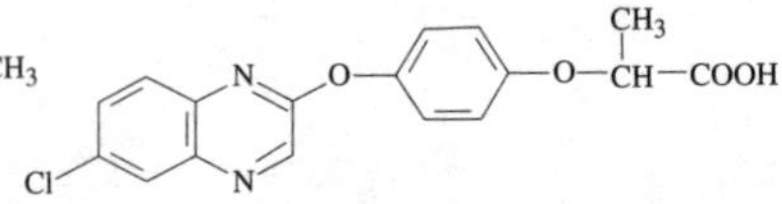

(quinoxaline structure) $-O-\bigcirc\!\!\!\!-O-\overset{CH_3}{\underset{}{CH}}-COOH$

11-5: Quizalotop-P

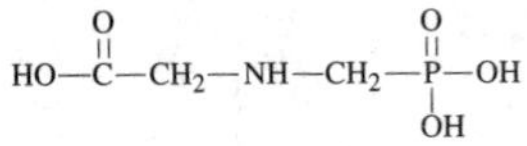

$$HO-\overset{O}{\overset{\|}{C}}-CH_2-NH-CH_2-\overset{O}{\overset{\|}{P}}-OH$$
(with OH)

11-6: Glyphosate

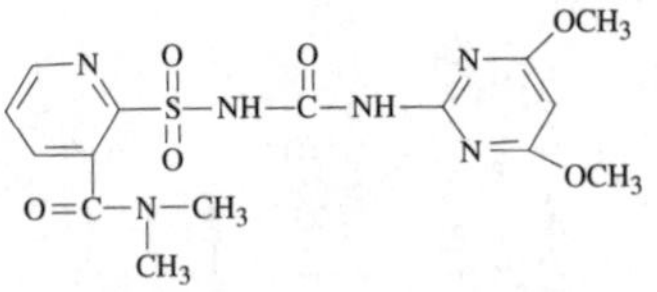

(pyridine sulfonyl structure) $\overset{O}{\overset{\|}{S}}-NH-\overset{O}{\overset{\|}{C}}-NH-$ (pyrimidine with OCH$_3$, OCH$_3$)

$O=\overset{}{\underset{}{C}}-N-CH_3$ with CH$_3$

11-7: Nicosulfuron

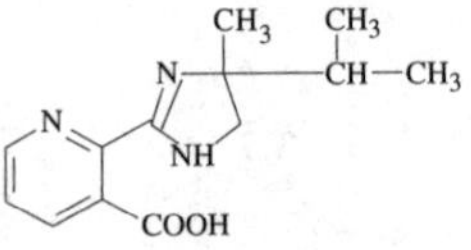

(imidazopyridine structure with CH$_3$, CH$_3$, CH—CH$_3$, NH, COOH)

11-8: Imazapyr

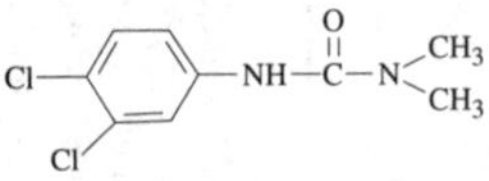

$$Cl-\bigcirc\!\!\!\!-NH-\overset{O}{\overset{\|}{C}}-N\overset{CH_3}{\underset{CH_3}{}}$$
(with Cl)

11-9: Diuron

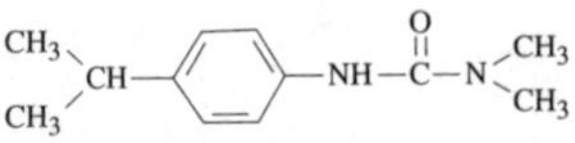

$$\overset{CH_3}{\underset{CH_3}{}}CH-\bigcirc\!\!\!\!-NH-\overset{O}{\overset{\|}{C}}-N\overset{CH_3}{\underset{CH_3}{}}$$

11-10: Isopsoturon

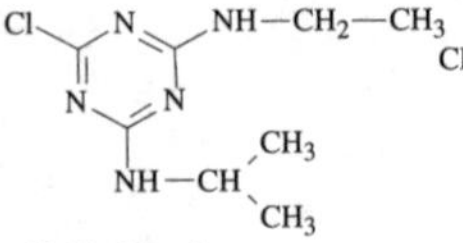

(triazine) $Cl-$ with $NH-CH_2-CH_3$ and $NH-CH\overset{CH_3}{\underset{CH_3}{}}$

11-11: Atrazine

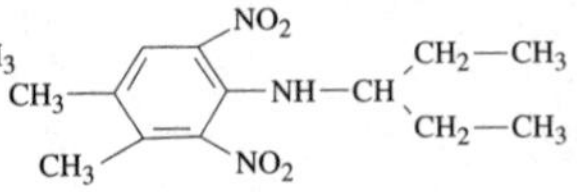

(benzene with NO$_2$, NO$_2$, CH$_3$, CH$_3$) $-NH-CH\overset{CH_2-CH_3}{\underset{CH_2-CH_3}{}}$

11-12: Pendimethalin

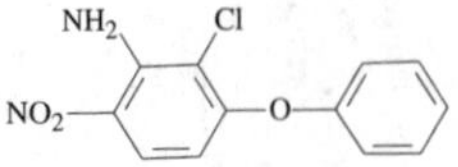

(benzene with NH$_2$, Cl, NO$_2$) $-O-\bigcirc$

11-13: Aclonifen

(benzene with CH$_2$—CH$_3$, CH$_2$—CH$_3$) $N\overset{\overset{O}{\overset{\|}{C}}-CH_2Cl}{\underset{CH_2-O-CH_3}{}}$

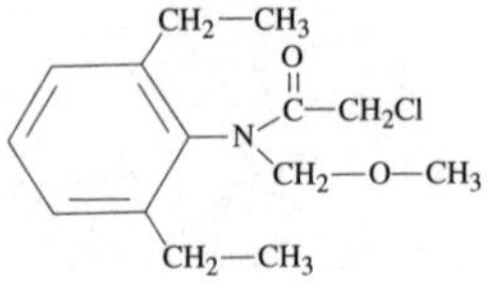

11-14: Alachlor

Scheme 1:   Chemical structures of herbicides.

11.2.2   Insecticides

Insecticides are the pesticides most commonly known by the public. Insects are not only a nuisance in everyday life; they pose a real danger to man, animals, crops, and the environment in general. In agriculture, insecticides are used widely to control insects in fruits, vegetables, rice, and other cereals. Other application areas are on farm animals, animal housing, and to control insects that are vectors of diseases. Mosquitoes (for malaria) and tsetse flies (for sleeping sickness) are just two examples. Control of these insects is a never-ending task, especially in hot and humid countries.

Insects have been a problem in all times. The first insecticides were plant extracts. Tobacco and garlic extracts were and still are particularly popular. Indeed, nicotine has some insecticidal properties and is still produced for this purpose. Its activity is low, however, and garlic extracts have no proven efficacy. At times, preparations containing arsenic and even mercury were used, but it became obvious that these toxic products caused more damage to the people who used them than to the insects.

Insecticides may act on direct contact with immediate effect or via the stomach of the insect with delayed efficacy. They can be deposited on surfaces or can be incorporated into the plant (systemic effect), killing the insects only when they feed on the plant material.

The first synthetic insecticides with demonstrated efficacy were polychlorinated compounds, with DDT as the best known example. DDT had many advantages such as low price, high efficacy, long-term effects, and relative low toxicity to mammals and man. Unfortunately, as most organochlorines, it is persistent in the environment. This has led to soil or water contamination. In addition, it accumulates in fat *via* the food chain. Therefore DDT and most other polychlorinated organic compounds were replaced by nonaccumulating molecules for agricultural use. Today DDT is still used in malarial areas as part of the WHO malaria eradication program but it has been phased out in most areas as it is considered highly toxic to human as well as wildlife populations, especially to those animals existing higher up in the food chain.

Another important class of insecticides is the organophosphates, with chlorpyriphos and parathion as typical examples. Organophosphates were also discovered in the 1940s. They are less persistent in the environment and food chain, but they are often toxic to humans and to nontarget species. They were widely used especially for large area spraying, and also as public area or household insecticides. Carbamate insecticides were developed in the 1960s. They are efficacious and nonpersistent. Carbofuran, for example, is degraded in soil microbiologically mainly to carbon dioxide. However, most carbamates, and especially carbofuran, are toxic to fish, birds, honey bees, and other nontarget species. Today there is a strong tendency to replace organophosphates and carbamates with less toxic alternatives. Pyrethrins (also called pyrethrum) are very active natural insecticides that have been used since ancient times. Pyrethrins are extracted from chrysanthemum flowers using methanol or supercritical carbon dioxide as the solvent. Pyrethrin I is a mixture of three esters of chrysanthemic acid, pyrethrin II of the corresponding pyrethrin acids. Pyrethrins bind to the sodium channels of the organisms prolonging their opening and thereby causing death. They paralyze the insects on contact nearly immediately (*knockdown effect*). Death occurs later. They are used to control a wide range of insects and mites in public health, stored products, fruit production, and on farm and domestic animals. Commercial products usually contain synergists, for example, piperonyl butoxide. Their purpose is to inhibit detoxification, thereby increasing the overall effect. Pyrethrins are not very toxic to nontarget species except aquatic species and bees. The main disadvantage is their sensitivity to sunlight, alkali, clay, and heat. This limits their use under real environmental conditions. From the chemical point of view, pyrethrins are interesting molecules, because they contain a cyclopropane ring and other structural features that can lead to isomers or enantiomers. Therefore, it is difficult to synthesize the molecules in the laboratory.

The obvious advantages of pyrethrins have challenged chemists to develop synthetic analogues that have a similar action mechanism, but avoid some of the disadvantages of the natural products. Synthetic compounds with pyrethrin-like properties are called pyrethroids. A well-known example is permethrin. It still contains the cyclopropylcarboxylic ester group, but has otherwise a simpler structure and is easier to synthesize than natural substances. Fenvalerate still contains the ester group, but no longer a cyclopropyl moiety. In etofenprox even the ester group is missing; therefore it belongs to the class of nonester pyrethroids. It has a very low toxicity for mammals, but its mode of action against insects is similar to that of other pyrethroids. Neonicotinoids is another class of modern insecticides with high efficacy against insects and low toxicity for humans. Imidacloprid and dinotefuran are two examples. They affect selectively the central nervous system of insects. They show contact and systemic activity and are readily absorbed by plants, which are then protected against insect attack. Examples of compounds with insecticidal activity are listed in Table 11.5 and in Scheme 2.

11.2.3 Fungicides

Fungi and similar microorganisms cause severe damage to fruits, cereals, vegetables, and other crops both before and after harvest. The effects are seen as mildew, decay, rotting, scorch, blight, rust, and many other plant diseases that reduce the value of agricultural products. Therefore, fungicides are probably pesticides with the highest economic value to the farmers. This is particularly true in humid climatic zones.

The first fungicides that were used in agriculture were inorganic compounds of copper, mercury, and sulfur. Copper is applied as oxychloride, sulfate, hydroxide, or octanoic acid salt. Cu^{++} ions kill the spore cells. They are sprayed or dusted onto leaves, where they have a protective action when applied before the fungal spores begin to germinate. Sulfur reacts with thiol groups in the organisms and interrupts respiration of the organism. Copper and sulfur are nonselective and are used on many crops, such as fruits, vines, vegetables, and flowers. Their specific activity is low, however, and frequent applications and

TABLE 11.5 Insecticides (Examples)

Common name or chemical name[*]	Chemical class	Mode of action	Typical uses	Structure
Nicotine or (S)-3-(1-methyl-2-pyrrolidin-2-yl) pyridine	Biopesticide	Predominantly respiratory action	Tobacco extracts against sucking insects	11–15
Imidacloprid or 1-[(6-chloro-3-pyridyl methyl]-N-nitro-2-imidazol idinimine	Neonicotinoid	Systemic uptake by plant and further distributed through the leaves, acts through central nervous system of insects	Control of sucking insects, termites, biting insects	11–16
Carbofuran or 2,3-dihydro-2.2-dimethyl-7-benzo furanylmethylcarbamate	Carbamate	Cholinesterase inhibitor, systemic action	Control of soil-dwelling and foliar feeding insects and nematodes in vegetables, nuts, cotton	11–17
Chlorpyrifos or O,O-diethyl O-3,5,6-trichloro-2-pyridyl phosphorothioate	Organophosphate	Cholinesterase inhibitor, nonsystemic contact, stomach and respiratory action	Control of soil-dwelling and foliar feeding insecticides in fruits and vegetables, also noncrop uses	11–18
Permethrin or (3-phenoxyphenyl) methyl 3-(2,2-dimethylcyclo-propanecarboxylate	Pyrethroid	Contact and stomach action	Control of a broad range of insects in agriculture, public areas, and storage facilities	11–19
Pyrethrum (Pyrethrin) or Plant extract	Pyrethroid	Opening of sodium channels, leading to paralysis and later death, nonsystemic, contact only action	Control of insects and spider mites	11–20
DDT or 1,1´-(2,2,2-trichloroethylidene)bis [4-chlorobenzene]	Organochlorine	Nerve poison, affecting sodium balance of nerve membranes, nonsystemic contact and stomach action	Mosquito control for malaria eradication; for crop use replaced by less persistent products	11–21

(Continued)

TABLE 11.5 Insecticides (Examples) (*Continued*)

Common name or chemical name[*]	Chemical class	Mode of action	Typical uses	Structure
Parathion or O,O-diethyl O-(4-nitrophenyl) phosphorothioate	Organophosphate	Cholinesterase inhibitor, nonsystemic contact, stomach and respiratory action, activated by oxidative desulfuration	Control of sucking and chewing insects and mites in field crops, fruits, and vegetables	11–22
Etofenprox or 2-(4-ethoxyphenyl)-2-methylpropyl-3-phenoxybenzyl ether	Nonester Pyrethroid	Contact and stomach action	Control of insects in rice fields, vegetables, fruit, oilseed rape, public health pests, and on animals	11–23
Fenvalerate or (RS)-α-cyano-3-phenoxybenzyl (RS)-2-(4-chlorophenyl)-3-methylbutyrate	Pyrethroid	Nonsystemic insecticide and acaricide with contact and stomach action	Control of biting and boring insects in crops and flying and crawling insects in public areas	11–24

[*] According to *Chemical Abstracts* or IUPAC.

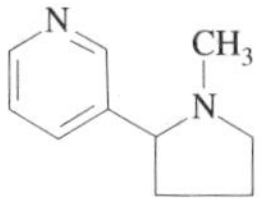

11-15: Nicotine 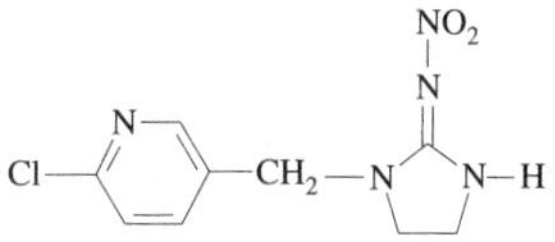 **11-16: Imiolacloprid** 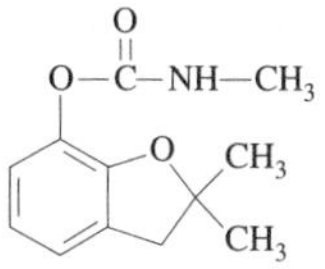 **11-17: Carbofuran**

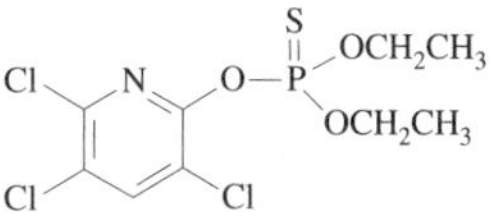

11-18: Chlorpyriphos 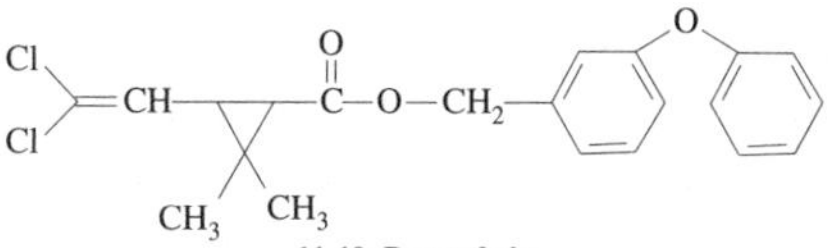 **11-19: Permethrin**

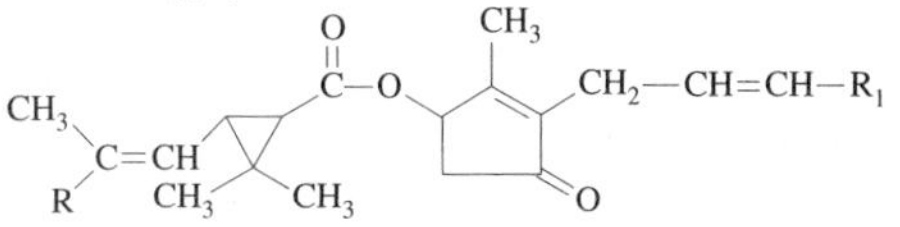

R = CH$_3$ORCOOCH$_3$ **11-20: Pyrethrum** **11-21: DDT**

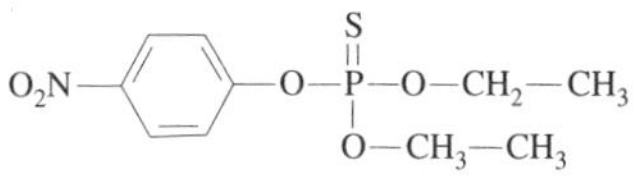

R$_1$ = – CH=CH$_2$ORCH$_3$

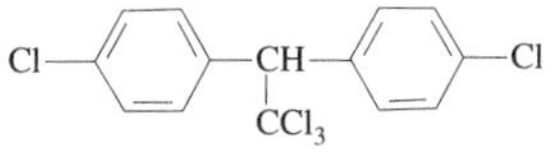

11-22: Parathion 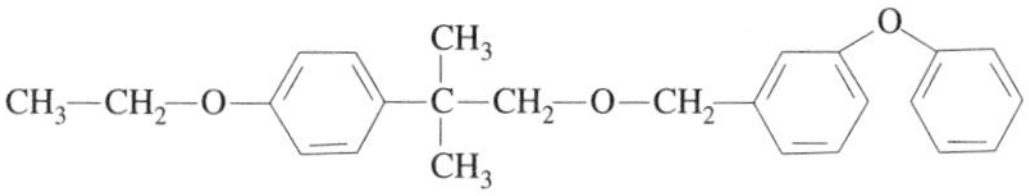 **11-23: Etofenprox**

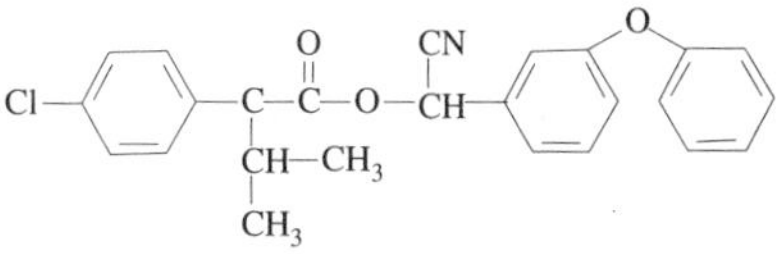

11-24: Fenvalerale 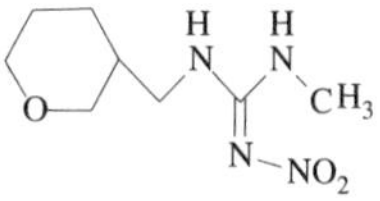 **11-25: Dinotefuran**

Scheme 2: Chemical structures of insecticides.

high rates of up to 6 kg/ha are required to achieve the desired protection. Copper compounds and sulfur are still used today on a large scale. They are allowed in *organic* farming with the justification that copper is an essential element that occurs naturally in the environment, and that it is, therefore, not harmful to animals and the environment. Unfortunately, this assumption overlooks that the high rate and frequent applications lead to much higher copper concentrations in the soil than are present naturally. As the element copper cannot degrade, this *organic* farming practice will lead to poisoning of the agricultural soil.

Dithiocarbamates were the next step of fungicide development. They are organic compounds containing several sulfur atoms and have a similar mode of action as sulfur. Often, they also contain other inorganic atoms, such as Mn, Zn, and Cu. Examples are mancozeb, maneb, thiram, or ziram. Folpet is a phthalimide containing sulfur. It is often applied in combination with other fungicides. Organic sulfur compounds also need high application rates in the kg/ha range. Although they are not very toxic themselves, there is some concern about the potential for formation of the possible metabolite ethylene thiourea (ETU), which is a suspected human carcinogen.

Later other compounds were discovered that have higher efficacy or selectivity against microorganisms. Imazalil inhibits the biosynthesis of ergosterol with systemic and protective action. It is used on vegetables, flowers, and fruits; and to protect seeds and crops during storage. Quinoxyfen is an example of a newly developed fungicide. It inhibits the cell growth and offers long-term protection against powdery mildew in cereals, sugar beets, vegetables, and so on, with application rates of 50 to 150 g/ha. This shows that it is nearly 50 to 100 times more active than the traditional sulfur and copper products. Strobilurins are derived from natural origin. They have become important modern fungicides. Azoxystrobin is an economically successful example. It inhibits mitochondrial respiration by blocking electron transfer between cytochromes. It inhibits spore germination and mycelical growth and is active against many pathogenic microorganisms, even those resistant to other fungicides. Toxicity for humans is low and no adverse effects on the environment have been observed. Examples of fungicides are listed in Table 11.6, their schemical structures in Scheme 3.

Fungicides influence the microorganism population on the crop. This may affect the fermentation pathways during processing of food, for example, the production of cheese, soy sauce, or wine. Because of these side effects, not all fungicides can be used for every purpose. Effects on the taste and fermentation must be tested before a new fungicide can be used on crops that are processed by biotechnological methods.

11.2.4. Miscellaneous compounds

There are many other substances on the market that are pesticides or are related to pesticides, such as repellents. Table 11.7 and Scheme 4 contain some examples.

Rodenticides are used to control rats and mice in fields, in storage areas, and in household environments. Rodents not only destroy harvested products, but are also vectors for contagious diseases. Most rodenticides belong to the coumarin group and act as anticoagulants. Bromadiolone is an example of a relative selective rodenticide that is highly toxic to rodents, but less toxic to domestic animals such as dogs and cats.

TABLE 11.6 Fungicides (Examples)

Common name or chemical name [*]	Chemical class	Mode of action	Typical uses	Structure
Copper salts	Inorganic	Prevents spore germination; nonsystemic	Control of powdery mildew, blights, and rust	—
Sulfur	Inorganic	Inhibits respiration; nonsystemic	Control of mildew, shot-hole, mites	—
Mancoceb or manganese ethylenebis(dithiocarbamate) polymeric, complex with Zn salt	Dithiocarbamate	Inhibits respiration, nonspecific with protective action	Control of many fungal diseases in field crops, fruit, flowers	11-26
Thiram or tetramethyl thiuram disulfide	Dithiocarbamate	Contact fungicide with protective action	Control of mildew, rust, scab, and the like on fruits and seeds	11-27
Folpet or N-(trichloromethylthio) phthalimide	Phthalimide	Inhibits respiration, foliar application with protective action	Control of mildew, leaf spot, scab, rot and the like on fruits, olives potatoes, and so on	11-28
Metalaxyl or methyl-N-(methoxyacyl)-N-(2,6-xylyl)-DL-alaninate	Acylalanine	Inhibits protein synthesis in fungi, systemic with protective action	Control of air- and soil-borne diseases on crops	11-29
Quinoxyfen or 5,7-dichloro-4-quinolyl-4-fluoro phenyl ether	Quinoline	Growth signal inhibitor, protectant, not an eradicant	Control of powdery mildew in many crops	11-30
Imazalil or Allyl 1-(2,4-dichlorophenyl)-2-imidazol-1-ylethyl ether	Imidazole	Inhibits ergosterol biosynthesis, systemic with protective action. Inhibits mitochondrial	Control of a wide range of fungal diseases	11-31
Azoxystrobin or Methyl (E)-2-{2-[6-(2-cyanophen oxy) pyrimidin-4-yloxy]phenyl}-3-methoxyacrylate	Strobilurin	respiration by blocking electron transfer between cytochromes, systemic with protective action	Control of a wide range of pathogens on cereals, vines, potato, rice, fruits, nuts, and so on	11-32

[*] According to *Chemical Abstracts* or IUPAC.

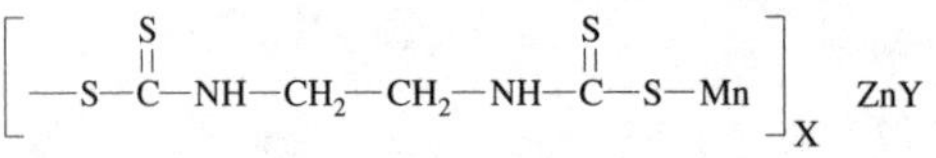

11-26: Mancozeb

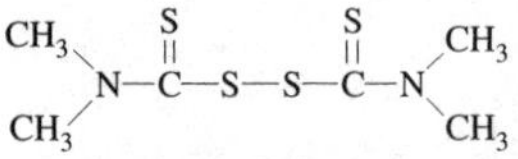

11-27: Thiram

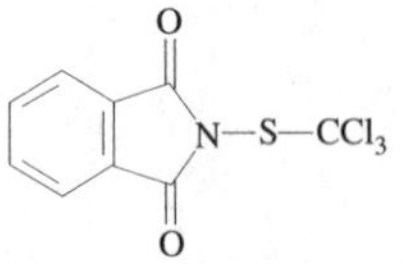

11-28: Folpet

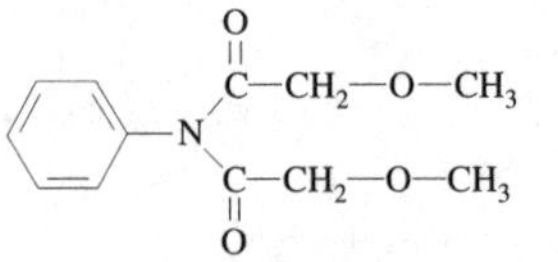

11-29: Metalaxyl

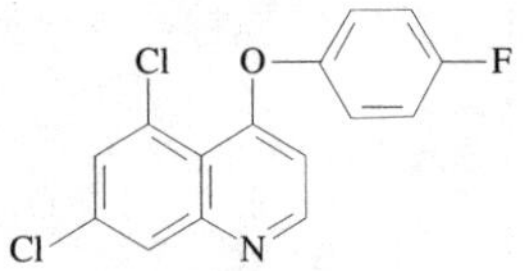

11-30: Quinoxyfen

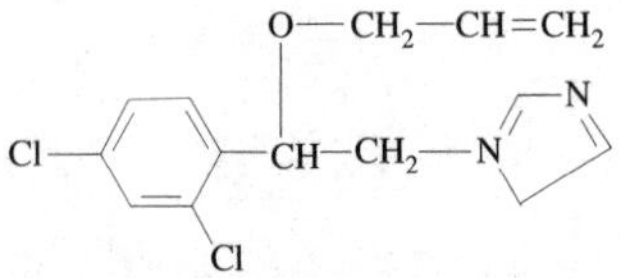

11-31: Imazotil

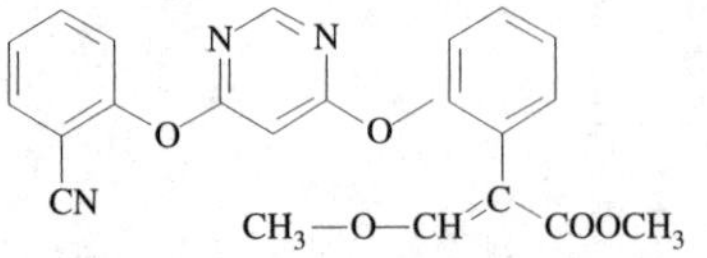

11-32: Azoxyshobin

Scheme 3:   Chemical structures of fungicides.

Nematodes (*worms*) attack fruits, vegetables, or plant roots, thereby causing losses to the crop during growth and in storage areas. Fumigants like 1,3-dichloropropene or methyl bromide control nematodes in soil or pests in mills, warehouses, grain elevators, ships, and in stored products in general. Because of their high toxicity and environmental risk these two fumigants cannot be used in the field or on animals. Here, other compounds with nematocidal properties are needed. However, safer, inexpensive alternatives are currently not available.

Repellents are sometimes added to pesticide formulations to keep nontarget species away from sprayed areas without killing them. Anthrachinone, for instance, repels birds. It is added to seeds to protect the seeds from being eaten and the birds themselves from being poisoned by toxic treated seeds. Some pyrethroids have a repelling effect on honey bees, which is very useful, because bees are kept away when the product is applied to flowering plants during the bee season. Insect repellents are also used by humans. The most famous active substance used in insect repellents is DEET. It was discovered by American scientists in the 1940s and no other substance with equal or better efficacy was discovered for over 50 years. Only recently a new substance came on the market with the trade name Bayrepel™ that is claimed to be superior to DEET.

Another topic is the protection of perishable fruits, vegetables, potted plants, and cut flowers. Many tropical fruits and vegetables are chill sensitive and cannot be transported or stored under low-temperature conditions. Other means for delaying deterioration are therefore needed.

TABLE 11.7 Other Pesticides and Repellents (Examples)

Common name or CA name*	Chemical class	Mode of action	Typical uses	Structure
Bromadialone or 3-[3-(4′-bromobiphenyl-4-yl)-3-hydroxy-1 phenylpropyl]-4-hydroxycoumarin	Coumarin anticoagulant	Anticoagulant, rodenticide	Control of rats and mice in storage areas, households, and industrial areas	11-33
1,3-dichloropropene	Fumigant	Soil fumigant	Controls nematodes in fruits, nuts, and berries	11-34
Methyl bromide or Bromemethane	Fumigant	Multipurpose fumigant	Controls a wide variety of pests in glass houses and storage areas	11-35
Anthrachinone	Bird repellent	Induces retching in birds	Used as seed treatment for cereals	11-36
DEET or N,N-diethyl-m-toluamide	Aromatic amine		Insect repellent for human use	11-37
Bayrepel or 1-Piperidinecarboxylic-acid, 2-(2-hydroxyethyl)-1-methylpropylester	Piperidine derivative		Insect repellent for human use	11-38
Ethylene	Hydrocarbon	Induction of ripening	Induction of ripening of fruits and vegetables	11-39
1-MCP or 1-Methylcyclopropene	Hydrocarbon	Blocks ethylene receptor sites	Delay of ripening, conservation of fruits, flowers, and so on	11-40

*According to *Chemical Abstracts* or IUPAC.

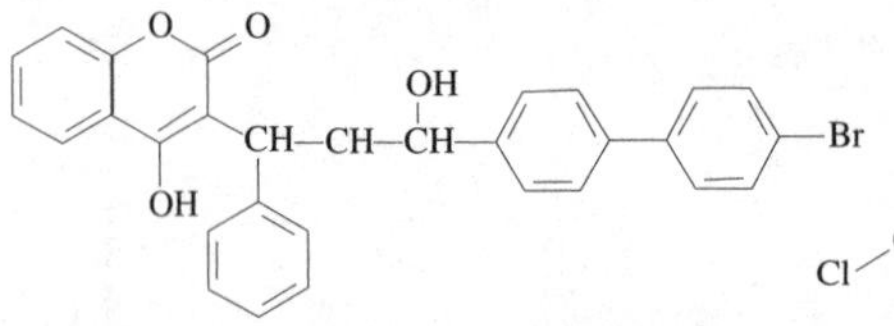

11-33: Bromadiolone 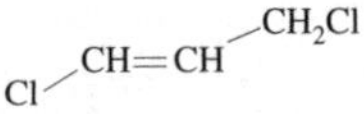   **11-34: (E) - 1.3 dichloropropene**   **11-35: Methyl bromide**

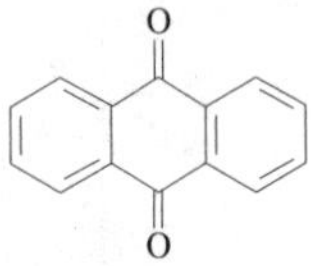

11-36: Anthrachinone 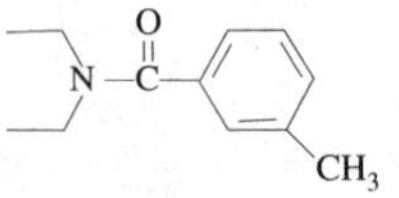   **11-37: DEET** 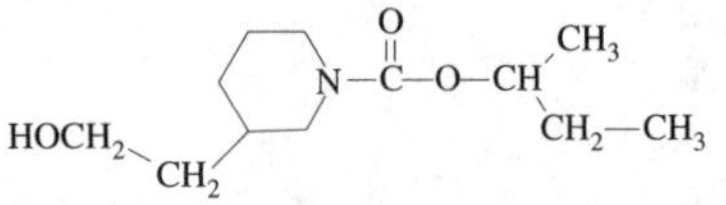   **11-38: Bayrepel**

CH$_2$=CH$_2$ CH$_3$—C=CH
                                         \ /
                                         CH$_2$

11-39: Ethylene   **11-40: 1-MCP**

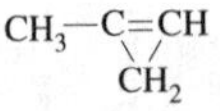

Scheme 4:   Chemical structures of other crop protection chemicals and repellents.

Traditionally, ethylene-induced ripening was the postharvest fruit management tool. Ethylene occurs naturally in plants. It starts and coordinates ripening processes (e.g., softening, color change, conversion of starch to sugars, loss of acidity, and so on) in fruits and vegetables. This has been used commercially to keep fruits edible after transportation and storage. Fruits are harvested well before ripening and are transported *green* to the far away destination and stored there. A few days before marketing they are exposed to an ethylene atmosphere. The ethylene facilitates a rapid ripening process, leading to products of high optical quality. The quality of taste, however, is compromised, because not all sugar, vitamin, and aroma components were developed by the time of harvest and are not improved by the ethylene exposure process.

Very recently, the opposite of the ethylene techniques has been developed to keep fruits fresh during transportation. Fruits are harvested after natural ripening and are protected from deterioration by compounds that suppress the natural ethylene response of the crop. Such a compound is 1-methylcyclopropene (1-MCP). It extends the postharvest shelf life and the quality of numerous fruits and vegetables, in particular, apple, tomato, and avocado fruits. In apples, 1-MCP maintains critical taste components including firmness (crunchiness), sugar content (sweetness), and acidity (tartness). It is also used to extend the lifetime of cut flowers and potted ornamental plants. 1-MCP acts by attaching to a site (receptor) in fruit tissues that normally binds to ethylene. If ethylene binding is prevented, ethylene no longer promotes ripening and senescence. This causes fruits to ripen and soften more

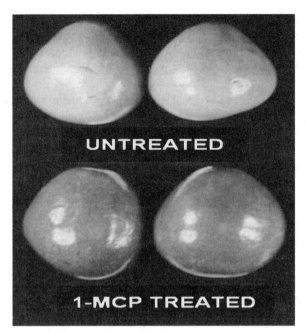

Figure 11.3   Effect of 1-MCP on the preservation of food. (*Source: Courtesy D. Huber, J. Jeong, and M. Ritenour, Horticultural Sciences Department, Florida Cooperative Extension Service, Institute of Food and Agricultural Sciences, University of Florida, January, 2003. EDIS.*)

slowly than in nature, thereby maintaining their edible condition for longer periods of time (Fig. 11.3).

Chemically 1-MCP is very interesting, because it is a highly unstable cyclopropene compound. A few years ago scientists thought about cyclopropenes as scientific curiosities without practical value. The 1-MCP proves that they were wrong. Being unstable and a gas as a neat substance, it is supplied commercially adsorbed to solid materials in powder form to make handling easier. When the powder is mixed with a specified amount of water, the 1-MCP gas is released to the gas phase where it interacts with the plants before it decomposes.

11.2.5   Chemical synthesis of pesticides

Synthesis routes of pesticides are very diverse because of the complex nature of the organic molecules. They are normally produced in batch processes on a scale of one to five tons per batch. *Old* pesticides are applied in relatively large quantities. Therefore, the price of the active ingredient is an important factor. This is illustrated by the examples of atrazine (Eq. 11.2) and glyphosate (Eq. 11.3) that are produced from readily available, inexpensive starting materials, such as cyanur chloride, diethylamine, 2-aminopropane and phosphoric acid, formaldehyde, and glycine, respectively. Modern pesticides are applied in much smaller amounts and the manufacturing cost is no longer the main factor. Sulfonylureas or pyrethroids, for instance, are manufactured from rather complex starting materials that themselves need several steps in the synthesis. This is illustrated by the example of nicosulfuron (Eq. 11.4)

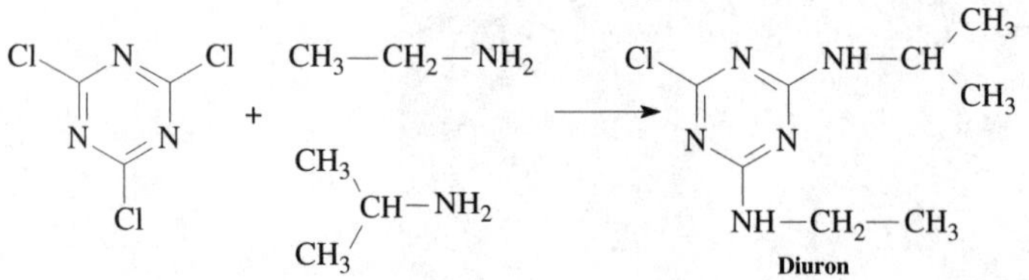

$$(11.2)$$

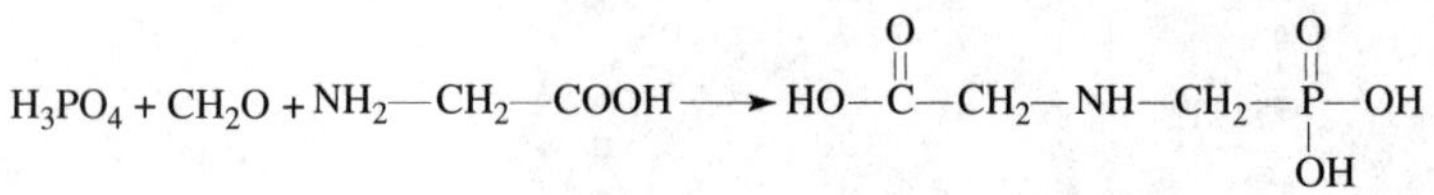

$$(11.3)$$

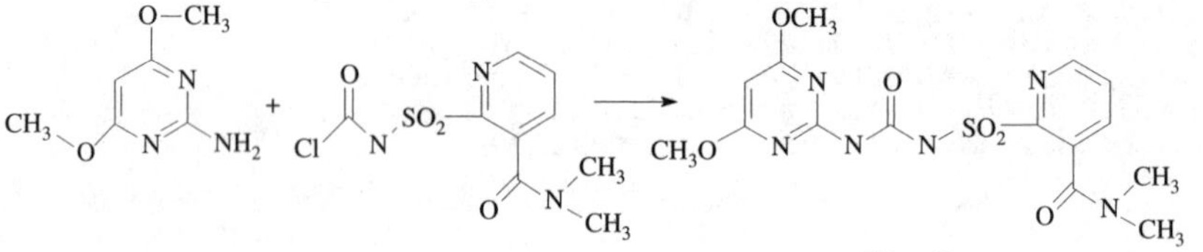

$$(11.4)$$

Another very important issue in synthesis is the formation of unwanted side products or the presence of highly toxic impurities in the starting materials. The best known example is the formation of dioxins during the production of 2,4-D (Eq. 11.5) and other similar compounds that use dichlorophenol or trichlorophenol as starting materials. Many different compounds belong to the dioxin group, depending on the number and pattern of chlorine atoms (e.g., TCDD = tetrachlorodibenzo-1,4-dioxins; HCDD = hexachlorodibenzo-1,4-dioxins). Some of these dioxin congeners are extremely toxic, even in very small concentrations. Between 1950 and 1970 there were several instances when dioxin-containing pesticides were sprayed on agricultural land or used in public hygiene. The best known example is *agent orange*, a mixture of 2,4-D and 1,2,4-T that was used during the Vietnam war on a large scale as a defoliant. Today the main risk for dioxin formation is during the incineration of old stocks of organochlorines. At low temperatures (e.g., 500 to 700°C) the oxidation is incomplete and dioxins may form. Incineration must be carried out at temperatures around 1200°C to provide a safe way of disposal of highly chlorinated compounds.

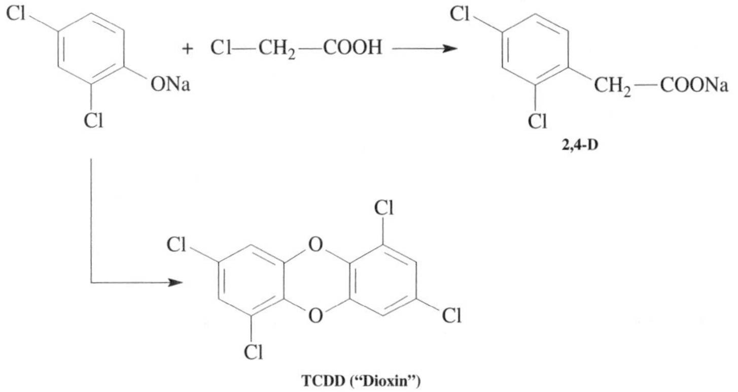

(11.5)

Nitrosamines is another class of highly toxic carcinogenic compounds that are of concern in pesticide chemistry. They are formed in nature, for instance, in rotten food or even during barbecuing of nitrite-containing meat, by nitrosylation of amino acids. Pesticides of the amino acid class, like glyphosate, are at risk in containing nitrosamines as impurities (Eq. 11.6). This potential is recognized and the main manufacturers have optimized their synthesis to avoid side reactions that may lead to nitrosamines. However, cheap imitation products are sometimes contaminated. Nitrosamine and dioxin analysis is a quality control element in industrialized countries, but the analytical procedures are very complicated and need sophisticated equipment, that is not available everywhere.

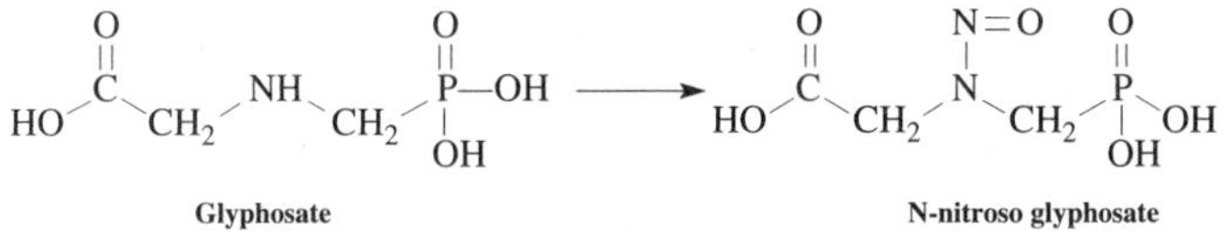

(11.6)

11.3 Formulated Products

The active substances are not used as neat chemicals. They are sold as formulated products. Products for small-scale use in house or garden are often ready-to-use formulas. That means they come in diluted form, often in spray cans, and can be used as they are. For large-scale use this is not economical, because a large part of the spray mix is simply water, needed to dilute the active substance. Therefore, pesticide products for use in agriculture or vector control are concentrates. They are diluted with water

TABLE 11.8   Definition of Terms for Pesticide Products

Term	Meaning	Remarks
a.i. or a.s.	Active ingredient or active substance	The chemical that controls the pest
Parent compound	Same as active ingredient	The parent compound degrades to metabolites
Formulation	Product containing the a.i., solvents, and additives. Some formulations contain . two or more a.i	(a) Concentrates for use on farms or public areas (b) Ready-to-use products for household application
Spray solution	Solution applied in the field	Prepared before use by dilution of the formulation with water
Tank mixes	Combination of two formulations in the same spray solution	Prepared before use in the spray tank

immediately before use, often in the tank of the sprayer. Thus, it is important to distinguish between the different stages of dilution. (See Table 11.8 for definitions.) Formulated pesticide products contain active and inert ingredients. Active ingredients kill or control the pest(s), whereas inert ingredients are designed to preserve the active ingredients, make them easier to apply, or improve their activity. Here are some examples.

Surfactants act as wetting agents. They help to disperse the droplets of the applied product on the hydrophobic surface of the leaves of plants and enhance the uptake through the membranes of the cells. Solvents are used to dissolve the active ingredient and other components. The most common solvents are water or oil, depending on the solubility characteristics of the substances. Oils (e.g., diesel oil) often enhance the product properties, as they facilitate the uptake through leaves or bark. Drift retardants are polymers, such as polyacrylamides, that reduce the apparent vapor pressure of volatile mixtures by adsorption and help to aggregate very fine droplets formed during the spraying process. They are added to a formulation to reduce drift to nontarget areas by evaporation or aerosol formation (*spray mist*) during application of the products. Foaming or antifoaming agents are added to facilitate or prevent foam formation during mixing of the products in the sprayers, depending on the properties of the products. The risk of accidental ingestion of toxic products can be minimized by the addition of emetic (vomiting causing) agents. Stenching agents produce strong odors to avoid mistaken consumption of pesticides as soft drinks, especially by children.

Pesticide products for large-scale use are usually concentrates that must be diluted with water. The water's properties, like hardness, salt content, pH, or temperature, can vary widely from region to region and formulations must be designed to cope with such differences. Incomplete mixing or precipitation during mixing in the spray tanks would render uniform application impossible and pose a risk to the applicator and the treated crop.

There are many different formulations on the market. The following are the most common types:

SC. Soluble concentrates are powders or liquids containing components that are completely soluble in water. They are easy to use but difficult to develop. They have a higher risk for leaching into the groundwater than other products.

EC. Emulsifiable concentrates are organic solutions (usually petroleum oil fractions) of the active ingredient. They contain emulsifying agents that facilitate a fine dispersion of the product upon mixing with water. They penetrate the waxy layer of leaves more easily, thereby increasing efficacy. The disadvantage is that they diffuse through the skin easily and may be more hazardous for the operator.

WP. Wettable powders containing compounds that are insoluble in water and organic solvents. The solid material is finely ground and coated with wetting and dispersing agents. They form suspensions when mixed with water.

DG. Dispersible granules that are small beads that disperse in water upon mixing. They are often used when highly water-soluble mobile substance should be retained in the soil for an extended time.

As active substances are often salts of acids or bases, their content in the formulation must be given as an acid or base equivalent to make products with different counter ions comparable. Glyphosate, for instance, is supplied as glyphosate sodium, glyphosate ammonium, glyphosate isopropyl ammonium, and glyphosate trimethylsulfonium (*trimesium*) salts, each of which has a different molar mass and consequently a different content of free glyphosate per kg formulated product.

A formulation can contain more than one active ingredient to increase the efficacy range per application. This reduces labor time, because fewer sprayings are necessary than with separate sprayings of single pesticide formulations. A similar result is achieved when several products are mixed directly before application (*tank mixes*). However, the products of the different formulations must be compatible. For instance, oil- and water-based products cannot be mixed and would lead to phase separation. Chemical incompatibility can occur as result of reaction of acids with bases or hydrolysis of pH sensitive compounds.

All commercial packages of pesticide formulations must have a label that is approved by competent authorities. The label is an official document and contains important information on the fields of use, the storage, safety precautions, application rates, incompatibilities, and so on,

that the applicator must know to apply the product safely and successfully. A short form of the label is permanently attached to the commercial container. The full label is more like a brochure. The U.S. label of
Roundup Custom™, a popular glyphosate formulation has 20 pages.
The following data are the minimum required:

An ingredient statement listing the names and amounts of the active
ingredients and the amount of inert ingredients

The net contents in the container

The name and address of the manufacturer and the supplier, along
with an establishment number telling which factory made the chemical

A registration number showing that the product has been registered
with a competent government authority for the uses listed on the label

The signal word and symbol on the label telling how toxic a product
is. Signal words are *danger* along with the word poison, and the skull
and crossbones symbol (highly toxic), *warning* (moderately toxic), and
caution (slightly toxic)

Emergency first aid measures and exposure conditions requiring medical attention

Statement about environmental toxicity, for example: "This product is
highly toxic to bees exposed to direct treatment or residues on crops."

Precautions to protect the environment, such as "Do not contaminate
water when cleaning equipment or when disposing of wastes." "Do not
apply where runoff is likely to occur."

List of physical and chemical hazards, like *flammable* or *corrosive*

The pests the product controls; the crops, animals, or other items the
product can be used on, legally

Directions on how the product should be applied; how much to use;
and where and when the product should be applied

The time that must pass from application until it is safe to harvest a
food crop; expressed as *days to harvest* or *pre-harvest interval* (*PHI*),
this is the time required for the residue to drop to safe levels: it is often
listed as a number in parentheses following the crop name

The misuse statement as a reminder that it is a violation of the law
to use a product in a manner inconsistent with its labeling

Storage and disposal directions that must be followed for environmental and human safety

It is important to comply with the label recommendations. First of all,
to ensure the safety of the operators, but also to protect bystanders and

consumers that would be at risk when too high pesticide residues are in the food. There is also a commercial component: If a product is applied incorrectly, for example, with high application rates leading to residues >MRL (maximum residue level), less the agricultural products produced by the guilty farmer or even by a whole country will be removed from the market. Residues of pesticides can be detected easily by modern analytical methods such as GC/MS or LC/MS/MS and food can be traced back to the growing area by bioanalytical methods.

11.4   Biological Pest Control

There were various attempts in history to control pests biologically. Often species (predators) were introduced that feed on the organism that needed control. A very simple and extremely successful example is the use of sheep to control grass and weeds. On the other hand, this example also illustrates the unwanted side effects that are associated with biological pest control. Sheep not only feed on grass but also on sprouts of trees and crops. They are *nonselective* and must be carefully managed to avoid damage of the existing ecosystems. Actually there are examples of attempts to control pests biologically that had disastrous consequences. To mention only one, the release of ferrets to control rabbits in New Zealand led to extinction of a number of species that were not adapted to these predators, which did not previously exist in New Zealand. The reason was very simple: the ferrets did not attack the target pest, but rather, fed on the eggs of birds.

Modern biological pest control uses more refined techniques. In this chapter, pheromones, living organisms, and genetically engineered plants are described in more detail. Pheromones are sexual attractants that affect species very selectively (see Table 11.9 for examples). They are used to lure insects into traps, where they can be killed, either with a contact insecticide or mechanically, for instance, with glue. Codlemone may serve as an example of how pheromones work. Chemically codlemone is a dodecanedienol, a substance with a low, but measurable vapor pressure. It is the sex pheromone of the codling moth and was isolated from virgin females. As it has a rather simple structure, it can be produced easily by chemical synthesis. If synthetic codlemone is applied in high concentrations, it interferes with the natural mating process. It leads the male onto the wrong track, one that does not lead to a female. This mechanism prevents the proliferation of the moths by disrupting mating without actually killing the insect. In practical application, slow release containers are placed in apple or peach orchards. There the pheromone evaporates slowly into the air and disperses throughout the orchard. Typical formulations contain 63 percent codlemone, 31 percent dodecanol, and 6 percent tetradecanol. Traps also contain a contact insecticide, such as

TABLE 11.9 Pheromones Used for Insect Control (Examples)

No.	Common name	Chemical name	Target pest	Structure
1	Codlemone	(E,E)-8,10-dodecadien-1-ol	Codling moth, hickory shuck worm	11–41
2	Disparlure	(7R,8S)-7,8-epoxy-2-methyloctadecane	Gypsy moth	11–42
3	Farnesol	(Z,E)-3,7,11-trimethyl-2,6,10-dodecatrien-1-ol	Spider mite	11–43
4	Grandlure	(+)-cis-2-isopropenyl-1-methylcyclobutaneethanol	Boll weevil	11–44
5	Melon fly attractant	4-(4-hydroxyphenyl)-2-butanone acetate	Melon fly	11–45

permethrin that kills the insect. Codlemone itself is not toxic and has no adverse effects on nontarget organisms.

An example of a *living pesticide* is *Bacillus thuringiensis* (Bt). It is a naturally occurring bacteria species with insecticidal properties. It lives in insect-rich environments, for instance in soil or food storage areas. The strains with the highest potency against a pest are selected from these natural populations. They are then used to produce large quantities of Bt in controlled fermentation. The insecticidal endotoxins and spores are harvested as water dispersible concentrates. The endotoxins are protein-like toxins that are produced inside the bacteria cells and are released after the cell walls are disrupted. After being ingested by the

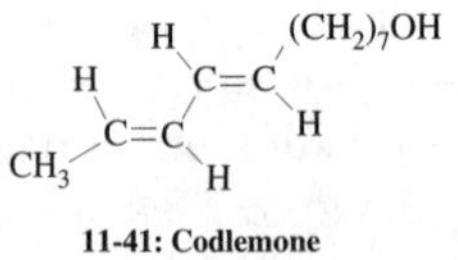

11-41: Codlemone

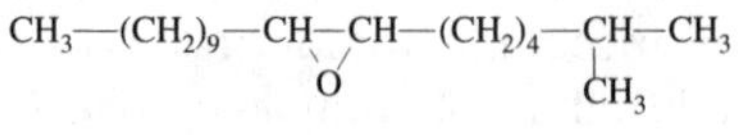

11-42: Disparlure

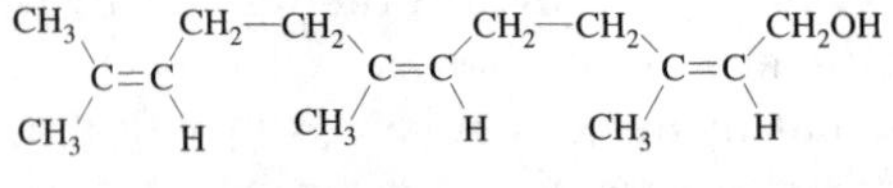

11-43: Farnesol

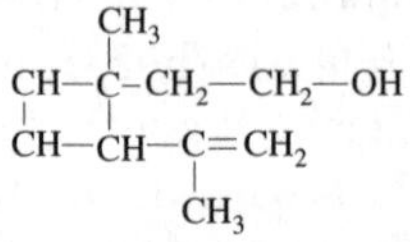

11-44: Grandlure

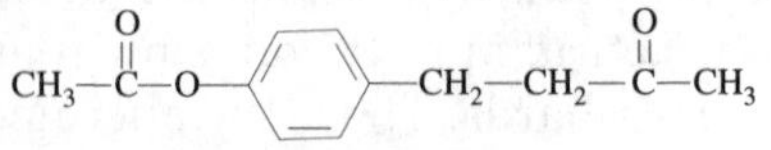

11-45: Melon fly pheromone

Scheme 5: Chemical structures of pheromones.

insects or larvae the endotoxins are hydrolyzed to smaller fragments. The fragments bind very specifically to selected receptor sites. Here, they cause widening of the channels in the cell membranes and increased water uptake and eventually cell rupture. Bt is active against moths, butterflies, potato beetles, and other related species. *Bacillus thuringiensis* is not infectious or toxic to humans and can be applied on food crops until the last day before harvest. Also, it has no ill effect on the environment. Its practical use, however, is limited by its short persistence in the field that makes frequent applications necessary. Another disadvantage is that it is efficacious only after the crop is infested and has already been damaged to a certain extent. Where this is acceptable, for instance, in forests, Bt is an interesting alternative to chemical pesticides. Other biological systems, such as nematodes, predatory mites, baculovirus, can be used to control pests. Some examples are listed in Table 11.10.

Most living organisms have the same advantages and disadvantages. The main advantages are high selectivity, low toxicity to nontarget species, and low environmental persistence. On the other hand, they are very sensitive to temperature and light, they are difficult to store, transport, and apply. Because of the high selectivity, they can control only one pest at a time. In integrated production (IP) farming, biological and chemical pesticides are used in combination and only when necessary. This is a very useful strategy to prevent resistance in the target pest and to reduce the negative impact of farming on the noncultivated environment.

Transgenic plants are another modern development to enhance food production. The principle here is that genes with special properties are isolated from the original species and introduced into a target species by genetic engineering techniques. A famous example is the gene that is responsible for the production of endoxines in *Bacillus thuringiensis*. When Bt genes are transferred into crop plants, the plants themselves exhibit insecticidal properties and become resistant against insects.

TABLE 11.10   Living Organisms Used to Control Pests (Examples)

No.	Species	Target pest or mode of action
1	*Bacillus thuringiensis* (Bt)	Bacteria produce endotoxins that kill insects and larvae (insecticide)
2	*Candida oleophila*	Fungus that inhibits infestation of plants by other fungi (fungicide)
3	*Heterorhabditis megidis*	Parasitic nematode that controls soil insects (insecticide)
4	*Phytophtera palmivora*	Fungus that controls strangler vines (herbicide)
5	*Aphidus ervi*	Parasitic wasp controls aphids in glasshouse grown crops

Bt genes have been introduced into potatoes, cotton, maize, and other plants. The action mechanism is similar as described for Bt itself.

The *class II EPSP synthase gene* also called *roundup ready gene* induces tolerance to glyphosate. When introduced into soybeans, cotton, maize, canola, and the like, these plants tolerate glyphosate. That means that glyphosate can be used to control weeds in these crop plants without damaging them, although it is originally a total herbicide that kills all plants.

The advantage for the farmer is that he needs only one product, instead of several different selective (and more expensive) herbicides. *Roundup ready* soybeans were launched in 1996 and today 50 percent of the soybean crop in the United States is derived from *roundup ready* seeds. Other glyphosate-resistant transgenic crops introduced by Monsanto are maize and oil seed rape. Competing companies also developed herbicide-resistant plants or plants genetically modified to be protected against certain pests, but none has achieved a commercial breakthrough, mainly because of political reasons.

One controversial issue is that the farmers must buy transgenic seeds that are controlled by one monopolist supplier. Local seed suppliers cannot produce seeds from harvested crops, because the transgenic plants are sterile and unable to produce new seeds for germination. Another potential problem is that the resistance genes of the transgenic plants are released into the environment in an uncontrolled way leading to problems that are presently not foreseen. The European Union has put a ban on all new transgenic plants and requires a clear labeling of all food that contains products from transgenic crops. Europeans feel that it is an unacceptable risk, as it is not proven that genetically modified food is safe. Many Americans on the other hand think that this is no problem at all, as there is no proof that genetically engineered food is not safe. This illustrates the different perceptions of risk in different parts of the world.

11.5   Testing Requirements for New Pesticides

11.5.1   General information and physical and chemical properties

In the past, pesticides were optimized for efficacy, but sometimes with harmful side effects to man or the environment. Today, modern pesticides have high selectivity for the target organism, and testing programs are mandatory to minimize risks and maximize benefits. Before a pesticide can be sold, the supplier must ask the competent authorities of the respective countries for a permit, the *marketing authorization*. The applicant must submit a dossier containing all information about the new substance that is needed to make a thorough risk or benefit analysis. A complete registration dossier contains around 200 different scientific reports with over 20,000 pages, the equivalent of 25 books. The studies are conducted under stringent quality control following the OECD guideline of GLP (Good Laboratory Practice). GLP is an internationally accepted quality standard. It assures that the personnel are qualified, that the methods are validated and that the instruments are properly calibrated. All data and reports are audited by an independent quality assurance unit.

The first part of a dossier contains basic information about the active substance and the applicant. Here are some examples: name, structure, and route of synthesis of the substance must be described. The typical purity of the technical material and the identity isomers and impurities is determined by the analysis of five production batches to assess the reproducibility of the process and to help to identify fake products. In the next chapter the chemical and physical properties are described, determined according to official guidelines published by the OECD and other organizations. This applies to the active substance, all significant metabolites, and all formulated products. Some of the required studies are listed in Table 11.11.

TABLE 11.11   Examples of Physical and Chemical Tests Required for the Registration of New Pesticides

No.	Description of test	Purpose of test or remarks
1	Appearance (physical state, color, odor)	Identification
2	Melting point, boiling point, density	Identification, information
3	Vapor pressure	Risk for evaporation
4	Flammability, explosivity, corrosivity, oxidative properties	Hazard during transport and storage
5	Surface tension	Risk to surface water
6	Solubility in water (at 3 pH values) and organic solvents	General information, risk for leaching
7	Spectra (UV/VIS, IR, NMR, Mass)	Identification
8	Partition coefficient n-octanol or water	Risk for bioaccumulation
9	Hydrolysis and photolysis in water	Risk for persistence
10	Stability and photochemical degradation in air	Risk for air pollution, persistence, and ozone depletion
11	Stability upon storage or shelf life	Quality of product
12	Analytical methods for purity and impurities in active substance and products	Quality and risk of toxic impurities
13	Analysis of five production batches	Reproducibility of process
14	Analytical methods for residues in food	Consumer safety
15	Analytical methods for residues in soil, water, air	Environmental monitoring

Appearance, melting and boiling points, density, spectra, and solubility data serve to establish unambiguously the identity of the substance. The vapor pressure value is an indication of the volatility of the substance. If a substance is volatile it may evaporate and be transported through air and inhaled by people, thereby increasing the risk to bystanders or nontarget plants. Flammability and explosivity are also safety parameters that show whether a substance may be dangerous when shipped or stored.

The partition coefficient is a very important property to assess possible accumulation of a substance in the food chain. It is defined as the ratio of the solubility of a substance in n-octanol and in water.

Log (Kow) = log [(concentration in octanol) / concentration in water)]

A high log (Kow) means that the substance dissolves better in octanol than in water (e.g., if log Kow = 6, a million times). In practical terms this means that there is a risk that the substance accumulates in fat. On the other hand, a log (Kow) < −2 shows that the substance is a hundred times more soluble in water than in octanol. This usually is a warning sign that the substance has a tendency to leach into ground water.

Hydrolysis and photolysis experiments allow conclusions on how stable a substance is and whether it may have a tendency to persist in the environment. Here the rate constants and the half life of a substance in water are measured at different pH values and under irradiation with simulated sunlight. The hydrolysis is tested at four different pH values: pH 4, pH 7, and pH 9 simulate the natural situation in soil and water, while a pH 1.2 is used to simulate the acidity in the stomach and gives some indication about hydrolysis after accidental oral ingestion. Organochlorines are often stable against hydrolysis. Other compounds undergo complicated reactions leading to a variety of products that are difficult to analyze. Methyl bromide is a compound that undergoes a straightforward hydrolysis resulting in the formation of methanol and hydrogen bromide (Eq. 11.7).

$$CH_3Br + H_2O \xrightarrow{pH4} CH_3OH + HBr \tag{11.7}$$

The storage stability (shelf life) of active substances and formulated products are studies under simulated climatic conditions to assure that they maintain their quality when stored or transported. A product must be stable for at least 2 years under the conditions on the climatic zone, where it is used.

The efficacy chapter contains the information directly related to the use of the active substance in the field, like mode of action, intended use, application rates, frequency of applications, efficacy against target organisms, and potential for development of resistance in target organisms. Efficacy must be demonstrated for all target organisms and for all formulations intended for marketing. Efficacy is evaluated in field studies under different climatic conditions. Studies are conducted by specialized field trial contractors or by government affiliated agricultural research stations.

In all tests, the chemical analysis of the samples is a crucial step. Modern chromatographic methods are applied, for example, HPLC, LC-MS-MS, or GC-MS and all methods must be carefully validated. This means that linearity, accuracy, precision, and selectivity must be proven. Validations are required for methods used during the development phase and also for monitoring after a marketing authorization is granted. *Enforcement methods* are intended for use by government laboratories to monitor product quality, worker exposure, or residues in food, feed, and the environment.

11.5.2 Toxicity

The toxicity of pesticides (and all other new chemical products) must be tested on animal models before a product can be sold in the market. Toxicity tests have different levels of complexity and different objectives. Acute toxicity studies are usually carried out on rats and the effect lasts up to 7 days. Their purpose is to determine how toxic a substance is after accidental exposure. Tests are conducted not only through oral intake and inhalation, but also through exposure to the skin and eye. The latter test was originally done on rabbits, but is now being replaced by in vitro alternatives. The final result (end point) of the acute oral tests used to be the LD-50 value, the dose after which 50 percent of the animals die. It is now being replaced by the acute reference dose (ARfD). This is the highest dose that is still safe for a human after a single exposure to the pesticide. It is calculated from the highest dose that caused no harm in the acute animal studies (NOAEL = no observed adverse effect level) and a safety factor (F). The safety factor is usually 100 and it accounts for the uncertainty that is associated with the comparison of the toxic responses in animals and humans.

$$\text{ARfD} = \text{NOAEL (in acute studies)} * \text{safety factor (F)}$$

Short-term toxicity studies simulate the exposure of workers or farmers to the products they use on a regular basis. They are performed with rats and dogs and last 1 to 6 months. Chronic and carcinogenicity studies last up to 2 years and simulate the long-term exposure of consumers to small concentrations of pesticides in food. Rats, mice, and dogs are used as animal models. The final result is the acceptable daily intake (ADI), the highest exposure level that is still safe for humans.

ADI = NOAEL (in chronic studies) * safety factor (F)

ADI and NOAL are given in mg/kg (body weight). That means they are normalized to the body weight to make the results of different species comparable.

Absorption, distribution, excretion, and metabolism (ADME) is a test to establish the fate of a substance in the body of a living organism. This study type is also used to investigate the transfer into secondary products such as milk and eggs. Animal models for ADME studies are rats, goats, and hens. Distribution of the original active substance (parent) in the different organs is determined and the formed metabolites are identified and quantified. The final result is a mass balance accounting for the fate of the applied substance, for example, how much is excreted via exhaled air, urine, or feces and how much stays in the body. C-14 labeled test substance is often used for this type of study, because radioactivity can be followed more easily than unlabeled substances.

11.5.3 Residues in food

Food safety is a key issue in modern society, because we are exposed to food from birth to death. Plant protection products are often used on food crops, biocides in food factories, or animal housing. Therefore, tests are required to determine the risk of pesticide residues in the raw agricultural commodity (RAC) and the processed food products.

The first step is to identify the critical crops and the compounds for which residue studies must be conducted. The breakdown and reaction products and metabolites in treated plants and products are often identified after treatment with a C-14 labeled test substance in *plant metabolism* studies. The objective is to determine the fate of the parent substance and its metabolites in the crop plants and their processing products. From the results of the studies, the relevant residues are defined (*residue definition*). For example, in plants glyphosate is converted to its main metabolite AMPA (= aminomethylphosphohonic acid), which was also studied as part of the residue assessment.

$$HO-\underset{\underset{CH_2}{}}{\overset{\overset{O}{\|}}{C}}-NH-\underset{\underset{\underset{OH}{|}}{CH_2}}{}-\underset{\underset{OH}{|}}{\overset{\overset{O}{\|}}{P}}-OH \xrightarrow[\text{[O}_2\text{]}]{\text{metabolism}} NH_2-CH_2-\underset{\underset{OH}{|}}{\overset{\overset{O}{\|}}{P}}-OH + 2CO_2 + H_2O$$

Glyphosate   AMPA

(11.8)

After the significant residues are identified, field studies are designed to determine experimentally the level of residues in crops that were treated according to the label recommendation and to good agricultural practice. Field trials are required in typical growth areas and in different climate zones. The program must be repeated in a second year to account for climatic fluctuations. Samples of the crop and the processed food are collected at various intervals after application of the pesticide and at the normal harvest day. They are deep-frozen to $< -20°C$ to avoid degradation of the residues and sent to the analytical laboratory, where they are analyzed for residues of all compounds included in the residue definition. The result is a maximum residue level (MRL), the concentration that will not be exceeded when a product is used exactly as recommended by the manufacturer. If a higher concentration is found in a crop, it can be concluded that the farmer did something wrong. However, residues >MRL do not mean that the food is unsafe, because the MRL is usually much lower than the level that would affect the health of the consumer. Only if a residue exceeds the ADI (acceptable daily intake), the food poses an unacceptable health risk.

11.5.4   Human safety risk assessment

Risk, in its scientific meaning, has two components, namely hazard and exposure. To swim in an ocean is hazardous, but people living inland are never exposed to swimming in the ocean. This means that their personal risk of being harmed by the ocean is very small.

$$Risk = hazard \times exposure$$

The hazard of pesticides comes from their toxicity, exposure from their use, or from residues in food. In principle every person can be exposed to pesticides, therefore, a large number of toxicity tests must be conducted before a pesticide gets a marketing authorization. (See Table 11.12 for examples.) The toxicity of pesticides varies widely, as can be seen from the examples in Table 11.13. On acute exposure glyphosate, nicosulfuron, and mancozeb are essentially not toxic, whereas carbofuran is very toxic. It is 50 times more toxic than imidacloprid, which has a similar use. On the other hand, the application rate of carbofuran is three times higher than that of imidacloprid. If we assume that the exposure of farm workers parallels the application rate, the exposure to carbofuran is three times that of imidacloprid. Three times higher exposure together with 50 times higher acute toxicity leads to 150 times higher risk for workers using carbofuran than for workers using imidacloprid.

Chronic toxicity does not parallel the acute toxicity. Chlorpyriphos and paraquat for instance, have medium acute toxicity, but a high

TABLE 11.12   Examples of Toxicological Tests Required for the Registration of a New Pesticide

No.	Description of test	Animal model	Purpose or remarks
1	Acute oral and inhalation toxicity	Rat	Accidental ingestion or inhalation
2	Acute dermal toxicity	Rat	Accidental exposure or skin
3	Skin and eye irritation, sensitization	Rat, rabbit in vitro	Damage of eye or skin after accidental exposure
4	Short term toxicity up to 6 months	Rat, dog	Risk after multiple exposure (workers)
5	Long-term toxicity and carcinogenicity up to 2 years	Rat, dog, mouse	Risk upon lifetime exposure (consumer via food)
6	Reproductive toxicity	Rat, rabbit	Risk for fertility and exposure during pregnancy
7	Neurotoxicity, endocrine disruption	Rat, hen	Risk to nerve and hormone systems
8	ADME (adsorption, distribution metabolism, and excretion studies)	Rat, goat hen	Assessment of fate in the body and the risk of transfer to milk or egg

chronic toxicity as can be seen from their low NOAEL values. A NOAEL value of 0.1 mg/kg/day means that test animal exposed to this dose did not show negative health effects. Only after the intake of higher doses were adverse reactions observed. The ADI is the maximum amount that is safe for human intake. It is derived from the NOAEL and a safety factor.

For example, the NOAEL of glyphosate is 400 mg/kg/day and the ADI is 0.1 mg/kg/day. This means that human beings with 60-kg body weight can eat 6 mg glyphosate without risk for their health. This is the toxicity

TABLE 11.13   Oral Toxicity of Some Common Pesticides (Approximate Values)

Compound	Acute tox LD50 (rat) (mg/kg)	Chronic tox NOAEL (rat) (mg/kg/day)	ADI or RfD (mg/kg/day)	Application rate (g/ha)
Glyphosate	>5000	400	0.1	3000
Nicosulfuron	>5000	1000	1.25	50
Mancozeb	>5000	5	0.03	8000
Imidacloprid	424	6	0.06	290
Carbofuran	8	20	0.002	1000
Chlorpyrifos	2680	0.1	0.01	600
Paraquat	112	0.6	0.005	500
2,4-D	640	5	0.01	1000

ADI = acceptable daily intake; RfD = reference dose
SOURCE: J. T. Stevens (2001).

consideration. Consumers are exposed to pesticide via food. Assuming a residue of 1 mg glyphosate per kg bread and a consumption of 0.5 kg bread per day, the exposure is 0.5 mg/day. This is well below the acceptable intake of 6 mg/day, indicating that the risk of consuming bread with glyphosate residues is low.

The situation is different for paraquat, which has an ADI of 0.005 mg/kg/day. Taking the 60-kg person as an example, the acceptable intake is 0.06 mg/day. On the exposure side, let us assume a residue of 0.2 mg/kg in bread and a daily consumption of 0.5 kg. This leads to an exposure of 0.1 mg/kg, which is about twice the acceptable intake. The conclusion is that, in this example, the risk associated with the residues of paraquat is not acceptable.

In general, risk values <1 are considered acceptable, while values >1 are not acceptable and require risk management. The risk management must reduce either exposure or toxicity. The latter would mean that a toxic substance is replaced by a different, less toxic compound. Exposure can be reduced in different ways. Protective clothes for farm workers, such as gloves, goggles, overalls, and boots, are well-known risk management tools that reduce exposure by preventing dermal or oral intake of the pesticide. Another way to manage risk is to limit the use of a pesticide. For instance, an insecticide may not be allowed on food crops, because high residues and chronic toxicity would lead to an unacceptable risk. Its use for vector control in nonagricultural areas could be acceptable, because exposure through food is not important.

The concept of *risk* is a valuable tool to make responsible and scientifically sound decisions. By definition *risk* is never zero. It must be balanced by a benefit to make it acceptable.

11.5.5 Environmental fate and environmental toxicology

All pesticides that can come into contact with the environment are subject to a risk assessment. The basis for this risk assessment is provided by data from environmental fate and environmental toxicity studies, which are carried out in the laboratory or under field conditions. The fate (adsorption, degradation, and mobility) of the active substance must be studied in soil, air, water, and sediments. The laboratory studies are frequently performed with ^{14}C-labeled substances to make the mass balance easier. It is important to know how a substance degrades in the environment, because sometimes the degradation products are more persistent than the parent substance. DDT, for instance, is converted to metabolites by stepwise dechlorination (Eq. 11.9). The metabolites (e.g., DDD or DDA) can be found in soil for many years after the DDT itself is degraded.

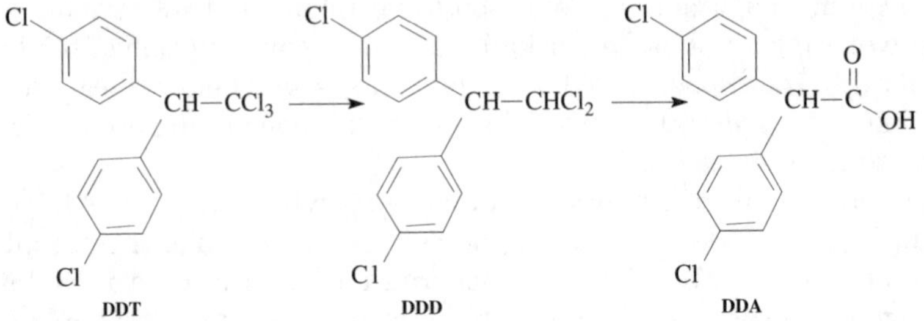

$$(11.9)$$

DDT   DDD   DDA

Examples of environmental fate studies:

Adsorption and desorption equilibrium constants are determined in different soils. The route and rate of degradation in water, sediments, and soil is studied under aerobic, anaerobic, and photolysis conditions and at different temperatures in the laboratory under standardized conditions. If the experimental results show that there is a risk for leaching, semi-field and field studies become necessary. *Lysimeter* is the name of a set-up, in which intact soil cores (e.g., 1-m diameter and 1-m depth) are installed in a container that allows collecting the water that leaches through the soil core. This water (leachate) is analyzed for content of parent substance and metabolites. If any substance exceeds a concentration of 100 ng/l (= 0.1 ppb), high risk for leaching is established and the substance cannot be used in the field. Also, here the formation of degradation products must be taken into account. They are often more polar than the parent and are therefore more soluble in water. This can lead to higher mobility. Atrazine is a typical example. The parent molecule has already a tendency to leach into groundwater where it has a half-life of 100 to 200 days. Its metabolite desethylatrazine (DEA, Eq. 11-10), however, is even more soluble and moves more readily through the soil. Its high mobility was so much of a concern that atrazine was banned from use, in which it can come into contact with groundwater.

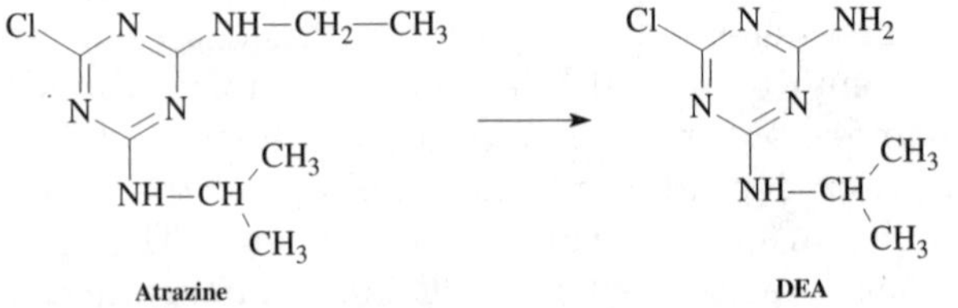

$$(11.10)$$

Atrazine   DEA

Field soil dissipation and accumulation studies are carried out to determine the lifetime of a substance in soil and to see whether it has a tendency to accumulate. The final result is the DT-90 value. This is the time after which 90 percent of the starting material has disappeared.

TABLE 11.14   Environmental Fate and Environmental Toxicity Studies Required for the Registration of Pesticides

No.	Description of test	Test system	Purpose or risk
1	Rate and route of degradation, laboratory studies	Soil, sediment, water	Persistence, formation of dangerous degradation products
2	Field soil dissipation and accumulation studies	Soil	Persistence, accumulation
3	Column leaching, lysimeter, field leaching	Soil, ground water	Mobility in soil, leaching into ground water
4	Toxicity to aquatic life	Fish, daphnia, algae, microorganisms	Interruption of food chain
5	Toxicity to birds	Quail, duck	Accumulation in food chain
6	Toxicity terrestrial life	Earthworm, microorganisms	Interruption of and accumulation in food chain
7	Effects on nontarget arthropods	Honeybees, predatory mites	Protection of nontarget species

If the DT-90 is longer than 3 months, there is a risk for accumulation and risk management procedures must be proposed. One such procedure could be to apply a product only every second year.

The second component of an environmental risk assessment for plant protection products that can come into contact with the environment are environmental toxicology tests (Table 11.14). Toxicity to aquatic life is tested on bacteria, daphnia, fish, and algae representing the food chain of aquatic species. Toxicity in terrestrial systems is studied with microorganisms, earthworms, nontarget plants, and other soil-dwelling organisms. Birds and mammals that feed on worms, fish or treated plants are the last step in the food chain. They are also included in the environmental toxicity program. Toxicity to beneficial insects is tested to assess the risk for nontarget species. Toxicity to honey bees is critical, as they are required to fertilize flowering plants and trees. Sometimes it is not possible to make insecticides so selective that they kill pests (insects) but not honeybees. In this case, the risk management provision is to avoid using the product during the honeybee season or on flowering plants.

The evaluation of the results from the environmental fate leads to a predicted environmental concentration (PEC) of the pesticide representing the exposure level. The accumulation in the food chain of fish is expressed as BAF (bioaccumulation factor in aquatic environment), that of mammals and birds as BCF (bioconcentration factor in terrestrial environment). Accumulation increases the exposure. The NOAEL (no observed adverse effect level) represents the hazard level. It is the result of the environmental toxicology studies. If the PEC is higher than the NOAEL multiplied by a safety factor, the risk is considered too high and the pesticide is not registered (see Table 11.14).

11.6   Social and Economic Aspects

11.6.1   Social consequences of pesticide use

The benefits of pesticides are social and economic. Social components are human health, availability of food, quality of life, and protection of the environment. Economic factors are loss of harvest, profitability of agriculture, and revenue of the industry. Some social factors are directly related to economic factors. If a farmer cannot make a living on his farm, he and his family must leave the land and migrate to the city, a process that occurs very frequently everywhere in the world. If the agrochemical industry does not make a profit, the factories must close down and the employees lose their jobs. If food becomes scarce prices go up and people with low incomes cannot afford a balanced diet or may even starve.

From these statements it seems clear that we need pesticides, if we want people to lead a decent life. There are, however, also a lot of social problems connected with pesticides. All pesticides are, by definition, toxic to some living organism—insecticides to insects, herbicides to plants, fungicides to fungi, and so on. In addition, they often have direct or indirect effects on nontarget organisms. Some are more toxic or longer lasting than others. Some accumulate in the environment and cause harm far away from the original site or purpose of application. No pesticide can be considered *safe*. Risks associated with the use of a pesticide must be outweighed by the benefits.

The perception of risks and benefits changes with time as the history of DDT illustrates. DDT was first synthesized in 1874 by the German chemist Zeidler. However, the insecticidal properties of DDT were discovered only later, in 1939, by Paul Mueller in Switzerland, who received the Nobel Prize for his discovery. After it became available on the market, DDT was accepted immediately and used on a large scale. For the first time in history, people could control insects effectively. Mothers could relieve their children from lice, farmers could protect their livestock and harvest. DDT has probably been responsible for saving more lives than any other synthetic chemical, perhaps with the exception of antibiotics. It is estimated that around 1940 ca 300 million people suffered from malaria. The mortality rate was about 1 percent. Thirty years later DDT and WHO (World Health organization) malaria program had eradicated malaria in many parts of the world. DDT also controlled the vectors of other diseases, such as louse-borne typhus, the plague, yellow fever, viral encephalitis, cholera, and the like.

Around 1970, people discovered that DDT benefits had their price. It is persistent in the environment and had accumulated over the years following its intensive use. Birds and aquatic life were directly affected, humans indirectly through DDT accumulation in the food chain. It deemed that risks outweighed the benefits and DDT was banned from most uses. Other seemingly less harmful insecticides had become available and replaced DDT. At the time organophosphates and carbamates were believed to be better alternatives, because they did not bioaccumulate. Today, we know that they are often very toxic to humans and animals. They harm the operators, when applied without proper protection, and they are the cause of many accidental poisonings. Actually, they are to a large extent responsible for the bad reputation that pesticides have today. The next step was the use of pyrethroids. They were effective and their use had relatively low risks, but the insects developed resistance, leaving us with a new problem. This shows how sometimes a known risk is simply replaced by another risk, often unknown at the time of the introduction of the product.

There is also a big gap between real and perceived risk. In Europe and the United States the public is mainly concerned about residues and food safety. With proper precautions, however, this is not a real problem, as residues in food are low and do not contribute significantly to our total intake of chemicals. Actually, there are no cases confirmed in which residues of modern pesticides in food were the cause of poisoning of humans. As there is a large overproduction of agricultural commodities in western countries, people worry about minor risks, like pesticide residues, as they do not have to worry about supply of food in principle. The picture is strikingly different in other parts of the world, where food production does not match the growth of the population and where malnutrition and famine are well known.

The largest real risk of pesticides is associated with their use. Farm workers and their families are often poisoned by pesticides as a consequence of improper protection during application. A picture of a farm worker spraying a field with bare feet and hands, without eye protection, wearing his everyday clothes illustrates this (Fig. 11.4). He holds the nozzle right in front of him and walks directly into the spray mist. The cotton face mask is not of much use, because it was designed to retain bacteria and other large particles, not vapors or aerosols. Also, there is skin exposure and inhalation. His clothes absorb large amounts of pesticide product. When he gets home, he takes home the pesticide and may expose his family and food supplies. What a difference to the application methods (see Fig. 11.5) used in modern farming. Here the worker is wearing protective clothing and spraying behind himself. He will

Figure 11.4 Worker is applying pesticide without protection. The face mask does not protect against vapors and aerosols. The skin is directly exposed and neither goggles nor protective clothes are worn. (*Source: Courtesy of Gerald R. Stephenson, University of Guelph.*)

(hopefully) remove his disposable garment before going home. Improper application methods are the main reason for acute and chronic poisoning by pesticides. The challenges are to develop better methods, applicable to hot climates and small farms, and to train the farm workers how to use them.

11.6.2   Economic aspects

The global crop protection market for the year 2000 was about US$32 billion. Herbicides accounted for about 45 percent, insecticides 22 percent, and fungicides 20 percent of worldwide sales. The top selling agrochemicals were glyphosate and paraquat as herbicides, chlorpyriphos and imidacloprid as insecticides, azoxystrobin and mancozeb as fungicides. The most important uses were on cereals (except rice) with

Figure 11.5   Proper application technique: The operator wears a gas mask, goggles and a protective coverall. (*Source: Courtesy T. Kratz, Landis Kane Consulting, Switzerland.*)

26 percent, fruits and vegetables 22 percent, rice 11 percent, soy beans 10 percent, cotton 10 percent, others 21 percent. In terms of mass, the production of organic crop protection chemicals grew until 1980. Since then it is more or less constant, although the treated areas and number of applications are still increasing. However, the amount of pesticide applied per hectare is decreasing, as the products become more efficacious and pest management procedures change. The average application rate in 1980 was 3 kg a.i./ha/treatment; for modern pesticides it is less then 0.3 kg a.i./ha/treatment (a.i. = active ingredient). Nevertheless, altogether about 1 million t of active ingredients are produced worldwide every year. The sales of the different pesticide classes is illustrated in Figs. 11.6, 11.7, and 11.8.

The development of new pesticides is very expensive and costs more than US$100 per substance and the failure rate is high. It takes 5 to 8 years from the time of discovery to bring a new substance to the market. On the other hand, only the best-selling products have sales exceeding US$100 per year. On average, new substances are patent protected for another 10 years after they are brought to the market. Thereafter they can be produced by any company with much lower development cost and therefore at lower prices. Off-patent compounds produced by nonresearch companies are called *generics* or generic products.

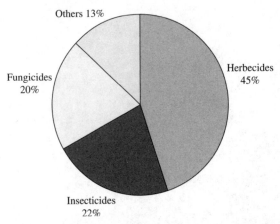

Figure 11.6 Sales of the most important crop protection chemicals. (*Source: Cramer, H.H., 2003.*)

Generics are low-cost alternatives to new substances and are widely used in many parts of the world. The problem is that many generics are still used in Asia, Africa, Eastern Europe, and South America, which are banned in economically advanced countries, because of unacceptable risks. According to UN Food and Agriculture Organization (FAO) and World Health organization (WHO), about one-third of the pesticides sold in the developing countries, representing a market value of US$900 million, do not comply with international regulations. Those low-quality pesticides often contain hazardous active compounds or are impure. In 2000, of the pesticide sales of US$32 billion worldwide, the

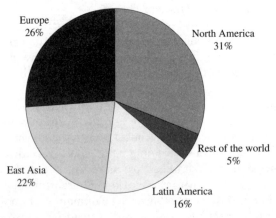

Figure 11.7 Regional distribution of crop protection product sales (2002). (*Source: Castle, D., 2003.*)

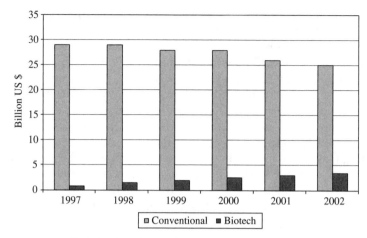

Figure 11.8   Global sales of conventional and biotech crop protection (in billion US dollars at distributor level). (*Source: Uttley, N., 2003.*)

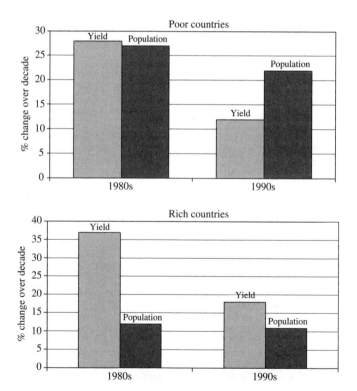

Figure 11.9   Comparison of agricultural productivity and population growth in *poor* and *rich* countries. In poor countries the population grows faster than the yield per hectare; in rich countries it is the opposite. (*Source: Castle, D., 2003.*)

market share of developing countries was just $US3 billion or about 10 percent. The reason for this imbalance is economic. Many farmers in poor countries cannot afford modern efficient pesticides, although they would need them urgently. This is illustrated in Fig. 11.9. In the poor countries the population and the yield per hectare grew at about an equal rate in the 1980s. In the 1990s, however, the population grew more rapidly than the agricultural production. If this does not change soon, the next food crisis is inevitable. In the *rich*, rapidly developing countries, on the other hand, productivity increases faster than population, leading to overproduction and to economic problems for farmers. The most obvious solution to the problem would be to import food from countries with overproduction to those with food deficit. However, this is not feasible, because the cost of transportation is too high.

Bibliography

Anderson, T. M., Industrial fermentation processes, *Encyclopedia of Microbiology*, Vol. 2, Academic Press, 2000.

Appleby, A. P., F. Müller, and S. Carpy, Weed control, in *Ullmann's Encyclopedia of Industrial Chemistry*, 6th ed., Vol. 39, p. 199, 2003.

Castle, D., in *Registration of Agrochemicals in an Enlarged Europe*, IIR Conference, Brussels 2003.

Copping, L. G. (Ed.), *The BioPesticide Manual*, British Crop Protection Council, 1998.

Cramer, H. H., Crop protection, in *Ullmann's Encyclopedia of Industrial Chemistry*, 6th ed., Vol. 9, p. 677–700, 2003.

Hirose, Y., Production and isolation of enzymes, Enzyme *Catalysis in Organic Synthesis*, Vol. 1, 41–66, Drauz, K., and Waldmann, H. (Eds.), Wiley-VCH, Weinheim, Germany, 2002.

Huber, D., J. Jeong, and M. Ritenour, *Use of 1-Methylcyclopropene (1-MCP) on Tomato and Avocado Fruits, Potential for Enhanced Shelf Life and Quality Retention*, Document HS-914, Florida Cooperative Extension Service, Institute of Food and Agricultural Sciences, University of Florida. Available at EDIS Web site, http://edis.ifas.ufl.edu, January 2003.

Maisch, W. F., Fermentation processes and products, *Corn Chemistry and Technology*, 2nd ed., White, P. J. and Johnson, L. A. (Eds.), AACC Monographic, 695–721, 2003.

Notermans, S. and F. Rombouts, (Eds.), *Frontiers in Microbial Preservation and Fermentation*, Elsevier Science Ltd., Oxford, 2002.

OECD, *Guidelines for the Testing of Chemicals*, ISSN 1607-310X, 2002, available at www.sourceoecd.org.

Praeve, P., U. Faust, W. Sittig, D. A. Sukatsch (Eds.), *Handbuch der Biotechnologie*, *Akademische Verlagsgesellschaft*, Wiesbaden, 1982.

Sanchez, S., and Demain, A. L., Metabolic regulation of fermentation processes, *Enzyme and Microbial* Technology, 31, 895–906, 2002.

Stephenson, G. R., Pesticides use and world food production: Risks and benefits, Chap. 15, 261–270, in *Environmental Fate and Effects of Pesticides*, J. R. Coats, Coats and H. Yamamoto (Eds.), Symposium Series 853, American Chemical Society, Washington, DC.

Stevens, J. T. and C. B., Breckenridge, Crop protection chemicals, in *Principles and Methods of Toxicology*, 4th ed., A. W. Hayes, (Ed.), Taylor & Francis, Philadelphia, 2001.

Synthesis and Chemistry of Agrochemicals VI, ACS Symposium Series No. 800, American Chemical Society, 2002.

Tomlin, C. D. S. (Ed.), *The Pesticide Manual*, 12th ed., British Crop Protection Council, 2000.

Uttley, N., in *Registration of Agrochemicals in an Enlarged Europe,* IIR Conference, Brussels, 2003.

Vert, M., Polymers from fermentation, Poly(lactic acid)s and their precursors, the lactic acids, *Actualité Chemique,* 11, 12, 79–82, 2002, (in French).

Waites, M. J., N. L. Morgan, J. S. Rockey and G. Higton, *Industrial Microbiology,* Blackwell Scientific, Oxford, 2001.

Internet sites with information about agrochemicals:

www.europa.eu.int/publications/en

www.epa.gov/pesticides

www.pesticides.gov.uk

www.wwf.org

www.oecd.org/document

www.fao.org

Chemical Explosives and Propellants

Bassam El Ali

12.1 Chemical Explosives

12.1.1 Introduction

Explosives are chemical compounds or their mixtures that rapidly produce a large volume of hot gases when properly initiated. Explosives are known to detonate at the rate of kilometers per second. Explosives are capable of exerting sudden high pressures, generating loud noise, and more or less destructive types of reactions that produce mechanical, chemical, or nuclear explosions. A mechanical explosive depends on physical reactions such as overloading a container with compressed air. Such a device has some application in mining, in which the release of gas from chemical explosives may be undesirable, but otherwise is used very little. A nuclear explosive is one in which a sustained nuclear reaction can be made rapidly, releasing large amounts of energy. Chemical explosives account for virtually all explosive applications.

The destructive effects of explosives are much more spectacular than their peaceful uses. However, it appears that more explosives have been used by industries for peaceful purposes than in all the wars [1, 2].

A distinguishing feature of explosives is the rapidity with which the chemical decomposition proceeds into the unreacted body of the explosive. Solid explosives are chemical compounds or mixtures of compounds that, when initiated by energy in the form of shock, impact, heat, friction, or spark, undergo very fast chemical decomposition reactions [3]. The decomposition takes place in the absence of an external supply of oxygen or other oxidizer. Thus the reaction releases large amounts of heat and gas that are used to perform work in the surroundings.

The production of explosives has increased in the last decade. In 2002, the U.S. explosive production was 2.51 million metric tons (Mt), 6 percent higher than in 2001. Coal mining, with 68 percent of the total consumption, was the dominant use of explosives in the United States. Figure 12.1 shows how sales for consumption of U.S. industrial explosives have changed since 1993.

12.1.2 Development of explosives

Among the important explosive materials, black powder, also known as gunpowder, was most likely the first explosive discovered accidentally by the Chinese alchemists. Black powder was formed during the process of separating gold from silver at a low temperature. They added potassium nitrate (KNO_3) and sulfur to gold ore in the furnace and they forgot to add charcoal to the mixture. However, they added the amount of charcoal at the last step of the reaction. As a result, a black powder was formed causing a strong explosion. The black powder was then introduced in the market as a mixture of potassium nitrate, charcoal,

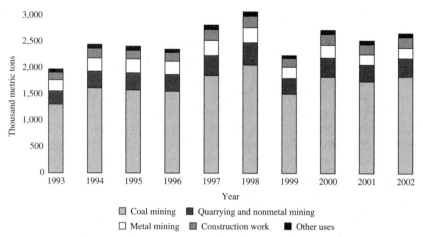

Figure 12.1 Sales for consumption of U.S. industrial explosives. (*Source: Institute of Makers of Explosives.*)

and sulfur. The composition and the order of addition determine the properties of this composite explosive.

Black powder contains a fuel and an oxidizer. The fuel is a powdered mixture of charcoal and sulfur. Potassium nitrate is the oxidizer.

Black powder was successfully introduced for blasting in 1627. The use of black powder then spread fast for mining, road building, and recovering ore, in copper mines, and other important industrial applications [4].

The limitations of black powder as a blasting explosive were apparent for difficult mining and tunneling operations. More efficient explosives were required. Liquid nitroglycerine [$C_3H_5O_3(NO_2)_3$], **1**, that was discovered by the Italian Professor Sobrero, was later studied and manufactured by the Swedish inventor, Immanuel Nobel in 1863. The major problem that the Nobel family faced was the transportation of liquid nitroglycerine that causes loss of life and property. The destruction of the Nobel factory in 1864 was one of many accidents caused by the explosion of nitroglycerine.

$$
\begin{array}{c}
\text{H} \\
| \\
\text{H}-\text{C}-\text{O}-\text{NO}_2 \\
| \\
\text{H}-\text{C}-\text{O}-\text{NO}_2 \\
| \\
\text{H}-\text{C}-\text{O}-\text{NO}_2 \\
| \\
\text{H}
\end{array}
$$

1

The discovery of mercury fulminate [Hg(CNO)$_2$] by Alfred Nobel in 1864 improved the initiation process. This chemical complex replaced the black powder in the initiation of nitroglycerine in boreholes.

In 1866, a major explosion had completely destroyed the nitroglycerine factory. Alfred Nobel had discovered the safety hazard of nitroglycerine during transportation. He reduced the sensitivity of nitroglycerine by mixing it with adsorbent clay known as *Kieselguhr*.

The chemistry of explosives was very active at that time as nitroglycerine was introduced. Nitrocellulose, known as gun cotton, was produced by the direct nitration of cellulose. The discovery and the use of nitrocellulose were associated with the names of two scientists, Schönbein and Böttger.

The synthesis of nitrocellulose was extremely difficult and many accidents took place, such as the destruction of the manufacturing plants in France, England, and Austria [4]. The stability of nitrocellulose was improved by the process of pulping, boiling, and washing. In 1868, a dry, compressed, highly-nitrated nitrocellulose was discovered to detonate by using a mercury fulminate detonator. Wet nitrocellulose could be exploded by adding a small amount of dry nitrocellulose.

An interesting discovery was made in 1875 by Alfred Nobel when an explosive gel was formed by mixing nitrocellulose and nitroglycerine. This gel was converted later in 1888 into gelatin dynamite and ballistile, known at first as smokeless powder. Ballistile was in fact a mixture of nitrocellulose, nitroglycerine, benzene, and camphor. The explosive properties of dynamite were improved by the Swedish chemists Ohlsson and Norrbin by adding ammonium nitrate (NH$_4$NO$_3$). Ammonium nitrate was not considered hazardous and explosive until disastrous accidents took place in 1947 in the harbor of Texas City in the SS Groundchamp and the SS High Flyer, both loaded with fertilizer grade ammonium nitrate (FGAN) [4].

Commercial explosives were very attractive, especially in coal mining. Black powder was mainly used in coal mining; however, many disastrous explosions occurred. The attempts to modify black powder by mixing it with *cooling agents* such as ammonium sulfate or paraffin were not very successful.

The development of dynamite and blasting gelatin by Nobel placed nitroglycerine-based explosives at a position where they could dominate the mining industries. Later, some governments prohibited the mining industries. Eventually, some governments prohibited the use of explosives in coal mining operations owing to a potential increase in dust and gases. After a thorough study and various experiences in explosives, several materials were recommended for use in the production of explosives for coal mines. These materials were mostly based on ammonium nitrate. Dynamite and black powder failed the tests set for this purpose and were replaced by ammonium nitrate. The explosives that passed the tests were called *permitted explosives* [4].

The development of military explosives started approximately since the discovery of black powder. In 1885, Turpin discovered picric acid or trinitrophenol, **2**, that was found to be a suitable replacement for black powder.

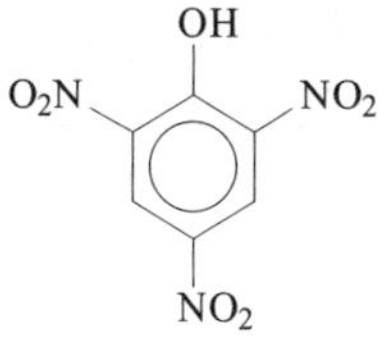

OH

O_2N NO_2

NO_2

2

Picric acid was adopted all over the world as the basic explosive for military uses. The major problems of picric acid are corrosion of the shells in the presence of water and high sensitivity of its salts. Also, picric acid melts only at a very high temperature.

Tetryl, **3**, was also discovered at the same time as picric acid. Tetryl was used for the first time as an explosive in 1906.

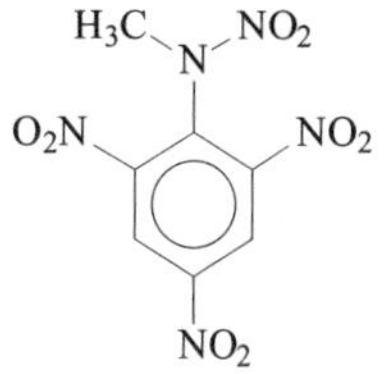

H_3C NO_2
       N

O_2N NO_2

NO_2

3

In 1863, Wilbrand prepared trinitrotoluene, **4** (TNT, **4**), for the first time. TNT was prepared in pure form in 1880 by Hepp and the structure was established in 1883 by Clams and Becker. In 1902, TNT, **4**, was adopted for use by the German army and in 1912 by the U.S. army.

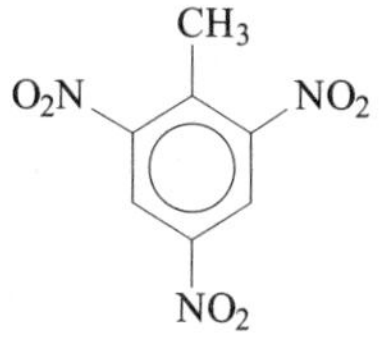

CH_3

O_2N NO_2

NO_2

4

The storage of TNT was made easier by the use of the mixture of TNT and ammonium nitrate, known as amatol.

After World War I, research in the area of explosives was concentrated on the development of new and more powerful explosive materials. For example, cyclotrimethylene trinitramine ($C_3H_6N_6O_6$), known as RDX, **5**, and pentaerythritol tetranitrate (PETN) ($C_5H_8N_4O_4$), **6**, were among the newly discovered explosives.

TABLE 12.1   Examples of PBX Compositions

HMX –	Cyclotetramethylene tetranitramine (Octogen)
HNS –	Hexanitrostilbene
PETN –	Pentaerythritol tetranitrate
RDX –	Cyclotrimethylene trinitramine (Hexogen)
TATB –	1,3,5- triamino-2,4,6-trinitrobenzene

SOURCE: *The Chemistry of Explosives,* 2nd ed. [4].

RDX was first prepared in 1899 by the German scientist Henning for medicinal use. It was discovered as a good explosive only in 1920 by Herz and in a higher yield by Hale in 1925 and by Buchmann in 1940.

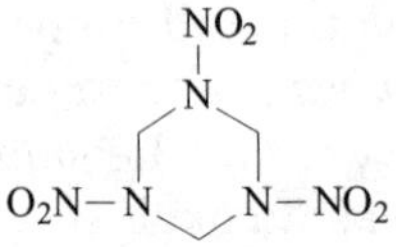

$$
\begin{array}{c}
NO_2 \\
| \\
N \\
\diagup \quad \diagdown \\
O_2N-N \qquad N-NO_2
\end{array}
$$

5

PETN, **6**, was discovered in 1894 and obtained by the nitration of pentaerythritol. During World War II, RDX was used more than PETN. The mixture of 50 percent PETN and 50 percent TNT was developed and used for foiling hand and antitank grenades, and detonators.

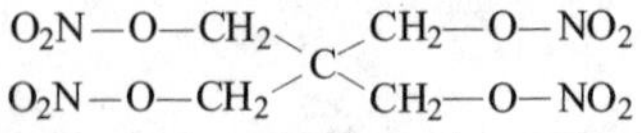

$$
\begin{array}{c}
O_2N-O-CH_2 \diagdown \qquad \diagup CH_2-O-NO_2 \\
C \\
O_2N-O-CH_2 \diagup \qquad \diagdown CH_2-O-NO_2
\end{array}
$$

6

Another class of explosives known as polymer bonded explosives (PBXs) was developed to reduce the sensitivity of the explosive crystals by embedding them in a rubber-like polymer, such as polystyrene. PBXs based on RDX and RDX/PETN, and also on HMX were developed (Tables 12.1 and 12.2). Energetic plasticizers have also been developed for PBXs production (Table 12.3).

Research and development is continuing in the chemistry of explosives to produce compounds that are insensitive to accidental initiation but still highly efficient as explosives. NTO (**7**), ADN (**8**), and TNAZ (**9**) are good examples of explosive compounds under development and testing.

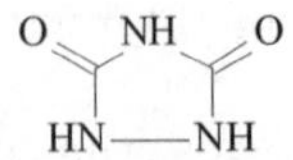

$$
\begin{array}{ccc}
O \quad NH \quad O & & \\
\diagup \quad | \quad \diagdown & NH_4^+ \; \overset{-}{N} \diagup NO_2 & \\
HN——NH & \diagdown NO_2 &
\end{array}
$$

$$
\begin{array}{c}
NO_2 \\
| \\
N \\
\diagup \quad \diagdown \\
O_2N \quad NO_2
\end{array}
$$

5-Nitro-1,2,4-triazol-3-one   Ammonium dinitramide   1,3,3-Trinitroazetidine

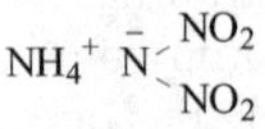

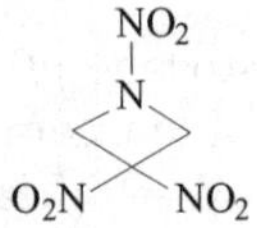

NTO (**7**)   ADN (**8**)   TNAZ (**9**)

TABLE 12.2   Examples of Energetic Polymers

Common name	Chemical name	Structure
GLYN (monomer)	Glycidyl nitrate	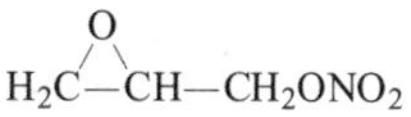 $H_2C-CH-CH_2ONO_2$ (epoxide O bridging H_2C and CH)
PolyGLYN	Poly (glycidyl nitrate)	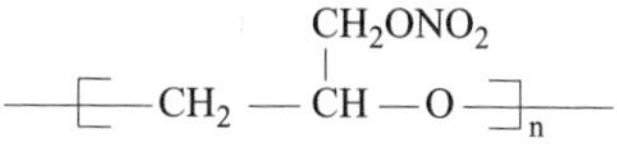 $-[-CH_2-CH(CH_2ONO_2)-O-]_n-$
NIMMO (monomer)	3-Nitratomethyl-3-methyl oxetane	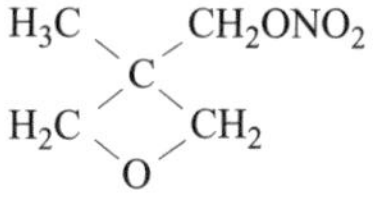 oxetane ring with H_3C and CH_2ONO_2 on C
polyNIMMO	Poly (3-nitratomethyl-3-methyl oxetane)	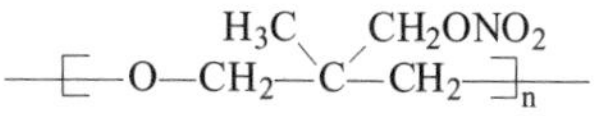 $-[-O-CH_2-C(CH_3)(CH_2ONO_2)-CH_2-]_n-$
GAP	Glycidyl azide polymer	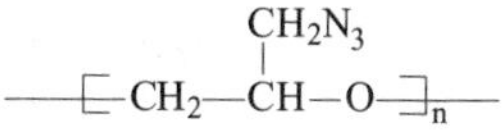 $-[-CH_2-CH(CH_2N_3)-O-]_n-$
AMMO (monomer)	3-Azidomethyl-3-methyl oxetane	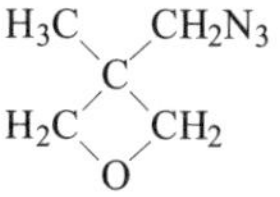 oxetane ring with H_3C and CH_2N_3 on C
PolyAMMO	Poly (3-azidomethyl-3-methyl oxetane)	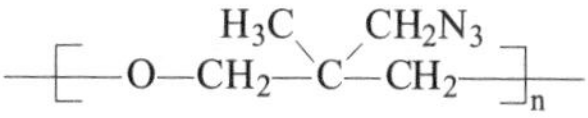 $-[-O-CH_2-C(CH_3)(CH_2N_3)-CH_2-]_n-$
BAMO (monomer)	3,3-bis-azidomethyl oxetane	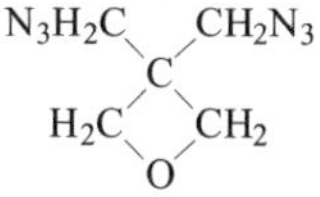 oxetane ring with N_3H_2C and CH_2N_3 on C
PolyBAMO (monomer)	Poly(3,3-bis-azidomethyl oxetane)	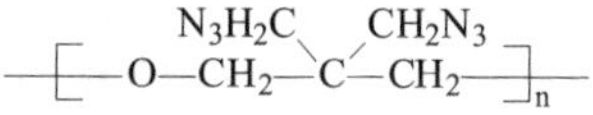 $-[-O-CH_2-C(CH_2N_3)(CH_2N_3)-CH_2-]_n-$

SOURCE: *The Chemistry of Explosives*, 2nd ed. [4].

12.1.3 Classification of explosives

Explosives were originally classified into two main types: low and high. This classification was based on the relative detonation speeds and the pressures produced by these reactions. Low explosives burn very rapidly and generate pressures below 50,000 psi. However, high explosives produce pressures of 5 million psi. The detonation rate of explosives has been characterized by comparing it with the speed of sound. High explosives are faster than the speed of sound, whereas low explosives detonate at a lower rate than the speed of sound. This classification is of limited use in this century and thus explosives have been classified in relation to their chemical nature and to their performance. The explosives have been divided into three classes (Fig. 12.2) [4, 5]:

TABLE 12.3   Examples of Energetic Plasticizers

Common name	Chemical name	Structure
NENAs	Alkyl nitratoethyl nitramines	$R-\underset{\underset{NO_2}{\vert}}{N}-CH_2-CH_2ONO_2$
EGDN	Ethylene glycol dinitrate	$O_2NOH_2C-CH_2ONO_2$
MTN	Metriol trinitrate	$H_3C-\underset{\underset{CH_2ONO_2}{\vert}}{\overset{\overset{CH_2ONO_2}{\vert}}{C}}-CH_2ONO_2$
BTTN	Butane-1,2,4-triol trinitrate	$O_2NOH_2C-\underset{\underset{ONO_2}{\vert}}{CH}-CH_2-CH_2ONO_2$
K10	Mixture of di- and tri nitroethylbenzene	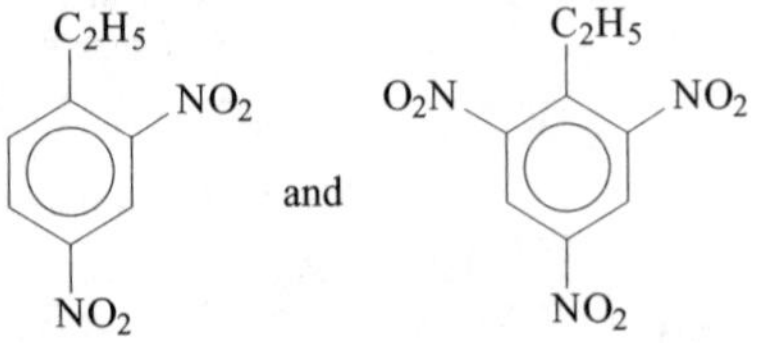 (nitroethylbenzene ring structures, and)
BDNPA/F	Mixture of bis-dinitropropylacetal and bis-dinitropropylformal	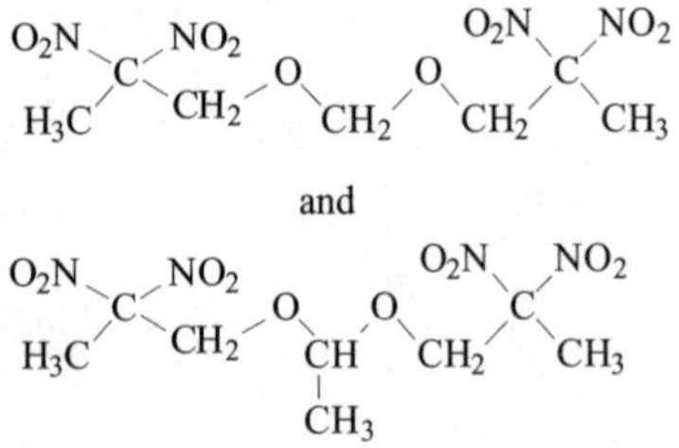 (bis-dinitropropyl acetal and formal structures)

SOURCE: *The Chemistry of Explosives*, 2nd ed. [4].

1. Primary explosives
2. Secondary explosives
3. Tertiary explosives or propellants

These explosives are also divided into two other groups based on the chemical nature of the materials [4, 5].

1. Pure compounds
2. Mixture of compounds

Explosives are compounds that have functions of groups that have explosive properties. The most common functional groups, both oxidizer

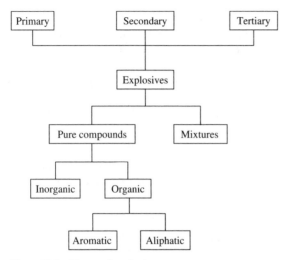

Figure 12.2. Types of explosives.

and fuel types, are as follows:

Nitro	NO_2
Nitrate	NO_3
Nitroso	NO
Azide	N_3
Amine	NH_2
Hydroxy	OH

The nitro group is present in the form of salts of nitric acid, nitrate esters (ONO_2 derivatives), aliphatic or aromatic nitro compounds (CNO_2 derivatives), and N-nitro compounds (NNO_2) such as nitramines, nitroureas, and so on.

Primary explosives. Primary or initiator explosives are most sensitive to heat, friction, impact, and shock. This class of explosives has been studied in detail and they are very reactive even when present in small quantities. Primary explosives are capable of transforming a low-energy into a high-intensity shock wave.

Primary explosives are used to initiate the materials with increasing mass and decreasing sensitivity. They are used mainly in military detonators and commercial blasting caps.

Primary explosives are characterized by their low detonation rates and less energy than secondary explosives. Most primary explosives are powders with a good flow transfer and pressing characteristics to permit high-speed automatic loading of detonators that are produced in very large quantities. Initiators are very sensitive to shock and are loaded into detonators in the production plant. The manufacture of detonators requires maximum precautions to prevent the accumulation of substantial masses of explosives, dusting, and accidental initiation. Few compounds can act as primary explosives and meet the restrictive military and industrial requirements for reliability, low cost, stability, and ease of manufacture. Most primary explosives are dense and inorganic compounds.

Mercury fulminate.   Mercury fulminate or mercuric cyanate, **10**, is a grey-white powder obtained by reacting mercuric nitrate with alcohol in nitric acid

$$Hg(CNO)_2$$

10

Mercury fulminate is the most sensitive among the initiating agents to impact and friction, although its sensitivity decreases as the density of the pressed mass increases. The sensitivity also decreases by the addition of water but increases in the presence of sunlight. Mercury fulminate decomposes when stored at elevated temperatures.

Mercury fulminate will easily detonate after initiation. It decomposes into stable products.

$$Hg(CNO)_2 \longrightarrow 2CO + N_2 + Hg$$

Mercury fulminate is no longer used as a primary explosive worldwide.

Lead azide.   Lead azide, **11**, is the primary explosive used in military detonators in the United States and has been intensively studied [6].

$$Pb(N_3)_2$$

11

Lead azide is a very stable compound at ambient and high temperatures, and has good flow characteristics. It is less sensitive to ignition compared to mercury fulminate. Therefore, lead styphnate, known to be readily ignitable material, is used as a cover charge to ensure initiation. Lead azide builds up to detonation rapidly on ignition. It has a good shelf life in dry conditions, but is unstable in the presence of moisture, oxidizing agents, and ammonia. Lead azide has a high initiating capacity for secondary explosives.

Lead azide can be easily hydrolyzed at high humidity or in the presence of moistures. The hydrazoic acid produced reacts with copper and its alloys to produce the very sensitive cupric azide. Serious protection must be provided by hermetic sealing and the use of noncopper or coated-copper metal [6].

Silver azide.   Silver azide, AgN_3 (**12**), is a potential replacement for lead azide because it can be used in smaller quantities as an initiator. It usually requires less energy for initiation than lead azide. It is less apt to hydrolyze and more sensitive to heat. Silver azide is not compatible with sulfur compounds, with tetrazene, and with some metals, such as copper [6].

Lead styphnate.   Lead styphnate, **13**, also known as 2,4,6-trinitro-soreinate, is an explosive compound used to start the ignition-to-detonation process in the explosive sequence. It is stable at elevated temperatures and noncorrosive. The addition of graphite enhances its electrical conductivity in systems designed for electrical initiation. Dry lead styphnate is the most sensitive of the primary explosives to electrostatic discharge [6].

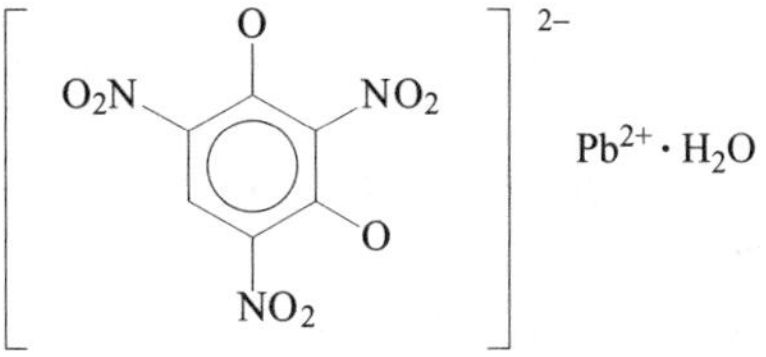

13

Lead styphnate is a weak primary explosive because of its high metal content (44.5 percent) and therefore is not used in the filling of detonators [4].

Tetrazene.   Tetrazene, **14**, is classified as a primary explosive that is a very hazardous material. An explosion can occur during the preparation that is based on sodium nitrite and amino guanidine sulfate (**15**).

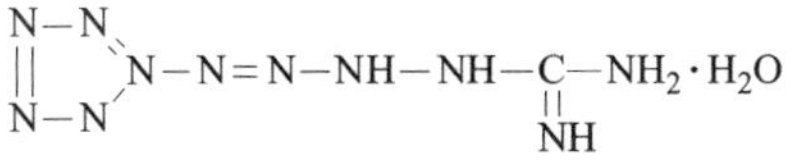

14 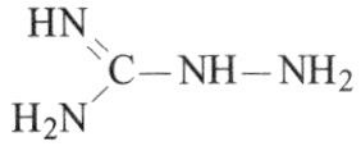   **15**

Tetrazene is hygroscopic and stable at ambient temperatures. The detonation property of tetrazene can detonate if it is not compacted. Tetrazene is not suitable for filling detonators [4, 6].

Secondary explosives.   Secondary explosives are usually more difficult to detonate and they differ from primary explosives in three basic ways [4, 6]:

1. Secondary explosives do not easily go from burning to detonation.

2. Electrostatic ignition is difficult with secondary explosives.

3. Secondary ignition requires large shocks.

Nitroglycerine.   Nitroglycerine, **1**, is still one of the most widely produced nitrate esters. The first process for the manufacture of nitroglycerine was developed in France and in England in 1882 [4, 6].

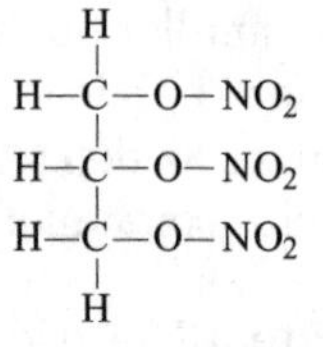

$$
\begin{array}{c}
H \\
| \\
H-C-O-NO_2 \\
| \\
H-C-O-NO_2 \\
| \\
H-C-O-NO_2 \\
| \\
H \\
\mathbf{1}
\end{array}
$$

Nitroglycerine is a very powerful secondary explosive with a high brisance, that is, shattering effect. It is used in dynamite. Nitroglycerine is also thickened or gelatinized by the addition of a small percentage of nitrocellulose. It provides a source of high energy in propellant compositions. Nitroglycerine is insoluble in water but readily dissolves in most organic solvents.

Nitrocellulose.   Nitrocellulose, **16**, is a generic term denoting a family of compounds. The composition is determined by the percentage by weight of nitrogen. The number of nitrogen groups present in nitrocellulose can be calculated by the following equation [4, 6]:

$$
n = \frac{162N}{1400 - 45N}
$$

where N is the percentage calculated from chemical analysis.

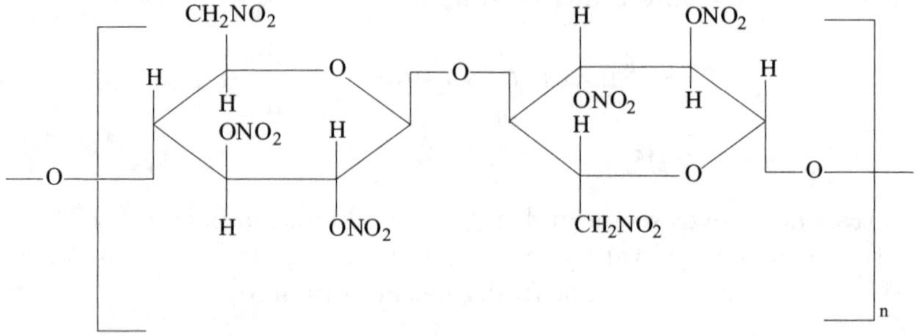

16

Nitrocellulose compositions used in explosive applications vary from 10 to 13.5 percent of nitrogen. The thermal stability of nitrocellulose decreases with increasing nitrogen content. It dissolves in organic solvents to form a gel. The gel is used as gun propellant, double-rocket propellant, and gelatine and semi-gelatine commercial blasting explosives.

Picric acid.   Picric acid, **2**, also known as 2,4,6-trinitrophenol, has been used in grenade and mine fillings and has a tendency to form impact-sensitive metal salts (picrates) with the metal walls of the shells [4–6].

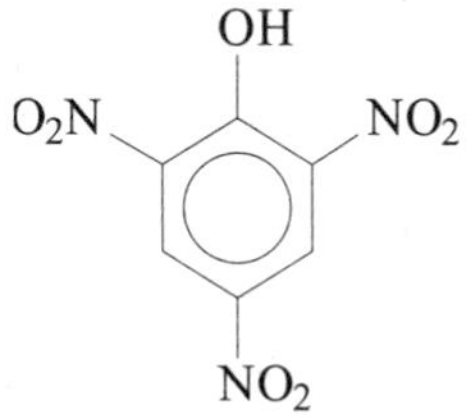

OH

O_2N NO_2

NO_2

2

Picric acid is a strong acid, very toxic, soluble in hot water, alcohol, ether, benzene, and acetone. The explosive power of picric acid is superior to TNT with regard to the velocity of detonation.

Tetryl.   Tetryl or 2,4,6-trinitrophenyl methylnitramine, **3**, was frequently used as a base charge in blasting caps, as the booster explosive in high-explosive shells, and as an ingredient of binary explosives. Tetryl is now replaced by PETN or RDX. Tetryl is a very toxic chemical [4–6].

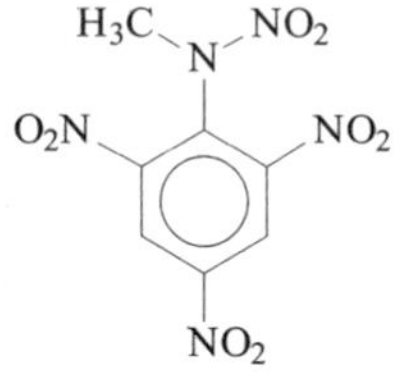

H_3C N NO_2

O_2N NO_2

NO_2

3

Tetryl is moderately sensitive to initiation by friction and percussion. It is more sensitive than picric acid and TNT.

TNT.   TNT, also known as trinitrotoluene, **4**, remains an important military explosive, particularly in mixtures with ammonium nitrate (NH_4NO_3). TNT has a low melting point (80°) that permits loading into bombs and shells in the molten state [4–6].

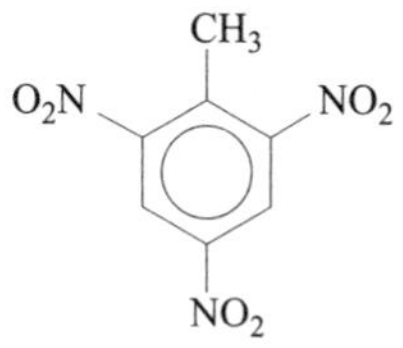

CH_3

O_2N NO_2

NO_2

4

TNT has the advantages of low manufacturing costs and cheap raw materials, safe handling, a low sensitivity to impact and friction, high-explosive power, and also has good chemical and thermal stability.

TNT is widely used in commercial explosives and is much safer to produce and handle than nitroglycerine and picric acid.

The important disadvantage of TNT is the exudation (leaching out) of the isomers of dinitrotoluenes and trinitrotoluenes. The exudation may result in the formation of cracks and cavities leading to premature detonation [4–6].

Nitroguanidine.   Nitroguanidine, **17**, also known as picrite ($CH_4N_4O_2$), has been used as an industrial explosive but not as a military explosive because of its relatively low energy content and difficulty of initiation [4–6].

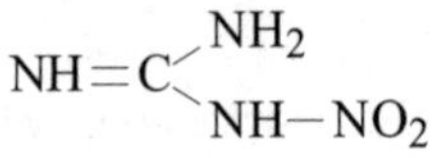

$$NH = C \underset{\textstyle NH-NO_2}{\overset{\textstyle NH_2}{\diagup}}$$

17

Nitroguanidine can also be used in flashes and propellants because of the low heat and temperature of explosion. Nitroguanidine combined with nitrocellulose, nitroglycerine, and nitrodiethyleneglycol forms a colloidal gel that is the basis of these propellants.

PETN.   Pentaerythritol tetranitrate (PETN), **6**, is known as one of the most sensitive military high explosives. It has a great shattering effect. PETN is not used in its pure form because of its high sensitivity to friction and impact [4, 6, 7].

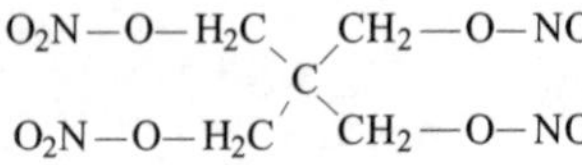

$$O_2N-O-H_2C \diagdown C \diagup CH_2-O-NO_2$$
$$O_2N-O-H_2C \diagup C \diagdown CH_2-O-NO_2$$

6

Pentolite is a military explosive composed of 20 to 50 percent PETN and TNT. However, PETN has been largely replaced by RDX in military applications.

RDX.   RDX, cyclonite or cyclotrimethylene trinitramine, **5**, is one of the most powerful explosives (Fig. 12.3). Pure RDX is very sensitive to initiation by impact and friction. However, the polymer bonded explosive (PBX) is less sensitive as a result of embedding the RDX crystals in a polymeric matrix [4, 6, 7].

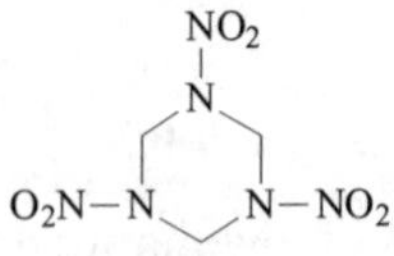

$$O_2N-N \underset{}{\diagdown} N-NO_2$$
with NO_2 on top nitrogen

5

Figure 12.3 HMX- and RDX-based explosives [9]. (*Source: http://www.army-technology.com/contractors/explosives/ dyno/dyno2.html.*)

RDX has a very high explosive power compared with TNT and picric acid. RDX has a high melting point that makes it difficult to use in casting [4, 6].

HMX. HMX, octogen, or cyclotetramethylene tetranitramine, **18**, is present in four different crystalline forms in their density and sensitivity [4, 6] (Fig. 12.3).

Crystal density	$(20°C/g \cdot cm^{-3}$
α-form	1.87
β-form	1.96
γ-form	1.82
δ-form	1.78

18

HMX has similar properties of RDX with respect to the chemical reactivity and solubility in organic solvent. HMX is superior to RDX as an explosive because of its high ignition temperature and chemical stability. The disadvantage of HMX compared to RDX is reflected in its lower explosive power [4, 6,].

Tertiary explosives or propellants. Tertiary explosives or propellants are combustible materials that contain, in addition to their components, the oxygen needed for their combustion. Propellants produce large amounts of gas upon combustion. Propellants only burn and do not explode. However, burning proceeds violently and is accompanied by a flame or spark. Propellants can be initiated by a flame or spark, and change from a solid to a gaseous state relatively slowly. Examples of these propellants include black powder, smokeless propellants, blasting explosives, and ammonium nitrate [4–6]. Propellants will be a subject of detailed investigation later in this chapter.

12.1.4 Chemistry of explosives

Nitration represents a major chemical reaction that plays an essential role in the production of most explosives. A variety of nitrocompounds ranging from C-nitrocompounds such as TNT to N-nitrocompounds such as RDX and HMX to O-nitrocompounds such as trinitroglycerol are considered among the most energetic compounds. The preparation of these explosives is thus of substantial importance [8].

Nitration is an electrophilic reaction of the addition of a nitro group to an organic compound with a nitrating agent. The introduction of the nitro group can take place on a carbon atom (C-nitration), oxygen atom (O-nitration), or nitrogen atom (N-nitration) [4, 8].

C-Nitration

$$\mathrm{-\!\!\overset{|}{\underset{|}{C}}\!\!-H + HNO_3 \longrightarrow -\!\!\overset{|}{\underset{|}{C}}\!\!-NO_2 + H_2O}$$

O-Nitration

$$\mathrm{-\!\!\overset{|}{\underset{|}{C}}\!\!-OH + HNO_3 \longrightarrow -\!\!\overset{|}{\underset{|}{C}}\!\!-ONO_2 + H_2O}$$

N-Nitration

$$\mathrm{\overset{\diagdown}{\underset{\diagup}{N}}\!\!-H + HNO_3 \longrightarrow \overset{\diagdown}{\underset{\diagup}{N}}\!\!-NO_2 + H_2O}$$

The nitrating agents for the manufacture of explosives via C-nitration, O-nitration, or N-nitration are mixtures of nitric and sulfuric acids, except for RDX and HMX where a mixture of nitric acid and ammonium nitrate is used [4].

C-Nitration.   Picric acid, **2**, can be synthesized by reacting phenol in sulfuric acid and then adding nitric acid (Eq. 12.1) [4].

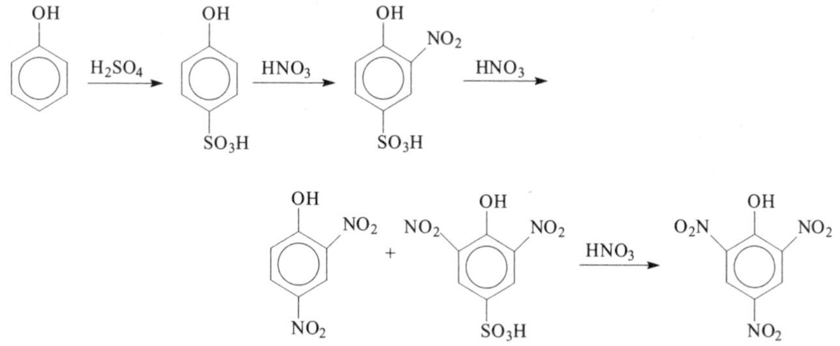

(12.1)

Tetryl, **3**, can be prepared by reacting dimethylaniline [$C_6H_5N(CH_3)_2$] with sulfuric acid followed by the addition of a mixture of nitric acid and sulfuric acid (Eq. 12.2) [4].

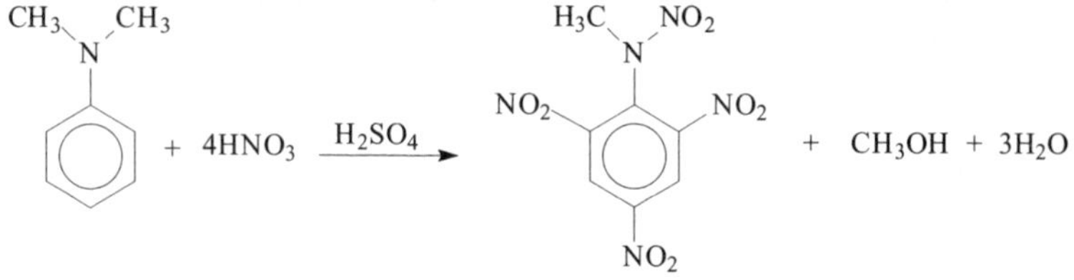

In this reaction, one methyl group is oxidized and the benzene ring undergoes an electrophilic nitration in the 2-,4- and 6-positions.

Another method for the synthesis of tetryl is the treatment of methylamine with 2,4-or 2,4-dinitrochlorobenzene to give dinitrophenylmethylamine. Later, this is treated with nitric acid to tetryl (Eq. 12.3) [4, 8].

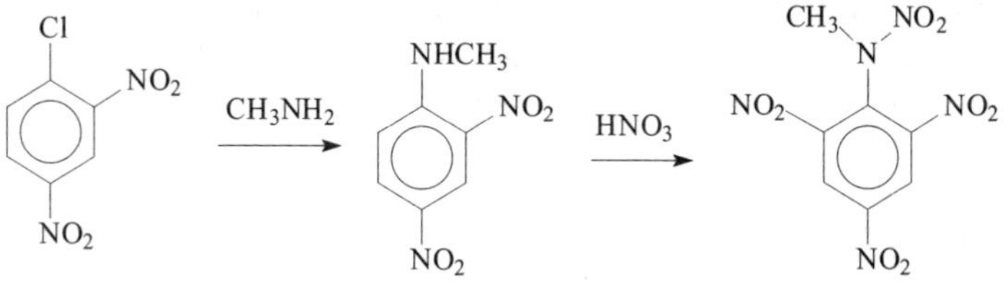

In both processes purification is carried out by washing in cold and boiling water. Finally, the tetryl is recrystallized by dissolving in acetone and precipitated with water, or recrystallized from benzene [4].

TNT, **4**, is produced by the nitration of toluene with a mixture of nitric and sulfuric acids. Toluene is first mononitrated and then dinitrated, and finally crude trinitrotoluene (TNT) is produced by the trinitration step. This step requires a high concentration of mixed acids with a free SO_3 group (Eq. 12.4) [4].

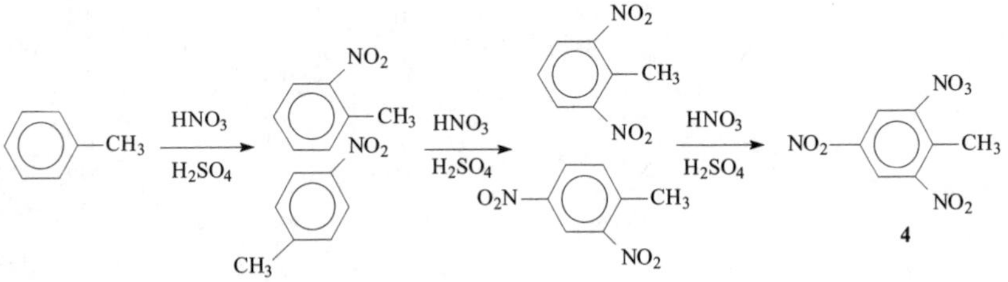

(12.4)

O-Nitration. Nitroglycerine, **1**, is an explosive prepared in a batch reactor or continuous process by mixing pure glycerine with a mixture of highly concentrated sulfuric acid and nitric acid at a controlled temperature (Eq. 12.5).

$$\begin{array}{l} CH_2\text{—}OH \\ CH\text{—}OH \\ CH_2\text{—}OH \end{array} + 3HNO_3 \xrightarrow{H_2SO_4} \underset{\mathbf{1}}{\begin{array}{l} CH_2\text{—}ONO_2 \\ CH\text{—}ONO_2 \\ CH_2\text{—}ONO_2 \end{array}} + 3H_2O \quad (12.5)$$

The mixtures were cooled with brine after the reaction and then poured into a separator, where the nitroglycerine was separated by gravity [4, 6].

The batch process has been widely replaced by continuous processes such as the Biazzi and the nitro Nobel injector processes.

Nowadays, nitroglycerine is not used for military explosives; it is only used in propellants and commercial blasting explosives [4, 6].

Nitrocellulose, **16**, obtained by the nitration of cellulose with a mixture of sulfuric and nitric acids, can be quite hazardous in dry form. Therefore, nitrocellulose is stored and transported in 30 percent water or ethanol.

The application of nitrocellulose depends strongly on the degree of nitration. For example, guncotton with 13.45 percent nitrogen is used in the manufacture of double-base and high-energy propellants [4].

PETN (pentaerythritol tetranitrate), **6**, is an old explosive made from acetaldehyde and formaldehyde that reacts by condensation under basic catalysis followed by a crossed Cannizaro reaction to produce pentaerythritol. This is a one-pot reaction (Eq. 12.6) [1].

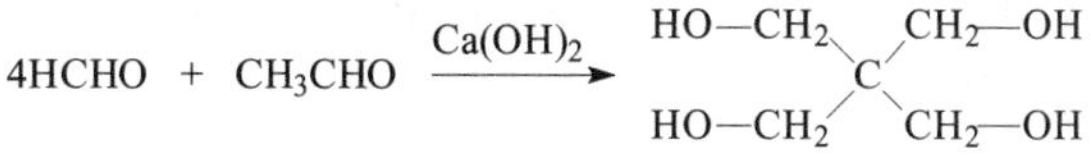

$$4HCHO + CH_3CHO \xrightarrow{Ca(OH)_2} \begin{array}{c} HO-CH_2 \\ \\ HO-CH_2 \end{array} C \begin{array}{c} CH_2-OH \\ \\ CH_2-OH \end{array} \qquad (12.6)$$

Pentaerythritol

The nitration of pentaerythritol takes place by adding concentrated nitric acid at 25–30°C to produce PETN (Eq. 12.7) [1, 4]. PETN is not very soluble in nitric acid or water and is easily filtered from solution.

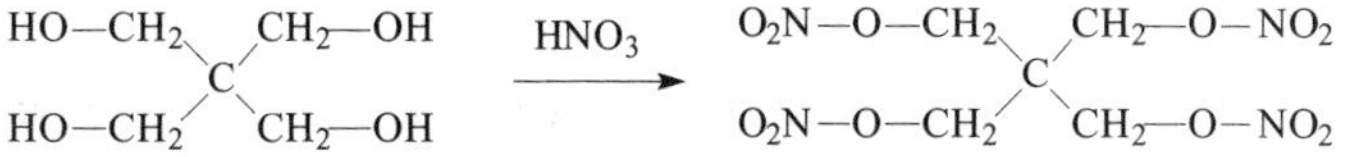

$$\begin{array}{c} HO-CH_2 \\ \\ HO-CH_2 \end{array} C \begin{array}{c} CH_2-OH \\ \\ CH_2-OH \end{array} \xrightarrow{HNO_3} \begin{array}{c} O_2N-O-CH_2 \\ \\ O_2N-O-CH_2 \end{array} C \begin{array}{c} CH_2-O-NO_2 \\ \\ CH_2-O-NO_2 \end{array} \qquad (12.7)$$

PETN, 6

PETN has been largely replaced by RDX, which is thermally more stable. However, PETN is widely used in industry as the major component in a cast booster for initiating blasting agents [1].

N-Nitration.   RDX, **5**, is a cyclic nitramine obtained by the nitration of hexamethylene tetramine (HMT) in the presence of an excess of concentrated nitric acid. RDX is formed by the nitration of the three outside nitrogen atoms of HMT with removal of the internal nitrogen and methylene ($-CH_2-$) groups. RDX is produced along with ammonium nitrate (NH_4NO_3) and formaldehyde (HCHO) as by-products. However, another molecule of RDX can be produced by adding NH_4NO_3, HCHO, and acetic acid (Eqs. 12.8 and 12.9) [1].

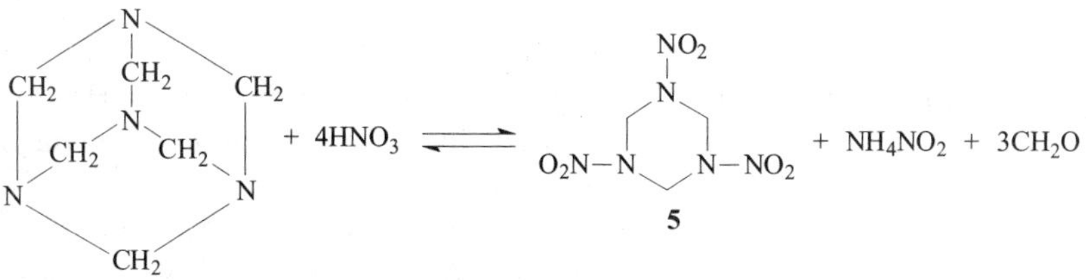

$$(12.8)$$

$$3CH_2O + 3NH_4NO_3 + 6(CH_3CO)O \rightleftharpoons RDX + 12CH_3CO_2H \qquad (12.9)$$

HMX, **18**, was identified as an impurity of the reaction of the production of RDX. HMX is an eight-membered ring that can be formed by adjusting the reaction conditions. The nitration of all four nitrogen atoms in HMT and the removal of two methylene groups is shown in Eq. 12.10 [1]. RDX that is formed as a product of the reaction of formation of HMX, must be removed by alkaline hydrolysis or by a difference in solubility in acetone.

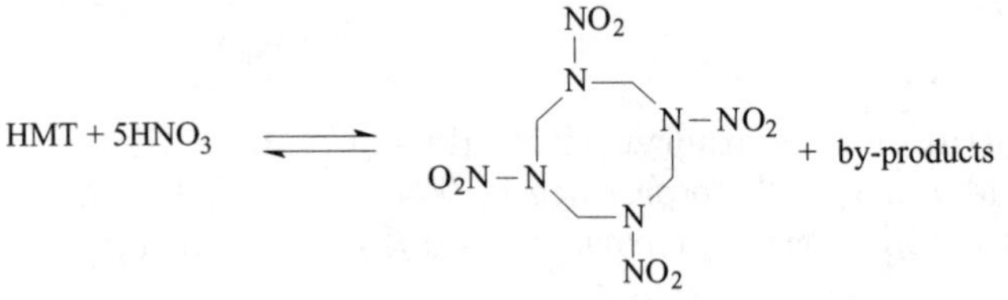

$$\text{HMT} + 5\text{HNO}_3 \rightleftharpoons \qquad + \text{by-products} \qquad (12.10)$$

18

Nitroguanidine, **17**, is another aliphatic nitramine that is used as a major constituent in triple-base propellants. It exists in two crystalline forms, the α- and β-forms. The α-form of nitroguanidine, **17**, can be prepared by reacting guanidine nitrate in concentrated sulfuric acid followed by the addition of an excess of water from which guanidine nitrate, **19**, is crystallized (Eqs. 12.11 and 12.12) [4].

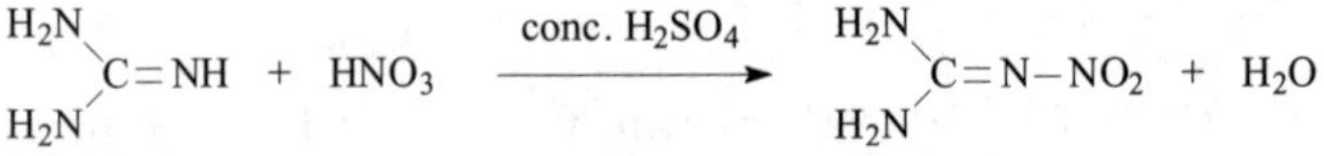

$$\begin{array}{c} H_2N \\ C=NH \\ H_2N \end{array} + HNO_3 \xrightarrow{\text{conc. } H_2SO_4} \begin{array}{c} H_2N \\ C=N-NO_2 \\ H_2N \end{array} + H_2O \quad (12.11)$$

Guanidine   Guanidinenitrate, **19**

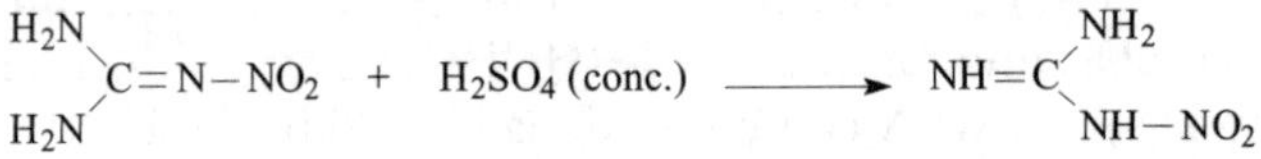

$$\begin{array}{c} H_2N \\ C=N-NO_2 \\ H_2N \end{array} + H_2SO_4 \text{ (conc.)} \longrightarrow NH=C\begin{array}{c} NH_2 \\ NH-NO_2 \end{array} \quad (12.12)$$

19   Nitroguanidine, **17**

The β-form may be prepared by nitrating a mixture of guanidine sulfate and ammonium sulfate and crystallizing it from hot water. However, the β-form can be converted into the α-form by dissolving it in concentrated sulfuric acid and pouring it into excess water.

Ammonium nitrate (NH_4NO_3), **20**, is manufactured by injecting the gaseous ammonia into 40 to 60 percent nitric acid at 150°C (Eq. 12.13) [1].

$$NH_3 + HNO_3 \longrightarrow NH_4NO_3 \qquad (12.13)$$

20

Dense ammonium nitrate crystals are formed by spraying droplets of molten ammonium nitrate solution (>99.6 percent) slowly in a short tower. The crystals obtained are usually used in conjunction with nitroglycerine [4]. Ammonium nitrate is the cheapest source of oxygen available for commercial explosives.

12.2   Propellants

12.2.1   Gun propellants

Gun propellants, chemically similar to explosives, can burn but do not detonate. Propellants are not pure substances but mixtures of chemical compounds capable of producing large volumes of gas and they give off heat [4, 5]. They are mainly used in launching projectiles from guns, rockets, and missile systems (Fig. 12.4). Propellants are divided into four different categories [5, 6]:

a. Single base

b. Double base

c. Triple base

d. Composite

Single-base propellants.   Single-base propellants are essentially formed from pure nitrocellulose that is produced from a cellulose monomer, **25**. Cellulose is produced industrially from cotton and wood pulp. The hydroxyl groups (−OH) in the cellulose are nitrated by a

Figure 12.4   Gun powder [10]. (*Source: http://www.nof.co. jp/E_nof/business%20activities/Explosives.htm.*)

reaction with a mixture of nitric and sulfuric acids. The control of the degree of nitration of the single hydroxyl group on the side may produce mono nitrocellulose, **26**, or all hydroxyl groups forming nitrocellulose, **27** [5, 6].

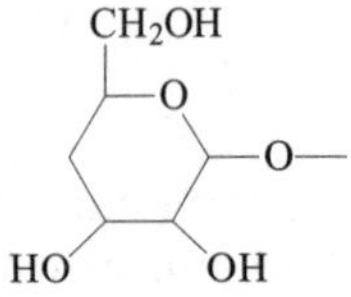

CH_2OH CH_2ONO_2 CH_2ONO_2

HO   OH   OH   OH O_2NO ONO_2

25 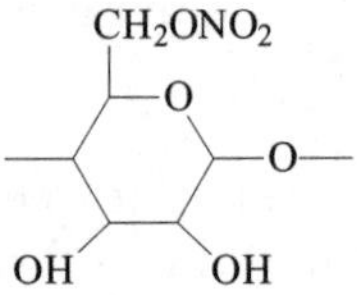   **26** 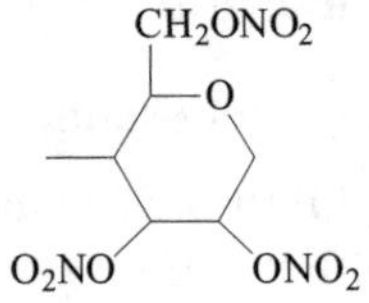   **27**

High-grade nitrocellulose or *guncotton* has a nitrogen weight percent of 13.4. Pyrocellulose is 12.6 percent nitrogen. Nitrocellulose of lower percentages on nitrogen (<12.0 percent) is not suitable for propellant use.

At high temperature, nitrocellulose slowly decomposes, producing nitric oxide (NO) and nitrogen dioxide (NO_2). The secondary reaction, such as the addition of NO_2 to the other parts of the chain, can be prevented by adding stabilizers such as diphenylamine, nitrodiphenylamine, ethylaniline, carbazole, and others [5, 6]. These stabilizers are generally present in quantities of less than 2 percent. Some examples of single-base propellants are given in Table 12.4.

Double-base propellants. Double-base propellants are made of gelatinized or plasticized nitrocellulose with another liquid propellant or explosive. This second type of propellant is nitroglycerine. Other nitrate esters may also be used in place of nitroglycerine. Also, dinitrotoluene

TABLE 12.4   Properties of Some Single-Base Propellants

	Propellant types		
Property	M10	M12	Navy pyro (pyrocellulose)
Nitrocellulose	98.0	97.7	99.5
% Nitration	13.15	13.15	12.95
Oxidizers			
Potassium sulfate	1.0	0.75	
Stabilizers			
Diphenylamine	1.0		0.5
Lubricants, antistats			
Graphite	0.1	0.8	
Tin		0.75	

TABLE 12.5   Properties of Some Double-Base Propellants

Property	M1	M2	M5	M6	M7	M8	M9
Nitrocellulose	85.0	77.45	81.95	87.0	54.6	52.15	57.75
% Nitration	13.5	13.25	13.25	13.15	13.15	13.25	13.25
Nitroglycerine		19.5	15.0		35.5	43.0	40.0
Dinitrotoluene	10.0			10.0			
Oxidizers							
Barium nitrate	1.4		1.4				
Potassium nitrate	0.75		0.75			1.25	1.5
Potassium perchlorate					7.8		
Plasticizers							
Dibutylphthalate	5.0			3.0			
Diethylphthalate						3.0	
Stabilizers							
Diphenylamine	1.0			1.0			0.75
Ethyl centalite		0.6	0.6		0.9	0.6	
Lubricants, antisats							
Graphite		0.3	0.3				
Carbon black					1.2		

is used as the second component. The second propellant is intimately bonded to nitrocellulose and does not separate [5, 6].

The function of the second component is also to adjust the oxygen balance that affects the energy output and reaction temperature. The oxygen balance is also adjusted by the addition of inorganic oxidizers such as nitrates, perchlorates, or sulfates. Some examples of double-base propellants are given in Table 12.5.

Triple-base propellants.   A third energetic material such as nitroguanidine is added in the double-base propellants to reduce the muzzle flash that is the result of a fuel-air explosion of the combustion products. The triple-base propellant is now formed of nitrocellulose, nitroglycerine, and nitroguanidine. The nitroguanidine is added in 50 percent to the propellant composition to adjust the gas output, energy, temperature, and burning rate. Triple-base propellants are used in tank guns and large caliber guns [4–6]. Some typical triple-base propellants are given in Table 12.6.

Composite propellants.   A composite propellant represents a group of propellants in which the composites, the fuels, and the oxidizers are separated materials. This type of propellant has replaced the sensitive propellants that suffer from the possibility of accidental initiation from fire, impact, electric spark, and others [4–6].

TABLE 12.6   Properties of Some Triple-Base Propellants

Property	Propellant types					
	M15	M16(T6)	M17	M30(T36)	M31(T34)	T20
Nitrocellulose	20.0	55.5	22.0	28.0	20.0	20.0
% Nitration	13.15	12.6	13.15	12.6	12.6	13.15
Nitroglycerine	19.0	27.5	21.5	22.5	19.0	13.0
Nitroguanidine	54.7		54.7	47.7	54.7	60.0
Dinitrotoluene		10.5				
Oxidizers						
Potassium sulfate		1.5				
Stabilizers						
Diphenylamine					1.5	
Ethyl centralite	6.0	4.0	1.5	1.5		2.0
Plasticizers						
Dibutylphthalate					4.5	5.0
Lubricants,						
Antistats						
Cryolite	0.3		0.3	0.3	0.3	
Graphite		0.5	0.1	0.1		
Lead carbonate						1.0

The composite propellants are used for rockets and for gas generators (Figure 12.5). The typical composite is a blend of a crystalline oxidizer and is amorphous or plastic fuel [5]. The typical composite propellant oxidizers are as follows:

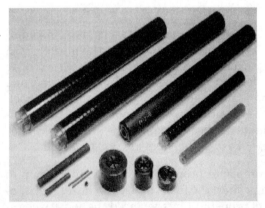

Figure 12.5 Extruded-type solid propellants [10]. (*Source: http://www.nof.co.jp/E_nof/business%20activities/Explosives.htm.*)

Sodium nitrate	$NaNO_3$
Potassium nitrate	KNO_3
Ammonium nitrate	NH_4NO_3
Ammonium perchlorate	NH_4ClO_4
Potassium perchlorate	$KClO_4$
Lithium perchlorate	$LiClO_4$

The fuel acts both as a fuel and a binder and it provides the mechanical strength and structural properties to the composite propellant. The typical composite propellant fuels are as follows:

Asphalt	Amorphous
Polyisobutylene	Polymer, amorphous
Polysulfide rubber	Polymer, thermosetting
Polyurethane rubber	Polymer, thermosetting
Polybutadiene-acrylicco-polymer	Polymer, thermosetting
Polyvinyl chloride	Polymer, thermoplastic
Cellulose acetate	Polymer, thermoplastic

The amorphous fuel systems can be mixed and pressed into molds or molded directly into a rocket motor. The thermosetting fuel systems are mixed as liquid monomers with the oxidizers and curing agents, poured into molds, and then heated and cured. These systems do not change their shape, even at very high temperatures. The thermoplastic system does soften and melt, but at much higher temperatures than the amorphous systems [5].

The oldest of the composite propellants is black powder that is produced by mixing the ingredients wet and then wheel-milling them. The tremendous pressures in the mill cause the sulfur to plasticize and flow. The product is then treated, dried, and screened to produce the various grades as a function of the particle sizes. The composite propellants are less vulnerable to initiate than nitrocellulose-base propellants [5].

Gun propellants contain additives that are necessary to impart certain required properties to the propellants. Examples of additives used in gun propellants are shown in Table 12.7, in which the additives are classified according to their functions [5].

12.2.2 Rocket propellants

Rocket propellants are similar to gun propellants as they burn smoothly and without detonation. Rocket propellants are required to burn at a

TABLE 12.7   Examples of Additives Used in Gun Propellants

Function	Additive	Action
Stabilizer	Carbamite (diphenyl diethyl), urea methyl centralite (diphenyl dimethyl urea), chalk, and diphenylamine	Increase shelf life of propellant
Plasticizer	Dibutyl phthalate, carbamite, and methyl centralite	Gelation of nitrocellulose
Coolant	Dibutyl phthalate, carbamite, methyl centralite, and dinitrotoluene	Reduce the flame temperature
Surface moderant	Dibutyl phthalate, carbamite, methyl centralite, and dinitrotoluene	Reduce burning rate of the grain surface
Surface lubricant	Graphite	Improve flow characteristics
Flash inhibitor	Potassium sulfate, potassium , nitrate potassium aluminum fluoride, and sodium cryolite	Reduce muzzle flash
Decoppering agent	Lead or tin foil, compounds containing lead or tin	Remove deposits of copper left by the driving band
Anti-wear	Titanium dioxide	Reduce erosion of gun barrel

pressure of 7 MPa compared to 400 Mpa for gun propellants. Rocket propellants must burn for a longer time to provide a sustained impulse [4]. There are two main types of solid rocket propellants: double-base and composite rocket propellants.

Double-base rocket propellants.   Double-base rocket propellants are homogeneous mixtures that contain nitrocellulose with nitroglycerine. The size of the propellant grain has an enormous effect on the manufacturing process. Smaller grains that are used in small rocket motors are obtained from the extruded double-base rocket propellant. A double-base rocket propellant produces larger grains and is used in large rocket motors [4].

A double-base rocket propellant is composed of nitroglycerine (13.25 percent N content), plasticizer, and other additives. The specific impulse is estimated as 2000 N.S.Kg^{-1} and the flame temperature at 2500 K [4].

Composite rocket propellants.   This type of propellant contains two-phase mixtures with a crystalline oxidizer in a polymeric fuel or binder matrix. The finely dispersed powder of ammonium perchlorate suspended in a fuel is the oxidizer. The fuel is a plasticized polymeric material such as hydroxy-terminated polybutadiene cross-linked with a diisocyanate (rubber type) or carboxy-terminated polybutadiene (plastic type) [4].

Extrusion is employed for the composite propellants of plastic properties, and cast technique is used for composite propellants of rubbery properties.

Composite rocket propellants that have a specific impulse of 2400 $N.S.Kg^{-1}$ and a flame temperature of 2850 K, contain ammonium perchlorate, hydroxy-terminated polybutadiene, aluminum, and other additives [4].

The composite propellant with carboxy-terminated polybutadiene as a plasticizer has a specific impulse of 2600 $N.S.Kg^{-1}$ and a flame temperature of 3500 K.

In addition to solid rocket propellants, a wide class of liquid rocket fuel is also available. Liquid propellants exist as monopropellants and bipropellants. Monopropellants are liquids that burn in the absence of external oxygen. They have low energy and a specific impulse and are used in small missiles that require low thrust. Hydrazine, NH_2-NH_2, is the most widely used monopropellant. However, other monopropellants such as hydrogen peroxide, ethylene oxide (C_2H_4O), and nitromethane (CH_3NO_2), are also used [4].

Bipropellants are formed of two components, a fuel and an oxidizer, that are stored in separate tanks. The two components are injected into the chamber and they ignite as they come into contact. The fuel component of the bipropellant includes methanol, kerosene, hydrazine, and others. The most used oxidizer is nitric acid [4].

12.3 Pyrotechnics [5]

Pyrotechnics are materials that differ from explosives and propellants in their lower reaction rate and less gas production. Pyrotechnics usually consist of a mixture of two ingredients: a fuel and an oxidizer. Generally, the fuels are metals and the oxidizers are either salts or metal oxides.

In both explosives and propellants the materials are relatively nonporous. The binders and plasticizers used effectively fill the pore spaces. Pyrotechnics are porous and the heat transfer related to the hot gas permeation into the reactant material mixture becomes important. In many pyrotechnics no binders or plasticizers are used. The explosives and propellants have burning (or detonation) rates that depend on density, temperature, and pressure. However, the burning rates of pyrotechnics are, in addition, affected by porosity, particle sizes, purity, homogeneity (degree of mixing), and stoichiometry (fuel or oxidizer ratio).

The big environmental and health concerns led to the avoidance of many substances that have been used in the past in pyrotechnic mixtures. These include beryllium, cadmium, mercury, chromates, lead compounds, and many others.

Pyrotechnics are divided into five major production categories: sound, light, heat, smoke, and delay.

12.3.1 Sound producers

Sound-producing pyrotechnics are primarily used in the fireworks industry and in military simulators. There are two types of sound producers: a composition that produces a loud, short-duration sound, like a bomb or grenade, and a composition that produces a shrill whistle of long duration. The short, loud, bang producers are sometimes made of black powder that is heavily confined in a cardboard tube. More typically, however, sound is produced by photoflash mixes. These consist of various blends containing aluminum, magnesium, potassium perchlorate, and other additives. Some of the blends are so fast that they require little or no confinement to explode. Typical sound producers are firecrackers, whistles, salutes, and military grenade, and ground-burst simulators.

12.3.2 Light producers

Basically, there are two subdivisions of light-producing pyrotechnics: flash powders and flares. The main difference in their performance is in the bulk burning rates. Flash powders burn very rapidly, some almost bordering on detonation velocities. The flash powders are generally loaded as loose or lightly pressed mixed dry powder. They are sometimes initiated by an ignitron and in some applications by a detonator. In military tactical use, they are employed as lighting for night reconnaissance photography.

The majority of photoflash mixes use magnesium, or aluminum, or both as the fuel and barium nitrate, or potassium perchlorate, or both as oxidizers (Eqs. 12.14 and 12.15).

$$8Al + 3KClO_4 \rightarrow 4Al_2O_3 + 3KCl \qquad (12.14)$$

$$8Al + 3BaNO_3 \rightarrow 2Al_2O_3 + 3BaO + 1.5N_2 \qquad (12.15)$$

Typical fuels are magnesium (sometimes mixed with aluminum), manganese, and silicon. Typical oxidizers include the nitrates of barium, sodium, potassium, and strontium. Binders used are castor and linseed oils and paraffin waxes. Another interesting flare mix is magnesium or teflon. The chlorine and fluorine from the teflon are the oxidizers in this mix. Magnesium or teflon flares burn several hundred degrees (°C) hotter than metal or salt flares and radiate very strongly in the infrared spectrum.

12.3.3 Heat producers

Heat-producing pyrotechnics are used for ignition mixtures, thermites, sparks, incendiaries, and heat pellets for thermally activated batteries. The first ignition mixes are generally either metal-salt or metal-metal oxide mixes with no binders. The metal-salt mixtures are very sensitive to impact, flame (or a concentrated heat source such as a glowing hot wire), and sparks, whereas the metal-metal oxide mixtures are not sensitive to impact, flame, or spark. Both types of mixtures are used as the *first fire* or the ignition element in a pyrotechnic train. Typical fuels are aluminum, zirconium, titanium, titanium hydrides, magnesium, boron, and in former times, beryllium. Typical oxidizers include the various nitrates already mentioned with the light producers, as well as calcium chromate, lead nitrate, iron oxides, copper oxide, and the perchlorates of sodium, potassium, and ammonia. The calcium chromate and lead nitrate are now considered to cause environmental problems and are falling out of use. A typical reaction of the metal-metal oxide type is the reduction-oxidation of aluminum mixed with copper oxide.

$$2Al + 3CuO \rightarrow Al_2O_3 + 3Cu \qquad (12.16)$$

This reaction, where an oxide of one metal is reduced and the other metal oxidized, is the same reaction as in the thermites. These materials, generally a mixture of aluminum and iron oxide, produce molten iron as one of the products.

$$2Al + Fe_2O_3 \rightarrow Al_2O_3 + 2Fe \qquad (12.17)$$

The molten iron is then the heat transfer medium that is used to perform the specific task of the thermite device, usually welding. Other fuels include nickel, and other oxidizers include Fe_3O_4 and Cu_2O.

12.3.4 Smoke producers

These pyrotechnics are also subdivided into organic and inorganic types and can be made in a variety of colors and optical densities. The earliest organic smokes were droplets of kerosene that condensed from vapors that were produced from smokes by boiling kerosene. This smoke has the obvious drawback of being extremely flammable. The other organic smokes are quite different. They are aerosol droplets of condensed organic dyes that are vaporized out of a relatively low-temperature smoke. Irritant agents such as tear gas are also formed in a similar way. The flame temperature of the burning mix is sufficiently low so the organic dye material does not thermally decompose. The pyrotechnic mixture frequently used to produce this low-temperature burn consists of potassium chlorate and powdered sugar diluted with bicarbonates to cool the reaction.

The inorganic smokes are from reactions that produce zinc chloride as the major obscurant. These mixtures, called HC smokes, are a blend of aluminum powder, zinc oxide, hexachlorethane (HC), and sometimes a pinch or two of magnesium and ammonium perchlorate. At low percentages of aluminum, the major smoke products are Al_2O_3 and $ZnCl_2$ with CO. At higher aluminum concentrations the products form a grey, instead of white, smoke that consists of Al_2O_3, $ZnCl_2$, and carbon.

12.4   Manufacturing of Explosives

12.4.1   TNT production

TNT may be prepared by either a continuous or a batch process, using toluene, nitric acid (HNO_3), and sulfuric acid as raw materials (Eq. 12.18) [7].

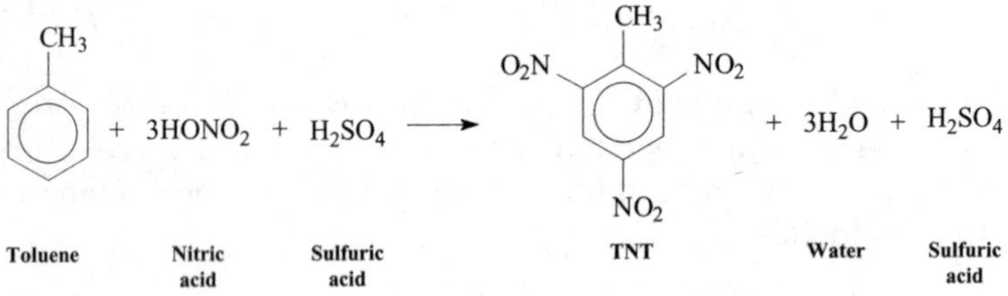

$$+ 3HONO_2 + H_2SO_4 \longrightarrow + 3H_2O + H_2SO_4 \tag{12.18}$$

| Toluene | Nitric acid | Sulfuric acid | | TNT | Water | Sulfuric acid |

The production of TNT by nitration of toluene is a three-stage process, performed in a series of reactors (Fig. 12.6). The mixed acid stream flows countercurrent to the organic stream. The fresh 60 percent HNO_3 solution and the spent acid are fed into the first reactor along with toluene. The organic layer that is formed in the first reactor is subjected to further nitration in the second reactor with fresh 60 percent HNO_3 and spent acid. The product obtained from the second reactor is a mixture of all possible isomers including dinitrotoluene (DNT) as a major one. DNT is pumped into the third reactor and treated with a fresh feed of 97 percent HNO_3 and oleum (a solution of sulfur trioxide, SO_3, in anhydrous sulfuric acid). The resulting products from the third reactor include mainly 2,4,6-trinitrotoluene (TNT). The crude TNT is purified by washing it with water to remove free acid. The TNT is then neutralized with soda ash and treated with a 16 percent aqueous sodium sulfite (Sellite) solution to remove the contaminating isomers. The wash water (yellow water) is recycled to the early nitration stages and the Sellite waste solution (red water) that is obtained from the purification

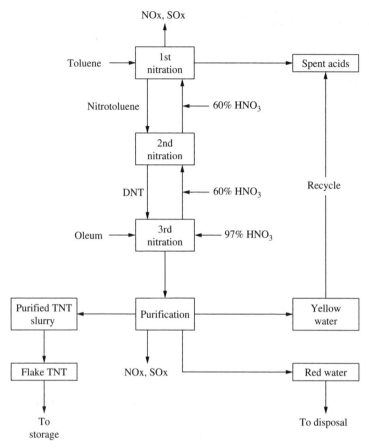

Figure 12.6   Nitration of toluene to form trinitrotoluene (TNT).

process is discharged directly as a liquid waste stream and then collected and sold or concentrated to a slurry and incinerated. The final TNT crystals are melted and passed through hot air dryers, where most of the water is evaporated. The dehydrated product is solidified, and the TNT flakes packaged for transfer to storage or loading area.

12.4.2   Black powder production [6]

Black powder is mainly used as an igniter for nitrocellulose gun propellants and to some extent in safety blasting fuses, delay fuses, and firecrackers. Potassium nitrate black powder (74 wt percent plus 15.6 wt percent carbon, 10.4 wt percent sulfur) is used for military applications. The slower-burning, less costly, and more hygroscopic sodium nitrate

TABLE 12.8   Approximate Composition of
Reaction Products of Black Powder

Component	Wt percent
Gases	
Carbon dioxide	49
Carbon monoxide	12
Nitrogen	33
Hydrogen sulfide	2.5
Methane	0.5
Water	1
Hydrogen	2
Total	44
Solids	
Potassium carbonate	61
Potassium sulfate	15
Potassium sulfide	14.3
Potassium thiocyanate	0.2
Potassium nitrate	0.3
Ammonium carbonate	0.1
Sulfur	9
Carbon	0.1
Total	56

black powder (71.0 wt percent plus 16.5 wt percent carbon, 12.5 wt percent sulfur) is used industrially.

The reaction products of black powder are complex and vary with initiation, confinement, and density (Table 12.8).

The conventional process for the production of black powder involves a mechanical mixing of the different components. Dry potassium nitrate is pulverized in a ball mill. The sulfur is milled into cellular charcoal to form a uniform mixture in a separate ball mill. The mixture of nitrate and the sulfur-charcoal is screened and then loosely mixed by hand or in a tumbling machine. Magnetic separators may be used to eliminate any ferrous metals. The preliminary mixture is transferred to an edge-runner wheel mill with large and heavy cast iron wheels. A clearance between the pan and the wheels is needed for safety purposes. The magnitude of this gap also contributes to the density of the black powder granules obtained. As the temperature increases during the milling operation and moisture evaporates, water is added to minimize dusting and improve incorporation of the nitrate into the charcoal. The milling operation requires approximately 3 to 6 h. The moist milled powder is transferred to a hydraulic press where it is consolidated in layers into cakes at pressures of approximately 41.3 MPa (6000 psi) applied for approximately 30 min. Each cake is about 2.5 cm thick and 60 cm square. The density of the powder increases to 1.6 to 1.8 g/cm^3, depending on the pressure applied. The cakes are then transferred to a corning mill consisting of adjustable corrugated rollers that are cascaded so that a series of crushing actions occur with each crushing, followed by automatic screening fines; and the cakes that have been screened are recycled to the press feed or used in fuse powder or fireworks. Coarse material is recycled. The grains are polished and dried. Graphite is added, and they are blended by tumbling in a large hardwood rotating drum. The powder is screened before placing it into air-tight metal drums.

12.4.3 RDX and HMX production [6]

Both RDX and HMX are white, stable, crystalline solids. Both are much less toxic than TNT and may be handled with no physiological effect if appropriate precautions are taken to assure cleanliness of operations. Both RDX and HMX detonate to form mostly gaseous, low molecular weight products with little intermediate formation of solids. The calculated molar detonation products of RDX are 3.00 H_2O, 3.00 N_2, 1.49 CO_2, and 0.02 CO. RDX has been stored for as long as 10 months at 85°C without perceptible deterioration.

At present, HMX is the highest-energy solid explosive produced on a large scale, primarily for military use. It exists in four polymorphic forms of which the beta form is the least sensitive and most stable and the type required for military use. The mole fraction products of detonation of HMX in a calorimetric bomb are 3.68 N_2, 3.18 H_2, 1.92 CO_2, 1.06 CO, 0.97 C, 0.395 NH_3, and 0.30 H_2.

Both RDX and HMX are substantially desensitized by mixing with TNT to form cyclotols (with RDX) and octols (with HMX) or by coating with waxes, synthetic polymers, and elastomeric binders.

The two most common processes for making RDX and HMX use hexamethylenetetramine (hexamine) as starting material. The Bachmann process, now used exclusively in the United States, is a simplification of a series of complex reactions that may be summarized as follows:

$$C_6H_{12}N_4 + 4HNO_3 + 2NH_4NO_3 + 6(CH_3CO)_2O \longrightarrow 2RDX + 12CH_3COOH$$

$$(12.19)$$

In the Bachmann process, the reactants are mixed, and the slurry aged to complete the reaction and increase the yield. The reaction vessels are stainless steel, jacketed, and temperature controlled. A solution of one part hexamine in 1.65 parts of acetic acid, and a solution of 1.50 parts of ammonium nitrate dissolved in 2.0 parts of nitric acid and 5.20 parts of acetic anhydride are added continuously to each of two reactors that are maintained at 65–72°C. The mixture is cycled in a loop in about 15 s from the reactor through a temperature-controlled jacketed reactor leg. The slurry containing RDX overflows continuously into the first of the three aging tanks, is kept at 65–72°C, and passes from one to the other over a period of about 24 min. The aged slurry is transferred to the first of a series of seven temperature-controlled simmer tanks. Dilution liquor is added to the first tank to produce an acetic acid concentration of about 63 percent in the simmer system. The hot slurry moves from one tank through the others in a gradual cooling process to hydrolyze undesirable by-products and precipitate the RDX from the solution.

The RDX-acetic acid slurry is filtered and water-washed and the spent acetic acid is processed for recovery. The RDX is recrystallized from cyclohexanone.

In the Bachmann process an 80 to 84 percent yield is obtained, about 10 percent of which is cyclotetramethylenetetranitramine (HMX).

A modification of the Bachmann process used to make RDX with the same starting materials and in similar equipment is employed for the manufacture of HMX. The reaction temperature is lower (44 ± 1°C as compared to 68°C for RDX) and the raw materials are mixed in a two-step process. The yield of HMX per mole of hexamine is about 55 to 60 percent, as compared to 80 to 85 percent in the manufacture of RDX.

A typical HMX batch process starts with a reactor that contains acetic acid. The proper amounts of hexamine in glacial acetic acid, ammonium nitrate in nitric acid, and acetic anhydride are added in stages. The HMX slurry is transferred to the aging tanks, held for 30 min at 44°C, and pumped to a simmer tank where sufficient dilution liquor is added to reduce the concentration to 80 percent acetic acid. After dilution the temperature is raised to 110°C, to decompose undesirable by-products and to improve the filtering characteristics of the HMX.

The slurry is filtered at about 60°C to retain RDX in solution. The filtered product is almost 99 percent HMX. The crystals are washed with cold water and crystallized from acetone, cyclohexanone, or both, depending on the particle distribution desired. Recrystallization also converts the more sensitive alpha crystals to the higher density beta form and reduces occluded acid to less than 0.02 percent. The RDX is recovered from the spent acid that is reclaimed.

12.5   Thermochemistry of Explosives

Thermochemistry is an important part of explosive chemistry: it provides information on the type of chemical reactions, energy changes, mechanisms, and kinetics that occur when a material undergoes an explosion [4].

Any explosive reaction will result in breaking the explosive molecule apart into its constituent atoms. Subsequently a series of new small and stable molecules are formed by the rearrangement of these atoms. These molecules are usually water (H_2O), carbon dioxide (CO_2), carbon monoxide (CO), and nitrogen (N_2). In addition, molecules of hydrogen (H_2), carbon (C), aluminum oxide (Al_2O_3), sulfur dioxide (SO_2), and so on, were also found in the products of some explosives. The nature of the products will depend upon the amount of oxygen available during the reaction. This supply of oxygen will depend in turn upon the quantity of oxidizing atoms that are present in the explosive molecule [4].

12.5.1 Oxygen balance [4]

The amount of oxygen in each molecule of nitroglycerine (1) and TNT (4) can be calculated and compared with the amount of oxygen required for complete oxidation of the fuel elements, that is, hydrogen and carbon.

If the amount of oxygen present in the explosive molecule is insufficient for the complete oxidation, a negative oxygen balance will result. This can be seen in the molecule TNT. Nitroglycerine, however, has a high proportion of oxygen, more than required for complete oxidation of its fuel elements and therefore has a positive oxygen balance. This oxygen balance can be defined as the amount of oxygen, expressed in weight percent, liberated as a result of the complete conversion of the explosive material to carbon dioxide, water, sulfur dioxide, aluminum oxide, and so on [4].

$$
\begin{array}{cc}
\begin{array}{l}
\quad\quad\text{H} \\
\quad\quad| \\
\text{H--C--O--NO}_2 \\
\quad\quad| \\
\text{H--C--O--NO}_2 \\
\quad\quad| \\
\text{H--C--O--NO}_2 \\
\quad\quad| \\
\quad\quad\text{H} \\
\quad\quad\mathbf{1}
\end{array}
&
\begin{array}{l}
\text{CH}_3 \\
\text{O}_2\text{N} \quad\quad \text{NO}_2 \\
\bigcirc \\
\text{NO}_2 \\
\quad\mathbf{4}
\end{array}
\end{array}
$$

When detonation of TNT ($C_7H_5N_3O_6$), **4**, takes place the explosive is oxidized to give a mixture of gases such as carbon dioxide, water, and nitrogen (Eq. 12.20):

$$C_7H_5N_3O_6 \rightarrow nCO_2 + nH_2O + nN_2 \qquad (12.20)$$

Equation (12.20) is not balanced. The introduction of oxygen atoms (O) is needed to balance the equation (Eq. 12.21):

$$C_7H_5N_3O_6 \rightarrow 7CO_2 + 2\tfrac{1}{2}H_2O + 1\tfrac{1}{2}N_2 - 10\tfrac{1}{2}O \qquad (12.21)$$

TABLE 12.9   Balanced Reaction Formulae for Some Explosives

Explosive substance	Balanced reaction formulae for complete combustion
Ammonium nitrate	$NH_3NO_3 \rightarrow 2H_2O + N_2 + 1O$
Nitroglycerine	$C_3H_5N_3O_9 \rightarrow 3CO_2 + 2^1/_2H_2O + 1^1/_2N_2 + {}^1/_2O$
EGDN	$C_2H_4N_2O_6 \rightarrow 2CO_2 + 2H_2O + N_2 + 0O$
PETN	$C_5H_8N_4O_{12} \rightarrow 5CO_2 + 4H_2O + 2N_2 - 2O$
RDX	$C_3H_6N_6O_6 \rightarrow 3CO_2 + 3H_2O + 3N_2 - 3O$
HMX	$C_4H_8N_8O_8 \rightarrow 4CO_2 + 4H_2O + 4N_2 + 4O$
Nitroguanidine	$CH_4N_4O_2 \rightarrow CO_2 + 2H_2O + 2N_2 - 2O$
Picric acid	$C_6H_3N_3O_7 \rightarrow 6CO_2 + 1^1/_2H_2O + 1^1/_2N_2 - 6^1/_2O$
Tetryl	$C_7H_5N_5O_8 \rightarrow 7CO_2 + 2^1/_2H_2O + 2^1/_2N_2 - 8^1/_2O$
TNT	$C_7H_5N_3O_6 \rightarrow 7CO_2 + 2^1/_2H_2O + 1^1/_2N_2 - 10^1/_2O$

SOURCE: *The Chemistry of Explosives*, 2nd ed. [4].

The balance of the reaction formula for the combustion of TNT gave a negative sign for oxygen. This therefore indicates that TNT has insufficient oxygen in its molecule to oxidize its reactants full to form water and carbon dioxide. This amount of oxygen as percent by weight can now be calculated as shown in Eq. 12.21 [4].

Tables 12.9 and 12.10 include balanced reaction formulae and oxygen balance for some explosives.

12.5.2   Heat of formation [4]

The heat of formation for a reaction containing explosive chemicals can be described as the total heat evolved when a given quantity of

TABLE 12.10   Oxygen Balance for Some Explosives

Explosive substance	Empirical formula	Oxygen balance/ percent weight
Ammonium nitrate	NH_3NO_3	+19.99
Nitroglycerine	$C_3H_5N_3O_9$	+3.50
EGDN	$C_2H_4N_2O_6$	0.00
PETN	$C_5H_8N_4O_{12}$	−10.13
RDX	$C_3H_6N_6O_6$	−21.60
HMX	$C_4H_8N_8O_8$	−21.62
Nitroguanidine	$CH_4N_4O_2$	−30.70
Picric acid	$C_6H_3N_3O_7$	−45.40
Tetryl	$C_7H_5N_5O_8$	−47.39
TNT	$C_7H_5N_3O_6$	−74.00

SOURCE: *The Chemistry of Explosives*, 2nd ed. [4].

a substance is completely oxidized in an excess amount of oxygen, resulting in the formation of carbon dioxide, water, and sulfur dioxide. The explosive substances that do not contain sufficient oxygen in its molecule for complete oxidation, that is, TNT, lead to the production of carbon monoxide, carbon, and hydrogen. The energy liberated during the formation of these products is known as the *heat of explosion*.

The value for the heat of formation can be negative or positive. If the value is negative, heat is liberated during the reaction and the reaction is exothermic; whereas, if the value is positive, heat is absorbed during the reaction and the reaction is endothermic. For reactions involving explosive components the reaction is always exothermic. In an exothermic reaction the energy evolved may appear in many forms, but for practical purposes it is usually obtained in the form of heat. The energy liberated when explosives deflagrate is called the *heat of deflagration*, whereas the energy liberated by detonating explosives is called the *heat of detonation* in $kJ \cdot mol^{-1}$ or the *heat of explosion* in $kJ \cdot kg^{-1}$.

In a chemical reaction involving explosives, energy is initially required to break the bonds of the explosive into its constituent elements as shown in Eq. (12.22) for RDX [4].

$$C_3H_6N_6O_6 \rightarrow 3C + 3H_2 + 3N_2 + 3O_2 \qquad (12.22)$$

These elements quickly form new bonds with the release of a greater quantity of energy as shown in Eq. (12.23) [4].

$$3C + 3H_2 + 3N_2 + 3O_2 \rightarrow 3CO + 3H_2O + 3N \qquad (12.23)$$

The molecules of an explosive are first raised to a higher energy level through input of the *heat of atomization* to break their interatomic bonds. Then the atoms rearrange themselves into new molecules, releasing a larger quantity of heat and dropping to an energy level lower than the original as shown in Fig. 12.7.

12.5.3 Heat of explosion [4]

When an explosive is initiated either to burning or detonation, its energy is released in the form of heat. The liberation of heat under adiabatic conditions is called the *heat of explosion*, denoted by the letter Q. The heat of explosion provides information about the work capacity of the explosive, where the effective propellants and secondary explosives generally have high values of Q. For propellants burning in the chamber of a gun, and secondary explosives in detonating devices, the heat of explosion is conventionally expressed in terms of constant volume

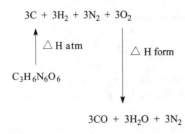

$$3C + 3H_2 + 3N_2 + 3O_2$$

$\triangle$ H atm

$\triangle$ H form

$$C_3H_6N_6O_6$$

$$3CO + 3H_2O + 3N_2$$

Figure 12.7   Energy is taken in to break the bonds of RDX into its constituent elements, then energy is released when new bonds are formed [4].

conditions Q_v. For rocket propellants burning in the combustion chamber of a rocket motor under conditions of free expansion to the atmosphere, it is conventional to employ constant pressure conditions. In this case, the heat of explosion is expressed as Q_p.

Consider an explosive that is initiated by a stimulus of negligible thermal proportions. The explosion can be represented by the irreversible process that includes the initiation followed by the explosion to give gaseous products with a heat Q ultimately lost to the surroundings.

Under constant volume conditions Q_v can be calculated from the standard internal energies of formation for the products $\Delta U^0_{f(products)}$ and the standard internal energies of formation for the explosive components $\Delta U^0_{f(explosive)(products)}$ as shown in Eq. (12.24).

$$Q_v = \sum \Delta U^0_{f(products)} - \sum \Delta U^0_{f(explosive)(components)} \qquad (12.24)$$

A similar expression is given for the heat of explosion under constant pressure conditions as shown in Eq. 12.25, here ΔU^0_f represents the corresponding standard enthalpies of formation:

$$Q_p = \sum \Delta U^0_f - \sum \Delta U^0_{f(explosive\,components)} \qquad (12.25)$$

In considering the thermochemistry of solid and liquid explosives, it is usually adequate, for practical purposes, to treat the state functions ÎH and ÎU as approximately the same. Consequently, heat or enthalpy terms tend to be used for both constant pressure and constant volume conditions.

Therefore, the heat of explosion Q can be calculated from the difference between the sum of the energies for the formation of the explosive components and the sum of the energies for the formation of the explosion products, as shown in Eq. 12.26.

$$Q = \Delta H_{(reaction)} = \sum \Delta H_{f(products)} - \sum \Delta H_{f(explosive\,components)} \qquad (12.26)$$

12.5.4   Explosive power and power index [4]

In an explosive reaction, heat and gases are liberated. The volume of gas V and the heat of explosion Q can both be calculated independently but these values can be combined to give the value for the explosive power as shown in Eq. 12.27.

$$\text{Explosive power} = Q \times V \tag{12.27}$$

The value for the explosive power is then compared with the explosive power of a standard explosive (picric acid) resulting in the power index, as shown in Eq. (12.28), where data for $Q_{(\text{picric acid})}$ are 3250 kJ g^{-1} and 0.831 dm^3, respectively.

$$\text{Power Index} = \frac{Q \times V}{Q_{(\text{picric acid})} \times V_{(\text{picric acid})}} \times 100 \tag{12.28}$$

As expected, the values for the power and power index of secondary explosives are much higher than the values for primary explosives.

12.6   Safety and Environmental Considerations

In view of the catastrophic effects and the increasingly severe legal and economic implications of a disastrous explosion, a large amount of effort is devoted to the safety of operations, while keeping control costs at an acceptable level. This work has been given increased impetus in the United States by the military plant modernization now underway and the development of a variety of new or improved processes for making explosives.

Many governmental regulations control the classification, shipping, and handling of explosive materials. The publications in the field of safety have increased greatly and a number of important symposia have been held on the subject, including the annual explosive safety seminars conducted by the Explosives Safety Board of the United Sates Department of Defense.

Standard safety procedures include attention to such matters as intraline distances to minimize explosion propagation, cleanliness, nonsparking equipment and explosion-proof motors, and location in selected areas away from the heavily populated areas. The development of continuous processes for explosive manufacture that use extensive automatic controls and minimize in-line quantities has contributed significantly to decreasing the hazards of operations.

Highly detailed and systematic methodologies have been developed to identify possible failure modes and to quantify their probability of occurrence and impact on operations, including hazards analyses, failure analyses, and risk or reliability assessment. They are characterized by the step-by-step detailed examination of a process, often using fault trees to specify and quantify the likelihood of undesirable events. Experiments are conducted simulating the actual application and environment of the explosives in the process to obtain valid quantitative data for calculation. The estimated probabilities of an accident occurring for each of the operations in a process are generally computerized to enable an assessment to be made of the overall hazard of the process and an attack launched on the most dangerous elements.

The concept of TNT equivalency has been increasingly used for evaluating the magnitude of an accidental explosion that may occur at some stage of an operation involving explosive materials. The TNT equivalency value expressed as percent is the relative weight of TNT, when tested as a hemispherical charge in the ground burst mode that yields the same peak pressure or positive impulse as is produced at a given distance by the explosive or explosive system tested. The information obtained enables the designer to convert the data in barricade design manuals that are usually presented in terms of surface bursts of TNT, to the explosive as it occurs in process operations.

A great deal of experimental work has been done to quantify the hazards of an unwanted explosion. The vulnerability of structures and people to shock waves and fragment impact has been well established. This effort has also led to the design of protective structures superior to the conventional barricades and that permit considerable reduction in allowable safety distances.

Environmental effects.   Pollutants from explosives are primarily produced during manufacture of explosives and the acids used in nitration. They may also be produced during incorporation in munitions or industrial explosives and in cleanup and disposal operations.

The wastewater effluent from a plant may be very acidic or very basic and high in oxygen demand, dissolved and particulate solids, soluble nitrates and sulfates, and oil and grease. Significant improvement may be made by good housekeeping practices, close process control, the separation of the contaminated from the clean process waters, and the application of the proven techniques for pollution abatement.

Many procedures are being studied and used to reduce the level of air, water, and solid waste contaminants. Specific approaches and principles of attack for decreasing water contamination include: (1) minimizing the quantity of water leaving the plant by the recycling process and cooling waters; (2) segregating and treating highly polluted waters before dilution; (3) using settling reservoirs for water treatment and removal of suspended particles by sedimentation; (4) centrifuging to remove suspended solids; (5) employing ion-exchange resins to concentrate pollutants; and (6) biologically denitrifying nitrates under anaerobic conditions.

12.7 Classification, Transportation, and Storage of Explosives

12.7.1 Explosives classification

The basis for the classification of explosives throughout the world is based upon conventions developed by the United Nations committee (UNO) on Transport of Dangerous Goods [5].

The transportation of dangerous materials is regulated to prevent accidents to personnel and property. However, the regulations are designed so as not to impede the movement of dangerous materials other than those that are too dangerous to be accepted for transportation. The regulations are addressed to all modes of transport. It should be noted that the numerical order of the classes does not necessarily indicate the degree of danger the recommended definitions of hazard classes in the UNO system are to indicate which materials are dangerous and in which class, according to their specific characteristics.

The United Nations system (UNO) divides hazardous materials into nine classes for the purpose of determining the degree of risk in shipping and transport. The order and number of the class are not meant to imply the degree of risk or danger. These classes are as follows:

1. Explosives

2. Gases

3. Flammable liquids

4. Flammable solids

5. Oxidizing substances, organic peroxides

6. Toxic and infectious substances

7. Radioactive materials

8. Corrosive substances

9. Miscellaneous dangerous substances and articles

Explosive substances and articles in Class 1 are assigned to one of six divisions, depending on the type of hazard. An explosive substance is defined as a solid or liquid substance (or mixtures of substances) that is in itself capable by a chemical reaction of producing gas at such a temperature and pressure and at such a speed as to cause damage to the surroundings. Pyrotechnic substances are included even when they do not evolve gases.

12.7.2 Transportation of explosives [5]

The UNO recommendations in the orange book specify in detail the restrictions on containers, road vehicles, and rail wagons that may be used for the transport of explosives. These are quoted in the following: "Freight containers, vehicles and wagons should not be offered for the transport of explosive substances and articles of Class I, unless the freight container, vehicle or wagon is structurally serviceable as witnessed by current International Convention for Safe Containers (CSC) approval plate (applicable to freight containers only) and a detailed visual examination as follows:

1. Prior to loading a freight container, vehicle, or wagon with explosives, it should be checked to ensure it is free of any residue of previous cargo and to ensure it is structurally serviceable and the interior floor and walls are free from protrusions.

2. *Structurally serviceable* means that the freight container, vehicle, or wagon is free from major defects in its structural components, top and bottom side rails, top and bottom end rails, door sills and headers, floor crass members, corner post, and corner fittings in a freight container. Major defects are dents or bends in structural members greater than 19 mm in depth, regardless of the length; cracks or breaks in structural members; more than one splice in top or bottom end rails, or door headers or more than two splices in any top or bottom side rail or any splice in a door sill or corner post; door hinges, and hardware that are seized, twisted, broken, missing, or otherwise inoperative, gaskets and seals that do not seal; freight containers, any distortion of the overall configuration great enough to prevent proper alignment of handling equipment, mounting and securing on chassis, vehicles or wagons, or insertion into ship's cells.

3. In addition, deterioration in any component of the container, vehicle or wagon, regardless of the materials of construction, such as rusted-out metal in side-walls or disintegrated fiberglass is unacceptable. Normal wear, however, including oxidization slight dents, scratches, and other damages that does not affect serviceability or the weather-tight integrity of the units is acceptable."

12.7.3 Storage of explosives [5]

Storage facilities for explosives must be constructed to meet a number of specific requirements. In addition, all storage facilities should have the fire symbol posted in a location that is visible to an approaching vehicle. All walk-in type storage facilities should have placards on or near the entry doors specifying personnel limits and the compatibility groups. Vegetation and other combustibles around storage facilities should be controlled to minimize any possible grass, brush, forest, or trash fires. All storage facilities should have fire extinguishers with a rating of 2A: 20BC (2A wood type; B, flammable liquid; C, electrically nonconductive), and all personnel handling explosives must know the location of the fire extinguishers in the storage facilities. All explosives should be properly packaged and labeled while in storage, with the exception of day-use magazines. Personnel are not allowed to work in storage magazines or igloos other than to open and inspect the contents of an individual package. A safe operating procedure is required for each facility to cover all aspects of facility use, including the items mentioned and any specific requirements.

References

1. Kent, J. A., *Riegel's Handbook of Industrial Chemistry*, 9th ed., Van Nostrand Reinhold, New York, N.Y., pp. 1186–1226, 1992.
2. Urbanski, T., *Chemistry and Technology of Explosives*, Vol. 1, Pergamon Press, New York, N.Y., 1964.
3. Morrow, S., Explosives, solid, *Encyclopedia of Physical Science and Technology*, by R. A. Meyers (Ed.), Vol. 6, Harcourt Brace, New York, N. Y., 291–300, 1992.
4. Akhovan, J., *The Chemistry of Explosives*, RSC, London, U.K., 2001.
5. Cooper, P. W. and S. R. Kurowski, *Introduction to the Technology of Explosives*, Wiley-VCH, New York, N.Y., 1996.
6. Lindner, V., *Kirk-Othmer, Encyclopedia of Chemical Technology*, 3rd ed., Vol. 9, Wiley-Interscience, New York, N.Y. , p. 561, 1980.
7. Austin, G. A., *Shreve's Chemical Process Industries*, McGraw-Hill Book Company, New York, N. Y., 1985.
8. Olah, C. A. and D. R. Squire, *Chemistry of Energetic Materials*, Academic Press, New York, N.Y., 139, 1991.
9. Available at *http://www.army-technology.com/contractors/explosives/dyno/dyno2.html* (DYNO ASA - Military Explosives).
10. Available at *http://www.nof.co.jp/E_nof/business%20activities/Explosives.htm* (NOF Corporation-Japan: Explosives & Propulsion).

13

Petroleum and Petrochemicals

Dr. James G. Speight

13.1 Introduction

Petroleum (also called *crude oil*), in the unrefined or crude form, like many industrial feedstocks has little or no direct use and its value as an industrial commodity is only realized after the production of salable products. Even then, the market demand dictates the type of products that are needed. Therefore, the value of petroleum is directly related to the yield of products and is subject to the call of the market.

As the basic elements of crude oil, hydrogen and carbon form the main input into a refinery, combining into thousands of individual constituents, and the economic recovery of these constituents varies with the individual petroleum according to its particular individual qualities, and the processing facilities of a particular refinery (Fig. 13.1). In general, crude oil, once refined, yields three basic groupings of products that are produced when it is broken down into cuts or fractions (Table 13.1).

The gas and gasoline cuts form the lower boiling products and are usually more valuable than the higher boiling fractions and provide gas (liquefied petroleum gas), naphtha, aviation fuel, motor fuel, and feedstocks, for the petrochemical industry. Naphtha, a precursor to gasoline and solvents, is extracted from both the light and middle range of distillate cuts and is also used as a feedstock for the petrochemical industry. The middle distillates refer to products from the middle boiling range of petroleum and include kerosene, diesel fuel, distillate fuel oil, and light gas oil; waxy distillate and lower boiling lubricating oils are sometimes included in the middle distillates. The remainder of the crude oil includes the higher boiling lubricating oils, gas oil, and residuum (the nonvolatile fraction of the crude oil). The residuum can also produce heavy lubricating oils and waxes but is more often used for asphalt production. The complexity of petroleum is emphasized insofar as the actual proportions of light, medium, and heavy fractions vary significantly from one crude oil to another.

In the early days of petroleum refining, the first processes were developed to extract kerosene for lamps. Any other products were considered to be unusable and were usually discarded. Thus, the first refining processes were developed to purify, stabilize, and improve the quality of kerosene. However, the invention of the internal combustion engine led (at about the time of World War I) to a demand for gasoline for use in increasing quantities as a motor fuel for cars and trucks. This demand on the lower boiling products increased, particularly when the market for aviation fuel developed. Thereafter, refining methods had to be constantly adapted and improved to meet the quality requirements and needs of car and aircraft engines.

Since then, the general trend throughout refining has been to produce more products from each barrel of petroleum and to process those products in different ways to meet the product specifications for use in modern engines. Overall, the demand for gasoline has rapidly expanded and demand has also developed for gas oils and fuels for domestic central heating, and fuel oil for power generation, as well as for light distillates and other inputs, derived from crude oil, for the petrochemical industries.

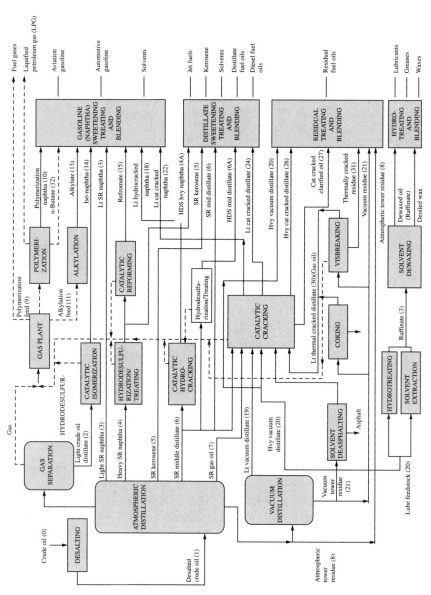

Figure 13.1 General schematic of a refinery.

TABLE 13.1   Crude Petroleum Is a Mixture of
Compounds That Can be Separated into Different
Generic Boiling Fractions

	Boiling range*	
Fraction	°C	°F
Light naphtha	−1–150	30–300
Gasoline	−1–180	30–355
Heavy naphtha	150–205	300–400
Kerosene	205–260	400–500
Light gas oil	260–315	400–600
Heavy gas oil	315–425	600–800
Lubricating oil	>400	>750
Vacuum gas oil	425–600	800–1100
Residuum	>510	>950

*For convenience, boiling ranges are converted to the
nearest 5°.

As the need for the lower boiling products developed, petroleum yielding the desired quantities of the lower boiling products became less available and refineries had to introduce conversion processes to produce greater quantities of lighter products from the higher boiling fractions. The means by which a refinery operates in terms of producing the relevant products depends not only on the nature of the petroleum feedstock but also on its configuration (i.e., the number of types of the processes that are employed to produce the desired product, slate) and the refinery configuration is, therefore, influenced by the specific demands of a market. Therefore, refineries need to be constantly adapted and upgraded to remain viable and responsive to the ever-changing patterns of crude supply and product market demands. As a result, refineries have been introducing increasingly complex and expensive processes to gain more and more lower-boiling products from the heavier and residual ends of a barrel.

To convert crude oil into desired products in an economically feasible and environmentally acceptable manner. Refinery processes for crude oil are generally divided into three categories: (1) separation processes, of which distillation is the prime example, (2) conversion processes, of which coking and catalytic cracking are prime examples, and (3) finishing processes, of which hydrotreating to remove sulfur is a prime example. However, before separation of petroleum into its various constituents can proceed, there is the need to clean the petroleum. This is often referred to as desalting and dewatering in which the goal is to remove water and the constituents of the brine that accompany the crude oil from the reservoir to the wellhead during recovery operations.

13.2   Desalting and Dewatering

Petroleum is recovered from the reservoir mixed with a variety of substances: gases, water, and dirt (minerals) (Burris and McKinney, 1992). Thus, refining actually commences with the production of fluids from the well or reservoir and is followed by pretreatment operations that are applied to the crude oil either at the refinery or prior to transportation. Pipeline operators, for instance, are insistent upon the quality of the fluids put into the pipelines; therefore, any crude oil to be shipped by pipeline or, for that matter, by any other form of transportation must meet rigid specifications in regard to water and salt content. In some instances, sulfur content, nitrogen content, and viscosity may also be specified.

Field separation, which occurs at a field site near the recovery operation, is the first attempt to remove the gases, water, and dirt that accompany crude oil coming from the ground. The separator may be no more than a large vessel that gives a quieting zone for gravity separation into three phases: gases, crude oil, and water containing entrained dirt.

Desalting is a water-washing operation performed at the production field and at the refinery site for additional crude oil cleanup (Fig. 13.2). If the petroleum from the separators contains water and dirt, water washing can remove much of the water-soluble minerals and entrained solids. If these crude oil contaminants are not removed, they can cause operating problems during refinery processing, such as equipment plugging and corrosion as well as catalyst deactivation.

The usual practice is to blend crude oils of similar characteristics, although fluctuations in the properties of the individual crude oils may cause significant variations in the properties of the blend over a period of time. Blending several crude oils prior to refining can eliminate the frequent need to change the processing conditions that may be required to process each of the crude oils individually.

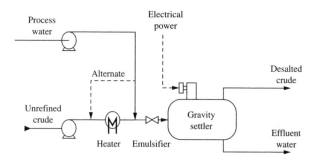

Figure 13.2   An electrostatic desalting operation.

However, simplification of the refining procedure is not always the end result. Incompatibility of different crude oils, which can occur if, for example, a paraffinic crude oil is blended with a heavy asphaltic oil, can cause sediment formation in the unrefined feedstock or in the products, thereby complicating the refinery process (Mushrush and Speight, 1995).

13.3   Evaluation

After petroleum recovery and prior to the commencement of refining, there needs to be an estimate of the potential behavior of petroleum during refining operations. Briefly, three frequently specified properties that are used to evaluate petroleum are density (specific gravity, API gravity), characterization factor, and sulfur content (Matar, 1992; Speight, 1999; 2000; 2001).

Specific gravity is the ratio of the weight of a given volume of oil to the weight of the same volume of water at a standard temperature, usually 60°F (15.6˚C). This method of measuring density and gravity first arose as a result of the need to define the character of products in more detail; it was natural to extend the measure to crude oils in general. The API (American Petroleum Institute) gravity bears an inverse relationship to the specific gravity (sp gr) and is used commonly.

$$\text{Gravity, }^{\circ}\text{API} = (141.5/\text{specific gravity}) - 131.5$$

Other factors, such as the Watson characterization factor, are also used. A highly paraffinic crude oil can have a characterization factor as high as 13, whereas a highly naphthenic crude oil can be as low as 10.5, and the breakpoint between the two types of crude oil is approximately 12. Sulfur content, the carbon residue, and distillation data are also valuable in petroleum evaluation (Speight, 2001).

13.4   Distillation

The first and the most fundamental step in the refining process (after the crude oil has been cleaned and any remnants of brine removed) is distillation (Nelson, 1958; Bland and Davidson, 1967; Speight, 1999, and references cited therein; Speight and Ozum, 2002, and references cited therein), which is often referred to as the *primary refining process*.

Distillation involves the separation of the different hydrocarbon compounds that occur naturally in a crude oil into a number of different fractions (a fraction is often referred to as a *cut*).

In the atmospheric distillation process (Fig. 13.3), heated crude oil is separated in a distillation column (distillation tower, fractionating

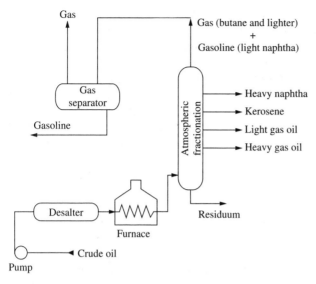

Figure 13.3   An atmospheric distillation unit.

tower, atmospheric pipe still) into streams that are then purified, transformed, adapted, and treated in a number of subsequent refining processes, into products for the refinery's market. The lighter, more volatile, products separate out higher up the column, whereas the heavier, less volatile, products settle out toward the bottom of the distillation column. The fractions produced in this manner are known as *straight run fractions* ranging from (atmospheric tower) gas, gasoline, and naphtha, to kerosene, gas oils, and light diesel, and to (vacuum tower) lubricating oil and residuum.

The feed to a distillation tower is heated by flow through pipes arranged within a large furnace. The heating unit is known as a pipe still heater or pipe still furnace, and the heating unit and the fractional distillation tower make up the essential parts of a distillation unit or pipe still. The pipe still furnace heats the feed to a predetermined temperature—usually a temperature at which a predetermined portion of the feed will change into vapor. The vapor is held under pressure in the pipe in the furnace until it discharges as a foaming stream into the fractional distillation tower. Here the unvaporized or liquid portion of the feed descends to the bottom of the tower to be pumped away as a bottom nonvolatile product, whereas the vapors pass up the tower to be fractionated into gas oils, kerosene, and naphtha.

Pipe still furnaces vary greatly and, in contrast to the early units where capacity was usually 200 to 500 bbl/day, can accommodate 25,000 bbl, or more of crude petroleum per day. The walls and ceiling are insulated with firebrick and the interior of the furnace is partially divided into two sections: a smaller convection section where the oil first enters the furnace and a larger section (fitted with heaters) where the oil reaches its highest temperature.

The atmospheric residuum is then fed to the vacuum distillation unit at the pressure of 10 mmHg where light vacuum gas oil, heavy vacuum gas oil, and vacuum residue are the products (Fig. 13.4).

The fractions obtained by vacuum distillation of the reduced crude (atmospheric residuum) from an atmospheric distillation unit depend on whether or not the unit is designed to produce lubricating or vacuum gas oils. In the former case, the fractions include (1) heavy gas oil, which is an overhead product and is used as catalytic cracking stock or, after suitable treatment, a light lubricating oil; (2) lubricating oil (usually three fractions—light, intermediate, and heavy), which is obtained as a side-stream product; and (3) asphalt (or residuum), which is the non-volatile product and may be used directly as, or to produce, asphalt, and which may also be blended with gas oils to produce a heavy fuel oil.

Operating conditions for vacuum distillation are usually 50 to 100 mmHg (atmospheric pressure = 760 mmHg). In order to minimize large fluctuations in pressure in the vacuum tower, the units are necessarily of a larger diameter than the atmospheric units. Some vacuum distillation

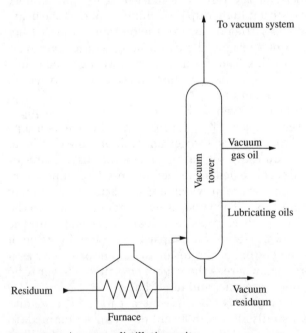

Figure 13.4 A vacuum distillation unit.

units have diameters on the order of 45 ft (14 m). By this means, a heavy gas oil may be obtained as an overhead product at temperatures of about 150°C (300°F), and lubricating oil cuts may be obtained at temperatures of 250–350°C (480–660°F), feed and residue temperatures being kept below the temperature of 350°C (660°F), above which cracking will occur. The partial pressure of the hydrocarbons is effectively reduced still further by the injection of steam. The steam added to the column, principally for the stripping of asphalt in the base of the column, is superheated in the convection section of the heater.

Fractions from the atmospheric and vacuum towers are often used as feedstocks to these second-stage refinery processes that break down the fractions, or bring about a basic chemical change in the nature of a particular hydrocarbon compound to produce specific products.

The atmospheric residuum or the vacuum residuum may be sent to a deasphalting unit to remove the very high molecular weight constituents (collectively called asphalt) and the deasphalted oil is used as feedstock to a catalytic cracking unit. Solvent deasphalting is a unique separation process in which the residue is separated by molecular weight (density), instead of by boiling point as in the distillation process. The purpose of a deasphalting unit is to produce asphalt as a final product and to produce a soluble stream (deasphalted oil that can be used in a catalytic cracking unit and have reduced coke deposition on the catalyst). The solvents used in a deasphalting unit vary from propane to pentane and the yields of asphalt and deasphalted oil vary with the hydrocarbon used in the unit.

13.5 Cracking, Coking, Hydrocracking, and Reforming

The basic processes introduced to bring about thermal decomposition of the higher boiling streams are known as *cracking*. In these processes, the higher boiling fractions are converted to lower boiling products. *Catalytic cracking* is the most common cracking process, in which heavy feedstock or cuts are broken down or changed by being heated, and reacted with catalysts.

The concept behind *thermal cracking* is the thermal decomposition of higher molecular weight constituents of petroleum to produce lower molecular weight, normally more valuable, products. The first commercial process was in 1913, which is known as the Burton Process. Even though catalyst cracking generally replaced thermal cracking in the 1940s, noncatalytic cracking processes using high temperature to achieve the decomposition are still in operation. Through catalytic processes more gasoline having higher octane, less heavy fuel oils, and light gases are produced. The light gases produced by catalytic cracking contain more olefins than those produced by thermal cracking.

In the thermal cracking process, a feedstock (e.g., gas oil) is fed to the fractionator with their thermal reactivity to separate gasoline and light and heavy oil. The light oil is then fed to the heater at 540–595°C (1000–1100°F) and a pressure of 350–700 psi, the light oil transforms to the vapor phase and is sent to the soaker. If the feedstock is heavy oil, temperatures on the order of 400–480°C (750–900°F) are used and higher pressures (350–700 psi) are used to maintain the feedstock in the liquid phase, then it is fed to the soaker. The liquid and vapor phase mix in the soaker and are sent to the separator, with the products coming out on the bottom as fuel oil and the light recycle back to the fractionator. Coking in the reactor is the main problem when heavy oil is heated at high temperatures.

In many refineries, coking processes, which use high temperatures and low pressures, have superseded thermal cracking processes.

The principal application of *catalytic cracking* (Table 13.2) is the production of high-octane gasoline, to supplement the gasoline produced by distillation and other processes. Catalytic cracking also produces heating oil components and hydrocarbon feedstocks, such as propene, and butene, for polymerization, alkylation, and petrochemical operations. The main process used in catalytic cracking involves the use of fluidized bed units (FCC).

TABLE 13.2   Summary of Catalytic Cracking Processes

Conditions
Solid acidic catalyst (silica-alumina, zeolite, and so on)
Temperature: 480–540°C (900–1000°F (solid or vapor contact))
Pressure: 10–20 psi
Provisions needed for continuous catalyst replacement with heavier feedstocks (residua)
Catalyst may be regenerated or replaced

Feedstocks
Gas oils and residua
Residua pretreated to remove salts (metals)
Residua pretreated to remove high molecular weight (asphaltic constituents)

Products
Lower molecular weight than feedstock
Some gases (feedstock and process parameters dependent)
Iso-paraffins in product
Coke deposited on catalyst

Variations
Fixed bed
Moving bed
Fluidized bed

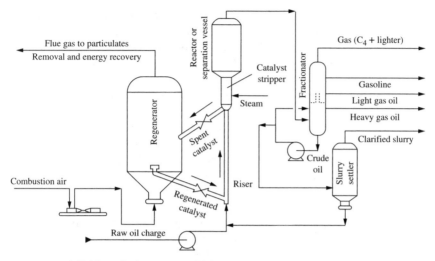

Figure 13.5   A fluid catalytic cracking (FCC) unit.

Fluid catalytic cracking (FCC) (Fig. 13.5) was first introduced in 1942 and uses a fluidized bed of catalyst with continuous feedstock flow. The catalyst is usually a synthetic alumina or zeolite used as a catalyst. Compared to thermal cracking, the catalytic cracking process (1) uses a lower temperature, (2) uses a lower pressure, (3) is more flexible, (4) and the reaction mechanism is controlled by the catalysts. Feedstocks for catalytic cracking include: straight-run gas oil, vacuum gas oil, atmospheric residuum, deasphalted oil, and vacuum residuum. Coke inevitably builds up on the catalyst over time and the issue can be circumvented by continuous replacement of the catalyst or the feedstock pretreated before it is used by deasphalting (removes coke precursors), demetallation (removes nickel and vanadium and prevents catalyst deactivation), or by feedstock hydrotreating (that also prevents excessive coke formation).

The catalyst, which may be an activated natural or synthetic material, is employed in bead, pellet, or microspherical form and can be used as a fixed bed, moving bed, or fluid bed. The fixed-bed process was the first process to be used commercially and uses a static bed of catalyst in several reactors that allows a continuous flow of feedstock to be maintained. Thus, the cycle of operations comprises (1) flow of feedstock through the catalyst bed, (2) discontinuance of feedstock flow and removal of coke from the catalyst by burning, and (3) insertion of the reactor on stream. The moving-bed process uses a reaction vessel (in which cracking takes place) and a kiln (in which the spent catalyst is regenerated) and catalyst movement between the vessels is provided by various means.

The fluid-bed process differs from the fixed-bed and moving-bed processes, insofar as the powdered catalyst is circulated essentially as a fluid with the feedstock. The several fluid catalytic cracking processes in use differ primarily in mechanical design. Side-by-side reactor-regenerator construction along with unitary vessel construction (the reactor either above or below the regenerator) are the two main mechanical variations.

Natural clays have long been known to exert a catalytic influence on the cracking of oils, but it was not until about 1936 that the process using silica-alumina catalysts was developed sufficiently for commercial use. Since then, catalytic cracking has progressively supplanted thermal cracking as the most advantageous means of converting distillate oils into gasoline. The main reason for the wide adoption of catalytic cracking is the fact that a better yield of higher-octane gasoline can be obtained than by any known thermal operation. At the same time the gas produced mostly consists of propane and butane with less methane and ethane. The production of heavy oils and tars, higher in molecular weight than the charge material, is also minimized, and both the gasoline and the uncracked *cycle oil* are more saturated than the products of thermal cracking.

Coking units convert heavy feedstock into a solid coke and lower boiling hydrocarbon products that are suitable as feedstock to other refinery units for conversion into higher value transportation fuels. From a chemical reaction viewpoint, coking can be considered as a severe thermal cracking process in which (unlike visbreaking, q.v.) the reactions are allowed to proceed to completion. As coke is a by-product of this process, the sulfur and metal contents of the coke are high (sometimes as high as 8 percent by weight).

There are three major coking processes: (1) delayed coking, (2) fluid coking, and (3) flexicoking.

Delayed coking (Fig. 13.6) involves the use of an insulated surge drum to accommodate the heater effluent that allows sufficient time for the coking reactions to proceed to completion. Delayed coking is a semi-continuous continuous, where as feed is introduced continuously into one of a pair of coking drums, the other is off-stream undergoing coke removal. Coke is removed from the drums by using high-pressure water jets (3000 psi) to bore a hole through the center of coke, and then inserting a probe to use high-pressure water jets to cut the coke radially. Feedstocks for delayed cokers include atmospheric residua, vacuum residua, heavy oils, and tar sand bitumen.

The *fluid coking* process (Fig. 13.7) is a continuous process in which coke is also made but only enough coke is burned to satisfy the heat requirements of the reactor and the feed preheating operations. The process is also (like delayed coking) a residuum conversion process that

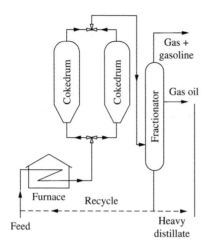

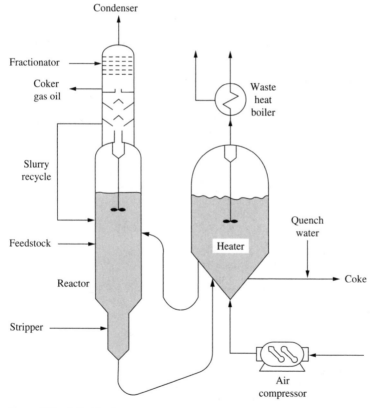

Figure 13.6 A delayed coker.

Figure 13.7 A fluid coker.

uses noncatalytic, thermal chemistry to achieve high conversion levels with even the heaviest refinery feedstock. Because a majority of the sulfur, nitrogen, and metals are rejected with the coke, the full-range of lighter products makes adequate feed for a fluid catalytic cracking unit. The process avoids process load swings and frequent thermal cycles that are typical of batch processes such as delayed coking. The configurations available with fluid coking include extinction recycle, once through, and once through with hydrocyclones.

The flexicoking process is an adaptation of the fluid coking process that uses the same reactor as a fluid coker but has an integrated gasification unit available for coke gasification to produce, in addition to the typical fluid coking slate of products, a low-BTU gas.

Visbreaking is a thermal cracking process in which the thermal reactions are not allowed to proceed to completion. It is used for partial conversion of the higher boiling fractions, usually the residua. Visbreaking, unlike the coking processes (Table 13.3), is a relatively mild thermal cracking operation mainly used to reduce the viscosity and pour point

TABLE 13.3   Comparison of Visbreaking with Delayed Coking and Fluid Coking

Visbreaking
Purpose: to reduce viscosity of fuel oil to acceptable levels
Conversion is not a prime purpose
Mild (470–495°C; 880–920°F) heating at pressures of 50–200 psi
   Reactions quenched before going to completion
   Low conversion (10%) to products boiling less than 220°C (430°F)
Heated coil or drum (soaker)

Delayed Coking
Purpose: to produce maximum yields of distillate products
Moderate (480–515°C; 900–960°F) heating at pressures of 90 psi
   Reactions allowed to proceed to completion
   Complete conversion of the feedstock
Soak drums (845–900°F) used in pairs (one on-stream and one off-stream being de-coked)
Coked until drum solid
Coke removed hydraulically from off-stream drum
Coke yield: 20–40% by weight (dependent upon feedstock)
Yield of distillate boiling below 220°C (430°F): ca. 30% (but feedstock dependent)

Fluid Coking
Purpose: to produce maximum yields of distillate products
Severe (480–565°C; 900–1050°F) heating at pressures of 10 psi
   Reactions allowed to proceed to completion
   Complete conversion of the feedstock
Oil contacts refractory coke
Bed fluidized with steam; heat dissipated throughout the fluid bed
Higher yields of light ends (<C_5) than delayed coking
Less coke made than delayed coking (for one particular feedstock)

of a vacuum residuum (vacuum tower bottoms) to meet No. 6 fuel oil specifications, or to reduce the amount of cutting stock required to dilute the residua to meet these specifications. Long paraffinic side chains attached to aromatic rings are the primary cause of high pour points and viscosities for paraffinic base residues. Visbreaking is carried out at conditions to optimize the breaking off of these long side chains and their subsequent cracking to shorter molecules with lower viscosities and pour point.

There are two types of visbreaker operations (time and temperature dependent): (1) coil cracking, and (2) soaker cracking. Coil cracking uses a higher furnace outlet temperature (475–500°C; 885–930°F) and reaction times from 1–3 min. Run times of 3–6 months are common before it shuts down for the cleanup of coke deposits. Soaker cracking (Fig. 13.8) may use a lower furnace outlet temperature (425–445°C; 800–830°F) and longer reaction times (from 5–8 min). The process has the advantages of lower energy consumption and longer run times of 6–18 months before having to shut down to remove coke deposits from the furnace tube.

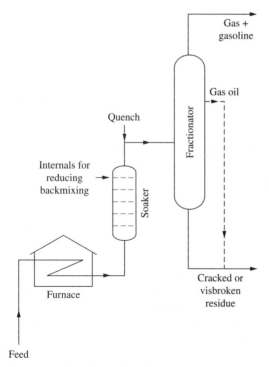

Figure 13.8   A soaker visbreaker.

TABLE 13.4   Summary of Hydrocracking Process Operations

Conditions
Solid acid catalyst (silica-alumina with rare earth metals, various other options)
Temperature: 260–450°C (500–845°F (solid or liquid contact)
Pressure: 1000–6000 psi hydrogen
Frequent catalyst renewal for heavier feedstocks
Gas oil: catalyst life up to three years
Heavy oil or tar sand bitumen: catalyst life less than 1 year

Feedstocks
Refractory (aromatic) streams
Coker oils
Cycle oils
Gas oils
Residua (as a full hydrocracking or hydrotreating option)
  In some cases, asphaltic constituents (S, N, and metals) removed by deasphalting

Products
Lower molecular weight paraffins
Some methane, ethane, propane, and butane
Hydrocarbon distillates (full range depending on the feedstock)
Residual tar (recycle)
Contaminants (asphaltic constituents) deposited on the catalyst as coke or metals

Variations
Fixed bed (suitable for liquid feedstocks
Ebullating bed (suitable for heavy feedstocks)

Hydroprocesses use the principle that the presence of hydrogen during a thermal reaction of a petroleum feedstock will terminate many of the coke-forming reactions and enhance the yields of the lower-boiling components such as gasoline, kerosene, and jet fuel.

Hydrogenation processes for the conversion of petroleum fractions and petroleum products are classified as *destructive* and *nondestructive* (Table 13.4). Destructive hydrogenation (hydrogenolysis or hydrocracking) is characterized by the conversion of the higher molecular weight constituents in a feedstock to lower-boiling products. Such treatment requires severe processing conditions and the use of high hydrogen pressures to minimize polymerization and condensation reactions that lead to coke formation. Nondestructive or simple hydrogenation is generally used for the purpose of improving product quality without appreciable alteration of the boiling range. Mild processing conditions are employed so that only the more unstable materials are attacked. Nitrogen, sulfur, and oxygen compounds undergo reaction with the hydrogen to remove ammonia, hydrogen sulfide, and water, respectively. Unstable compounds that might lead to the formation of gums, or insoluble materials, are converted to more stable compounds.

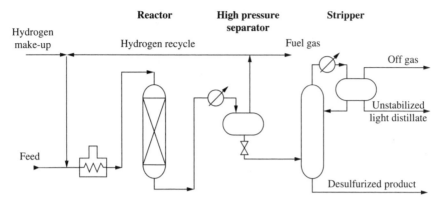

Figure 13.9   A distillate hydrotreater for hydrodesulfurization.

Hydrotreating (Fig. 13.9) is carried out by charging the feed to the reactor, together with hydrogen in the presence of catalysts such as tungsten-nickel sulfide, cobalt-molybdenum-alumina, nickel oxide-silica-alumina, and platinum-alumina. Most processes employ cobalt-molybdena catalysts that generally contain about 10 percent of molybdenum oxide and less than 1 percent of cobalt oxide supported on alumina. The temperatures employed are in the range of 300–345°C (570–655°F), while the hydrogen pressures are about 500–1000 psi (Scott and Bridge, 1971).

The reaction generally takes place in the vapor phase but, depending on the application, may be a mixed-phase reaction. Generally it is more economical to hydrotreat high-sulfur feedstocks prior to catalytic cracking than to hydrotreat the products from catalytic cracking. The advantages are that: (1) sulfur is removed from the catalytic cracking feedstock, and corrosion is reduced in the cracking unit; (2) carbon formation during cracking is reduced so that higher conversions result; and (3) the cracking quality of the gas oil fraction is improved.

Hydrocracking (Fig. 13.10) is similar to catalytic cracking, with hydrogenation superimposed and with the reactions taking place either simultaneously or sequentially. Hydrocracking was initially used to upgrade low-value distillate feedstocks, such as cycle oils (high aromatic products from a catalytic cracker, which are usually not recycled to extinction for economic reasons), thermal and coker gas oils, and heavy-cracked and straight-run naphtha. These feedstocks are difficult to process by either catalytic cracking or reforming, because they are characterized usually by a high polycyclic aromatic content, or by high concentrations, or by both the two principal catalyst poisons—sulfur and nitrogen compounds.

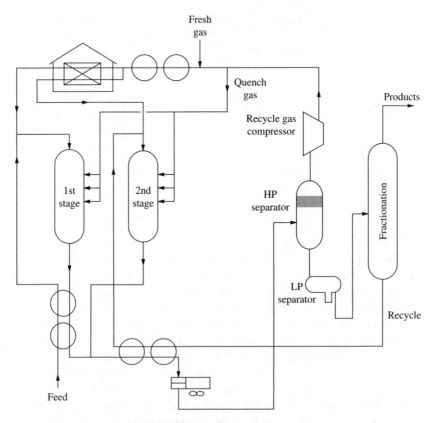

Figure 13.10   A two-stage hydrocracking unit.

The older hydrogenolysis type of hydrocracking practiced in Europe during, and after, World War II used tungsten or molybdenum sulfides as catalysts and required high reaction temperatures and operating pressures, sometimes in excess of about 3000 psi (203 atmospheres), for continuous operation. The modern hydrocracking processes were initially developed for converting refractory feedstocks (such as gas oils) to gasoline and jet fuel, but process and catalyst improvements and modifications have made it possible to yield products from gases and naphtha to furnace oils and catalytic cracking feedstocks.

A comparison of hydrocracking with hydrotreating is useful in assessing the parts played by these two processes in refinery operations. Hydrotreating of distillates may be defined simply as the removal of nitrogen-, sulfur-, and oxygen-containing compounds by selective hydrogenation. The hydrotreating catalysts are usually cobalt plus molybdenum or nickel plus molybdenum (in the sulfide form) impregnated on an alumina base. The hydrotreated operating conditions are such that appreciable hydrogenation of aromatics will not occur—1000–2000 psi hydrogen and about 370°C (700°F). The desulfurization reactions are usually accompanied by small amounts of hydrogenation and hydrocracking.

The commercial processes for treating, or finishing, petroleum fractions with hydrogen all operate in essentially the same manner. The feedstock is heated and passed with hydrogen gas through a tower or reactor filled with catalyst pellets. The reactor is maintained at a temperature of 260–425°C (500–800°F) at pressures from 100–1000 psi, depending on the particular process, the nature of the feedstock and the degree of hydrogenation required. After leaving the reactor, excess hydrogen is separated from the treated product and recycled through the reactor after the removal of hydrogen sulfide. The liquid product is passed into a stripping tower where steam removes dissolved hydrogen and hydrogen sulfide and, after cooling, the product is taken to product storage or, in the case of feedstock preparation, pumped to the next processing unit.

Reforming processes are used to change the inherent chemical structures of the hydrocarbons that exist in distillation fractions crude oil into different compounds. Catalytic reforming (Fig. 13.11) is one of the most important processes in a modern refinery, altering straight-run fraction or fractions from a catalytic cracker into new compounds through a combination of heat and pressure in the presence of a catalyst.

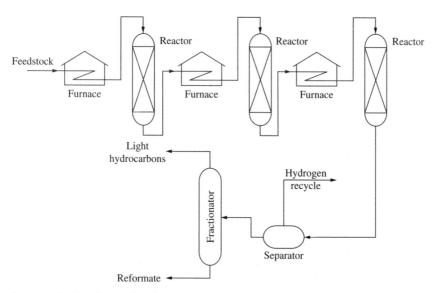

Figure 13.11   Catalytic reforming.

Reforming processes are particularly important in producing high-quality gasoline fuels. Reforming processes are classified as continuous, cyclic, or semi-regenerative, depending upon the frequency of catalyst regeneration.

In carrying out thermal reforming, a feedstock such as 205°C (400°F) end-point naphtha or a straight-run gasoline is heated to 510–595°C (950–1100°F) in a furnace, much the same as a cracking furnace, with pressures from 400–1000 psi (27–68 atmospheres). As the heated naphtha leaves the furnace, it is cooled or quenched by the addition of cold naphtha. The material then enters a fractional distillation tower where any heavy products are separated. The remainder of the reformed material leaves the top of the tower to be separated into gases and reformate. The higher octane of the reformate is due primarily to the cracking of longer-chain paraffins into higher-octane olefins.

The products of thermal reforming are gases, gasoline, and residual oil or tar, the latter being formed in very small amounts (about 1 percent). The amount and quality of the gasoline, known as reformate, is very dependent on the temperature. A general rule is: the higher the reforming temperature, the higher the octane number, but the lower the yield of reformate.

Thermal reforming is less effective and less economical than catalytic processes and has been largely supplanted. As it used to be practiced, a single-pass operation was employed at temperatures in the range of 540–760°C (1000–1140°F) and pressures of about 500–1000 psi (34–68 atmospheres). The degree of octane number improvement depended on the extent of conversion but was not directly proportional to the extent of crack per pass. However, at very high conversions, the production of coke and gas became prohibitively high. The gases produced were generally olefinic; and the process required either a separate gas polymerization operation or one in which C_3 to C_4 gases were added back to the reforming system.

More recent modifications of the thermal reforming process because of the inclusion of hydrocarbon gases with the feedstock are known as gas reversion and polyforming. Thus, olefinic gases produced by cracking and reforming can be converted into liquids boiling in the gasoline range by heating them under high pressure. As the resulting liquids (polymers) have high octane numbers, they increase the overall quantity and quality of gasoline produced in a refinery.

Like thermal reforming, catalytic reforming converts low-octane gasoline into high-octane gasoline (reformate). Whereas thermal reforming could produce reformate with research octane numbers of 65 to 80 depending on the yield, catalytic reforming produces reformate with octane numbers on the order of 90 to 95. Catalytic reforming is conducted in the presence of hydrogen over hydrogenation-dehydrogenation catalysts, which may be supported on alumina or silica-alumina. Depending on the catalyst, a definite sequence of reactions takes place, involving structural changes in the feedstock. This more modern concept actually rendered thermal reforming somewhat obsolescent.

The commercial processes available for use can be broadly classified as the moving-bed, fluid-bed, and fixed-bed types. The fluid- and moving-bed processes use mixed nonprecious metal oxide catalysts in units equipped with separate regeneration facilities. Fixed-bed processes use predominantly platinum-containing catalysts in units equipped for cycle, occasional, or no regeneration.

Catalytic reformer feeds are saturated (i.e., not olefinic) materials; in the majority of cases that feed may be a straight-run naphtha but in others by-product low-octane naphtha (e.g., coker naphtha) can be processed after treatment to remove olefins and other contaminants. Hydrocracker naphtha that contains substantial quantities of naphthenes is also a suitable feed.

Dehydrogenation is a main chemical reaction in catalytic reforming, and hydrogen gas is consequently produced in large quantities. The hydrogen is recycled though the reactors where the reforming takes place to provide the atmosphere necessary for the chemical reactions and also prevents the carbon from being deposited on the catalyst, thus extending its operating life. An excess of hydrogen above whatever is consumed in the process is produced, and, as a result, catalytic reforming processes are unique in that they are the only petroleum refinery processes to produce hydrogen as a by-product.

Catalytic reforming usually is carried out by feeding a naphtha (after pretreating with hydrogen if necessary) and hydrogen mixture to a furnace where the mixture is heated to the desired temperature, 450–520°C (840–965°F), and then passed through fixed-bed catalytic reactors at hydrogen pressures of 100–1000 psi (7–68 atmospheres). Normally pairs of reactors are used in series and heaters are located between adjoining reactors to compensate for the endothermic reactions taking place. Sometimes as many as four or five reactors are kept on-stream in series while one or more being regenerated.

The on-stream cycle of any one reactor may vary from several hours to many days, depending on the feedstock and reaction conditions.

The composition of a reforming catalyst is dictated by the composition of the feedstock and the desired reformate. The catalysts used are principally molybdena-alumina, chromia-alumina, or platinum on a silica-alumina or alumina base. The nonplatinum catalysts are widely used in regenerative process for feeds containing, for example, sulfur, which poisons platinum catalysts, although pretreatment processes (e.g., hydrodesulfurization) may permit platinum catalysts to be employed.

The purpose of platinum on the catalyst is to promote dehydrogenation and hydrogenation reactions, that is, the production of aromatics, participation in hydrocracking, and rapid hydrogenation of carbon-forming precursors. For the catalyst to have an activity for isomeriza-tion of both paraffins and naphthenes—the initial cracking step of hydrocracking—and to participate in paraffin dehydrocyclization, it must have an acid activity. The balance between these two activities is most important in a reforming catalyst. In fact, in the production of aro-matics from cyclic saturated materials (naphthenes), it is important that hydrocracking be minimized to avoid loss of the desired product and, thus, the catalytic activity must be moderated relative to the case of gasoline production from a paraffinic feed, where dehydrocyclization and hydrocracking play an important part.

Other processes to maximize the production of gasoline products include *isomerization* (Fig. 13.12) that is used for reforming or recom-bining lighter cuts into new products, *polymerization* (not in the true sense of the word) processes (Fig. 13.13), and *alkylation* processes (Fig. 13.14).

Present *isomerization* applications in petroleum refining are used with the objective of providing additional feedstock for alkylation units or high-octane fractions for gasoline blending. Straight-chain paraffins (n-butane, n-pentane, n-hexane) are converted to respective iso-hydrocarbons by continuous catalytic (aluminum chloride, noble metals) processes. Natural gasoline or light straight-run gasoline can provide feed by first fractionating as a preparatory step. High volumetric yields

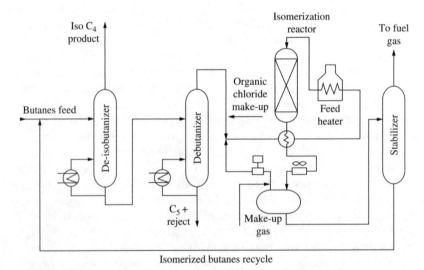

Figure 13.12 A butane isomerization unit

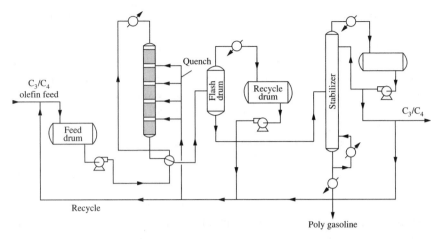

Figure 13.13   A polymerization unit.

(>95 percent) and 40–60 percent conversion per pass are characteristic of the isomerization reaction.

During World War II aluminum chloride was the catalyst used to isomerize butane, pentane, and hexane. Since then, supported metal catalysts have been developed for use in high-temperature processes, which operate in the range 370–480°C (700–900°F) and 300–750 psi (20–51 atmospheres),

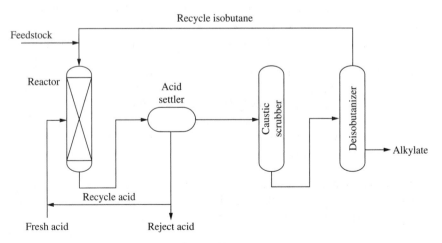

Figure 13.14   An alkylation unit (sulfuric acid catalyst).

while aluminum chloride plus hydrogen chloride are universally used for the low-temperature processes.

A nonregenerable aluminum chloride catalyst is employed with various carriers in a fixed-bed or liquid contactor. Platinum or other metal catalyst processes use fixed-bed operations and can be regenerable or nonregenerable. The reaction conditions vary widely depending on the particular process and feedstock, 40–480°C (100–900°F) and 150–1000 psi (10–68 atmospheres).

The purpose of the polymerization reaction is to polymerize low molecular weight olefins to form a high-octane product boiling in the gasoline boiling range. Phosphoric acid on an inert support is used as the catalyst but sulfur in the feedstock poisons the catalyst and basic compounds neutralize the acid. Oxygen dissolved in the feedstock strongly affects the reactions and must be removed. Propene and butene are used as feedstock and the reaction is highly exothermic, hence, the temperature is controlled either by propane quench or by generating steam.

Polymerization may be accomplished thermally or in the presence of a catalyst at lower temperatures. Thermal polymerization is regarded as not being as effective as catalytic polymerization but has the advantage that it can be used to *polymerize* saturated materials that cannot be induced to react by catalysts. The process comprises vapor-phase cracking of, for example, propane and butane followed by prolonged periods at high temperature (510–595°C, 950–1100°F) for the reactions to proceed to near completion.

Olefins can also be conveniently polymerized by means of an acid catalyst. Thus, the treated, olefin-rich feed stream is contacted with a catalyst (sulfuric acid, copper pyrophosphate, phosphoric acid) at 150 to 220°C (300–425°F) and 150–1200 psi (10–81 atmospheres), depending on the feedstock and product requirement.

Phosphates are the principal catalysts used in polymerization units; the commercially used catalysts are liquid phosphoric acid, phosphoric acid on kieselguhr, copper pyrophosphate pellets, and phosphoric acid film on quartz. The latter is the least active, but the most used and the easiest one to regenerate simply by washing and recoating; the serious disadvantage is that tar must occasionally be burned off the support. The process using the liquid phosphoric acid catalyst is far more responsible for attempts to raise production by increasing temperature compared to other processes.

The purpose of the alkylation reaction is to produce an iso-paraffin by the reaction of low molecular weight olefin with an iso-paraffin to produce constituents for high-octane gasoline. By proper choice of operating conditions, most of the product can be made to fall within the gasoline boiling range. Although alkylation is possible without catalysts, commercial processes use aluminum chloride, sulfuric acid, or hydrogen fluoride as catalysts, when the reactions can take place at low temperatures, minimizing undesirable side reactions, such as polymerization of olefins.

Alkylate is composed of a mixture of iso-paraffins, which have octane numbers that vary with the olefins from which they were made. Butenes produce the highest octane numbers, propene the lowest, and pentenes the intermediate values. All alkylates, however, have high octane numbers (>87) that make them particularly valuable.

The alkylation reaction as now practiced in petroleum refining is the union, through the agency of a catalyst, of an olefin (ethylene, propene, butene, and pentene) with iso-butane to yield high-octane branched-chain hydrocarbons in the gasoline boiling range. Olefin feedstock is derived from the gas produced in a catalytic cracker, whereas iso-butane is recovered by refinery gases or produced by catalytic butane isomerization.

To accomplish this either ethylene or propene is combined with iso-butane at 50–20°C (125–450°F) and 300–1000 psi (20–68 atmospheres) in the presence of metal halide catalysts such as aluminum chloride. Conditions are less stringent in catalytic alkylation; olefins (propene, butene, or pentene) are combined with iso-butane in the presence of an acid catalyst (sulfuric acid or hydrofluoric acid) at low temperatures and pressures (1–40°C, 30–105°F, and 14.8–150 psi, 1–10 atmospheres).

Sulfuric acid, hydrogen fluoride, and aluminum chloride are the general catalysts used commercially. Sulfuric acid is used with propene and higher-boiling feeds, but not with ethylene, because it reacts to form ethyl hydrogen sulfate. The acid is pumped through the reactor and forms an air emulsion with reactants, and the emulsion is maintained at 50 percent acid. The rate of deactivation varies with the feed and iso-butane charge rate. Butene feeds cause less acid consumption than the propene feeds.

Aluminum chloride is not widely used as an alkylation catalyst but when employed, hydrogen chloride is used as a promoter and water is injected to activate the catalyst as an aluminum chloride or hydrocarbon complex. Hydrogen fluoride is used for alkylation of higher-boiling olefins and the advantage of hydrogen fluoride is that it is more readily separated and recovered from the resulting product.

The most important variables in the process are: (1) reaction temperature, (2) acid strength, (3) iso-butane concentration, and (4) olefin space velocity.

13.6 Treating Processes

Petroleum contains impurities that have to be removed at different stages of the refinery. Some of the impurities can be harmful to the refinery processes, corrosive, odiferous, or degrade the final product unless removed. The most common and significant nonhydrocarbon impurity in petroleum is sulfur and the sulfur content can affect the refinery configuration. Strict pollution control in many markets also makes it necessary for refiners to be careful of the level of sulfur in final products such as gasoline, diesel fuel, and fuel oil. The removal of sulfur through desulfurization processes can add significantly to refinery costs both financially and in terms of energy consumption.

When the sulfur content of the crude oil is low (usually less than one percent by weight), the crude oil is known as a *sweet crude*, while crude oil with higher concentrations of sulfur is called *sour crude*. Removal of sulfur and other impurities form part of the treating processes and sulfur itself can form a valuable by-product in a refinery as an input into the chemical industry. Other impurities include nitrogen, oxygen, and salt, as well as small quantities of metals such as vanadium and nickel that are common in certain of the heavier crude oils. As well as extraction processes to purify oil and its products of impurities, specific additives are also used to react with corrosive or odiferous constituents to produce harmless and odorless substances. Such processes are generally termed *sweetening processes*.

Another process that is used for treating is the solvent dewaxing process (Fig. 13.15) that accepts lubricating oil and, on occasion, deasphalted oil as the feedstock.

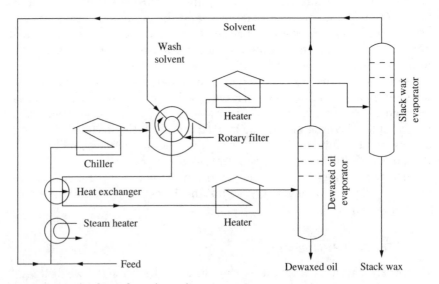

Figure 13.15   A solvent dewaxing unit.

In solvent dewaxing, the oil is diluted with a solvent that has a high affinity for oil, chilled to precipitate the wax, filtered to remove the wax, stripped of solvent, and dried. The solvents (principally propane, naphtha, methyl ethyl ketone-MEK) act as diluents for the high molecular weight oil fractions to reduce the viscosity of the mixture and provide sufficient liquid volume to permit pumping and filtering. Wax produced by the solvent dewaxing process is used to make: (1) paraffins for candle wax, (2) microwax for cosmetics, and (3) wax for petroleum jelly.

Catalytic dewaxing is a process in which the chemical composition of the feed is changed. The process involves catalytic cracking of long paraffin chains into shorter chains to remove the wax and produces lower molecular weigh products suitable for other uses. As an example, the feedstock is contacted with hydrogen at elevated temperature and pressure over a catalyst (such as a zeolite) that selectively cracks the normal paraffins to methane, ethane, and propane. This process is also known as hydrodewaxing. There are two types of catalytic dewaxing: (1) single catalyst process that is used for pour point reduction and to improve the oxygen stability of the product, and (2) a two-catalyst process that uses a fixed-bed reactor, and essentially no methane or ethane is formed in the reaction.

13.7   Petroleum Products

A petroleum refinery produces a wide variety of products; the types of products and the respective volumes are dependent on the nature (properties and composition) of the petroleum feedstock and also on the configuration of the refinery. The products vary from gases to solids and the amount of each produced is dependent upon the requirements of the market.

13.8   Fuel Gas (Refinery Gas) and Liquefied Petroleum Gas

Fuel gas or refinery gas is produced in considerable quantities during the different refining processes and is used as fuel for the refinery itself and as an important feedstock for the petrochemical industry.

Liquid petroleum gas (LPG) is frequently used as domestic bottled gas for cooking and heating, and forms an important feedstock for the petrochemical industry. It is also used in industry for cutting metals.

13.9   Gasoline

Gasoline (also known as *petrol* in many parts of the world) is one of the more important refinery products. The final gasoline product as a

TABLE 13.5   Streams for Gasoline

Stream	Producing process	Boiling range	
		°C	°F
Paraffinic			
Butane	Distillation	0	32
	Conversion		
Iso-pentane	Distillation	27	81
	Conversion		
	Isomerization		
Alkylate	Alkylation	40–150	105–300
Isomerate	Isomerization	40–70	105–160
Naphtha	Distillation	30–100	85–212
Hydrocrackate	Hydrocracking	40–200	105–390
Olefinic			
Catalytic naphtha	Catalytic cracking	40–200	105–390
Cracked naphtha	Steam cracking	40–200	105–390
Polymer	Polymerization	60–200	140–390
Aromatic			
Catalytic reformate	Catalytic reforming	40–200	105–390

transport fuel is a carefully blended mixture having a predetermined octane value (Table 13.5).

Gasoline is a complex mixture of hydrocarbons that boils below 200°C (390°F). The hydrocarbon constituents in this boiling range are those that have four to twelve carbon atoms in their molecular structure. Gasoline varies widely in composition, even those with the same octane number may be quite different. For example, low-boiling distillates with high aromatics content (above 20 percent) can be obtained from some crude oils. The variation in aromatics content as well as the variation in the content of normal paraffins, branched paraffins, cyclopentanes, and cyclohexanes all involve characteristics of any one individual crude oil and influence the octane number of the gasoline.

Because of the differences in composition of various gasolines, gasoline blending is necessary. The physical process of blending the components is simple but determination of how much of each component to include in a blend is much more difficult. The operation is carried out by simultaneously pumping all the components of a gasoline blend into a pipeline that leads to the gasoline storage, but the pumps must be set to deliver automatically the proper proportion of each component. Baffles in the pipeline are often used to mix the components as they travel to the storage tank.

Octane number is the important property in gasoline. Octane number measures the resistance offered by the fuel to compression. A higher octane implies higher resistance to compression (to promote high antiknock quality, ease of starting, quick warm-up, low tendency to vapor lock, and low engine deposits. High-octane constituents, necessary for modern engines, are produced during the second stage of the refining processes. The gasoline group of products also includes special solvents and industrial spirits used in paints dry-cleaning and rubber solvents.

The objective of *product blending* is to allocate the available blending components in such a way as to meet product demands and specifications. In the blending process, product streams from other units are collected and blended to produce the desired product. For example, in gasoline blending, product streams from hydrotreating units, reforming units, polymerization units, and alkylation units are blended to produce specification gasoline.

Aviation gasoline, usually found in use in light aircraft and older civil aircraft, have narrower boiling ranges than conventional (automobile) gasoline, that is, 38–170°C (100–340°F), compared to −1 to 200°C (30–390°F) for automobile gasoline. The narrower boiling range ensures better distribution of the vaporized fuel through the more complicated induction systems of aircraft engines. As aircraft operate at altitudes where the prevailing pressure is less than the pressure at the surface of the earth (pressure at 17,500 ft is 7.5 psi [0.5 atmosphere] compared to 14.8 psi [1.0 atmosphere] at the surface of the earth), the vapor pressure of aviation gasoline must be limited to reduce boiling in the tanks, fuel lines, and carburetors.

13.10 Solvents

Petroleum naphthas have been available since the early days of the petroleum industry. They are valuable as solvents because of their non-poisonous character and good dissolving power. The wide range of naphtha available, and the varying degree of volatility possible, offer products suitable for many uses.

Petroleum naphtha is a generic term, which is applied to refined, partly refined, or unrefined petroleum products. Naphthas are prepared by any one of several methods including (1) fractionation of distillates or even crude petroleum, (2) solvent extraction, (3) hydrogenation of distillates, (4) polymerization of unsaturated (olefinic) compounds, and (5) alkylation processes. The naphtha may also be a combination of product streams from more than one of these processes.

The main uses of petroleum naphtha fall into the general areas of (1) solvents (diluents) for paints, , (2) dry-cleaning solvents, (3) solvents for cutback asphalts, (4) solvents in rubber industry, and (5) solvents for industrial extraction processes. Turpentine, the older and more conventional solvent for paints, has now been almost completely replaced by the cheaper and more abundant petroleum naphtha.

13.11   Kerosene

Kerosene was the major refinery product before the onset of the *automobile age*, but now kerosene might be termed as one of several other petroleum products after gasoline. Kerosene originated as a straight-run (distilled) petroleum fraction that boiled between approximately 205–260°C (400–500°F). In the early days of petroleum refining some crude oils contained kerosene fractions of very high quality, but other crude oils, such as those having a high proportion of asphaltic materials, must be thoroughly refined to remove aromatics and sulfur compounds before a satisfactory kerosene fraction can be obtained.

The kerosene fraction is essentially a distillation fraction of petroleum. The quantity and quality of the kerosene vary with the type of crude oil; some crude oils yield excellent kerosene but others produce kerosene that requires substantial refining. Kerosene is a very stable product, and additives are not required to improve the quality. Apart from the removal of excessive quantities of aromatics, kerosene fractions may need only a lye (alkali) wash if hydrogen sulfide is present.

Diesel fuel also forms part of the kerosene boiling range (or middle distillate group of products). Diesel fuels come in two broad groups, for high-speed engines in cars and trucks requiring a high-quality product, and lower-quality heavier diesel fuel for slower engines, such as in marine engines or for stationary power plants. An important property of diesel is the cetane number (analogous to the gasoline octane number), which determines the ease of ignition under compression.

13.12   Fuel Oil

Fuel oil is classified in several ways but generally may be divided into two main types: distillate fuel oils and residual fuel oils. Distillate fuel oils are vaporized and condensed during a distillation process; they have a definite boiling range and do not contain high-boiling oils or asphaltic components. A fuel oil that contains any amount of the residue from crude distillation or thermal cracking is a residual fuel oil. The terms distillate fuel oil and residual fuel oil are losing their significance, because fuel oils are now made for specific uses and may be distillates, residuals, or mixtures of the two. The terms domestic fuel oils, diesel fuel oils, and heavy fuel oils are more indicative of the uses of fuel oils.

Domestic fuels are those used primarily in the home and include kerosene, stove oil, and furnace fuel oil. Diesel fuel oils are also distillate fuel oils, but residual oils have been successfully used to power marine diesel engines, and mixtures of distillates and residuals have been used on locomotive diesels. Heavy fuel oils include a variety of oils ranging from distillates to residual oils that must be heated to 260°C (500°F) or higher before they can be used. In general, heavy fuel oils consist of residual oils blended with distillates to suit specific needs. Included among heavy fuel oils are various industrial oils; when used to fuel ships, heavy fuel oil is called bunker oil.

Stove oil is a straight-run (distilled) fraction from crude oil whereas other fuel oils are usually blends of two or more fractions. The straight-run fractions available for blending into fuel oils are heavy naphtha, light and heavy gas oils, and residua. Cracked fractions such as light and heavy gas oils from catalytic cracking, cracking coal tar, and fractionator bottoms from catalytic cracking may also be used as blends to meet the specifications of the different fuel oils.

Heavy fuel oils usually contain residuum that is mixed (cut back) to a specified viscosity with gas oils and fractionator bottoms. For some industrial purposes where flames or flue gases contact the product (ceramics, glass, heat treating, open hearth furnaces) fuel oils must be blended to contain minimum sulfur contents; low-sulfur residues are preferable for these fuels.

The manufacture of fuel oils at one time largely involved using what was left after removing the desired products from crude petroleum. Now fuel oil manufacture is a complex matter of selecting and blending various petroleum fractions to meet definite specifications.

Fuel oil that is used for heating is graded from No. 1 Fuel Oil to No. 6 Fuel Oil, and covers light distillate oils, medium distillate, heavy distillate, a blend of distillate and residue, and residue oil. For instance No. 2 and No. 3 Fuel Oils refer to medium-to-light distillate grades used in domestic central heating. Fuel oil refers to a medium heavy residual oil used for heating large commercial premises.

13.13 Lubricating Oil

Lubrication oil comes in a huge variety of products, ranging from very fine fluid lubricants, to heavy viscous oils and greases.

Lubricating oil is distinguished from other fractions of crude oil by a usually high (>400°C, >750°F) boiling point, as well as high viscosity. Materials suitable for the production of lubricating oils principally comprise hydrocarbons containing from 25 to 35 carbon atoms per molecule, whereas residual stocks may contain hydrocarbons with 50 to 80 carbon atoms per molecule.

The development of vacuum distillation provided the means of separating lubricating oil fractions with predetermined viscosity ranges and removed the limit on the maximum viscosity that might be obtained in distillate oil. Vacuum distillation prevented residual asphaltic material from contaminating lubricating oils but did not remove other undesirable materials such as acidic components or components that caused the oil to thicken excessively when cold and become very thin when hot.

Lubricating oil may be divided into many categories according to the types of service it is intended to perform. However, there are two main groups: (1) oils used in intermittent service, such as motor and aviation oils and (2) oils designed for continuous service such as turbine oils.

13.14   Petroleum Wax

Petroleum wax is of two general types: the paraffin waxes in petroleum distillates and the microcrystalline waxes in petroleum residua. The melting point of wax is not directly related to its boiling point, because waxes contain hydrocarbons of different chemical structure. Nevertheless, waxes are graded according to their melting point and oil content.

Paraffin wax is a solid crystalline mixture of straight-chain (normal) hydrocarbons ranging from C20 to C30 and higher. Wax constituents are solid at ordinary temperatures (25°C, 77°F) whereas petrolatum (petroleum jelly) does contain both solid and liquid hydrocarbons.

Wax production by *wax sweating* was originally used in Scotland to separate wax fractions with various melting points from the wax obtained from shale oils. Wax sweating is still used to some extent but is being replaced by the more convenient wax recrystallization process. In wax sweating, a cake of slack wax is slowly warmed to a temperature at which the oil in the wax and the lower-melting waxes become fluid and drip (or sweat) from the bottom of the cake, leaving a residue of higher melting wax.

The amount of oil separated by sweating is now much smaller than it used to be because of the development of highly efficient solvent dewaxing techniques. Wax sweating is now more concerned with the separation of slack wax into fractions with different melting points. A wax sweater consists of a series of about nine shallow pans arranged one above the other in a sweater house or oven, and each pan is divided horizontally by a wire screen. The pan is filled to the level of the screen with cold water. Molten wax is then introduced, allowed to solidify, and the water is then drained from the pan leaving the wax cake supported on the screen. A single sweater oven may contain more than 600 barrels of wax, and steam coils arranged on the walls of the oven slowly heat the wax cakes, allowing oil and the lower-melting waxes to sweat from the cakes and drip into the pans. The first liquid removed from the pans is called *foots* oil which melts at 38°C (100°F) or lower and is followed by *interfoots oil* that melts in the range of 38–44°C (100–112°F). Crude scale wax next drips from the wax cake and consists of wax fractions with melting points over 44°C (112°F).

Wax recrystallization, like wax sweating, separates wax into fractions but, instead of relying upon differences in melting points, the process makes use of the different solubility of the wax fractions in a solvent such as a ketone. When a mixture of ketone and wax is heated, the wax usually dissolves completely, and if the solution is cooled slowly, a temperature is reached at which a crop of wax crystals is formed. These crystals will all be of the same melting point, and if they are removed by filtration, a wax fraction with a specific melting point is obtained. If the clear filtrate is cooled further, a second batch of wax crystals with a lower melting point is obtained. Thus, alternate cooling and filtration can subdivide the wax into a large number of wax fractions, each with different melting points.

This method of producing wax fractions is much faster and more convenient than sweating and results in a much more complete separation of the various fractions. Furthermore, recrystallization can also be applied to the microcrystalline waxes obtained from intermediate and heavy paraffin distillates, which cannot be sweated. Indeed, the microcrystalline waxes have higher melting points and differ in their properties from the paraffin waxes obtained from light paraffin distillates, and thus, wax recrystallization has made new kinds of waxes available.

13.15   Asphalt

The nonvolatile residuum is used to produce road asphalt (sometimes referred to as bitumen) as well as a variety of asphalt grades for roofing and waterproofing. It is produced to certain standards of hardness or softness in controlled vacuum distillation processes. Asphalt is a residuum and cannot be distilled even under the highest vacuum because the temperatures required to volatilize the residuum promote the formation of coke. Asphalts have complex chemical and physical compositions that usually vary with the source of the crude oil.

Asphalt is a product of many petroleum refineries (Barth, 1962) and may be residual asphalt, which is made up of the nonvolatile hydrocarbons in the feedstock, along with similar materials produced by thermal alteration during the distillation sequences, or asphalt may be produced by air-blowing an asphaltic residuum.

Asphalt manufacture is, in essence, a matter of distilling everything possible from crude petroleum until a residue with the desired properties is obtained. This is usually done by stages; crude distillation at atmospheric pressure removes the lower-boiling fractions and yields a reduced crude that may contain higher-boiling (lubricating) oils, asphalt, and even wax. Distillation of the reduced crude under vacuum removes the oils (and wax) as volatile overhead products and the asphalt remains as a bottom (or residual) product. At this stage the asphalt is frequently (and incorrectly) referred to as pitch.

There are wide variations in refinery operations and in the types of crude oils so different asphalts will be produced. Asphalts of intermediate softening points may be made by blending with higher and lower softening point asphalts. If lubricating oils are not required, the reduced crude may be distilled in a flash drum that is similar to a distillation tower but has few, if any, trays. Asphalt descends to the base of the flasher as the volatile components pass out of the top. Asphalt is also produced by propane deasphalting and can be made softer by blending the hard asphalt with the extract obtained in the solvent treatment of lubricating oils. On the other hand, soft asphalts can be converted into harder asphalts by oxidation (air blowing).

Road oils are liquid asphalt materials intended for easy application to earth roads. They provide a strong base or a hard surface and will maintain a satisfactory passage for light traffic. Liquid road oils, cutbacks, and emulsions are of recent date, but use of asphaltic solids for paving goes back to the European practices of the early 1800s.

Cutback asphalts are mixtures in which hard asphalt has been diluted with a lighter oil to permit application as a liquid without drastic heating. They are classified as rapid, medium, and slow curing, depending on the volatility of the diluent, which governs the rate of evaporation and consequent hardening.

An asphaltic material may be emulsified with water to permit application without heating. Such emulsions are normally of the oil-in-water type. They reverse or break on application to a stone or earth surface, so that the oil clings to the stone and the water disappears. In addition to their usefulness in road and soil stabilization, they are useful for paper impregnation and waterproofing. The emulsions are chiefly (1) the soap or alkaline type and (2) the neutral or clay type. The former break readily on contact, but the latter are more stable and probably lose water mainly by evaporation. Good emulsions must be stable during storage or freezing, suitably fluid, and amenable to control for speed of breaking.

Recently, asphalt has grown to be a valuable refinery product. In the post-1980 period, a shortage of good quality asphalt has developed. This is, in no short measure, because of the tendency of refineries to produce as much liquid fuels (e.g., gasoline) as possible. Thus, residua that would have once been used for asphalt manufacture are now being used to produce liquid fuels (and coke).

13.16   Coke

Petroleum coke is the residue left by the destructive distillation of petroleum residua. The coke formed in catalytic cracking operations is usually nonrecoverable because of adherence to the catalyst as it is often employed as fuel for the process.

The composition of coke varies with the source of the crude oil, but in general, large amounts of high molecular weight complex hydrocarbons (rich in carbon but correspondingly poor in hydrogen) make up a high proportion. The solubility of coke in carbon disulfide has been reported to be as high as 50 to 80 percent, but this is, in fact, a misnomer, because the coke is an insoluble, honeycomb-type material that is the end product of thermal processes.

Coke is employed for a number of purposes, but the major use is in the manufacture of carbon electrodes for aluminum refining that requires a high-purity carbon—low in ash and sulfur free. In addition, petroleum coke is employed in the manufacture of carbon brushes, silicon carbide abrasives, and structural carbon (pipes, Rashig rings, and so on), as well as calcium carbide manufacture from which acetylene is produced.

The use of coke as a fuel must proceed with some caution with the acceptance by refiners, of the heavier crude oils as refinery feedstocks. The higher contents of sulfur and nitrogen in these oils mean a product coke containing substantial amounts of sulfur and nitrogen. Both of these elements will produce unacceptable pollutants—sulfur oxides and nitrogen oxides—during combustion. These elements must also be regarded with caution in any coke that is scheduled for electrode manufacture and removal procedures for these elements are continually being developed.

13.17 Petrochemicals

The petrochemical industry began in the 1920s as suitable by-products became available through improvements in the refining processes. It developed in parallel with the oil industry and has expanded rapidly since the 1940s, with the oil refining industry providing plentiful cheap raw materials.

A *petrochemical* is any chemical (as distinct from fuels and petroleum products) manufactured from petroleum (and natural gas) and used for a variety of commercial purposes (Table 13.6). The definition, however, has been broadened to include the whole range of aliphatic, aromatic, and naphthenic organic chemicals, as well as carbon black and inorganic materials such as sulfur and ammonia. Petroleum and natural gas are made up of hydrocarbon molecules, which comprise one or more carbon atoms, to which hydrogen atoms are attached. Currently, oil and gas are the main sources of the raw materials because they are the least expensive, most readily available, and can be processed most easily into the primary petrochemicals (Table 13.7). Primary petrochemicals include: olefins (ethylene, propene, and butadiene), aromatics (benzene, toluene, and the isomers of xylene), and methanol. Thus, petrochemical feedstocks can be classified into three general groups: olefins, aromatics, and methanol; a fourth group includes inorganic compounds and synthesis gas (mixtures of carbon monoxide and hydrogen). In many instances, a

TABLE 13.6 Hydrocarbon Intermediates Used in the Petrochemical Industry

Carbon number	Hydrocarbon type		
	Saturated	Unsaturated	Aromatic
1	Methane		
2	Ethane	Ethylene	
		Acetylene	
3	Propane	Propene	
4	Butanes	n-butenes	
		Isobutene	
		Butadiene	
5	Pentanes	Isopentenes	
		(Iso pentenes)	
		Isoprene	
6	Hexanes	Methylpentenes	Benzene
	Cyclohexane		
7		Mixed heptenes	Toluene
8		di-isobutene	Xylenes
			Ethylbenzene
			Styrene
9			Cumene
12		Propene tetramer	
		tri-isobutene	
18			Dodecylbenzene
6–18		n-olefins	
11–18	n-paraffins		

TABLE 13.7 Sources of Petrochemical Intermediates

Hydrocarbon	Source
Methane	Natural gas
Ethane	Natural gas
Ethylene	Cracking processes
Propane	Natural gas, catalytic reforming, cracking processes
Propene	Cracking processes
Butane	Natural gas, reforming, and cracking processes
Butene(s)	Cracking processes
Cyclohexane	Distillation
Benzene	Catalytic reforming
Toluene	Catalytic reforming
Xylene(s)	Catalytic reforming
Ethylbenzene	Catalytic reforming
Alkylbenzenes	Alkylation
$>C_9$	Polymerization

specific chemical included among the petrochemicals may also be obtained from other sources, such as coal, coke, or vegetable products. For example, materials such as benzene and naphthalene can be made from either petroleum or coal, whereas ethyl alcohol may be of petrochemical or vegetable origin.

As stated above, some of the chemicals and compounds produced in a refinery are destined for further processing and as raw material feedstocks for the fast growing petrochemical industry. Such nonfuel uses of crude oil products are sometimes referred to as its nonenergy uses. Petroleum products and natural gas provide two of the basic starting points for this industry; methane from natural gas, and naphtha and refinery gases.

Petrochemical intermediates are generally produced by chemical conversion of primary petrochemicals to form more complicated derivative products. Petrochemical derivative products can be made in a variety of ways: directly from primary petrochemicals; through intermediate products that still contain only carbon and hydrogen; and through intermediates that incorporate chlorine, nitrogen, or oxygen in the finished derivative. In some cases, they are finished products; in others, more steps are needed to arrive at the desired composition.

Of all the processes used, one of the most important is polymerization. It is used in the production of plastics, fibers, and synthetic rubber, the main finished petrochemical derivatives. Some typical petrochemical intermediates are vinyl acetate for paint, paper and textile coatings, vinyl chloride for polyvinyl chloride (PVC), resin manufacture, ethylene glycol for polyester textile fibers, and styrene that is important in rubber and plastic manufacturing. The end products number in the thousands, some going on as inputs into the chemical industry for further processing. The more common products made from petrochemicals include adhesives, plastics, soaps, detergents, solvents, paints, drugs, fertilizer, pesticides, insecticides, explosives, synthetic fibers, synthetic rubber, flooring, and insulating materials.

Bibliography

Barth, E. J., *Asphalt Science and Technology,* Gordon and Breach, New York, N.Y., 1962.

Bland, W. F. and R. L. Davidson, *Petroleum Processing Handbook.* McGraw-Hill, New York, N.Y., 1967.

Burris, D. R. and J. D. McKinney, *Petroleum Processing Handbook.* J. J. McKetta (Ed.) Marcel Dekker, Inc., New York, N.Y., p. 666, 1992.

Matar, M. S., *Petroleum Economics and Engineering,* 2nd ed. H. K. Abdel-Aal, B. A. Bakr, and M. A. Al-Sahlawi (Eds.). Marcel Dekker, Inc., New York, N.Y., p. 33, 1992.

Mushrush, G. W. and J. G. Speight, *Petroleum Products: Instability and Incompatibility,* Taylor & Francis, Washington, D.C., 1995.

Nelson W. L., *Petroleum Refinery Engineering.* McGraw-Hill, New York, N.Y., 1958.

Scott J. W. and A. G. Bridge, *Origin and Refining of Petroleum*. American Chemical Society, Washington, DC, 1971.
Speight, J. G., *The Chemistry and Technology of Petroleum*, 3rd ed. Marcel Dekker, Inc., New York, N.Y., 1999.
Speight, J. G., *The Desulfurization of Heavy Oils and Residua*, 2nd ed. Marcel Dekker, Inc., New York, N.Y., 2000.
Speight, J. G., *Handbook of Petroleum Analysis*. John Wiley & Sons Inc., New York, 2001.
Speight, J. G. and B. Ozum, *Petroleum Refining Processes*. Marcel Dekker Inc., New York, N.Y., 2002.

Synthetic Polymers

Hasan A. Al-Muallem

14.1 Basic Concepts and Definitions

A polymer is a giant molecule made up of a large number of repeating units joined together by covalent bonds. The simple compounds from which polymers are made are called *monomers*. The word *polymer* is derived from the Greek words *poly* (many) and *meros* (parts). Polymer molecules have molecular weight in the range of several thousands or more, and therefore, are also referred to as *macromolecules*. This is illustrated by the following equation, which shows the formation of the polymer polystyrene.

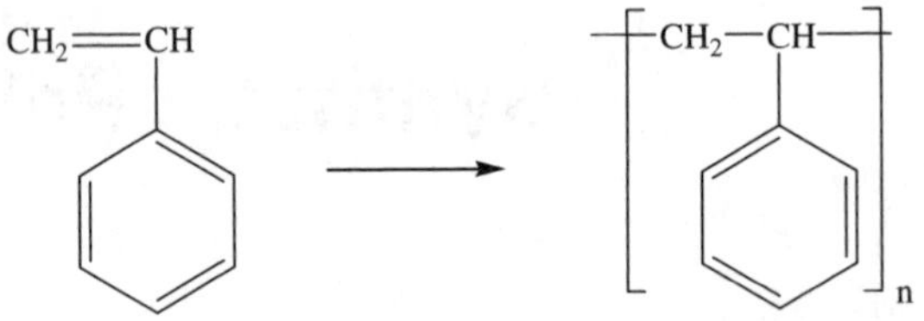

Styrene
(monomer)

Polystyrene
(polymer)

(1)

The styrene molecule is the monomer, and the resulting structure, enclosed in square brackets, is the polymer polystyrene. The unit in square brackets is called the *repeating unit*. Some polymers are derived from the mutual reaction of two or more monomers. For example, poly(hexamethylene adipamide) or nylon-6,6 is made from the reaction of hexamethylenediamine and adipic acid, as shown in the following equation:

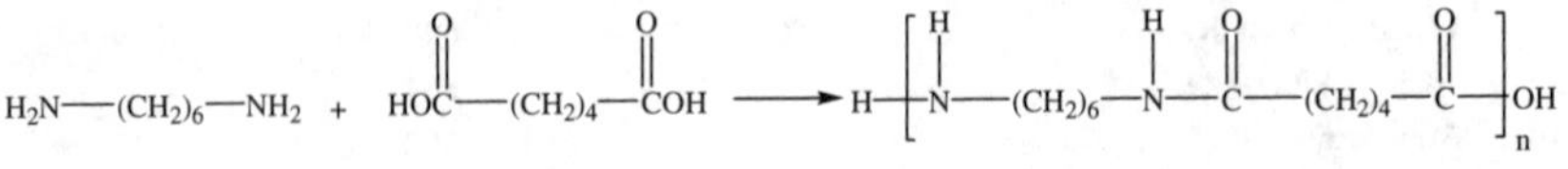

Hexamethylenediamine   Adipic acid   Poly(hexamethylene adipamide)
(nylon-6,6)

(2)

For a molecule to be a monomer, it must be at least bifunctional. The *functionality* of a molecule refers to its interlinking capacity, or the number of sites it has available for bonding with other molecules. Reactions between monofunctional molecules use up the reactive groups completely and render the product incapable of further reactions, whereas the presence of two condensable groups in both hexamethylenediamine ($-NH_2$) and adipic acid ($-COOH$) makes each of these monomers bifunctional with the ability to form polymers. In this respect, styrene is also a bifunctional monomer because the extra pair of electrons in the double bond can form two bonds with vinyl groups in other molecules.

The number of repeating units in the polystyrene structure (1) is indicated by the index n. This is known as the *degree of polymerization* (*DP*). It specifies the length of the polymer chain. *Oligomer* is a very low

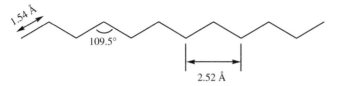

1.54 Å

109.5°

2.52 Å

Figure 14.1   Schematic structure of fully stretched polyethylene.

molecular weight polymer usually with less than 10 repeating units. The word oligomer is derived from the Greek word *oligos* meaning *a few*. Oligomers exhibit different thermal and mechanical properties compared to the corresponding high-molecular weight polymer. It is sometimes useful to prepare oligomers with certain functional groups at the end that can be used in further chemistry. The degree of polymerization represents one way of quantifying the molecular weight or size of a polymer. For example, a linear polyethylene consisting of one thousand ethylene units will have a molecular weight of 28,000, and an extended length of 2520 angstroms (Å) (Fig. 14.1). However, because of rotation of the carbon-carbon bonds, the polymer chains are seldom extended to their full contour length but are present in many different shapes or conformations.

14.2   Classification of Polymers

Polymers can be classified in many ways according to various criteria such as:

a. *Origin of the polymer*: Polymers can be classified as being natural or synthetic based on the origin of the polymer. Certain polymers, such as nucleic acids, proteins, cellulose, natural rubber, wool, and silk are found in nature. Clays, sands, graphite, and diamond are also naturally occurring polymers. On the other hand, thousands of polymers have been synthesized and more are likely to be produced in the future. In some cases, naturally occurring polymers can be produced by synthetic routes. For example, polyisoprene is the synthetic version of natural rubber (Hevea).

b. *Functional groups present in the repeating unit*: In this respect, polymers can be grouped in families like polyesters, polyamides, polyimides, polycarbonates, polyurethanes, polyureas, polyethers, polysulfides, and so on.

c. *Polymer structure*: A polymer can be described as:
   1. Linear, branched, cross-linked, ladder, star-shaped, comb-shaped, dendritic, and the like (Fig. 14.2)

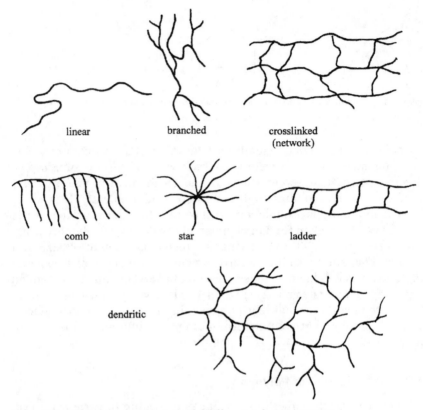

Figure 14.2   Schematic representation of different polymer structures.

2. Amorphous or crystalline based on absence or presence of long-range ordered pattern among polymer chains
3. Homopolymer or copolymer with different types of copolymer
4. Fibers, plastics, or elastomers. Polymers (synthetic or natural) can be divided into various families; fibers, elastomers, plastics, adhesives, and each family itself has subgroups

d. Polymerization mechanism: Based on the polymer-forming reaction; condensation versus addition or step-growth versus chain-growth polymerization reactions

e. Preparative technique: Bulk, solution, suspension, emulsion, or precipitation

f. Thermal behavior: Thermoplastics or thermosets

g. End use: Such as diene polymers (rubber industry); olefin polymers (sheet, film, and fiber industries); and acrylics (coating and decorative materials)

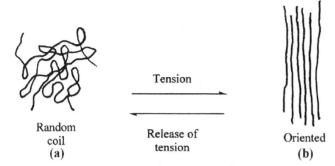

Figure 14.3 Illustration of rubbery elastomeric property. (a) Relaxed: high entropy; (b) Stretched: low entropy.

Elastomers.   Elastomers are polymeric materials with irregular structure and weak intermolecular attractive forces. Elastomers are capable of high extension (up to 1000%) under ambient conditions. That is, they have the particular kind of elasticity characteristic of rubber. The elasticity is attributed to the presence of chemical and/or physical crosslinks in these materials. In their normal state, elastomers are amorphous, and as the material is stretched, the random chains are forced to occupy more ordered positions. Releasing the applied force allows the elongated chains to return to a more random state. Thus, the restoring force after elongation is largely because of entropy. (Fig. 14.3)

In addition to natural rubber, there are synthetic elastomers such as diene elastomers (e.g., polybutadiene, polyisoprene, polychloroprene, and so on,), nondiene elastomers (e.g., polyisobutylene, polysiloxanes, polyurethanes), and nitrile and butyl rubber. Elastomers can also be made from block copolymers containing hard or rigid segments of polyurethane and soft or flexible segments derived from the polyester or polyether diols with degrees of polymerization generally above 15. Polyurethane elastomeric materials exhibit good abrasion resistance, chemical resistance, and good tear strength with a wide variation of flexibility available. These polyurethanes are also used in fabrics and sporting goods. (Fig. 14.4)

Fibers.   A fiber is often defined as an object with a length-to-diameter ratio of at least 100. Fibers (synthetic or natural) are polymers with high molecular symmetry and strong cohesive energies between chains that usually result from the presence of polar groups. Fibers possess a high degree of crystallinity characterized by the presence of stiffening groups in the polymer backbone, and of intermolecular hydrogen bonds. Also, they are characterized by the absence of branching or irregularly spaced

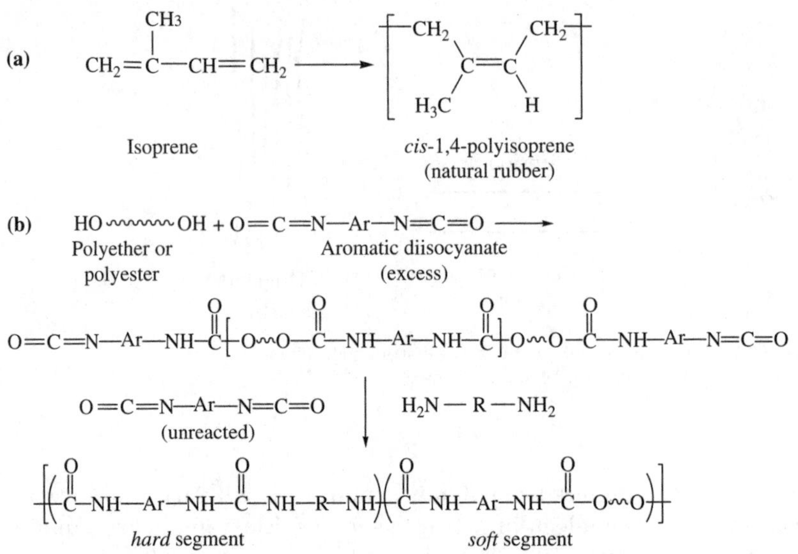

(a)

$$CH_2{=}C{-}CH{=}CH_2 \xrightarrow{\hspace{1cm}}$$

(with CH₃ above the second carbon)

Isoprene

cis-1,4-polyisoprene
(natural rubber)

(b) HO〜〜〜〜OH + O=C=N—Ar—N=C=O ⟶
Polyether or   Aromatic diisocyanate
polyester   (excess)

O=C=N—Ar—NH—C(=O)[O〜O—C(=O)—NH—Ar—NH—C(=O)]O〜O—C(=O)—NH—Ar—N=C=O

O=C=N—Ar—N=C=O   H₂N — R —NH₂
(unreacted)

$$\left[\left(C{-}NH{-}Ar{-}NH{-}C{-}NH{-}R{-}NH\right)\left(C{-}NH{-}Ar{-}NH{-}C{-}O{\sim}O\right)\right]$$

hard segment   soft segment

Figure 14.4 Examples of synthetic elastomers. (a) polyisoprene, and (b) polyurethane elastomer.

pendant groups that will otherwise disrupt the crystalline formation. Fibers are normally linear and drawn in one direction to make them long, thin, and threadlike, with great strength along the fiber. These characteristics permit formation of this type of polymer into long fibers suitable for textile applications. Typical examples of fibers include polyesters, nylons, and acrylic polymers such as polyacrylonitrile, and naturally occurring polymers such as cotton, wool, and silk. (Fig. 14.5)

Plastics. Plastics are the polymeric materials with properties intermediate between elastomers and fibers. In spite of the possible differences in chemical structure, the demarcation between fibers and plastics may sometimes be blurred. Polymers such as polypropylene and polyamides can be used as fibers and plastics by a proper choice of processing conditions. Plastics can be extruded as sheets or pipes, painted on surfaces, or molded to form countless objects. A typical commercial plastic resin may contain two or more polymers in addition to various additives and fillers. Additives and fillers are used to improve some property such as the processability, thermal or environmental stability, and mechanical properties of the final product.

Thermoplastics and thermosets. All polymers can be divided into two major groups (*thermoplastics* and *thermosets*) based on their thermal processing behavior. Thermoplastic polymers soften and flow under the action

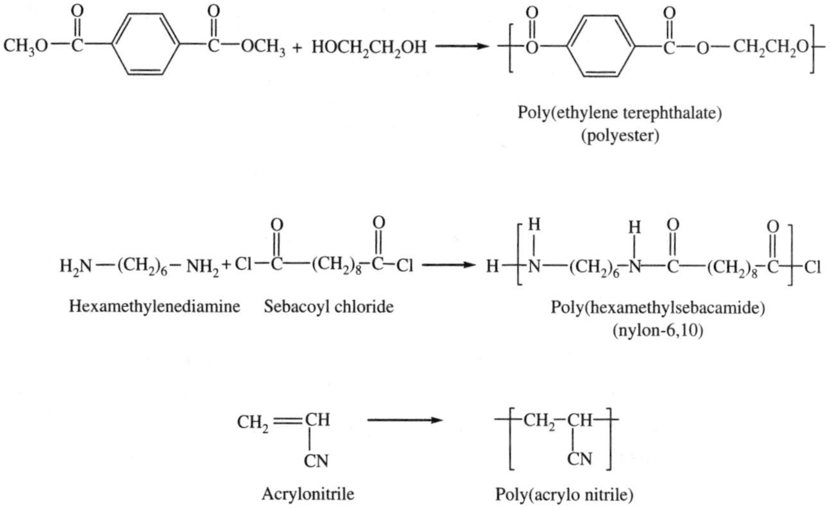

$$CH_3O-\overset{\overset{\displaystyle O}{\|}}{C}-\underset{}{\text{⬡}}-\overset{\overset{\displaystyle O}{\|}}{C}-OCH_3 + HOCH_2CH_2OH \longrightarrow \left[O-\underset{}{\text{⬡}}-\overset{\overset{\displaystyle O}{\|}}{C}-O-CH_2CH_2O \right]$$

Poly(ethylene terephthalate)
(polyester)

$$H_2N-(CH_2)_6-NH_2 + Cl-\overset{\overset{\displaystyle O}{\|}}{C}-(CH_2)_8-\overset{\overset{\displaystyle O}{\|}}{C}-Cl \longrightarrow H \left[\overset{\overset{\displaystyle H}{|}}{N}-(CH_2)_6-\overset{\overset{\displaystyle H}{|}}{N}-\overset{\overset{\displaystyle O}{\|}}{C}-(CH_2)_8-\overset{\overset{\displaystyle O}{\|}}{C} \right] Cl$$

Hexamethylenediamine   Sebacoyl chloride   Poly(hexamethylsebacamide)
(nylon-6,10)

$$CH_2=\underset{\underset{\displaystyle CN}{|}}{CH} \longrightarrow \left[CH_2-\underset{\underset{\displaystyle CN}{|}}{CH} \right]$$

Acrylonitrile   Poly(acrylo nitrile)

Figure 14.5   Examples of synthetic fibers: polyester, nylon and poly(acrylonitrile).

of heat and pressure. Upon cooling, the polymer hardens and assumes the shape of the mold (container). *Thermoplastics,* when compounded with appropriate ingredients, can usually withstand several heating and cooling cycles without suffering any structural breakdown. Examples of commercial thermoplastics are polystyrene, polyolefins (e.g., polyethylene and polypropylene), nylon, poly(vinyl chloride), and poly(ethylene terephthalate) (Fig. 14.6). Thermoplastics are used for a wide range of

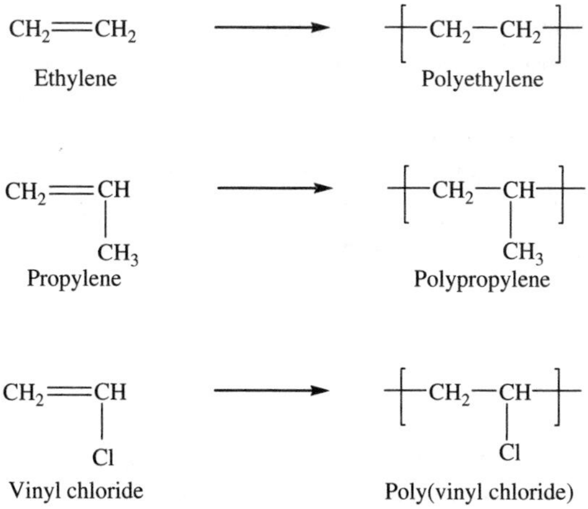

$$CH_2=CH_2 \longrightarrow \left[CH_2-CH_2 \right]$$

Ethylene   Polyethylene

$$CH_2=\underset{\underset{\displaystyle CH_3}{|}}{CH} \longrightarrow \left[CH_2-\underset{\underset{\displaystyle CH_3}{|}}{CH} \right]$$

Propylene   Polypropylene

$$CH_2=\underset{\underset{\displaystyle Cl}{|}}{CH} \longrightarrow \left[CH_2-\underset{\underset{\displaystyle Cl}{|}}{CH} \right]$$

Vinyl chloride   Poly(vinyl chloride)

Figure 14.6   Examples of commercial thermoplastics.

applications, such as film for packaging, photographic, and magnetic tape, beverage and trash containers, and a variety of automotive parts and upholstery. Advantageously, waste thermoplastics can be recovered and refabricated by application of heat and pressure.

Thermosets are polymers whose individual chains have been chemically linked by covalent bonds during polymerization or by subsequent chemical or thermal treatment during fabrication. The thermosets usually exist initially as liquids called prepolymers; they can be shaped into desired forms by the application of heat and pressure. Once formed, these cross-linked networks resist heat softening, creep and solvent attack, and cannot be thermally processed or recycled. Such properties make thermosets suitable materials for composites, coatings, and adhesive applications. Principal examples of thermosets include epoxies, phenol-formaldehyde resins, and unsaturated polyesters. Vulcanized rubber used in the tire industry is also an example of thermosetting polymers. Thermosetting polymers are usually insoluble because the crosslinking causes a tremendous increase in molecular weight. At most, thermosetting polymers only swell in the presence of solvents, as solvent molecules penetrate the network. Examples of the reactions of phenol and formaldehyde to yield phenol-formaldehyde resins are shown in Fig. 14.7.

Properties of a specific polymer can often be varied by means of controlling molecular weight, end groups, processing, cross-linking, additives, and so on. Therefore, it is possible to classify a single polymer in more than one category. For example, nylon can be produced as fibers in the crystalline forms, or as plastics in the less crystalline forms. Also, poly(vinyl chloride) and siloxanes can be processed to act as plastics or elastomers.

Commodity and engineering polymers. On the basis of end use and economic considerations, polymers can be divided into two major classes: *commodity plastics* and *engineering polymers*. Commodity plastics are characterized by high volume and low cost. They are used frequently in the form of disposable items such as packaging film, but also find application in durable goods. Commodity plastics comprise principally of four major thermoplastic polymers: polystyrene, polyethylene, polypropylene, and poly(vinyl chloride).

Engineering plastics refer to those polymers that are used in the manufacture of premium plastic products where high temperature resistance, high impact strength, chemical resistance, or other special properties are required. They compete with metals, ceramics, and glass in a variety of applications. Engineering plastics are designed to replace metals. Compared to commodity plastics, engineering plastics are specialty polymers that provide outstanding properties such as superior mechanical

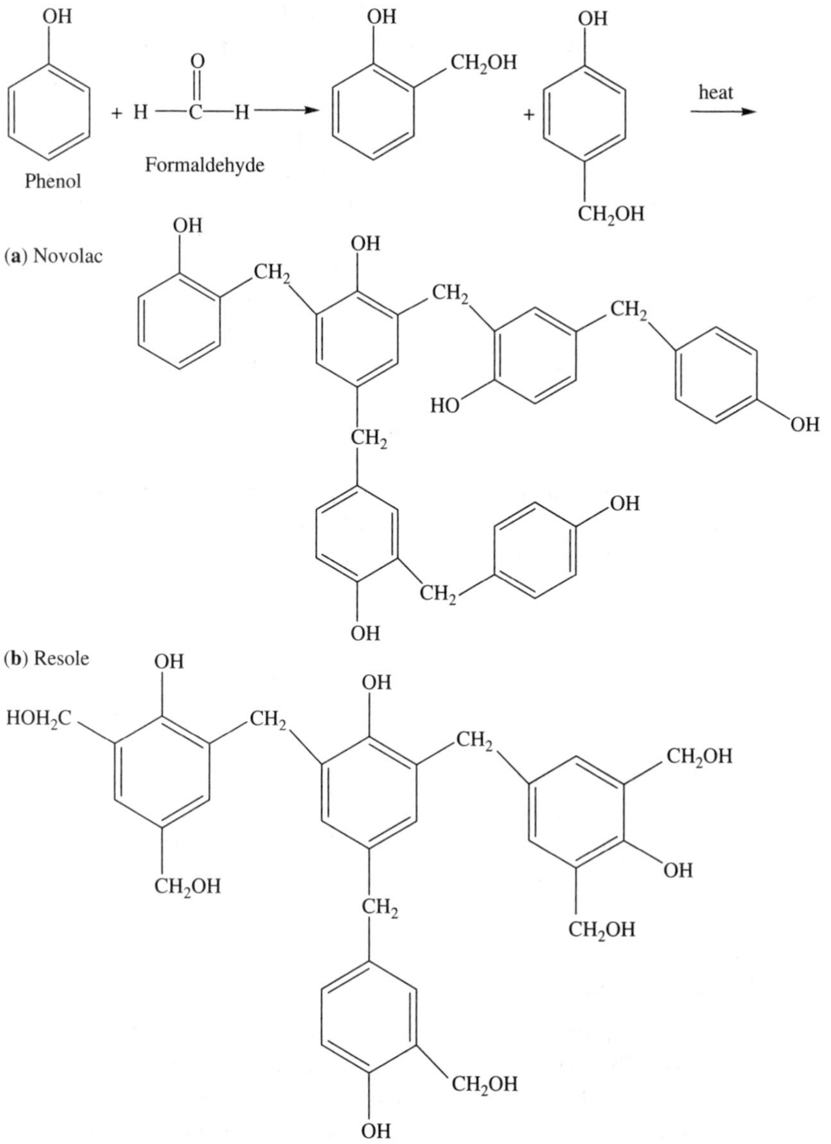

Figure 14.7 Representative structures of phenol-formaldehyde resins: (*a*) novolac (formed under acidic conditions), and (*b*) resole (formed under basic conditions).

properties, high thermal stability, excellent chemical resistance, low creep compliance, and high tensile, flexural, and impact strength. In contrast to commodity plastics, however, engineering and specialty polymers are produced at lower volume and higher cost. Examples of engineering plastics include aliphatic polyamides (such as nylon-6,6), aromatic polyamides, acrylonitrile-butadiene-styrene (ABS) resin, polyacetal, polycarbonate, polysulfones, poly(phenylene oxide) resins, poly(phenylene sulfide), and fluoroplastics such as teflon. Structures of some of these polymers are shown in Fig. 14.8.

Specialty polymers achieve very high performance and find limited but critical use in aerospace composites, in electronic industries, as membranes for gas and liquid separations, as fire-retardant textile fabrics for firefighters and race-car drivers, and for biomedical applications (as sutures and surgical implants). The most important class of specialty plastics is polyimides. Other specialty polymers include polyetherimide, poly(amide-imide), polybismaleimides, ionic polymers, polyphosphazenes, poly(aryl ether ketones), polyarylates and related aromatic polyesters, and ultrahigh-molecular-weight polyethylene (Fig. 14.9).

14.3   Polymers Industry

Today, polymeric materials are used in nearly all areas of daily life that it is right to claim that we live in a polymer age. The evidence is all around us. Polymers are used in shelter, clothing, health, food, transportation, and almost every facet of our lives. In fact, it is hard to imagine what the world would be like without synthetic polymers. Production and fabrication of polymers are major worldwide industries. The data presented in Tables 14.1 to 14.8 indicate the size of the polymers' marketplace in the 1992 to 2002 decade [1].

14.4   Polymer Structure

The properties of polymers are strongly influenced by details of the chain structure. The structural parameters that determine properties of a polymer include the overall chemical composition and the sequence of monomer units in the case of copolymers, the stereochemistry or tacticity of the chain, and geometric isomerization in the case of diene-type polymers.

Homopolymer.   It is a macromolecule consisting of only one type of repeating unit. The repeating unit may consist of a single species as in PS (1), or may contain more species of monomer unit. Polymers such as nylon-6,6 (2) that have repeating units composed of more than one monomer are considered homopolymers.

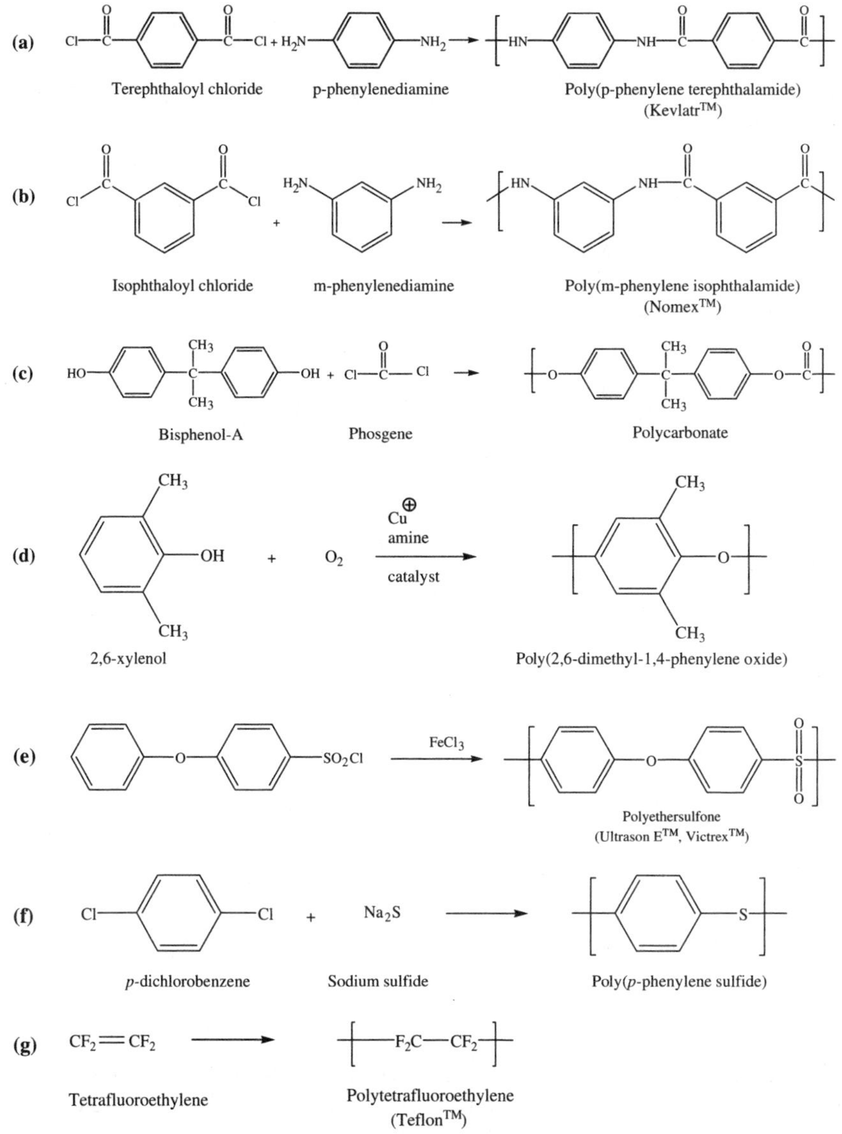

Figure 14.8 Examples of engineering plastics.

Copolymer. It is a macromolecule consisting of more than one type of building unit or mer. Copolymerization will be discussed in Sec. 14.10.

Head-to-head vs. head-to-tail. Substituted vinyl monomers can join in one of two arrangements; head-to-tail or head-to-head configurations

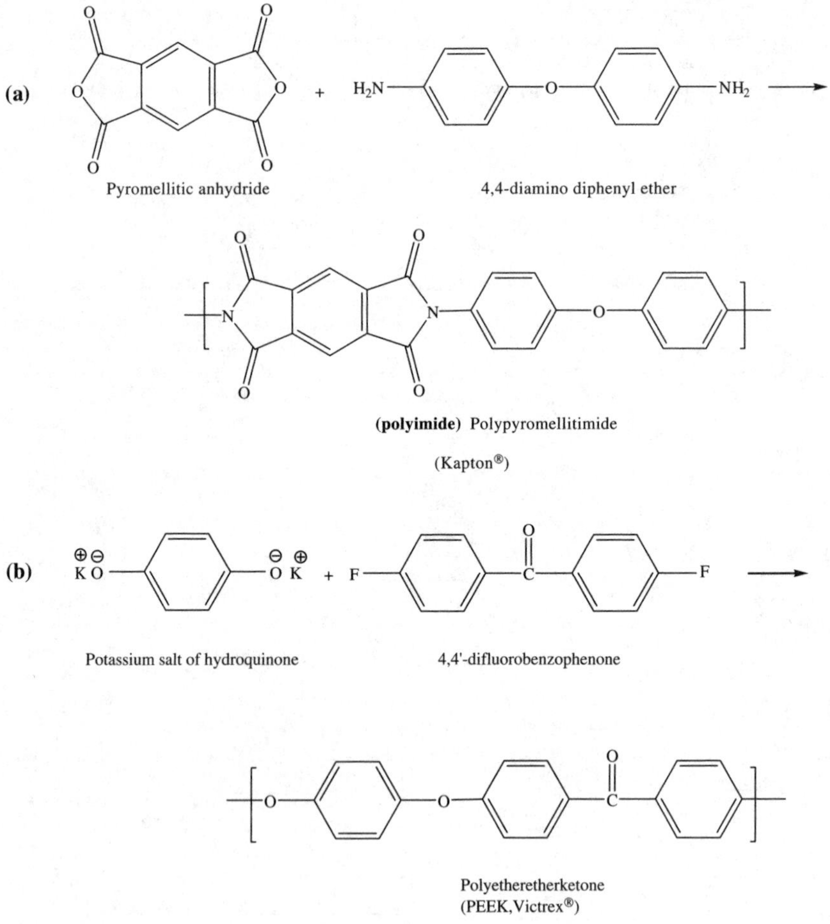

(a)

Pyromellitic anhydride   4,4-diamino diphenyl ether

(polyimide) Polypyromellitimide

(Kapton®)

(b)

Potassium salt of hydroquinone   4,4'-difluorobenzophenone

Polyetheretherketone
(PEEK,Victrex®)

Figure 14.9   Examples of specialty polymers.

(Fig. 14.10). Due to steric and electronic considerations, the usual arrangement is head-to-tail, so that the pendant groups are usually on every other carbon atom in the chain.

Linear polymers.   A linear polymer is a polymer molecule in which the atoms are more or less arranged in a long chain called the *backbone*. This is best illustrated with the structure of polyethylene (Fig. 14.1). However, polymers such as polypropylene or poly(1-pentene) in which a small chain called *pendant groups* presents in the repeating unit, are also designated as a linear polymer. The chains of pendant groups are much

TABLE 14.1   U.S. Plastics and Synthetic Rubber

U.S. Plastics

Production (Thousands of metric tons[a])

	1992	1993	1994	1995	1996	1997	1998	1999	2000	2001	2002
Thermoplastic resins											
Polyethylene											
Low density[b,c]	3,299	3,278	2,077	3,467	3,531	3,489	3,437	3,493	3,436	3,491	3,647
Linear low density[b,c]	2,106	2,196	2,278	2,378	2,885	3,124	3,278	3,677	3,607	4,659	5,139
High density[c,d]	4,449	4,509	5,043	5,085	5,612	5,696	5,862	6,289	6,336	6,933	7,243
Polypropylene[e]	3,820	3,914	4,327	4,940	5,439	6,042	6,271	7,028	7,139	7,228	7,691
Styrene polymers											
Polystyrene[f]	2,312	2,442	2,653	2,566	2,751	2,894	2,829	2,935	3,104	2,773	3,025
Styrene-acrylonitrile[g]	51	48	63	59	55	44	55	56	58	58	59
Acrylonitrile-butadiene-styrene and other styrene polymers[g,h]	1,184	1,326	1,465	1,319	1,347	1,359	1,447	1,406	1,415	1,240	1,323
Polyamine, nylon type	303	348	428	463	500	554	583	612	581	517	578
Polyvinyl chloride and copolymers[e]	4,531	4,652	5,312	5,577	5,996	6,388	6,578	6,764	6,551	6,467	6,939
Total	22,055	22,713	23,644	25,853	28,117	29,590	30,341	32,259	32,227	33,365	35,644
Thermosetting resins											
Epoxy[i]	207	232	273	287	300	297	290	298	314	273	297
Urea and melamine	807	914	1,005	955	1,104	1,197	1,302	1,354	1,437	1,379	1,460
Phenolic	1,326	1,396	1,465	1,453	1,577	1,694	1,787	1,990	1,974	1,979	2,013
Total	2,341	2,542	2,742	2,695	2,981	3,187	3,379	3,642	3,726	3,630	3,770
Grand Total, Plastics	24,396	25,255	26,386	28,548	31,098	32,777	33,720	35,901	35,953	36,995	39,414
U.S. Synthetic Rubbers											
Styrene-butadiene rubber	796	817	851	878	907	932	908	910	874	782	768
Polybutadiene	465	473	505	523	535	564	561	588	605	562	583
Ethylene-propylene	207	227	262	270	289	309	320	339	349	306	301
Nitrile, solid	74	78	84	83	85	85	87	88	89	84	84
Polychloroprene	72	70	76	70	70	73	69	66	64	57	54
Other[j]	381	400	408	427	437	449	447	478	484	424	425
Total	1,995	2,065	2,186	2,251	2,323	2,412	2,392	2,469	2,465	2,215	2,215

NOTE: Totals for plastics are for those products listed and exclude some small-volume plastics. Synthetic rubber data include Canada. [a]Dry-weight basis unless otherwise specified; [b]Density 0.940 and below; [c]Data include Canada from 2001; [d]Density above 0.940; [e]Data include Canada from 1995; [f]Data include Canada from 2000; [g]Data include Canada from 1994; [h]Includes styrene-butadiene copolymers and other styrene-based polymers; [i]Unmodified; [j]Includes butyl styrene-butadiene rubber latex, nitrile latex, polyisoprene, and miscellaneous others.

SOURCES: American Plastics Council, International Institute of Synthetic Rubber Producers.

TABLE 14.2 Canada Plastics

						Production (Thousands of metric tons[a])					
	1992	1993	1994	1995	1996	1997	1998	1999	2000	2001	2002
Polyesters, unsaturated	40	45	60	58	61	71	82	108	120	115	113
Polyethylene[a]	1660	1742	1908	2073	2194	2195	2283	2485	2751	3035	3330
Polystyrene[b]	167	158	177	189	209	181	180	200	203	186	195

[a]Includes high-, low-, and linear low-density polyethylene; [b]includes acrylonitrile-butadiene-styrene.
SOURCE: Statistics Canada.

TABLE 14.3 Europe Plastics and Synthetic Rubber

| | Production (Thousands of metric tons) | | | | | | | |
	1995	1996	1997	1998	1999	2000	2001	2002[a]
Polyethylene	2,832	3,000	8,508	9,731	10,223	10,579	5,481	4,778
Polystyrene	891	1,044	1,117	1,090	675	331	592	764
Acrylonitrile-butadienestyrene	643	604	762	859	971	1,038	466	121
Polyvinyl chloride	3,905	4,322	4,792	2,651	3,209	4,893	3,902	3,792
Epoxy resins	340	282	373	334	393	419	89	54
Polypropylene	NA	NA	NA	4,158	6,524	6,984	5,644	3,380
Polyamides	441	843	1,652	1,494	766	1,412	1,209	616
Synthetic rubber	1,753	1,946	2,419	2,245	2,239	2,342	2,691	2,068

[a]C&EN estimates based on partial reporting; NA = not available.
SOURCES: European Union and national government statistics offices, EuroChlor, Association of Petrochemicals Producers in Europe.

TABLE 14.4 Asia Plastics and Synthetic Rubber

In Japan

Production (Thousands of metric tons)

	1992	1993	1994	1995	1996	1997	1998	1999	2000	2001	2002
Polyethylene, low-density	NA	1573	1645	1748	1830	1839	1760	1856	1892	1852	1789
Polyethylene, high-density	NA	1024	1113	1238	1271	1313	1168	1301	1246	1240	1181
Polyethylene terephthalate	NA	1213	1279	1377	1360	1398	1300	1281	1308	1243	1211
Polypropylene	2038	2031	2225	2502	2730	2854	2520	2626	2721	2696	2641
Polyvinyl chloride	1981	1980	2112	2274	2511	2626	2457	2460	2410	2195	2225
Polystyrene	2005	1966	2099	2149	2178	2201	1975	2037	2024	1810	1837
Epoxy	NA	170	181	194	201	222	204	225	243	192	201
Phenolic resins	356	328	330	327	294	303	259	250	262	232	242
Polycarbonate	NA	149	171	227	251	292	317	351	354	370	386
Synthetic rubber	1388	1310	1349	1498	1520	1592	1520	1577	1590	1466	1522

Na = not available.
SOURCE: Ministry of Economy, Trade & Industry.

In South Korea

Production (Thousands of metric tons)

	1992	1993	1994	1995	1996	1997	1998	1999	2000	2001	2002
Acrylonitrile-butadiene-styrene	310	350	409	491	560	596	636	784	777	858	1,055
Polyethylene, low-density	761	853	923	970	1256	1394	1518	1642	1576	1614	1620
Polyethylene, high-density	1019	1189	1294	1232	1340	1549	1615	1756	1706	1839	1868
Polypropylene	1223	1436	1607	1619	1738	2056	2355	2440	2413	2422	2622
Polystyrene	754	809	869	905	1000	1104	1038	1105	1185	1224	1361
Polyvinyl chloride	726	760	791	914	1005	1087	1013	1170	1191	1238	1221

SOURCE: National Statistical Office, Republic of Korea.

In Taiwan

	Production (Thousands of metric tons)										
	1992	1993	1994	1995	1996	1997	1998	1999	2000	2001	2002
Acrylonitrile butadiene-styrene	569	642	742	757	911	979	901	1020	1059	982	1077
Polyethylene, low-density	211	217	224	214	233	235	224	236	243	472	411
Polyethylene, high-density	149	152	188	212	241	243	275	395	306	507	513
Polyester filament	903	1033	1179	1249	1298	1429	1615	1671	1632	1584	1603
Polypropylene	225	220	341	417	448	420	418	515	561	751	831
Polystyrene	528	593	626	671	808	790	784	784	748	911	892
Polyvinyl chloride	1043	1078	1114	976	1105	1149	1158	1397	1387	1434	1484
Synthetic rubber											
Styrene-butadiene	64	65	152	161	157	167	173	175	159	174	173
Polybutadiene	45	43	46	52	51	55	56	55	51	53	66

SOURCE: Taiwan Ministry of Economic Affairs.

TABLE 14.5   China Plastics and Synthetic Rubber

	Production (Thousands of metric tons)					
	1997	1998	1999	2000	2001	2002
Plastics	6474	7029	8418	10,794	12,038	13,665
Polyvinyl chloride	1536	1599	1894	2397	2877	3389
Polyethylene	2152	2292	2714	3000	3122	3547
Polypropylene	1881	2075	2722	3239	3225	3742
Polyester fiber	1274	2503	1898	2210	2932	3139
Synthetic rubber	623	589	761	836	1045	1168

SOURCE: China National Chemical Information Center.

smaller than the backbone chain, which usually has hundreds of thousands of atoms.

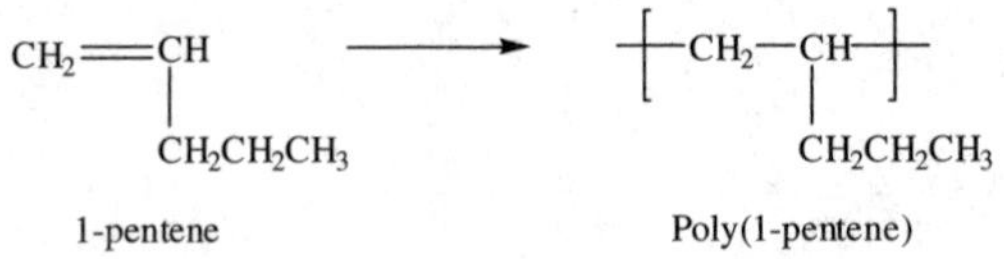

1-pentene   Poly(1-pentene)

Branched polymers.   In branched polymers (Fig. 14.2), chain extensions or branches are present on branch points, irregularly spaced along the polymer chain. The number of branches in nonlinear polyethylene (low-density polyethylene, LDPE) may vary from 1.5 per 20 methylene groups to 1 per 2000 methylene groups. This branching increases the specific volume and thus reduces the density of the polymer.

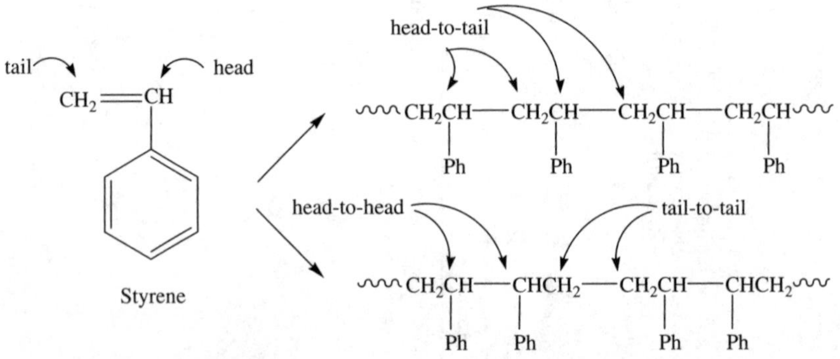

Figure 14.10   Illustration of head-to-tail and head-to-head configurations.

TABLE 14.6 U.S. Paints and Coatings

						Shipments (Millions of liters)					
	1992	1993	1994	1995	1996	1997	1998	1999	2000	2001	2002
Architectural	2180	2301	2441	2350	2422	2483	2392	2498	2464	2339	2566
Product[a]	1181	1351	1412	1423	1510	1609	1620	1665	1715	1540	1567
Special purpose	655	678	734	738	791	689	655	659	689	613	590
Total	4016	4330	4587	4512	4724	4780	4667	4822	4868	4493	4724

[a]For original equipment manufacturers.
SOURCE: Department of Commerce.

TABLE 14.7 U.S. Synthetic Fibers

	Production (Thousands of metric tons)										
	1992	1993	1994	1995	1996	1997	1998	1999	2000	2001	2002
Noncellulosic fibers											
Acrylic[a]	199	196	200	196	211	209	157	143	154	137	150
Nylon	1159	1206	1243	1226	1270	1286	1218	1217	1215	1019	1106
Olefin	894	959	1083	1085	1095	1216	1326	1395	1461	1342	1352
Polyester	1622	1613	1750	1763	1736	1855	1768	1736	1775	1456	1471
Total	3874	3975	4276	4270	4311	4567	4469	4491	4605	3954	4080
Cellulosic fibers											
Acetate[b] and rayon	225	229	227	226	215	208	166	134	158	103	81
Total fibers	4099	4204	4503	4496	4527	4775	4635	4625	4764	4057	4160

[a]Includes modacrylic; [b]Includes diacetate and triacetate; excludes production for cigarette filters.
SOURCE: Fiber Economics Bureau.

TABLE 14.8 Europe Synthetic Fibers

						Production (Thousands of metric tons)					
	1992	1993	1994	1995	1996	1997	1998	1999	2000	2001	2002
Acrylic	745	686	714	623	677	705	650	614	623	842	861
Polyester	955	895	982	973	895	995	959	909	968	1472	1480
Polyamide	650	591	641	632	632	673	641	595	636	708	698
Cellulosics	713	713	682	736	766	722	715	651	627	457	450

SOURCE: International Rayon & Synthetic Fibers Committee.

Cross-linked polymers. In cross-linked polymer systems (Fig. 14.2), polymer chains become chemically linked to each other resulting in a network. Network structures are formed when the average functionality of a mixture of monomers is greater than 2. Network polymers can also be made by chemically linking linear or branched polymers. For example, in a tire, the rubber polymer chains are interconnected with sulfur linkages in a process called vulcanization (Fig. 14.11).

Both linear and branched polymers are *thermoplastics*. However, cross-linked three-dimensional, or network-polymers are *thermoset* polymers. The cross-linked density may vary from the low cross-linked density in vulcanized rubber to high cross-linked density observed in ebonite (hard rubber; highly cross-linked natural rubber).

Ladder polymers. A ladder polymer consists of two parallel linear strands of molecules with a regular sequence of cross-links. Ladder polymers have only condensed cyclic units in the chain. The molecular structure of ladder polymers is more rigid than that of conventional liner polymers. Typical examples of ladder polymers are shown in Fig. 14.12.

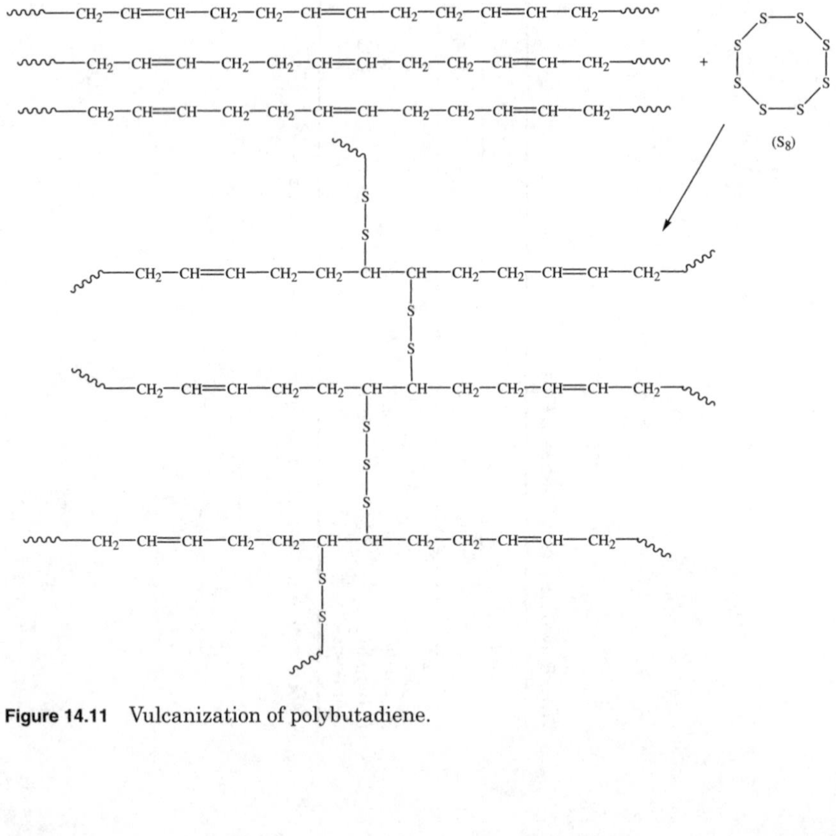

Figure 14.11 Vulcanization of polybutadiene.

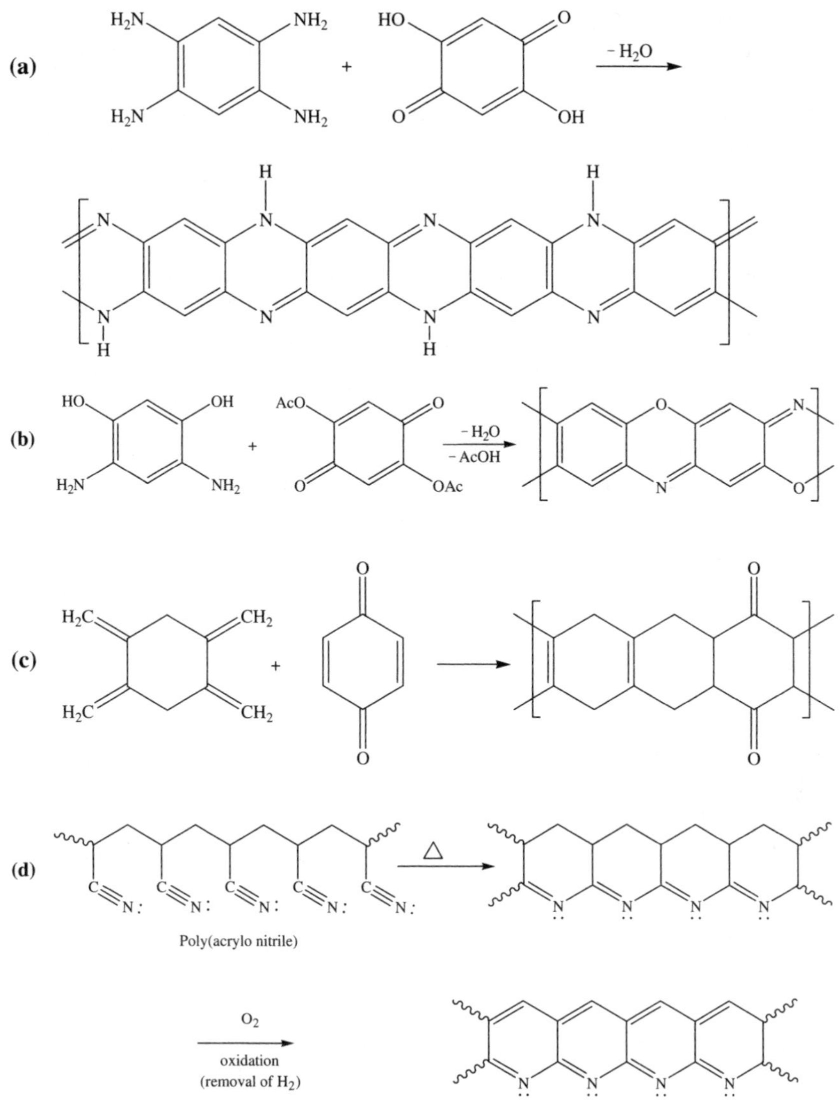

Figure 14.12   Examples of ladder polymers.

Star-shape polymers.   Star polymers contain three or more polymer chains emanating from a central structural unit (Fig. 14.2).

Comb polymers.   Comb polymers contain pendant chains (which may or may not be of equal length) and are related structurally to graft copolymers. Such a polymer might be formed, for example, by polymerizing a long-chain vinyl monomer (i.e., macromonomer).

Tacticity.   In addition to the type, number, and sequential arrangement of monomers along the chain, *tacticity* or the spatial arrangement of substituent groups is also important in determining properties. Three types of steric configurations are possible from polymerizing asymmetric vinyl monomers, as represented with polypropylene in Fig. 14.13:

1. All the methyl groups could be protruding from the chain in a random fashion; this is named an atactic polymer.
2. The methyl groups could alternate from one side of the chain to the other; this is called a syndiotactic polymer.
3. The methyl groups could all be on the same side; then the polymer is said to be isotactic polymer.

   Because of their orderly arrangements, the chains of the tactic polymers (syndiotactic and isotactic) can lie closer together and the polymers are partially crystalline, whereas atactic polymers are amorphous and soft indicating the absence of all crystalline order. Isotactic polypropylene is highly crystalline with a melting point of 160°C, whereas the atactic isomer is an amorphous (noncrystalline) soft polymer with a melting

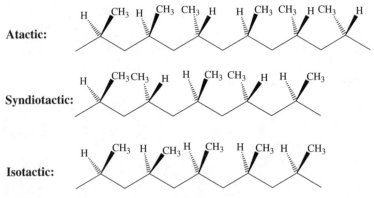

Figure 14.13   Three types of polypropylene.

point of 75°C. In addition to crystallinity, other polymer properties, such as thermal and mechanical behavior, can be significantly affected by the tacticity of the polymer.

Until 1953, most addition polymers were made by free-radical paths, which produce atactic polymers. In that year, however, the Nobel laureates Karl Ziegler and Giulio Natta introduced a new technique for polymerization using a type of catalyst that permits control of the stereochemistry of a polymer during its formation.

Geometric isomerism. When there are unsaturated sites along a polymer chain, several different isomeric forms are possible. As illustrated in Fig. 14.14, conjugated dienes such as isoprene and chloroprene can be polymerized to give either 1,2-, 3,4, or 1,4-polymer. In the case of 1,4-polymers, both *cis* and *trans* configurations are possible. Also, stereoregular (i.e., isotactic and syndiotactic) polybutadienes can be produced in case of 1,2- and 3,4-polymerization.

It is possible to control the type of polymer produced. This is best illustrated with polymerization of the monomer isoprene; 2-methyl-1,3-butadiene. The Ziegler-Natta catalyst, consisting of a titanium tetrachloride

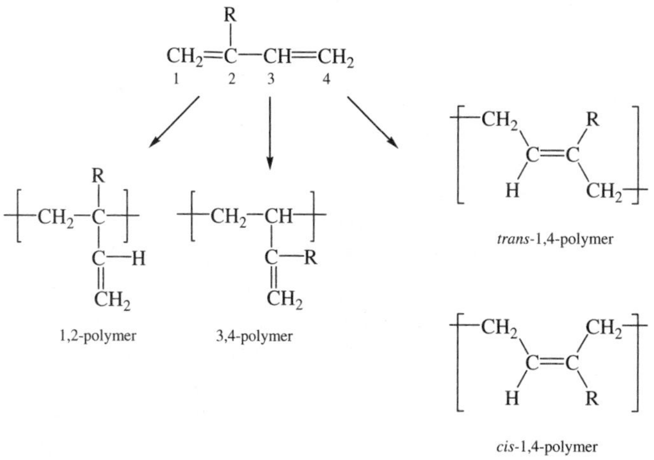

R	Monomer	Polymer
H	1,3-butadiene	polybutadiene
Cl	2-chloro-1,3-butadiene	polychloroprene
CH_3	2-methyl-1,3-butadiene	polyisoprene

Figure 14.14 Possible polymer structures from the polymerization of conjugated dienes.

catalyst, an alkyl aluminium cocatalyst ($TiCl_4/AlR_3$ or AlH_3) produces 96 percent *cis*-poly-1,4-isoprene, while VCl_3/AlR_3 system produces essentially 97 percent *trans*-poly-1,4-isoprene. When $Ti(OR)_4/AlR_3$ is used as the Ziegler-Natta catalyst/cocatalyst system, the product is 60 to 70 percent poly-3,4-isoprene [2]. When chromium hexacyanobenzene [$Cr(C_6H_5(CN)_6)$] is used as the catalyst for the polymerization of 1,3-butadiene, stereospecific polymers are obtained. A ratio of cocatalyst to catalyst of 2:1 yields syndiotactic poly1,2-butadiene, and a ratio of 10:1 yields isotactic poly-1,2-butadiene [3, 4].

Polymer molecular interaction.   Atoms in individual polymer molecules are joined to each other by primary forces, which are relatively strong covalent bonds. The bond energies of the carbon-carbon bonds are in the order of 80 to 90 kcal/mol. In addition, polymer molecules are attracted to each other (and for long-chain polymer chains even between segments of the same chain) by intermolecular secondary forces (or van der Waals forces). Secondary intermolecular forces include London dispersion forces, induced permanent forces, and dipolar forces, including hydrogen bonding. The strength of these forces is dependent on the distance between the interacting molecules. These forces collectively play a major role in determining the polymer properties and performance. Polymer behavior is affected by the proximity one chain can have relative to another. An amorphous polymer, for example, is more flexible than its crystalline counterpart. Thus, many physical properties of polymers are quite dependent on both the conformation (arrangements related to the rotation about single bonds) and configuration (arrangements related to the actual chemical bonding about a given atom).

Nonpolar polymers such as polyethylene are attracted to each other by weak London or dispersion forces that result from induced dipole-dipole interaction. This arises from the instantaneous fluctuations in the density of the electron clouds along the polymer chains. The energy range of these forces is about 2 kcal/mol repeat unit and is independent of temperature. London forces are the dominant forces between chains in largely nonpolar polymers present in elastomers and soft plastics.

Polar polymers, such as poly(vinyl chloride) are attracted to each other by dipole-dipole interactions resulting from the electrostatic attraction of a chlorine atom in one molecule to a hydrogen atom in another molecule (Fig. 14.15). These forces range from 2 to 6 kcal/mol repeat unit, however, they are reduced as the temperature is increased. These forces are characteristic of many plastics. London forces are also present in polar molecules. An especially strong type of dipole-dipole interaction (called hydrogen bonding) occurs between strong polar molecules in which the nitrogen, oxygen, or fluorine atoms in one molecule are attracted to the hydrogen atoms attached to a highly electronegative atom in another molecule (Fig. 14.15). Hydrogen bonds are the strongest

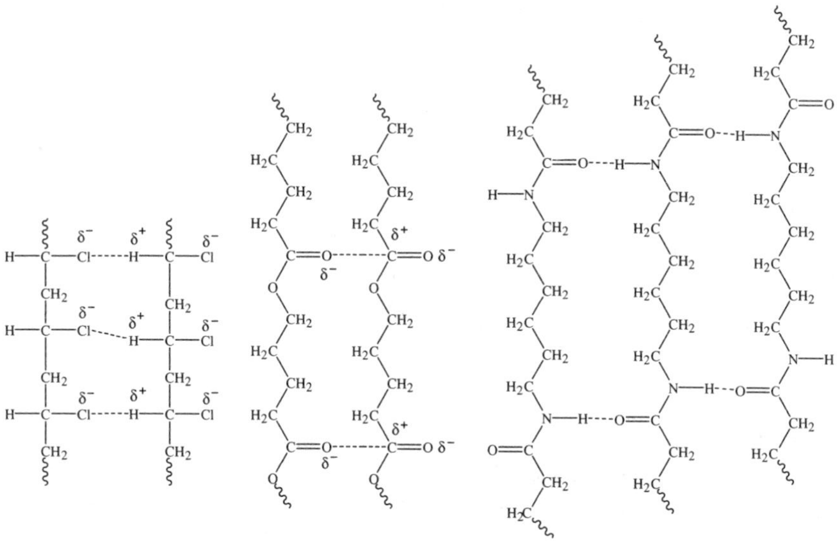

Figure 14.15 Typical dipole-dipole interaction between molecules of (a) poly(vinyl chloride), (b) polyester, and (c) nylone-6,6 (H-bonding).

intermolecular forces and may have energies as high as 10 kcal/mol repeat unit. Intermolecular hydrogen bonds are usually present in fibers, such as cotton, wool, silk, nylon, polyacrylonitrile, polyesters, and polyurethanes. Isotactic polypropylene, which has no hydrogen bonds, is also a strong fiber as a result of the good fit of the regularly spaced methyl pendant groups on the chain. Atactic polypropylene does not have this molecular geometry, and thus it is not a fiber.

In addition to the chemical interactions discussed above, physical interactions (namely chain entanglement) between polymer chains play an important role in determining the physical properties of polymers. Polymer chains that are sufficiently long can form stable, flow-restricting entanglements. Entanglements have significant importance in relation to viscoelastic properties, melt viscosity, and mechanical properties such as stress relaxation, creep, and craze formation. Paraffin wax has the same chemical composition as that of high density polyethylene (HDPE). However, the length of paraffin is too short to permit entanglement and hence it lacks the strength and other characteristic properties of HDPE.

The minimum polymer chain length or critical molecular weight, M_c, for the formation of stable entanglements depends upon the flexibility of a polymer chain which is affected by the polarity and shape of the polymer. Relatively flexible polymer chains, such as polystyrene, have a

TABLE 14.9 Entanglement Molecular Weight
for Linear Polymers [5]

Polymer	M_c
1,4-polybutadiene	5900
cis-polyisoprene	10,000
Polyisobutylene	15,200
Polydimethylsiloxane	24,400
Poly(vinyl acetate)	24,500
Poly(methyl methacrylate)	27,500
Poly(α-methylstyrene)	28,000
Polystyrene	31,200

high M_c, whereas more rigid-chain polymers, such as those with aromatic backbone have a relatively low M_c. For example, the melt viscosity (η) of a polymer exponentially increases with the critical chain length (z) as related in Eq. (1), regardless of the structure of the polymer. K is a temperature-dependent constant. Table 14.9 gives the molecular weight critical for entanglement for some polymers.

$$\eta = Kz^{3.4} \tag{1}$$

Flexibility.   The flexibility of amorphous polymers is impeded by chain entanglement, high intermolecular forces, and other factors such as the presence of reinforcing agents and cross-links. The flexibility of amorphous polymers above the glassy state is increased when flexibilizing groups such as methylene groups (CH_2), oxygen atoms, ethylene oxides, and dimethylsiloxane are present in the polymer backbone. Thus, the flexibility of aliphatic polyesters usually increases as m is increased in the polyester with the following structure:

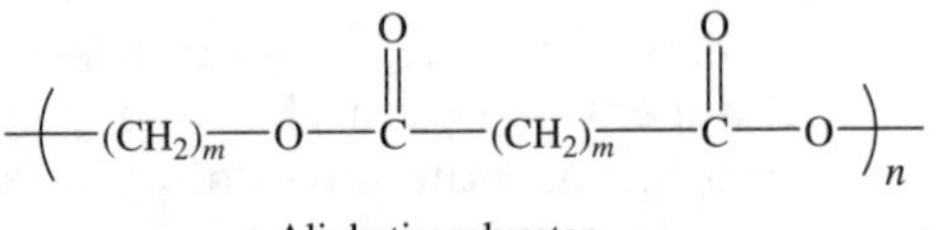

Aliphatic polyester

In contrast, presence of certain groups in the polymer backbone renders the polymer chain less flexible. Examples of stiffening groups are:

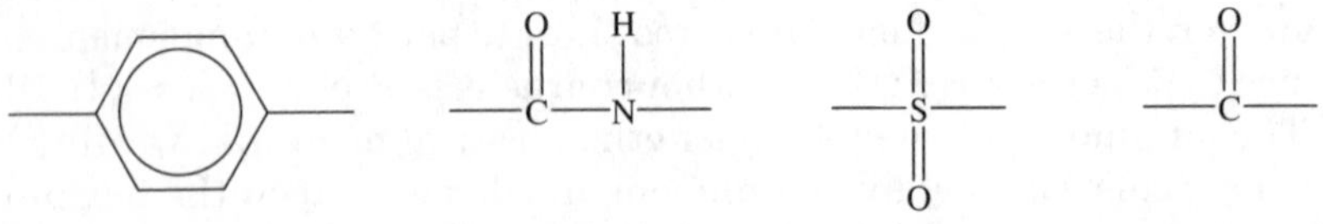

p-phenylene   Amide   Sulfone   Carbonyl

Poly(ethylene terephthalate) (PET), for example, is stiffer than both poly(butylene terephthalate) and poly(ethylene adipate), because of the presence of phenylene groups, and fewer methylene groups in PET.

Poly(ethylene adipate)   Poly(ethylene terephthalate)   Poly(butylene terephthalate)

Crystalline and amorphous state.   Polymer chains could be likened to spaghetti strands in a pot of spaghetti. Individual chains are randomly coiled and intertwined with no molecular order or structure. The lack of order in the arrangement of polymer chains produces an *amorphous* state. Commercial-grade (atactic) polystyrene and poly(methyl methacrylate) are examples of polymers that are amorphous in the solid state. On the other hand, polymer chains with very regular structures, such as linear polyethylene and polytetrafluoroethylene, which have symmetrically substituted repeating units, are highly *crystalline*. Also, stereoregular polymers (e.g., *isotactic*-polypropylene), can be arranged with well-defined crystalline morphology consisting of chain-folded lamella joined in highly regular structures.

The regular packing of chains forms regions of crystallinity called crystallites. This phenomenon is demonstrated qualitatively by the development of opacity when a rubber band is stretched and by the abnormal stiffening and whitening of unvulcanized rubber when it is stored for a long time at low temperature. Both observations are attributed to the side-by-side packing arrangements of individual chains (or different segments of the same chain) in a regular manner. The stretching of a polymer film or fiber results in a roughly parallel alignment of the macromolecular chains, which in turn facilitates crystallization (see Fig. 14.3).

The chemical structure of a polymer determines whether it will be crystalline or amorphous in the solid state. Both tacticity (i.e., syndiotactic or isotactic) and geometric isomerism (i.e., *trans* configuration) favor crystallinity. In general, tactic polymers with their more stereoregular chain structure are more likely to be crystalline than their atactic counterparts. For example, isotactic polypropylene is crystalline, whereas commercial-grade atactic polypropylene is amorphous. Also, *cis*-polyisoprene is amorphous, whereas the more easily packed *trans*-polyisoprene is crystalline. In addition to symmetrical chain structures that allow close packing of polymer molecules into crystalline lamellae, specific interactions between chains that favor molecular orientation, favor crystallinity. For example, crystallinity in nylon is enhanced because of the occurrence of specific interactions (i.e., hydrogen bonding) between an amide carbonyl group on one chain and the hydrogen atom of an amide group on an adjacent chain. Such bonding is illustrated in Fig. 14.15 for nylon-6,6, poly(hexamethylene adipamide). *Atactic*-poly(vinyl alcohol) is partly crystalline demonstrating the importance of specific interactions (hydrogen bonding) in favoring crystallinity in polymers.

It is important, however, to recognize that the regular packing arrangements usually exist only in small domains within the polymer. Order polymers are seldom 100 percent crystalline, and no bulk polymer is completely crystalline. A crystalline polymer really consists of microcrystallites embedded in a matrix of amorphous polymer. Therefore, the structure of a crystalline polymer is generally considered to consist of microcrystalline domains separated by amorphous, random coil regions, and that a single polymer chain may traverse several microcrystalline and amorphous regions (Fig. 14.16). The degree of crystallinity, that is, the fraction of the total polymer in the crystalline regions may vary from a few percentage points to about 90 percent. In addition to the crystallization of the backbone of polymers, crystallization may also occur in regularly placed bulky groups even when an amorphous structure is maintained in the backbone. In general, the pendant group must contain at least 10 carbon atoms for the side chain crystallization to occur.

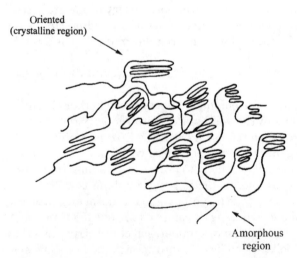

Oriented
(crystalline region)

Amorphous
region

Figure 14.16 Illustration of polymer structure consisting of a mixture of crystalline and amorphous regions.

Physically, amorphous polymers exhibit a glass transition temperature but not a melting temperature, and do not give a clear X-ray diffraction pattern. On the other hand, crystalline polymers may exhibit both glass transition temperature (T_g) corresponding to long-range segmental motions in the amorphous regions and a crystalline-melting temperature, or T_m, at which crystallites are destroyed and an amorphous, disordered melt is formed. The combination of crystalline and amorphous regions is important for the formation of materials that have both good strength (contributed largely by the crystalline portions) and some flexibility or softness (derived from the amorphous portions).

14.5 Polymer Structure-Property Relationships

The ultimate properties and the final uses of a polymer are intimately related to the chemical composition and structure of the polymer. The physical properties of polymers are related to the strength of the covalent bonds, the stiffness of the segments in the polymer backbone, and the strength of the intermolecular forces between the polymer molecules. However, polymer properties are also influenced by a multitude of other factors that include extent and distribution of crystallinity, molecular weight and molecular weight distribution, the nature and amount of additives such as fillers, reinforcing reagents, and plasticizers. For linear and lightly branched polymers, the contribution of end groups to the polymer structure is significant only at relatively low molecular weight. The effects of end groups on polymer properties, such as density, refractive index, and spectroscopic absorption, progressively decrease with the increase in molecular weight of polymers, and become negligible beyond a certain molecular weight.

14.5.1 Thermal properties

Thermal stability.   For a polymer to be considered *thermally stable* or *heat resistant*, it should not decompose below 400°C, and should retain its useful properties at temperatures near the decomposition temperatures. Table 14.10 lists some representative thermally stable polymers. A common feature among the polymers listed in Table 14.10 is the presence of aromatic moieties. Aromatic polymers have found uses in applications that involve exposure to high temperatures and aerospace industries (as fabrics for astronauts' clothing for example) because of being resistant to high temperatures. Thermal degradation or decomposition occurs as a result of the temperature causing the bonds to vibrate to the rupture point. In the case of cyclic repeating units, breaking of one bond in a ring does not lead to decrease in molecular weight,

TABLE 14.10 Examples of Thermally Stable Polymers [6]

Type	Structure	Decomposition temperature (°C)
Poly(p-phenylene)	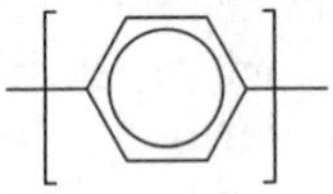	660
Polybenzimidazole	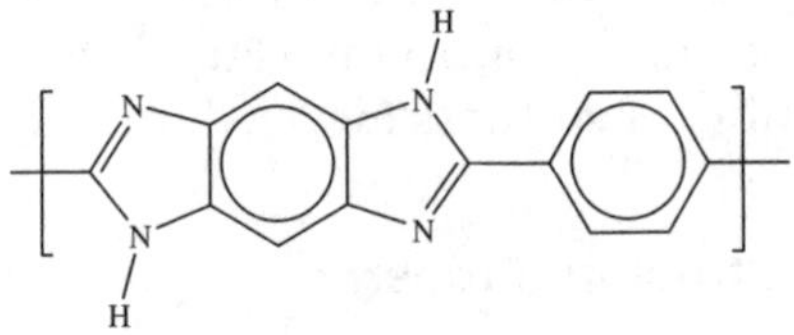	650
Polybenzimiazole	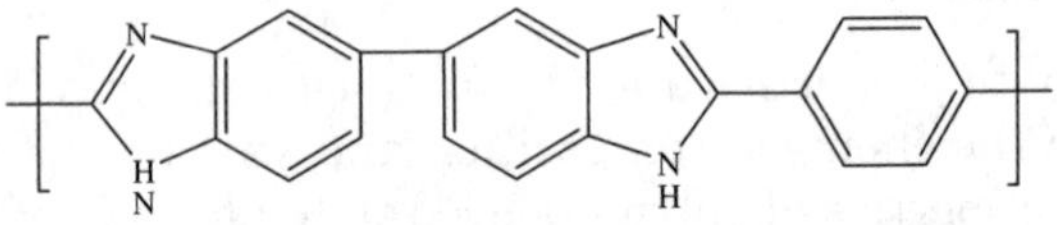	650
Polyquinoxaline	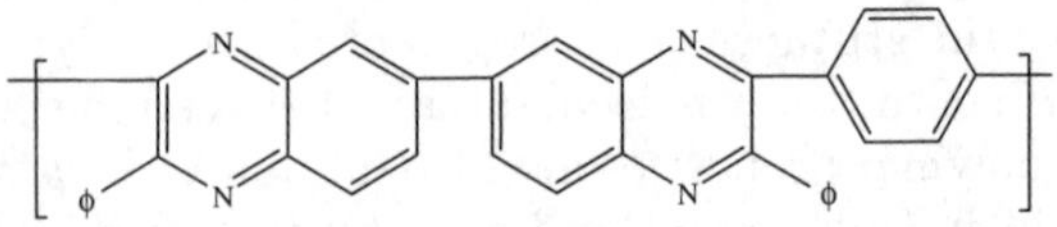	640
Polyoxazole	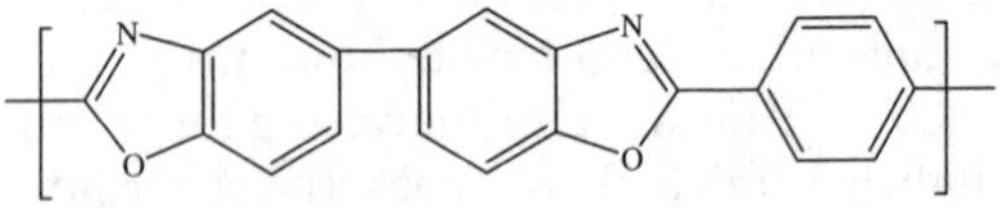	620
Polyimide	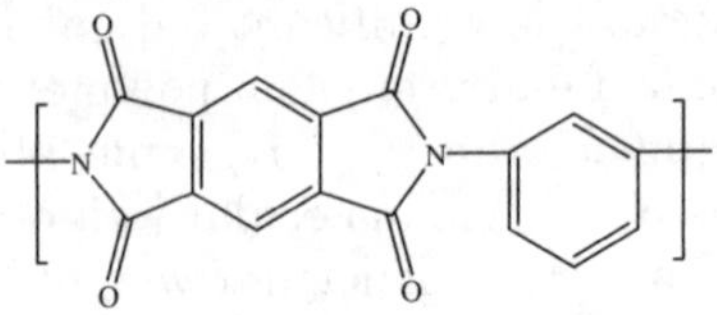	585
Poly(phenylene oxide)		570
Polybenzamide	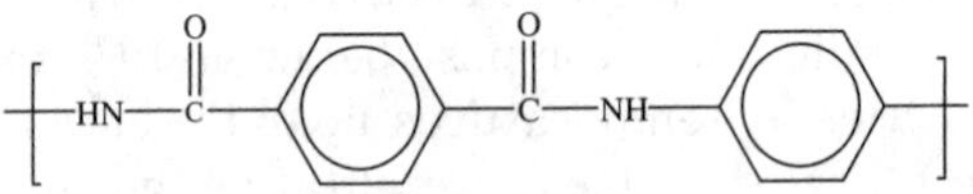	500
Polythiadizole	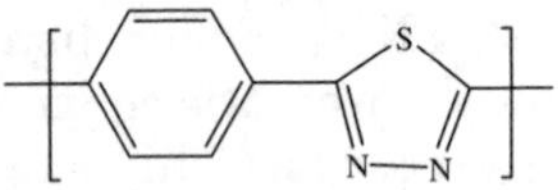	490
Poly(phenylene sulfide)	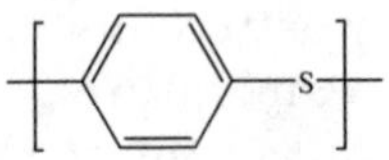	490

(*Continued*)

TABLE 14.10   Examples of Thermally Stable Polymers [6] (*Continued*)

Type	Structure	Decomposition temperature (°C)
Aromatic polyester	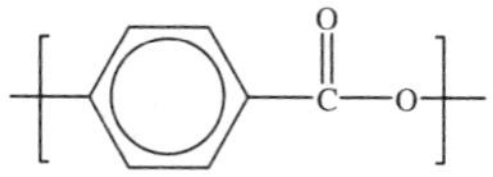	480
Polypyrrole	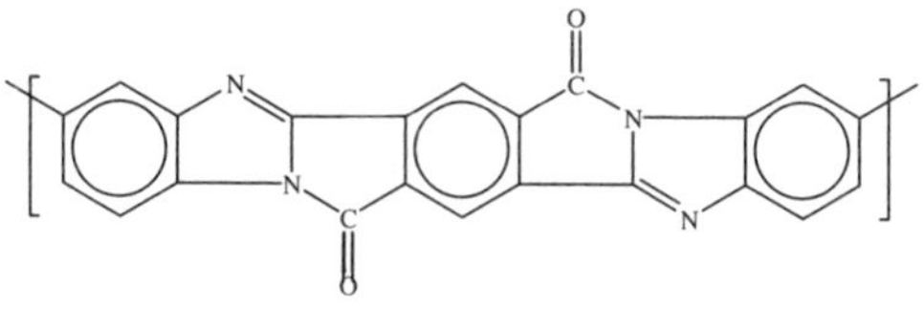	660

and the probability of two bonds breaking within one ring is low. Thus, ladder or semiladder polymers usually have higher thermal stabilities than open-chain polymers.

Glass transition temperature.   Glass transition temperature, T_g, is a characteristic temperature that marks the transition from the glassy state (hard and brittle amorphous state) to the rubbery or flexible state because of the onset of segmental motion. At temperatures below T_g, there is no segmental motion, except for the stretching or distortion of covalent bonds. As such, the flexibility of amorphous polymer is reduced drastically when they are cooled below T_g. The melting point, T_m, is the temperature range where total or whole polymer chain mobility occurs. For many polymers, T_g is approximately one-half to two-thirds of T_m (expressed in Kelvin). The glass transition temperature is characteristic of a particular polymer in much the same way that a melting point is characteristic of ordinary low molecular weight compounds.

The glass transition temperature of amorphous polymers is a function of the chemical structure of the polymer chain. It varies widely with the types of skeletal atoms present, with the types of side groups, and with the tacticity of side groups along the polymer backbone. Table 14.11 demonstrates the effects of structural variations on the crystalline melting temperature and glass transition temperature for several polymers.

In general, polymers with flexible backbone and small substituent groups have low T_g, whereas those with rigid backbones, such as polymers containing main-chain aromatic groups have high T_g. Thus, poly(dimethylsiloxae) has a very low T_g (−123°C), presumably because the silicon-oxygen bonds have considerable flexibility. On the other hand, polymers with highly rigid aromatic backbones such as the

TABLE 14.11   Approximate Temperatures of Thermal Transitions for Selected Semicrystalline Polymers [7,8]

Polymer	Structure	T_g (°C)	T_m (°C)
Polydimethylsiloxane (silicon rubber)	$\left[\!\!\begin{array}{c}CH_3\\|\\Si-O\\|\\CH_3\end{array}\!\!\right]$	−123	−29
Polyethylene	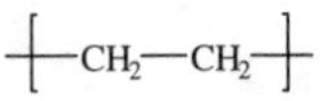 $-CH_2-CH_2-$	−120	135
Polypropylene (*atactic*)	$\begin{array}{c}-CH_2-CH-\\|\\CH_3\end{array}$	−20	176
Polypropylene (*isotactic*)		100	
Poly(vinyl chloride)	$\begin{array}{c}-CH_2-CH-\\|\\Cl\end{array}$	89	285
Polyacrylonitrile	$\begin{array}{c}-CH_2-CH-\\|\\C\equiv N\end{array}$	85	317
Polystyrene	$\begin{array}{c}-CH_2-CH-\\|\\C_6H_5\end{array}$	100	240
Poly(vinylidene fluoride)	$\begin{array}{c}F\\|\\-CH_2-C-\\|\\F\end{array}$	−45	172
Poly(vinyl acetate)	$\begin{array}{c}-CH_2-CH-\\|\\OCCH_3\\||\\O\end{array}$	32	
Poly(vinyl alcohol)	$\begin{array}{c}-CH_2-CH-\\|\\OH\end{array}$	85	258
Poly(methyl methaacrylate)	$\begin{array}{c}CH_3\\|\\-CH_2-C-\\|\\COCH_3\\||\\O\end{array}$	105	200

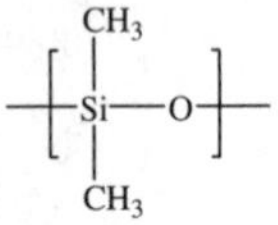

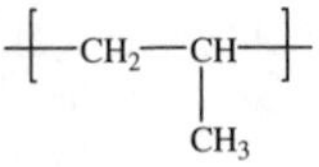

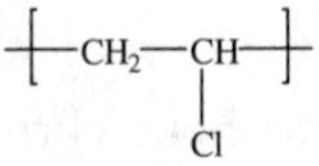

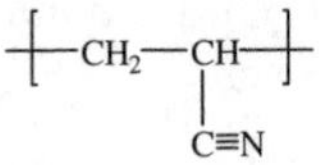

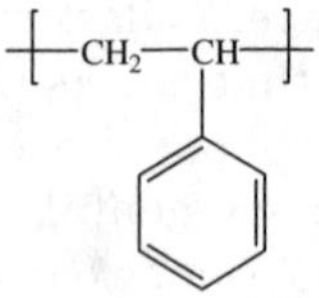

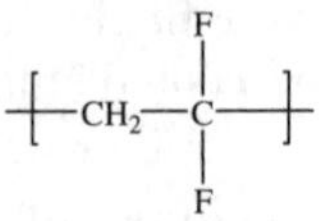

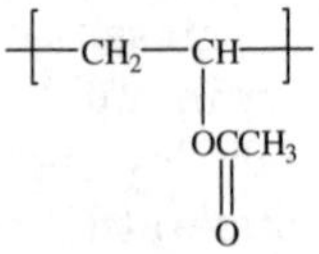

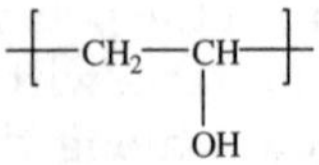

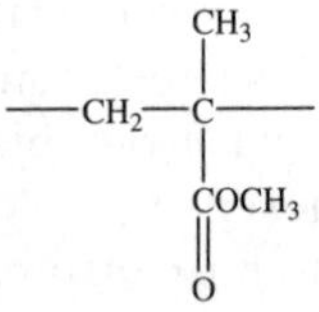

(*Continued*)

TABLE 14.11   Approximate Temperatures of Thermal Transitions for Selected Semicrystalline Polymers [7,8] (Continued)

Polymer	Structure	T_g (°C)	T_m (°C)
Polyoxymethylene	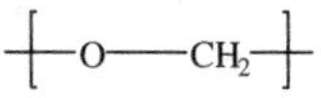 —[O—CH$_2$]—	−85	195
Polycaprolactone	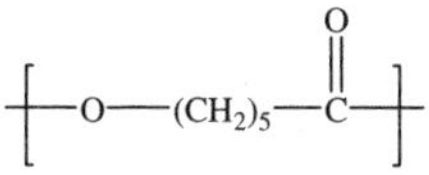 —[O—(CH$_2$)$_5$—C(=O)]—	−60	61
Poly(ethylene adipate)	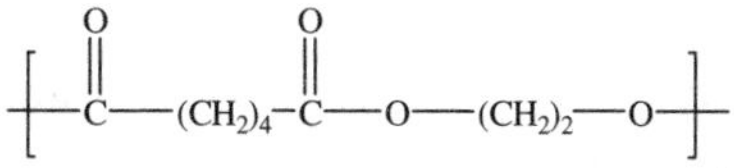 —[C(=O)—(CH$_2$)$_4$—C(=O)—O—(CH$_2$)$_2$—O]—	−63	
Poly(butylenes adipate)	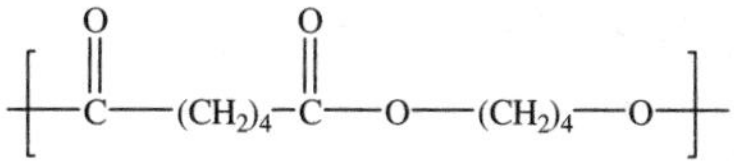 —[C(=O)—(CH$_2$)$_4$—C(=O)—O—(CH$_2$)$_4$—O]—	−118	
Poly(ethylene terephthalate) (Dacron)	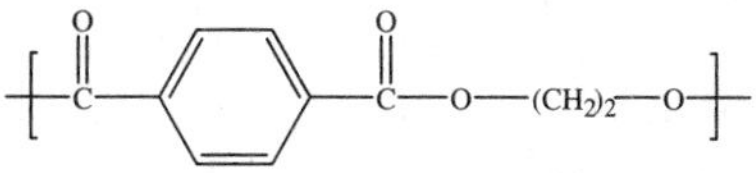 —[C(=O)—C$_6$H$_4$—C(=O)—O—(CH$_2$)$_2$—O]—	69	270
Polycarbonate (Lexan)	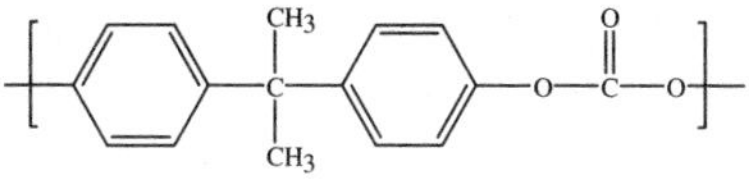 —[C$_6$H$_4$—C(CH$_3$)$_2$—C$_6$H$_4$—O—C(=O)—O]—	150	267
Nylon-6,6	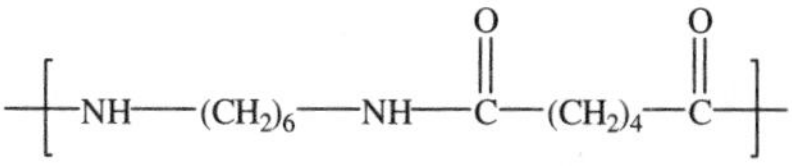 —[NH—(CH$_2$)$_6$—NH—C(=O)—(CH$_2$)$_4$—C(=O)]—	50	265
Nylon-6,10	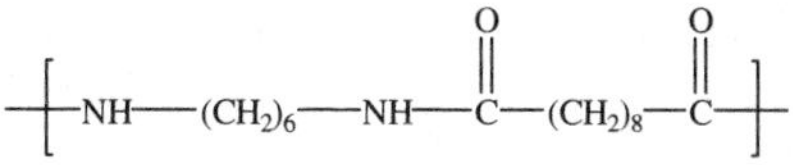 —[NH—(CH$_2$)$_6$—NH—C(=O)—(CH$_2$)$_8$—C(=O)]—	50	165 (226)
Nylon-6	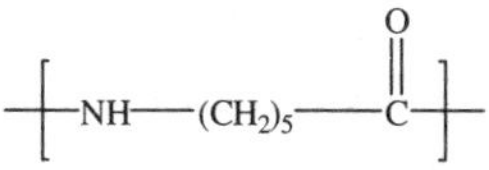 —[NH—(CH$_2$)$_5$—C(=O)]—	77	223
Nylon-11	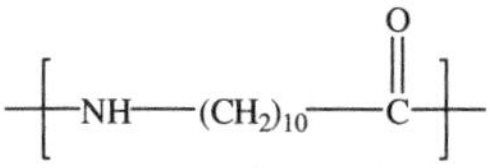 —[NH—(CH$_2$)$_{10}$—C(=O)]—	46	198
Poly(2,6-dimethyl-1,4-phenylene oxide) (Parlene)	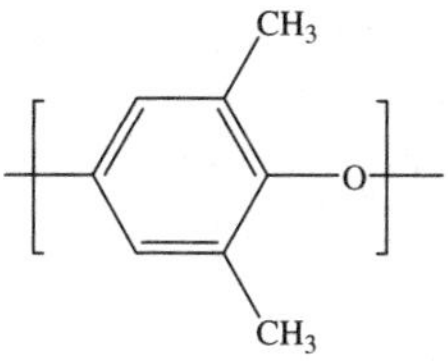 —[2,6-(CH$_3$)$_2$C$_6$H$_2$—O]—	220	338
Poly[2,2'-(m-phenylene)-5,5'-bibenzimidazole]	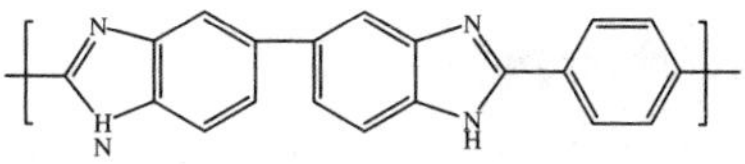 (bibenzimidazole–m-phenylene structure)		429

TABLE 14.12   Glass Transition Temperature (T_g) of Diene Polymers

Polymer	T_g (°C) cis	trans
1,4-Polybutadiene[9]	−102	−58
1,4-Polyisoprene[10]	−67	−70
1,4-Polychloroprene[9]	−20	−40

high-modulus fiber poly[2,2′-(m-phenylene)-5,5′-bibenzimidazole] has a reported T_g in the range from 427 to 500°C. Likewise, aromatic nylons, called aramids, have greater T_g values than aliphatic nylons counterparts. Bulky substituent groups hinder rotation and therefore raise T_g. For example, the bulky phenyl group in polystyrene restricts rotation, and hence its T_g is higher than that for polyethylene. Also, as shown in Table 14.11, T_g decreases as the size of the ester groups increases in polyacrylates and polymethacrylates. The T_g of poly(ethylene terephthalate) is 69°C, which is higher than that of poly(ethylene adipate) by 132°C demonstrating the effect of the phenylene stiffening group. Chain mobility is lowered by the presence of charged structures on the same chain, which will repel or attract each other. Also, polar interactions between neighboring chains could raise the T_g. As shown in Table 14.11, T_g increases as the degree of crystallinity increases; the T_g value of isotactic polypropylene is 100°C, whereas atactic polypropylene (less crystalline) has a T_g value of −20°C. The presence of other nonrotational groups such as unsaturated groups in the main chain decreases flexibility of the polymer chain. *Trans* geometric isomers have higher T_g than *cis* isomers, as for example, in the case of *cis*-polybutadiene ($T_g = $ −102°C) compared to *trans*-polybutadiene ($T_g = $ −58°C), or in the case of *cis*-polyisoprene ($T_g = $ −67°C) compared to *trans*-polyisoprene ($T_g = $ −70°C) (Table 14.12). Even the chain length has an effect. For low polymers, the T_g generally rises with increasing chain length until a limiting value is reached where further increase in molecular weight has very little effect on T_g. To a very large extent, the practical utility of polymers and their different properties depend heavily on their glass transition temperatures. Amorphous plastics perform best below T_g, but elastomers must be used above the brittle point or T_g.

14.5.2   Mechanical properties

Mechanical properties of a polymer depend on many variables— molecular weight, molecular weight distribution, morphology, additives, temperature, time, and so on. Mechanical properties of polymers are much more dependent on molecular weight over a very broad range, yet they too level off at high molecular weight. The *leveling off* molecular weight depends on the polymer structure. For very polar polymers, such as polyamides, the leveling off may occur at a molecular weight in the range 20,000 to 50,000. Whereas polyolefins such as polyethylene, in which mechanical strength is mainly because of dispersion forces, the leveling off occurs at much higher molecular weight (above 10^5). The dependence of properties on molecular weight is shown for a hypothetical polymer in Fig. 14.17.

Mechanical properties of a polymer are mainly determined by how much stress a sample will withstand before the sample *fails*. At low strain (i.e., <1 percent), the deformation of most polymers is elastic where the deformation is homogeneous and full recovery can occur over a finite time. Among the mechanical properties that are of fundamental interests in commercial polymers are:

1. *Tensile strength*: It is a measure of resistance to stretching. Anything that contributes to chain stiffening such as the presence of bulky side groups, cyclic units, cross-linking, and crystallinity, will increase the tensile strength, but decrease the tensile elongation.

2. *Compressive strength*: It is the extent to which a sample can be compressed before it fails.

3. *Flexural strength*: It is a measure of resistance to breaking, or snapping, when a sample is bent (flexed).

4. *Impact strength*: It is a measure of toughness, that is, how well a sample will withstand the sudden onset of stress, like a hammer blow. Usually, chain stiffening lowers the impact strength and increases embrittlement.

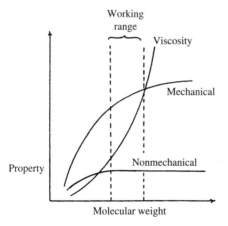

Figure 14.17 Dependence of properties on molecular weight (hypothetical polymers) [11].

5. *Fatigue*: It is a measure of how well a sample will withstand repeated application of tensile, flexural, or compressive stress.

6. *Creep*: It is the cold flow of a polymer. It is a measure of the change in strain when a polymer sample is subjected to a constant stress (such as gravity).

7. *Stress relaxation*: It refers to the decrease in stress when a sample is elongated rapidly to a constant strain.

14.5.3 Solubility

Solvents are frequently used during the polymerization processes, during fabrication (i.e., film casting, fiber formation, and coatings), and for the determination of molecular weight and molecular weight distribution. Dissolution of a polymer sample in a solvent occurs in two stages. The first stage is a solvation process in which the solvent molecules diffuse slowly through the polymer matrix to form a swollen, solvated mass called a *gel*. The second stage of dissolution consists of a breakdown of the swollen gel to give an actual solution of polymer molecules in the given solvent. In general, linear and branched polymers dissolve in the second stage, whereas cross-linked polymers remain in a swollen condition. However, lack of solubility does not necessarily mean that the polymer is of the network type; some linear polymers defy all attempts to dissolve them.

General rules for polymer solubility. A number of structural parameters of the polymer affect the ease and speed of the polymer solution. These include:

1. *Chemical structure*: Polar polymers tend to dissolve in polar solvents, and nonpolar polymers tend to dissolve in nonpolar solvents (like dissolves like). Therefore, chemical similarity of a polymer and solvent is a fair indication of solubility. For example, ethanol can dissolve poly(vinyl alcohol) but not polystyrene, whereas toluene can dissolve polystyrene but not poly(vinyl alcohol).

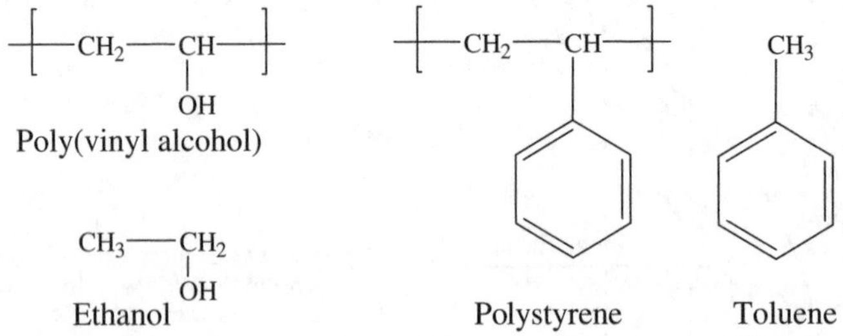

2. *Molecular weight*: In a given solvent at a particular temperature, the solubility of a polymer decreases as its molecular weight increases.

3. *Cross-linking*: If a polymer is cross-linked to a significant degree, it will only swell as a result of taking liquid into the network, but complete dissolution does not occur.

4. *Crystallinity*: In general, solubility of a polymer decreases with an increase in the degree of crystallinity, but it is possible to find solvents capable of overcoming the crystalline bonding forces and hence dissolving the polymer. Also, solubility is enhanced by heating the polymer to a temperature close to its crystalline melting point.

5. *Branching*: In general, solubility increases with increasing polymer branching.

Predictions of polymers solubilities.   Dissolution of polymer in a solvent occurs if the solvent-solvent and the polymer-polymer cohesive forces are similar to adhesive forces between the solvent and polymer molecules. Solubility of polymers is usually predicted using a parameter called *solubility parameter, δ*. The solubility parameter is related to cohesive energy density, or the molar energy of vaporization of a pure liquid. Cohesive energy density is defined as the energy needed to remove a molecule from its nearest neighbors. The magnitude of the cohesive forces may be described simply by the amount of energy needed to vaporize a certain volume of a substance, and is defined by Eq. 2.

$$\text{Cohesive energy density (CED)} = \delta^2 = \frac{E_0}{V_0} \qquad (2)$$

where E_0 = molar latent heat of vaporization
V_0 = molar volume
δ = solubility parameter

Hildebrand [12] showed that the enthalpy change per unit volume for endothermic dissolution (the usual case for polymers) is expressed as

$$\Delta H_{\text{dissolution}} = \phi_1 \phi_2 (\delta_1 - \delta_2)^2 \qquad (3)$$

where ϕ_1 = volume fraction of solvent in the solution
ϕ_2 = volume fraction of polymer in the solution
δ_1 = solubility parameter of the solvent
δ_2 = solubility parameter of the polymer

From the thermodynamic point of view, the solution process is governed by the free energy relationship (Eq. 4)

$$\Delta G = \Delta H - T\Delta S \tag{4}$$

where ΔG = the change in Gibbs free energy
ΔH = the change in enthalpy
T = the absolute temperature
ΔS = the change in entropy

For a polymer to dissolve spontaneously, it is essential that the free energy of solution, ΔG, is negative. The absolute temperature must be positive, and the entropy of solution, ΔS, is generally positive because of the increase in conformational mobility of the polymer chains. Therefore, the magnitude of the enthalpy of solution, ΔH, determines the sign of ΔG. A positive ΔH means the pure materials of solvent and the polymers are each in a lower energy state and hence dissolution is disfavored, whereas a negative ΔH indicates that the solution is the lower energy state. In general, the solution process is an endothermic process, and then ΔH must be less than $T\Delta S$ if the polymer is to be soluble. However, negative ΔH's are usually encountered when specific interactions such as hydrogen bondings are possible between polymer and solvent molecules. The best one can expect in a good solvent is that $\Delta H_{\text{solution}}$ should be close to zero (it may not be negative because dissolution of a polymer is usually an endothermic process); therefore, $(\delta_1 - \delta_2)^2$ must also be small. Hildebrand showed that the polymer will dissolve in a solvent or a mixture of solvents having solubility parameters, δ, that do not differ by more than ± 1.8 $(\text{cal/cm}^3)^{1/2}$, that is, $\delta_1 - \delta_2$ should be less than ± 1.8 H [Hildebrand unit, H = $(\text{cal/cm}^3)^{1/2}$].

$$\Delta H_{\text{dissolution}} = \phi_1\phi_2(\delta_1 - \delta_2)^2 \approx 0 \quad \text{if} \quad \delta_1 = \delta_2 \tag{5}$$

The solubility parameters of various solvents can be determined in the following way:

$$\delta = \left(\frac{E_0}{V_0}\right)^{1/2} = \left[\frac{\Delta H_v - RT}{V_0}\right]^{1/2} = \left[\frac{D(\Delta H_0 - RT)}{M}\right]^{1/2} \tag{6}$$

where D = density; M = molar mass; $V = M/D$; H_v = enthalpy of evaporation. The use of the above equation can be illustrated by calculating the solubility parameter for n-heptane that has a molecular weight of 100 g/mol, and at 298 K has a density of 0.68 g/cm^3 and heat of vaporization equals to 8700 cal/mol, δ is calculated to be 7.4 H.

The solubility parameters for polymers, however, cannot be determined from their energies of evaporation, because polymers are usually solids of high molecular weight and decompose before reaching their melting point. A more practical approach is the use of *group molar attraction constants*. These constants (designated G) are derived from studies of low-molecular weight compounds and represent the intermolecular forces for various molecular groupings. Two sets of G values have been suggested, one by Small [13], derived from heat of vaporization, and the other by Hoy [14], based on vapor pressure measurements. The set to be used is determined by the method used to determine δ_1 for the solvent. Typical values for G are shown in Table 14.13. G values are additive for a given structure, and are related to δ by

$$\delta = \frac{D\Sigma G}{M} \tag{7}$$

where G = molar attraction constant for a particular group in the repeating unit. D and M represent the density and molecular weight, respectively, of the repeating unit. For example, the solubility parameter of amorphous poly(vinyl acetate) (D = 1.2, M = 86) is calculated according to Small's G values as

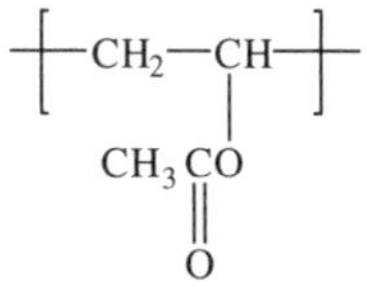

$$\delta = \frac{1.2(133 + 214 + 28 + 310)}{86} = 9.55 \text{ H}$$

or according to Hoy's values as

$$\delta = \frac{1.2(131.5 + 147.3 + 85.99 + 326.6)}{86} = 9.65 \text{ H}$$

The solubility parameter can be used as a guide in choosing solvents for a polymer. The smaller the difference $|\delta_{\text{solvent}} - \delta_{\text{polymer}}|$, the higher is the solubility of the particular polymer in that particular solvent. This generalization, however, should be taken with caution, because, as discussed before, factors such as strong polymer-solvent interactions, H-bonding, crystallinity, branching, and molar mass may contribute to solubility. Various polymers and solvents and their solubility parameters are listed in Tables 14.14 and 14.15.

The law of mixtures applies to the solubility parameter, and thus it is possible to mix two or more solvents to form a mixture with a certain

TABLE 14.13   Representative Group Molar Attraction Constants

Group	$G[\text{cal cm}^3]^{1/2}\text{ mol}^{-1}$	
	Small[13]	Hoy[14]
CH_3——	214	147.3
——CH_2——	133	131.5
⟍CH——	28	86
⟩C⟨	−93	32
==CH_2	190	126.5
——CH==	111	121.5
⟩C==	19	84.5
——CH == (aromatic)		117.1
——C_6H_5 (phenyl)	735	
——H (variable)	80–100	
HC==C——	285	
——C==N	410	
——N==C==O	359	358.5
——F	122	
——Cl (primary)	270	205
——Br (primary)	340	
——O—— (ether)	70	115
——S—— (sulfide)	225	209.5
⟩C==O (ketone)	275	262.7
——CO_2—— (ester)	310	326.6

value of solubility parameter. For example, a mixture of n-pentane ($\delta = 7.1$ H) and n-octane ($\delta = 7.6$ H) in a respective molar ratio of 70 to 30 will have a solubility parameter value given by

$$\delta = (0.70)(7.1) + (0.30)(7.6) = 7.25 \text{ H}$$

TABLE 14.14   Typical Values of the Solubility Parameter 'δ' for Some Common Polymers and Solvents [15]

Solvent	H $(\text{cal/cm})^{1/2}$	Polymer	H $(\text{cal/cm})^{1/2}$
Acetone	10.0		
Acetonitrile	11.9	Cellulose	15.5
Benzene	9.2	Cellulose diacetate	10.0
n-Butanol	11.4	Ethyl cellulose	10.3
2-Butanone	9.0	Nylon-6,6	13.6
Carbon tetrachloride	8.6	Polyacrylonitrile	15.4
Chlorobenzene	9.5	Polybutadiene	8.6
Chloroform	9.3	Polycarbonate	10.0
Cyclohexane	8.2	Poly(dimethyl siloxane)	7.3
Cyclohexanol	11.4	Polyethylene	7.9
Cyclohexanone	9.9	Poly(ethylene terephthalate)	10.7
Dichloromethane	9.7	Polyisobutylene	8.1
Dimethylformamide	12.1	Poly(methyl methacrylate)	9.5
Dimethyl sulfoxide	13.0		
Dioxane	9.9		
Ethanol	12.4		
Ether	7.4	Polypropylene	8.1
Ethyl acetate	9.0		
Ethylene glycol	14.6	Polystyrene	9.1
Formamide	19.2		
n-Heptane	7.4	Polytetrafluoroethylene	6.2
n-Hexane	7.2	Polyurethane	10.0
Methanol	14.5	Poly(vinyl acetate)	9.4
Styrene	9.3	Poly(vinyl alcohol)	12.5
Tetrahydrofuran	9.9	Poly(vinyl chloride)	9.7
Toluene	8.9		
Water	23.4		

TABLE 14.15   The More Important Solvents of Some Polymers [16]

Polymer	Solvent
Poly(acrylonitrile)	Dimethyl formamide, nitrophenol, ethylene carbonate
Polyethylene	Aromatic and chlorinated solvents, above 70°C
Poly(methyl methacrylate)	Ester, chlorinated and aromatic hydrocarbons
Polystyrene	Aromatic and chlorinated hydrocarbons
Poly(tetrafluoro ethylene)	Not known
Poly(vinyl acetate)	Aromatic and chlorinated hydrocarbons, low molecular mass esters, methanol
Poly(vinyl alchohol)	Water (if polymer is fully hydrolyzed)
Poly(vinyl chloride)	Tetrahydrofuran, toluene, dioxane (at lower molecular mass)
Poly(ethylene terphthalate)	Benzyl alcohol (warm), nitrobenzene (warm)
Linear polyester	Hydrocarbons, chlorinated hydrocarbones, low molecular
Alkyd resins	Ketons, esters mass alcohols
Polypropylene	Above 135°C decaline, tetraline, paraffins
Cellulose nitrate	Low molecular mass alcohols, ketones, esters
Cellulose acetate	Ketones, low molecular mass esters

It is very useful to know what solvents will dissolve a particular polymer. Most physical measurements of the polymer, such as determining molecular weight or molecular weight distribution, are carried out using a solution of the polymer in a suitable solvent. Also polymer-solvent interaction can be of special importance in some practical applications. Stiff polymers may be flexibilized by the addition of plasticizers, which are nonvolatile compatible liquids with the $\Delta\delta$ values between the polymer and the plasticizer being ≤ 1.8 H. Plasticizers work by reducing the intermolecular attractions (CED and δ) of polymers that permits slippage of polymer chains and thus reduces the T_g and modulus of the polymer. This makes processing the polymer less difficult. For example, poly(vinyl chloride) (PVC) itself is a hard, brittle, and rigid material. In this form it is often used to make pipes, rods, and compact discs. Poly(vinyl chloride), softened with a plasticizer such as esters, is used for making vinyl leather (used for handbags, briefcases, and inexpensive shoes), plastic raincoats, shower curtains, garden hoses, floor covering, and automobile upholstery.

14.5.4   Viscosity

Viscosity of a material is a measure of its resistance to flow. As a material's resistance to flow increases, its viscosity increases. Viscosities are usually reported in a unit called *poise*, which is dyne seconds per square centimeter (1 poise = 10^{-1} Nsm^{-2}). Another widely employed unit in reporting viscosity values is *Pascal*, which is Newton seconds per square centimeter (1 Pas = 10 poise). The flow of polymer (i.e., viscosity) is impeded by chain entanglement, high intermolecular forces, and other factors such as the presence of reinforcing agents and cross-links.

Viscosity measurements constitute an important method for determining *relative* molecular weights of polymers. Viscosities are measured at dilute concentrations (about 0.5 g/100 mL of solvent) by determining the flow time of volume of a solution through a capillary of fixed length. Also, viscosity measurements are run at constant temperature. The method is unique for being rapid, easy, and requires only minimal instrumentation. Relatively small amounts of dissolved polymer could cause tremendous increase in the viscosity of the resulting solution relative to that of the pure solvent. In general, the viscosity of a polymer solution increases with the polymer molecular weight. Other factors that control the viscosity of a polymer solution include the particular polymer and solvent, solute concentration, and temperature.

Viscosity can be expressed in several ways (Table 14.16). *Relative viscosity* (or viscosity ratio) (η_{rel}) is the ratio of solution viscosity to solvent viscosity that is proportional to a first approximation for dilute solutions to the ratio of the corresponding flow times. *Specific viscosity* (η_{sp}) is the fractional increase in viscosity. The *reduced viscosity* or (viscosity number)

TABLE 14.16   Commonly Used Viscosity Designations

Common name	IUPAC name	Definition
Relative viscosity	Viscosity ratio	$\eta_{rel} = \dfrac{\eta}{\eta_0} = \dfrac{t}{t_0}$
Specific viscosity	–	$\eta_{sp} = \dfrac{\eta - \eta_0}{\eta_0} = \dfrac{t - t_0}{t_0} = \eta_{rel} - 1$
Reduced viscosity	Viscosity number	$\eta_{red} = \dfrac{\eta_{sp}}{C} = \dfrac{\eta_{rel} - 1}{C}$
Inherent viscosity	Logarithmic viscosity number	$\eta_{inh} = \dfrac{\ln \eta_{rel}}{C}$
Intrinsic viscosity	Limiting viscosity number	$[\eta] = \left(\dfrac{\eta_{sp}}{C}\right)_{C=0} = (\eta_{inh})_{C=0}$

(η_{red}) is obtained by dividing η_{sp} by the concentration of the solution, C. Viscosity increases as the concentration increases. In order to eliminate the concentration effects, the specific viscosity is divided by concentration and extrapolated to zero concentration to give the *intrinsic viscosity* (or limiting viscosity number), $[\eta]$. Viscosity determined at a single concentration, *inherent viscosity* (η_{inh}), extrapolates to the same $[\eta]$, and can also be used as an approximate indication of molecular weight (Fig. 14.18).

Intrinsic viscosity is the most useful of the various viscosity expressions because it can be related to molecular weight by the Mark-Houwink-Sakurada equation:

$$[\eta] = K\overline{M}_v^{\alpha} \qquad (8)$$

where $\overline{M}_v$ is the viscosity-average molecular weight.

The exponent α is a measure of the interaction of the solvent and polymer. It is a function of the shape of the polymer coil in a solution, and usually has a value between 0.5 (for a randomly coiled polymer in a θ solvent) and 0.8 (when the polymer coils expand in good solvents). The α value is 0 for spheres, about 1 for semicoils, and is between 1.8 and 2.0 for a rigid polymer chain extended to its full contour length. The proportionality constant K is characteristic of the polymer and solvent. The constants K and α are the intercept and slope, respectively, of a plot of log $[\eta]$ versus log M of a series of fractionated polymer samples. Viscosity average molecular weights lie between those of the corresponding

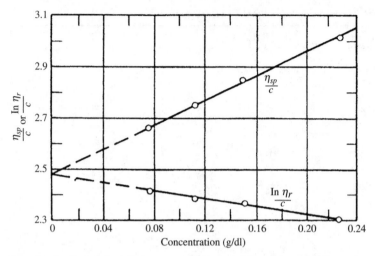

Figure 14.18 Reduced and inherent viscosity-concentration curves for a polystyrene in benzene [17].

weight-average molecular weight ($\overline{M}_w$) and number-average molecular weight ($\overline{M}_n$), but closer to the former. Hence, better results are obtained if K and α are determined with fractionated samples of measured $\overline{M}_w$. A partial list of K and α values is given in Table 14.17.

Viscosity measurements alone cannot be directly used in the Mark-Houwink-Sakurada equation to relate *absolute* viscosity and polymer molecular weight, since additional unknowns, K and α must be determined. Therefore, viscometry does not yield absolute molecular weight values; it rather gives only a relative measure of polymer's molecular weight. Viscosity measurements based on the principle of mechanical shearing are also employed, most commonly with concentrated polymer solutions or undiluted polymer; these methods, however, are more applicable to flow properties of polymers, not molecular weight determinations.

14.6   Rheology

Polymers are viscoelastic materials, meaning they exhibit both liquid-like properties (*visco*) and solid-like properties (*elastic*). Whether a material behaves more as a viscous or more as an elastic material depends upon the temperature, the particular polymer and its prior treatment, polymer structure, and the time scale of deformation. The particular property demonstrated by a polymer under given conditions allows polymers to act as solid or viscous liquids, as plastics, elastomers, or fibers, and so on.

Many of the viscoelastic properties of polymers are determined using stress-strain measurements. Several models were employed to

TABLE 14.17   Representative Viscosity-Molecular Weight Constants [18]

Polymer	Solvent	Temperature (°C)	Molecular weight range $\times 10^{-4}$	$K \times 10^3$	α
Polystyrene	Cyclohexane	35	8–42	80	0.50
(*atactic*)	Cyclohexane	50	4–137	26.9	0.599
	Benzene	25	3–61	9.52	0.74
Polyethylene	Decalin	135	3–100	67.7	0.67
(*low pressure*)					
Poly(vinyl chloride)	Benzyl alcohol	155.4	4–35	156	0.50
	Cyclohexane	20	7–13	13.7	1.0
Polybutadiene					
98% *cis*-1,4,	Toluene	30	5–50	30.5	0.725
2%-1,2					
97% *trans*-1,4,	Toluene	30	5–16	29.4	0.753
3%-1,2					
Polyacrylonitrile	DMF	25	5–27	16.6	0.81
	DMF	25	3–100	39.2	0.75
Poly(methyl methacrylate*co*–styrene)					
30–70 mol%	1-Chlorobutane	30	5–55	17.6	0.67
71–29 mol%	1-Chlorobutane	30	4.8–81	24.9	0.63
Poly(ethylene terephthalte)	*m*-Cresol	25	0.04–1.2	0.77	0.95
Nylon-6,6	*m*-Cresol	25	1.4–5	240	0.61

demonstrate the deformation of polymers resulting from the application of stress. In these models, the fluid or liquid part of the behavior is usually described in terms of a Newtonian dash pot or shock absorber; an element that represents chain and local segmental movement. The viscous behavior is more pronounced for polymers at temperatures above their T_g, and is related to the slow disentanglement and slippage of polymer chains past each other. On the other hand, the elastic or solid part of the behavior is described in terms of a Hookean or ideal elastic spring, which would represent bond flexing. The spring-like behavior is more important for polymers below their T_g. Hookes's law (Eq. [9]) states that the applied stress (s) is proportional to the resultant strain (γ), but is independent of the rate of this strain ($d\gamma/dt$). Stress is equal to force per unit area, and strain is the extension per unit length.

$$s = E\gamma \tag{9}$$

where E is Young's modulus of elasticity, which for polymers is usually replaced by shear modulus (G), where $E \approx 2.6G$.

Newton's law, which describes the flow properties, states that the applied stress (s) is proportional to the rate of strain ($d\gamma/dt$), but is independent of the strain (Eq. [10])

$$s = \eta \frac{d\gamma}{dt} \tag{10}$$

where η is viscosity.

The two modeling elements, spring and dash pot, are combined in various ways to demonstrate the deformation of a polymer subjected to the application of stress as shown in Fig. 14.19.

Figure 14.20 illustrates typical stress-strain behavior based on which polymers can be described as being soft, hard, tough, brittle, weak, strong, and so on. Polymers in class (a) are described as soft and weak, and are characterized by a low modulus of elasticity, low yield point, and moderate time-dependent elongation. Polyisobutylene is an example of this type of polymers. Yield point is the point on a stress-strain curve below which there is reversible recovery. Class (b) polymers are hard and brittle; they are characterized by a high modulus of elasticity, poorly defined yield point, and little elongation before failure. An example of this class is polystyrene. Polymer hardness is a property related to resist scratching, marring and abrasion, penetration and indentation. Soft and tough polymers (class [c]) such as plasticized PVC have a low modulus of elasticity, high elongation, and a well-defined yield point. Rigid PVC, on the other hand, is hard and strong (class [d]), characterized by a high modulus of elasticity and high yield strength. Toughness is measured as the area under the entire stress-strain curve. Hard and tough polymers (class [e]), such as the copolymer of acrylonitrile-butadiene-styrene, exhibit moderate elongation prior to yield point followed by nonrecoverable elongation.

Mechanical properties are also temperature dependent. Figure 14.21 demonstrates the effect of temperature on tensile modulus of an amorphous thermoplastic. The temperature effect on mechanical properties will also be a function of other structural parameters such as molecular weight, extent of cross-linking, and degree of crystallinity. The higher the molecular weight, the higher is the temperature necessary to overcome the increased molecular weight entanglements. Similarly, the higher the cross-link density, the less is the flow, and hence the greater will be the modulus. Semicrystalline polymers behave much like cross-linked polymers below the melting temperature; however, they flow above the melting temperature. Some mechanical properties are rather time-dependent such as creep and stress relaxation.

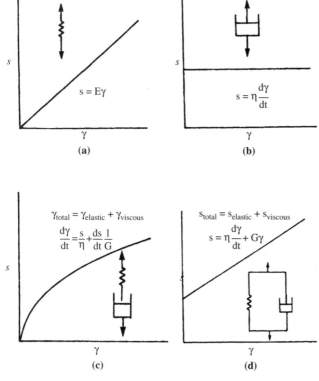

(a)

$$s = E\gamma$$

(b)

$$s = \eta\frac{d\gamma}{dt}$$

(c)

$$\gamma_{total} = \gamma_{elastic} + \gamma_{viscous}$$
$$\frac{d\gamma}{dt} = \frac{s}{\eta} + \frac{ds}{dt}\frac{1}{G}$$

(d)

$$s_{total} = s_{elastic} + s_{viscous}$$
$$s = \eta\frac{d\gamma}{dt} + G\gamma$$

Figure 14.19 Stress-strain plots for (*a*) a Hookean spring where *E* is the slope; (*b*) a Newtonian dash pot where s is constant, (*c*) stress-time plot stress for relaxation in the Maxwell model, and (*d*) stress-time plot stress for a Voigt-Kelvin model.

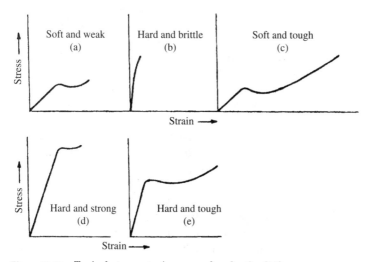

Soft and weak (a)

Hard and brittle (b)

Soft and tough (c)

Hard and strong (d)

Hard and tough (e)

Figure 14.20 Typical stress-strain curves for plastics [19].

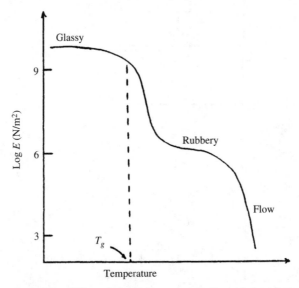

Figure 14.21   Effect of temperature on tensile modulus of an amorphous thermoplastic; log E, modulus scale; T_g, glass transition temperature [11].

14.7   Molecular Weight of Polymers

An important parameter that characterizes a polymer is its size. Molecular weight relates directly to a polymer's physical properties. Because of the massive size, polymer chains act as a group so that when one part of the chain moves the other parts are affected, and when one polymer chain moves, surrounding chains are affected by that movement. Also, the massive size of polymers allows cumulative effects of secondary interaction forces among polymer chains to become dominant factors in some behavior. Some of the unique properties that depend on the molecular weight and molecular weight distribution of polymers include melt viscosity, tensile strength, modulus, impact strength or toughness, and resistance to heat and corrosives. In contrast, properties like density, specific heat capacity, and refractive index are essentially independent of the molecular weight values above the critical molecular weight. Figure 14.17 depicts the relationship of polymer properties to molecular weight. For most practical applications, a particular polymer has a threshold molecular value (TMWV) that is the lowest molecular weight where the desired property value is achieved. Extremely high molecular weight might lead to processing difficulties. Accordingly, it is customary to establish a commercial polymer range above the TMWV that affords good mechanical properties, but below the extremely high molecular weight range affording a workable viscosity. What defines an optimum molecular weight depends largely on the chemical structure of the polymer and the application for which it is intended. The value of TMWV depends on T_g, the extent of crystallinity in crystalline polymers, the cohesive energy density (CED) of amorphous polymers, and the effect of reinforcements in polymeric composites.

Molecular-weight averages and molecular-weight distribution. Some naturally occurring polymers such as specific proteins (e.g., enzymes, muscle fibers, and natural silk) have a fixed number of building units, and thus have a discrete molecular weight, and are said to be monodisperse with respect to molecular weight. Commercial synthetic polymers, on the other hand, are usually produced with heterogeneous molecular weights. That is, they are made up of molecules of different numbers of repeating units, and are said to be polydisperse. The exact breadth of the molecular-weight distribution depends upon the specific conditions of polymerization. In addition to differing in size, polymer chains may vary in the extent and frequency of branching, the distribution of branching, and the length of branching. Therefore, it is necessary to define an *average* molecular weight to characterize an individual polymer sample. In this context, the numerical value of repeating units (termed *degree of polymerization*, DP) should be considered an average degree of polymerization, $(\overline{DP})$. As a result, the average molecular weight of a polydisperse polymer will equal the product of the average degree of polymerization $(\overline{DP})$ and the molecular weight of the repeating unit. There are different ways to report an average polymer molecular weight:

1. Number-average molecular weight $(\overline{M}_n)$: This is obtained from methods that depend on end-group analysis or colligative properties (freezing point depression, boiling point elevation, vapor pressure lowering, osmotic pressure, and so on) where the number of molecules of each weight in the sample is counted. In measurements of colligative properties, each molecule contributes equally irrespective of weight. $\overline{M}_n$ is simply the weight of sample per mole, and can be calculated like any other numerical average by dividing the sum of the individual molecular weight values by the number of molecules. Mathematically, $\overline{M}_n$ is shown as

$$\overline{M}_n = \frac{\text{total weight of sample}}{\text{number of molecules of } N_i} = \frac{W}{\sum_{i=1}^{\infty} N_i} = \frac{\sum_{i=1}^{\infty} N_i M_i}{\sum_{i=1}^{\infty} N_i} \qquad (11)$$

Most thermodynamic properties of polymers are best related to $\overline{M}_n$.

2. Weight-average molecular weight ($\bar{M}_n$): This is obtained from measurements in which each molecule or chain makes a contribution to the measured result relative to its size. Analytical methods based on light scattering, diffusion, sedimentation, and ultracentrifugation, for example, determine molecular weight based on mass of the species. The greater the mass, the greater is the contribution to the measurement. In light scattering, for example, the larger molecules contribute more because they scatter light more effectively. $\bar{M}_n$ is expressed mathematically as

$$\bar{M}_w = \frac{\sum_{i=1}^{\infty} w_i M_i}{\sum_{i=1}^{\infty} w_i} = \frac{\sum_{i=1}^{\infty} N_i M_i^2}{\sum_{i=1}^{\infty} N_i M_i} \tag{12}$$

where w_i is the weight fraction of each species. Bulk properties associated with large deformations, such as viscosity and toughness, are particularly affected by $\bar{M}_w$.

3. Viscosity-average molecular weight ($\bar{M}_v$): This is obtained from viscosity measurement and is given by the expression:

$$\bar{M}_v = \frac{\sum_{i=1}^{\infty} N_i M_i^{1+\alpha}}{\sum_{i=1}^{\infty} N_i M_i^{\alpha}} \tag{13}$$

where $0.75 < \alpha < 1$. Determination of the intrinsic viscosity, $[\eta]$, is widely used as a method for routine molecular-weight determination.

Molecular weight is related to $[\eta]$ by the *Mark-Houwink-Sakurada equation* given as

$$[\eta] = K\bar{M}_v^{\alpha}$$

Both K and α are empirical (Mark-Houwink) constants that are specific for a given polymer, solvent, and temperature.

4. Z-average molecular weight ($\bar{M}_z$): This is also determined by ultracentrifugation technique. The molecular weight depends both on size and mass of the molecules. However, the contribution of the mass of the particle are weighed further in this type of molar mass. Mathematically, $\bar{M}_z$ is given as

$$\bar{M}_z = \frac{\sum_{i=1}^{\infty} N_i M_i^3}{\sum_{i=1}^{\infty} N_i M_i^2} \tag{14}$$

Melt elasticity of a polymer is closely dependent on $\bar{M}_z$.

Also, the molecular weight type and value obtained depend, in large measure, on the method of measurement. Molecular weights obtained from methods based on colligative properties, end-group analysis, light scattering photometry, and ultracentrifugation are referred to as *absolute molecular weights*. These molecular weights are related to measured parameter(s) through basic mathematical relationships. On the other hand, methods like gel permeation chromatography (GPC) and viscometry yield what is referred to as *relative molecular weight*, as they require calibration with polymers of known molecular weight determined from an absolute molecular weight technique.

For a polymer that is heterogeneous with respect to molecular weight, $\overline{M}_z$ is always greater than $\overline{M}_w$, and $\overline{M}_w$ is always greater than $\overline{M}_v$ which in turn is greater than $\overline{M}_n$. When the exponent in the Mark-Houwink equation is equal to 1, then $\overline{M}_v$ is equal to $\overline{M}_w$. However, typical values of α are 0.5 to 0.8, and therefore, $\overline{M}_w$ is usually greater than $\overline{M}_v$ (Fig. 14.22).

The ratio $\overline{M}_w/\overline{M}_n$ is a measure of the breadth of the molecular weight range in a polymer sample, and is called polydispersity index, PDI, or molecular weight distribution, MWD. Molecular weight distribution is an important characteristic of polymers, and can significantly affect polymer properties. The molecular weight distributions of commercial

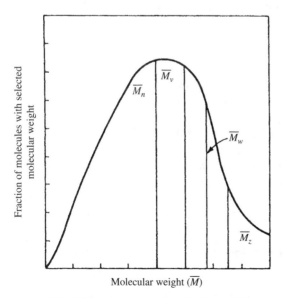

Figure 14.22   Molecular weight distributions [20].

TABLE 14.18   Typical Molecular Weight Determination Methods [20]

Method	Type of molecular weight average	Applicable weight range[a]
Light scattering	$\overline{M}_w$	To ∞
Membrane osmometry	$\overline{M}_n$	2×10^4 to 2×10^6
Vapor phase osmometry	$\overline{M}_n$	To 40,000
Electron and X-ray microscopy	$\overline{M}_{n,w,z}$	10^2 to ∞
Isopiestic method (isothermal distillation)	$\overline{M}_n$	To 20,000
Ebuliometry (boiling point-elevation)	$\overline{M}_n$	To 40,000
Cryoscopy (melting point depression)	$\overline{M}_n$	To 50,000
End-group analysis	$\overline{M}_n$	To 20,000
Osmodialysis	$\overline{M}_n$	500 to 25,000
Centrifugation		
Sedimentation equilibrium	$\overline{M}_z$	To ∞
Archibald modification	$\overline{M}_{z,w}$	To ∞
Trautman's method	$\overline{M}_w$	To ∞
Sedimentation velocity	Gives a real M only for monodisperse systems	To ∞
Chromatography	Calibrated	To ∞
Small-angle X-ray scattering	$\overline{M}_w$	
Mass spectrometry		To 10^6
Viscometry	Calibrated	To ∞
Coupled chromatography-Light scattering		To ∞

[a]"To ∞"means that molecular weight of the largest particles soluble in a suitable solvent can be determined in theory.

polymers vary widely. For example, commercial grades of polystyrene with $\overline{M}_n$ of over 100,000 have MWD between 2 and 5, whereas polyethylene synthesized in the presence of a stereospecific catalyst may have a MWD as high as 30. In contrast, the MWD of some vinyl monomers prepared by *living* polymerization can be as low as 1.06. Such polymers with nearly monodisperse molecular-weight distributions are useful as molecular weight standards for the determination of molecular weights and molecular weight distributions of commercial polymers. Typical techniques for molecular weight determination are given in Table 14.18.

14.8   The Synthesis of High Polymers

Polymerization is the process of joining together small molecules by covalent bonds. The small molecules (monomers) must be at least difunctional. Polymer-forming reactions can be classified into two categories: *condensation* versus *addition* and *stepwise* versus *chainwise*. The terms condensation and addition polymers were first proposed in 1929 by W. H. Carothers. Condensation reactions are those in which some part of the reacting system is eliminated as a small molecule. Thus, the condensation polymers contain fewer atoms within the polymer repeat unit than the reactants from which they are formed (or to which they can be degraded). For example, polyamides (such as nylon-6,6 (**2**)) are produced from condensation reactions between diamines and diacarboxylic acid. Also, most natural polymers such as cellulose, starch, wool, and silk are classified as condensation polymers. In contrast, an addition polymer has the same atoms as the monomer in its repeat unit. The most important group of addition polymers includes those derived from unsaturated vinyl monomers, such as ethylene, propylene, styrene, vinyl chloride, methyl acrylate, and vinyl acetate. While the atoms in the backbone of addition polymers are usually carbon atoms, the backbone of condensation polymers usually contains atoms of more than one element.

Carothers' classification (condensation *vs.* addition) is primarily based on the composition or structure of polymers. The second classification (chainwise *vs.* stepwise) was proposed by P. J. Flory, and is based on the kinetic scheme or mechanism governing the polymerization reactions. *Step reactions* are those in which the chain growth occurs in a slow, stepwise manner. Two monomer molecules react to form a dimer. The dimer can then react with another monomer to form a trimer, or with another dimer to form tetramer. Thus, the average molecular weight of the system increases slowly over a period of time. This is exemplified by the following polyesterification:

$$\text{HO}-\underset{\overset{\parallel}{O}}{C}-\text{R}-\underset{\overset{\parallel}{O}}{C}-\text{OH} \ + \ \text{HO}-\text{R}'-\text{OH} \ \xrightarrow{-\text{H}_2\text{O}} \ \text{HO}-\underset{\overset{\parallel}{O}}{C}-\text{R}-\underset{\overset{\parallel}{O}}{C}-\text{O}-\text{R}'-\text{OH}$$

$$\text{HO}-\underset{\overset{\parallel}{O}}{C}-\text{R}-\underset{\overset{\parallel}{O}}{C}-\text{O}-\text{R}'-\text{OH} \ + \ \text{HO}-\underset{\overset{\parallel}{O}}{C}-\text{R}-\underset{\overset{\parallel}{O}}{C}-\text{OH} \ \xrightarrow{-\text{H}_2\text{O}} \ \text{HO}-\underset{\overset{\parallel}{O}}{C}-\text{R}-\underset{\overset{\parallel}{O}}{C}-\text{O}-\text{R}'-\text{O}-\underset{\overset{\parallel}{O}}{C}-\text{R}-\underset{\overset{\parallel}{O}}{C}-\text{OH}$$

$$+ \ \text{HO}-\text{R}'-\text{OH} \ \xrightarrow{-\text{H}_2\text{O}} \ \text{HO}-\text{R}'-\text{O}-\underset{\overset{\parallel}{O}}{C}-\text{R}-\underset{\overset{\parallel}{O}}{C}-\text{O}-\text{R}'-\text{OH}$$

A high molecular-weight polymer is formed only near the end of the polymerization when most of the monomer has been depleted.

On the other hand, *chain polymerizations* require an ionic or radical initiation to begin chain growth which then takes place by rapid addition of olefin molecules to a growing chain end. The growth continues until some termination reaction renders the chain inactive.

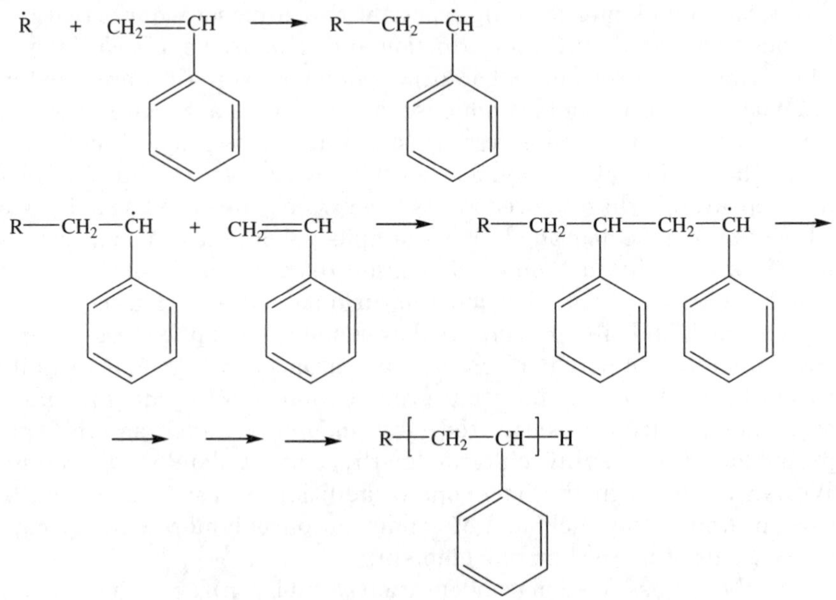

$$\dot{R} \;+\; CH_2{=}CH \longrightarrow R{-}CH_2{-}\dot{C}H$$

$$R{-}CH_2{-}\dot{C}H \;+\; CH_2{=}CH \longrightarrow R{-}CH_2{-}CH{-}CH_2{-}\dot{C}H \longrightarrow$$

$$\longrightarrow \longrightarrow \longrightarrow R{+}CH_2{-}CH{+}H$$

Polystyrene

The two classifications arise from two different bases of classification, yet there is a large but not total overlap between the two classifications. Condensation polymers are usually formed by the stepwise intermolecular condensation of reactive groups; and addition polymers ordinarily result from chain reactions involving some sort of active centers (radical, ionic, or metal-coordinated). With some exceptions, polymers made in chain reactions often contain only carbon atoms in the main chain (*homochain polymers*), whereas polymers made in step reactions may have other atoms, originating in the monomer functional groups, as part of the chain (*heterochain polymers*). Table 14.19 shows the main distinguishing features of chainwise and stepwise polymerizations. The following examples show that the two classifications cannot always be used interchangeably.

1. Polyurethanes and polyureas are produced from the reaction of diisocyanates with a diol or a diamine, respectively.

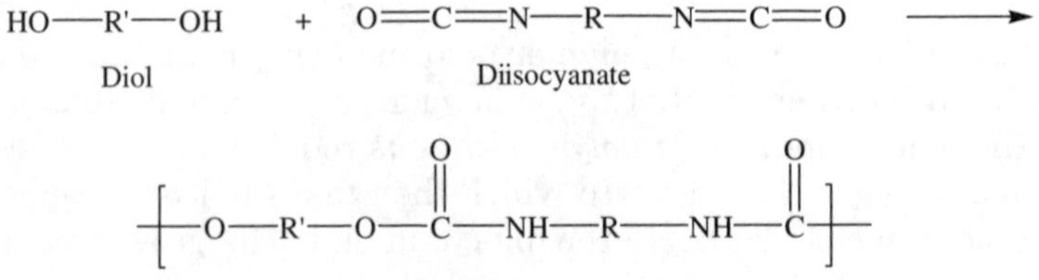

$$HO{-}R'{-}OH \;+\; O{=}C{=}N{-}R{-}N{=}C{=}O \longrightarrow$$

Diol   Diisocyanate

$$\left[O{-}R'{-}O{-}\overset{\overset{\displaystyle O}{\|}}{C}{-}NH{-}R{-}NH{-}\overset{\overset{\displaystyle O}{\|}}{C} \right]$$

Polyurethane

TABLE 14.19   Comparison of Step-Reaction and Chain-Reaction Polymerization

Step-reaction polymerization	Chain-reaction polymerization
Growth occurs through the reaction of any two molecular units with proper functional groups.	Growth occurs only by addition of one unit at a time.
Monomer consumed early in the reaction.	Monomer concentration decreases steadily throughout the reaction.
At any stage all molecular species are present in a calculable distribution.	Reaction mixture contains almost only monomer, high polymers, and very little growing chains.
Polymer chain length increases steadily during the polymerization.	Polymer chains are formed from the beginning of the polymerization and throughout the process.
$\overline{DP}$ is low to moderate.	$\overline{DP}$ can be very high.
Polymerization rate decreases steadily as functional groups consumed.	Polymerization rate increases initially as initiator units generated; remains relatively constant until monomer is depleted.
Long reaction times and high extents of reactions are essential to obtain high molecular weight.	Long reaction times give high yields but have little effect on molecular weight.

The reaction does not involve elimination of any small molecules, and thus according to Carothers could be classified as addition polymers. However, the polymers are structurally more similar to condensation polymers than to addition polymers. The repeating unit contains functional groups (or is heteroatomed). The formation of the two polymers also proceeds through stepwise kinetics.

2. Ladder polymers produced from Diels-Alder reactions are formed through a stepwise kinetic process, yet no small molecules are eliminated.

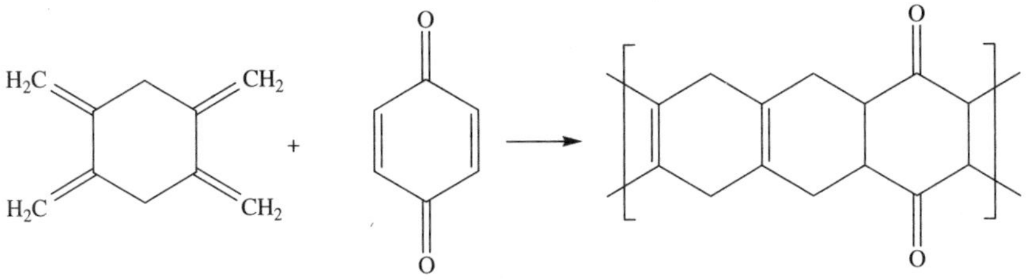

3. Polyester, a condensation polymer, can be produced by chainwise, acid-catalyzed ring openings of cyclic ester (lactone) without expulsion of small molecules, and also by stepwise polycondensation of ω-hydroxycarboxylic acid.

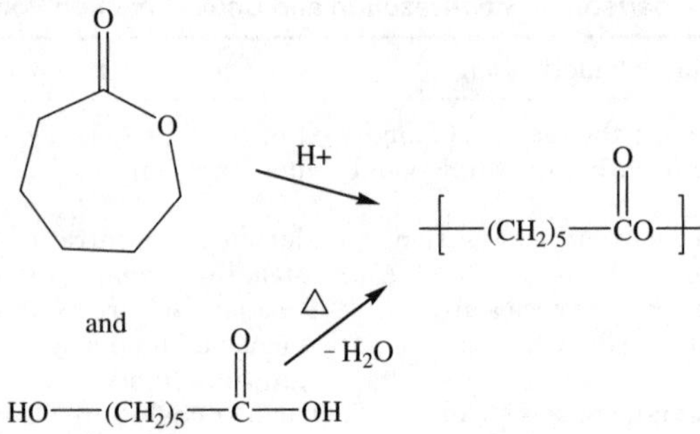

HO—(CH$_2$)$_5$—C—OH

4. Nylon-6, a condensation polymer, can be produced by chainwise, ring openings of cyclic amide (lactam) without expulsion of small molecules, and also by stepwise polycondensation of ω-amino acid.

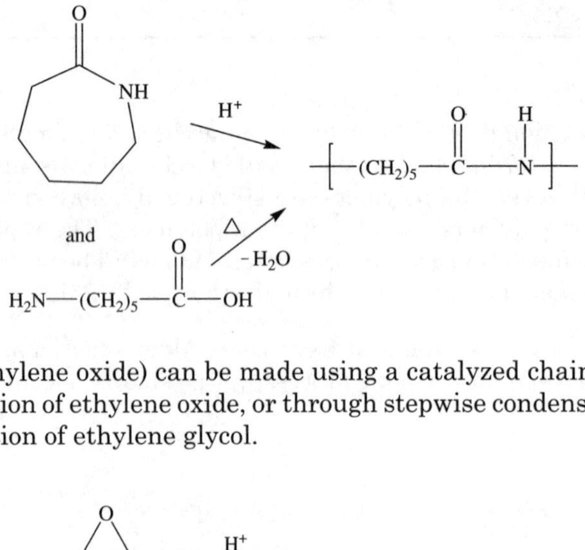

H$_2$N—(CH$_2$)$_5$—C—OH

5. Poly(ethylene oxide) can be made using a catalyzed chainwise polymerization of ethylene oxide, or through stepwise condensation polymerization of ethylene glycol.

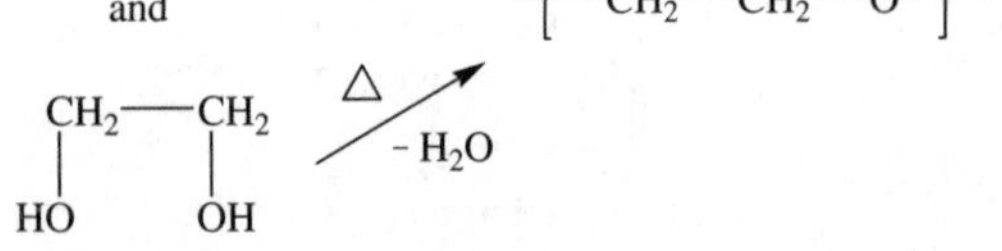

6. Hydrocarbon polymers can be made by the typical chainwise polymerization from ethylene and by the stepwise polymerization from 1,8-dibromooctane.

$$CH_2\!\!=\!\!CH_2 \longrightarrow \left[\!-CH_2\!-\!CH_2\!-\!\right]$$

$$BrCH_2\!-\!(CH_2)_6\!-\!CH_2Br \xrightarrow{\text{2 Na}} \!-\!\!(CH_2)_n\!\!-\! + \ 2\,NaBr$$

The boron trifluoride-catalyzed polymerization of diazomethane illustrates a chain-growth polymerization that is also a condensation reaction.

$$CH_2N_2 \xrightarrow{\text{BF}_3} \!-\!\!(CH_2)_n\!\!-\! + \ N_2$$

Polymers having identical repeating units but formed by entirely different reactions do not necessarily have identical properties. Physical and mechanical properties may differ markedly because different polymerization processes may give rise to differences in molecular weight, end groups, stereochemistry, or possibly chain branching.

14.8.1 Condensation or step-reaction polymerization

In condensation polymerization, polymer formation takes place through the condensation between two complementary functional groups with possible elimination of a small molecule such as water or HCl. The molecule participating in a polycondensation reaction may be a monomer, oligomer, or higher-molecular weight intermediate each having complementary functional end units, such as carboxylic acid or hydroxyl groups. The two cross-reacting functional groups can be in one molecule.

$$A\!-\!B \longrightarrow \left(\!A\!-\!B\!\right)$$

Another approach is to start with two difunctional molecules.

$$A\!-\!A + B\!-\!B \longrightarrow \left(\!A\!-\!A\!-\!B\!-\!B\!\right)$$

The reaction continues until one of the reagents is almost completely used up; equilibrium is established that can be shifted at will at high temperatures by controlling the amounts of reactants and products. In step-growth polymerization, the monomer molecules are consumed rapidly, and chains of any length x and y combine to form longer chains.

$$x\text{-mer} + y\text{-mer} \longrightarrow (x+y)\text{-mer}$$

An example of a condensation polymerization is the synthesis of nylon-66 by condensation of adipic acid and hexamethylene diamine as shown earlier in the equation.

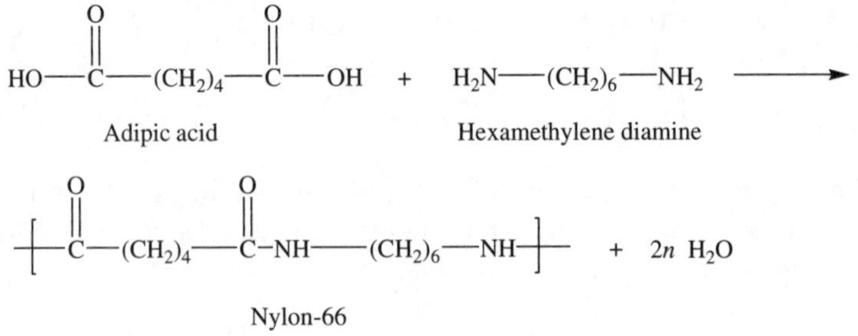

Nylon-66

This polymerization is accompanied by the liberation of two molecules of water for each repeating unit.

Molecular weight in a step-growth polymerization. One way to express molecular weight is through *degree of polymerization, DP,* that normally represents the number of repeating units in the polymer. Carothers developed a simple equation for relating molecular weight to percent conversion of monomer. The reaction conversion, p, is given by the expression:

$$p = \frac{N_0 - N}{N_0} \qquad (15)$$

or

$$N = N_0(1 - p) \qquad (16)$$

where N_0 refers to the total number of molecules present *initially*, and N refers to total molecules present after a given reaction period. The average number of repeating units in all molecules present, that is, $\overline{DP}$, is equal to N_0/N, which can then be expressed as

$$\overline{DP} = \frac{1}{1 - p} \qquad (17)$$

This simple equation demonstrates one fundamental aspect of step-reaction polymerizations—that very high conversions are necessary to achieve practical molecular weight. At 98 percent conversion, for example, $\overline{DP}$ is only 50. For $\overline{DP} = 100$, the monomer conversion must be 99 percent.

The number-average molecular weight is given as

$$\overline{M}_n = (\overline{DP})m = \frac{m}{1 - p} \qquad (18)$$

and the weight-average degree of polymerization is given as

$$\overline{M}_w = \frac{1+p}{1-p} m \tag{19}$$

where m is the molecular weight of a repeating unit. Thus, the molecular weight distribution for the most probable molecular weight distribution becomes $1 + p$, as shown below:

$$\frac{\overline{M}_w}{\overline{M}_n} = \frac{m(1+p)/(1-p)}{m/(1-p)} = 1+p \tag{20}$$

Therefore, when p is equal to 1 (i.e., 100 percent conversion), the polydispersity for the most probable distribution for step-reaction polymers is 2.

In general, high molecular-weight polymers can be obtained in a step-growth polymerization only under conditions of high monomer conversion, high monomer purity, high reaction yield, and stoichiometric equivalence of functional groups (in AA/B-B polymerization). Often, the later requirement can be achieved by preparing an intermediate low-molecular weight salt. Sometimes, a slight excess of one monomer may be used to control molecular weight.

Gel formation.   Bifunctional monomers give essentially linear polymers, whereas polyfunctional monomers, with more than two functional groups per molecule, give branched or cross-linked polymers. If a step-growth polymerization is carried out with monomer(s) of functionality $f > 2$, and if the reaction is carried out to a high conversion, a cross-linked network or a *gel* may be formed. A gel could be looked at as a molecule of essentially infinite molecular weight, extending throughout the reaction mass. In the production of thermosetting polymers, the reaction must be terminated short of the conversion at which gel is formed, or the product could not be molded or processed further (cross-linking is later completed in the mold). Hence, the prediction of *gel point* conversion is of great practical importance. Useful commercial polyesters, called *glyptals*, are produced by heating glycerol and phthalic anhydride. As the secondary hydroxyl is less active than the terminal primary hydroxyls in glycerol, the first product formed at conversions less than about 70 percent is linear polymer. A cross-linked product is produced by further heating.

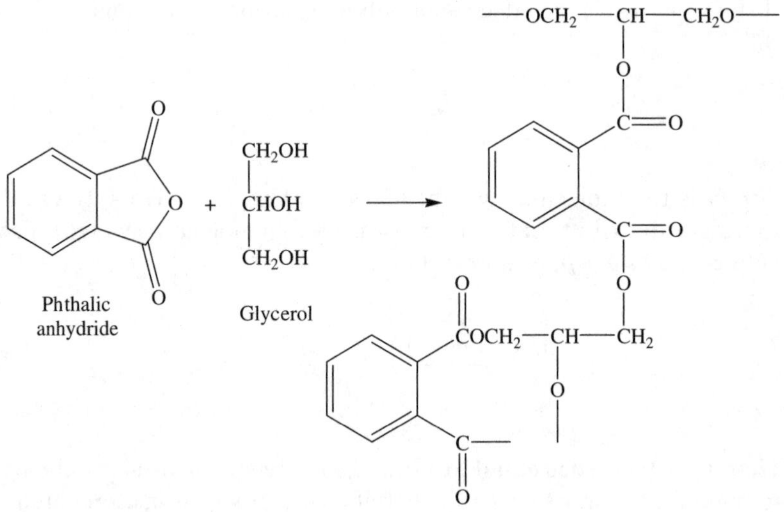

Phthalic anhydride + Glycerol → Cross-linked polyester

14.8.2 Addition or chain-reaction polymerization

Addition-reaction polymerization involves joining monomers together without the splitting of small molecules. The polymer formation involves three distinct kinetic steps: *initiation, propagation*, and *termination*. The *initiation* step constitutes the start of the reaction and requires an initiator to begin polymerization of a monomer. The initiator might be an anion, a cation, or a free radical ($R^{\bullet}$). The polymerization reaction can also be started using complex coordination compounds, which act as catalysts that are regenerated at the end of reaction. The type of mechanism best suited for polymerization of a particular vinyl monomer is related to the substituent(s) on the monomer that determines the polarity of the monomer and the acid-base strength of the ion formed. Vinyl monomers containing electron-withdrawing substituents form stable anions and polymerizes mainly with anionic polymerizations, whereas vinyl monomers that contain electron-donating groups form stable carbenium ions and best undergo cationic polymerization. Free radical polymerizations occur for vinyl monomers that are typically intermediate between electron-poor and electron-rich. Some monomers with a resonance-stabilized substituent-group such as a phenyl ring may be polymerized by more than one pathway. For example, styrene can be polymerized by both free-radical and ionic methods.

Growth of the polymer chain (*propagation*) occurs through continuous addition of monomer to the reactive chain end. Because polymerization

TABLE 14.20   Types of Chain Polymerization Suitable for Common Monomer

Monomer	Polymerization mechanism[a]			
	Radical	Cationic	Anionic	Coordination
Ethylene	+	+	−	+
Propylene and α-olefins	−	−	−	+
Isobutylene	−	+	−	−
Dienes	+	−	+	+
Styrene and α-methyl styrene	+	+	+	+
Vinyl chloride	+	−	−	+
Vinylidene chloride	+	−	+	−
Vinyl fluoride	+	−	−	−
Tetrafluoroethylene	+	−	−	+
Vinyl ethers	−	+	−	+
Vinyl esters	+	−	−	−
Acrylic and methacrylic esters	+	−	+	+
Acrylonitrile	+	−	+	+

[a] + = high polymer formed; − = no reaction or oligomers only.

occurs at the chain end, molecular weight increases rapidly even though large amounts of monomer remain unreacted. This constitutes a fundamental difference from step-reaction polymerization in which molecular weight increases slowly whereas monomer is consumed rapidly. The chain polymerization reaction propagates at a reactive chain end and continues until *termination* reaction render the chain end inactive (e.g., combination of radicals), or until monomer is completely consumed.

By bulk, almost all vinyl polymers are made by four processes: free radical (more than 50 percent), complex coordinate (12 to 15 percent), anionic (10 to 15 percent), and cationic (8 to 12 percent) [20]; Table 14.20 contains some common monomers and the suitable type of chain polymerization.

Table 14.21 list some commercially important polymers along with production techniques.

14.8.3   Free radical polymerization

Free radical polymerization offers a convenient approach toward the design and synthesis of special polymers for almost every area. In a free radical addition polymerization, the growing chain end bears an unpaired electron. A free radical is usually formed by the decomposition of a relatively unstable material called *initiator*. The free radical is capable of reacting to open the double bond of a vinyl monomer and add to it, with an electron remaining unpaired. The energy of activation for the propagation is 2–5 kcal/mol that indicates an extremely fast reaction (for condensation reaction this is 30 to 60 kcal/mol). Thus, in a very short time (usually a few seconds or less) many more monomers add successively

TABLE 14.21   Major Techniques Used in the
Production of Important Vinyl Polymers

Free radical
Low-density polyethylene (LDPE)
Poly(vinyl chloride)
Poly(vinyl acetate)
Polyacrylonitrile and acrylic fibers
Poly(methyl methacrylate)
Polyacrylamide
Polychloroprene
Poly(vinyl pyridine)
Styrne-acrylonitrile copolymers (SAN)
Polytetrafluoroethylene
Poly(vinylidene fluoride)
Acrylonitrile-butadiene-styrene copolymer (ABS)
Ethylene-methacrylic acid copolymers
Styrene-butadiene copolymer (SBR)
Nitrile rubber
Polystyrene

Cationic
Polyisobutylene and polybutenes
Isobutylene-isoprene copolymer (Butyl rubber)
Polyacetal
Poly(vinyl ether)s
Isobutylene-cyclopentadiene copolymer
Hydrocarbon and polyterpene resins

Anionic
Polyacetal
Thermoplastic olefin elastomers
(copolymers of butadiene, isoprene, and styrene)
cis-1,4-polybutadiene
cis-1,4-polyisoprene
Styrene-butadiene block and star copolymers
ABA block copolymers
(A = styrene, B = butadiene or isoprene)
Polycarbonates

Complex
High-density polyethylene (HDPE)
Polypropylene
Polybutadiene
Polyisoprene
Ethylene-propylene elastomers

to the growing chain. Free-radical polymerization of styrene, for example, involves adding about 1330 units of styrene to the polymer backbone in a single second. The average chain length in free radical polymerization remains constant throughout the reaction, and the maximum chain length is attained very rapidly. Finally, two free radicals react to annihilate each other's growth activity and form one or more polymer molecules. This can take place by *coupling* of two macroradicals or by *disproportionation*. Termination by disproportionation involves chain transfer of a hydrogen atom from one chain end to the free radical chain end of another growing chain, resulting in one of the *dead* polymers having an unsaturated chain end. Free radical polymerization can be accomplished in bulk, solution, suspension, or emulsion.

Free radical initiators.   Certain monomers, notably styrene and methyl methacrylate and some strained ring cycloalkenes, undergo polymerization on heating by free radical initiating species that are generated in situ. Thus, to prevent spontaneous polymerization of olefins on storage, certain compounds called *inhibitors* are added to the system to stabilize the olefin. Most monomers, however, require some kind of initiator. A large number of free radical initiators are available; they may be classified into the following four major types:

1. *Organic peroxides* (*ROOR*) *and hydroperoxides* (*ROOH*). These compounds are thermally unstable and decompose thermally by cleavage of the oxygen bond to yield RO• and HO• radicals. Examples of this type of initiators include

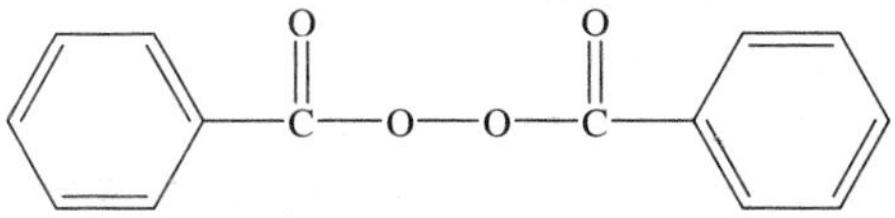

Benzoyl peroxide 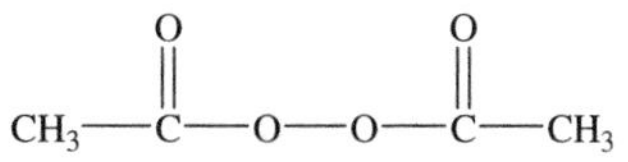   Diacetyl peroxide

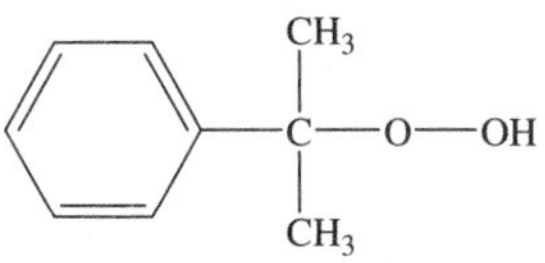

Cumyl hydroperoxide 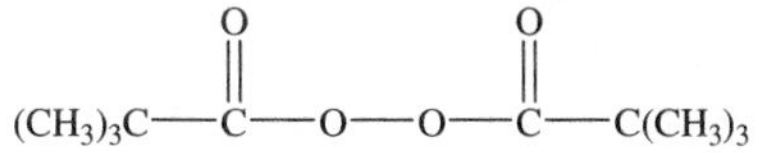   di-*t*-butyl peroxide

2. *Azo compounds.* *RN*=*NR*. Compounds such as α,α′azobis(isobutyronitrile) abbreviated as AIBN, decompose thermally to give nitrogen and two alkyl radicals.

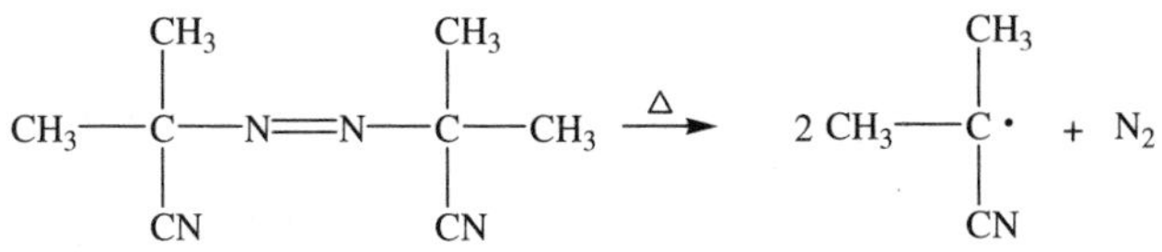

3. *Redox initiators.* Free radicals are produced in redox initiators by one-electron transfer reactions. This type of initiator is particularly useful in initiation of low-temperature polymerization and emulsion polymerization. Some typical examples are: persulfate + reducing agents; hydroperoxides + ferrous ion.

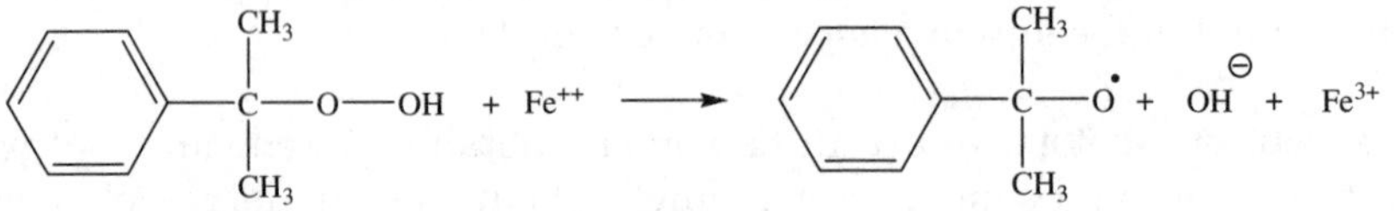

$$HOOH + Fe^{++} \longrightarrow HO^{\cdot} + OH^{\ominus} + Fe^{3+}$$

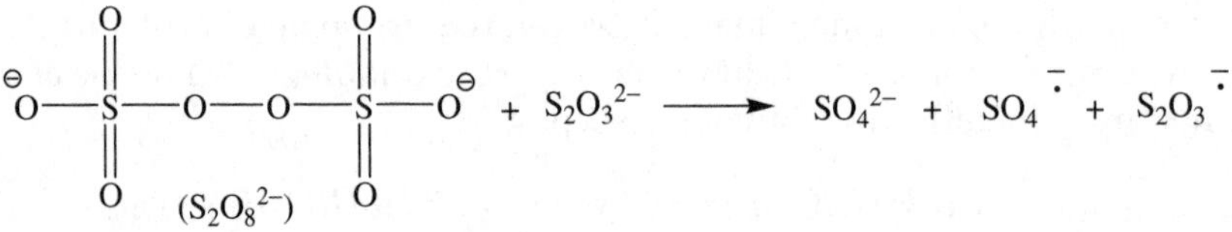

$$HSO_3^{\ominus} + Fe^{3+} \longrightarrow HSO_3^{\cdot} + Fe^{2+}$$

4. *Photoinitiators.* These are compounds that dissociate under the influence of light to form radicals. Peroxides and azo compounds dissociate photolytically as well as thermally. Advantageously, photoinitiation is independent of temperature; thus polymerization may be conducted at low temperatures. Furthermore, better control of the polymerization reaction is generally possible because narrow wavelength bands may be used to initiate decomposition, and the reaction can be stopped simply by removing the light source. A wide variety of photolabile compounds are available, including disulfide, benzil, and benzoin.

$$RS-SR \xrightarrow{h\nu} 2RS\cdot$$
Disulfide

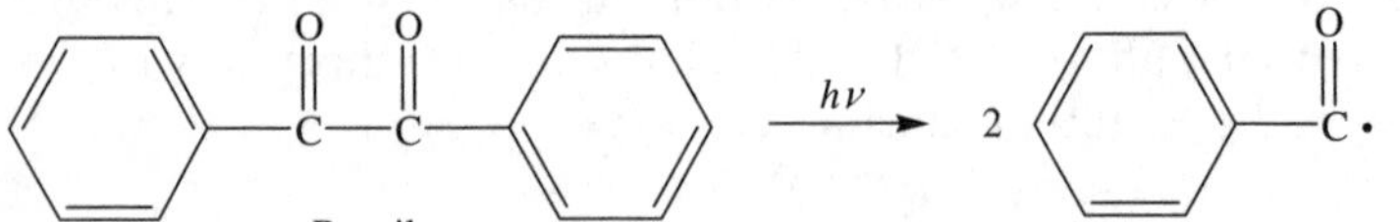

Benzil

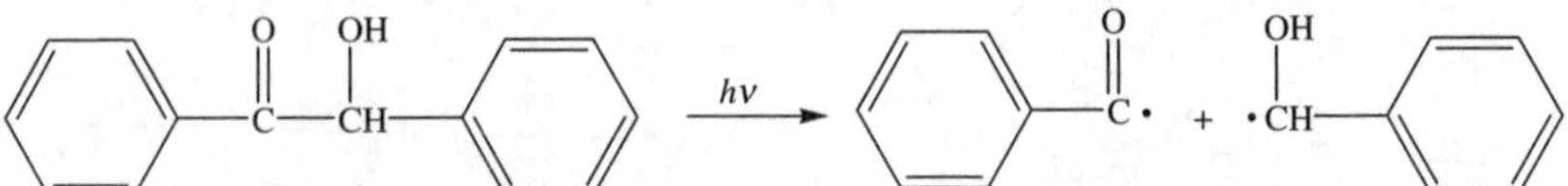

Benzoin

Mechanism and kinetics of free-radical polymerization

Initiation. The initiation of polymerization occurs in two consecutive steps. In the first step, the initiator molecule, represented by I, undergoes a first-order decomposition with a rate constant k_d to give two free radicals, $R^\bullet$.

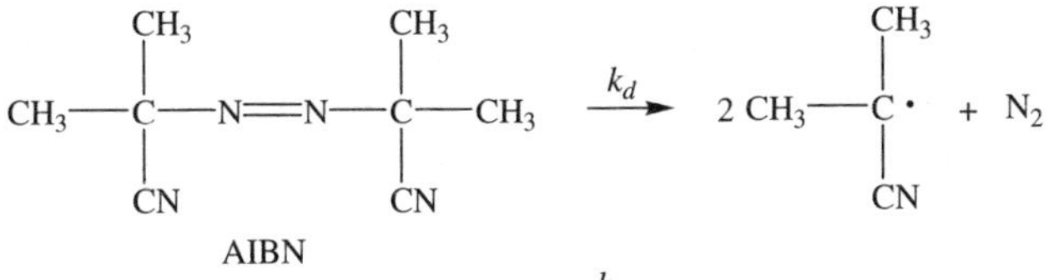

$$I \xrightarrow{k_d} 2R^\bullet$$

$$R_d = -d[I]/dt = k_d[I] \tag{21}$$

where R_d is the rate of decomposition, and k_d is decomposition rate constant. Actual initiation of a free radical chain then takes place by the addition of a free radical ($R^\bullet$) to a vinyl monomer (M) as illustrated by styrene.

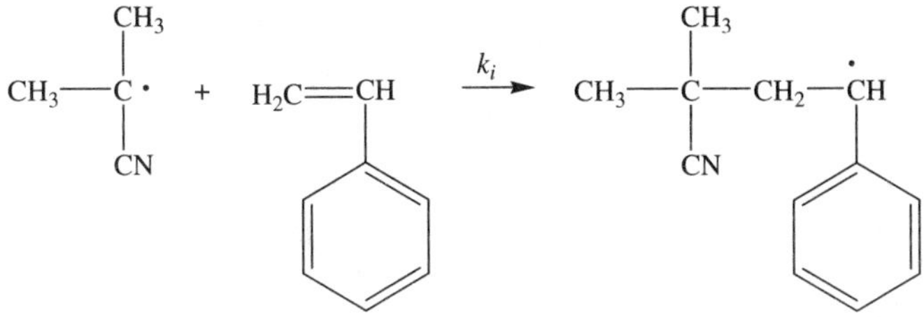

This may be abbreviated by

$$\underset{\text{Free radical}}{R^\bullet} + \underset{\text{Monomer}}{M} \xrightarrow{k_i} \underset{\text{New free radical}}{RM^\bullet}$$

$$R_i = d[RM^\bullet]/dt = k_i[R^\bullet][M] \tag{22}$$

where R_i is the rate of initiation, and k_i is initiation rate constant. The rate of decomposition of I is the rate-controlling step in the free radical polymerization. Thus, the overall expression describing the rate of initiation can be given as

$$R_i = 2k_d f[I] \tag{23}$$

where f is the efficiency factor and is a measure of the fraction of initiator radicals that produce growing radical of chains, that is, are able to react with monomer.

$$f = \frac{\text{radicals that initiate a polymer chain}}{\text{radicals formed from initiator}}$$

Propagation.   Chain propagation involves the addition of a free radical to the double bond of a vinyl monomer. The product itself is a free radical ($RM^{\bullet}$) to which more monomer molecules (M) add successively.

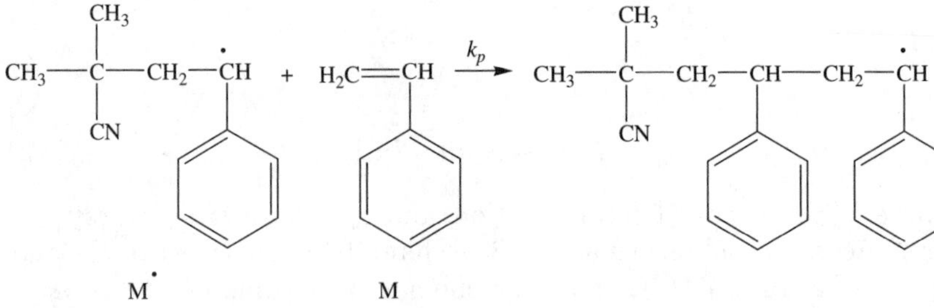

Propagation is a bimolecular reaction for which the rate expression is given by

$$R_p = k_p[M][M^{\bullet}] \qquad (24)$$

The above expression incorporates an approximation; a single rate constant, k_p, is used to describe all the propagation steps. This assumption agrees with the experimental finding that the specific rate constants associated with propagation are approximately independent of chain length; thus the specific rate constants for each propagation step (i.e., each successive addition of monomer) are considered to be the same.

Unsymmetrical vinyl monomers can add to each other in one of two ways; *head-to-tail* or *head-to-head*. The former is the most likely form of monomer addition. The principal reason for the preference of head-to-tail addition lies in the greater thermodynamic stability of the free radical and perhaps also in the steric inhibition encountered in the head-to-head addition (see Fig. 14.10).

Termination.   Free radical chains can be terminated by any reaction that destroys the active chain centers. This can happen to a small extent through the reaction with initiator radicals as shown in the reaction:

$$R\!-\!\!\!\sim\!\!\!\sim\!\!\!\sim\!\!-\!CH_2\!-\!\underset{\underset{X}{\displaystyle |}}{\overset{\overset{H}{\displaystyle |}}{C}}\bullet \; + \; \bullet R' \longrightarrow R\!-\!\!\!\sim\!\!\!\sim\!\!\!\sim\!\!-\!CH_2\!-\!\underset{\underset{X}{\displaystyle |}}{\overset{\overset{H}{\displaystyle |}}{C}}\!-\!R'$$

The more important means of termination, however, occur either by combination of the two growing free radicals or by disproportionation where an atom (usually hydrogen) is transferred from one polymer radical to another.

Combination

$$R\!-\!\!\!\sim\!\!\!\sim\!\!-\!CH_2\!-\!\underset{\underset{X}{|}}{\overset{\overset{H}{|}}{C}}\bullet \; + \; \bullet\underset{\underset{X}{|}}{\overset{\overset{H}{|}}{C}}\!-\!CH_2\!-\!\!\!\sim\!\!\!\sim\!\!-\!R \longrightarrow R\!-\!\!\!\sim\!\!\!\sim\!\!-\!CH_2\!-\!\underset{\underset{X}{|}}{\overset{\overset{H}{|}}{C}}\!-\!\underset{\underset{X}{|}}{\overset{\overset{H}{|}}{C}}\!-\!CH_2\!-\!\!\!\sim\!\!\!\sim\!\!-\!R$$

Disproportionation

$$R\!-\!\!\!\sim\!\!\!\sim\!\!-\!CH_2\!-\!\underset{\underset{X}{|}}{\overset{\overset{H}{|}}{C}}\bullet \; + \; \bullet\underset{\underset{X}{|}}{\overset{\overset{H}{|}}{C}}\!-\!CH_2\!-\!\!\!\sim\!\!\!\sim\!\!-\!R \longrightarrow R\!-\!\!\!\sim\!\!\!\sim\!\!-\!CH_2\!-\!\underset{\underset{X}{|}}{\overset{\overset{H}{|}}{C}}\!-\!H + \underset{\underset{X}{|}}{\overset{\overset{H}{|}}{C}}\!\!=\!\!CH\!-\!\!\!\sim\!\!\!\sim\!\!-\!R$$

$$R_t = -\mathrm{d}[M^\bullet]/\mathrm{dt} = 2k_t[M^\bullet][M^\bullet] = 2k_t[M^\bullet]^2 \qquad (25)$$

The relative proportion of each termination mode depends on the nature of the monomer and on the reaction temperature. In most cases, one or the other predominates, but combination will normally be preferred at low temperatures, and disproportionation becomes more significant at high temperatures. The results for several polymer systems are shown in Table 14.22. Termination via combination will yield a polymer with molecular weight twice that of the growing chain prior to termination, and with two initiator fragments (R) per molecule, and there is a head-to-head configuration at the juncture of the two macroradicals in the dead polymer, whereas termination by disproportionation will give a polymer with only one initiator fragment per molecule. Also, termination by disproportionation yields polymer • molecules that are essentially of the same molecular weight as the growing chain prior to termination, and results in one of the *dead* polymers having an unsaturated chain end.

Polymerization rate expression. The equations describing kinetics of free-radical polymerization steps contain a term for the concentration of radicals $[M^\bullet]$, which exists at very low concentration ($\sim 10^{-8}$ M) and thus

TABLE 14.22   Termination of Free Radical Polymerization at 60°C [21]

Monomer	Formula	Disproportionation	Combination
Acrylonitrile	$CH_2\!=\!CH\!-\!CN$	~0	~100
Methyl methacrylate	$CH_2\!=\!\underset{\displaystyle}{\overset{\displaystyle CH_3}{C}}\!-\!\underset{\displaystyle O}{\overset{\displaystyle}{C}}\!-\!OCH_3$	79	21
Styrene	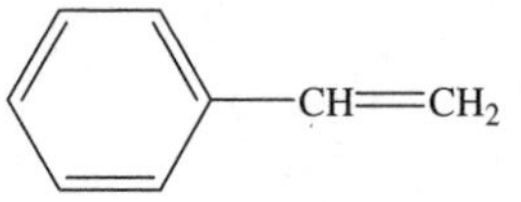 ⬡$-CH\!=\!CH_2$	23	77
Vinyl acetate	$CH_2 = CH\!-\!O\overset{\displaystyle O}{\overset{\displaystyle \parallel}{C}}CH_3$	~100	~0

is difficult to determine experimentally. Therefore, it is more useful to develop an expression involving more experimentally accessible terms. The rate of polymerization is nothing but the rate of monomer disappearance. Monomer disappears in the initiation steps as well as in the propagation reactions. Therefore,

$$-d[M]/dt = R_i + R_p = k_i\,[M][M^\bullet] + k_p[M][M^\bullet] \tag{26}$$

However, the number of monomer molecules consumed in the initiation reaction is extremely lower than the number of monomer molecules consumed in the propagation reactions. Therefore, the rate of polymerization in the above equation can be approximated as

$$-d[M]/dt \approx R_P \approx k_P[M][M^\bullet] \tag{27}$$

where $[M]$ is the monomer concentration and $[M^\bullet]$ is the total concentration of all chain radicals. The monomer radical change is given by

$$d[M^\bullet]/dt = k_i[R^\bullet][M] - 2k_t[M^\bullet]^2 \tag{28}$$

It is experimentally found that the concentration of radicals increases initially, but almost instantaneously reaches a constant, and that the number of growing chains is approximately constant over a large extent of reaction, that is, steady-state condition where $d[M^\bullet]/dt = 0$ and

$$k_i[R^\bullet][M] = 2k_t[M^\bullet]^2 \tag{29}$$

Also, a steady-state condition for $R^\bullet$ (dot is superscript) exists, yielding

$$d[R^\bullet]/dt = 2k_d f[I] - k_i[R^\bullet][M] = 0 \tag{30}$$

Solving for $[M^\bullet]$ and $[R^\bullet]$ gives

$$[M^\bullet] = \left(\frac{k_i[R^\bullet][M]}{2k_t} \right)^{1/2} \tag{31}$$

and

$$[R^\bullet] = \frac{2k_d f[I]}{k_i[M]} \tag{32}$$

Substituting the above expression for $[R^\bullet]$ into the expression for $[M^\bullet]$ in Eq. (31) gives

$$[M^\bullet] = \left(\frac{k_d f[I]}{k_t} \right)^{1/2} \tag{33}$$

which when substituted in the equation for R_p (Eq. [27]) yields an expression for the rate of polymerization.

$$R_p = k_p[M][M^\bullet] = k_p[M]\left(\frac{k_d f[I]}{k_t} \right)^{1/2} = k'[M][I]^{1/2} \tag{34}$$

where $\quad k' = \left(\dfrac{k_p^2 k_d f}{k_t} \right)^{1/2}$

Degree of polymerization is governed by the rate of polymerization compared to the rate of termination, and thus can be expressed as

$$\overline{DP} = \frac{R_p}{R_t} = \frac{k_p[M](k_d f[I]/k_t)^{1/2}}{2k_t[M^\bullet]^2} = \frac{k_p[M]}{2(k_d k_t f[I])^{1/2}} = k'' \frac{[M]}{[I]^{1/2}} \tag{35}$$

where $k'' = \dfrac{k_p}{(2k_d k_t f)^{1/2}}$

Thermodynamics of free-radical polymerization.   The free energy of polymerization, ΔG_p, is given by the first and second laws of thermodynamics for a reversible process as

$$\Delta G_{\mathrm{p}} = \Delta H_{\mathrm{p}} - T\Delta S_{\mathrm{p}} \qquad (36)$$

where ΔH_{p} is the *heat of polymerization* and defined as

$$\Delta H_{\mathrm{p}} = E_{\mathrm{p}} - E_{\mathrm{dp}} \qquad (37)$$

and E_{p} and E_{dp} are the activation energies for propagation (i.e., polymerization) and depolymerization, respectively. Both ΔH_{p} and ΔS_{p} are negative, and, therefore, ΔG_{p} will also be negative (i.e., polymerization is favored at low temperatures). At temperature, called the *ceiling temperature* (T_{c}), the polymerization reaches equilibrium. In other words, the rates of polymerization and depolymerization become equal and

$$\Delta G_{\mathrm{p}} = 0$$

The ceiling temperature is therefore defined as the temperature at which the rates of propagation and depolymerization are equal. For that reason, T_{c} is a threshold temperature above which a specific polymer cannot exist. Representative values of T_{c} for some common monomers are given in Table 14.23.

14.8.4 Ionic polymerization

Anionic polymerization.   Anionic polymerization is an addition polymerization in which the growing chain end bears a negative charge. The monomers suitable for anionic polymerization are those that have substituent groups capable of stabilizing a carbanion through resonance or induction. Typical monomers that can be polymerized by ionic mechanisms include styrene, acrylonitrile, and methyl methacrylate (Table 14.20).

Initiation.   Initiation of anionic polymerization is brought about by species that undergo nucleophilic addition to a monomer. The most typically used anionic initiators can be classified into two basic types:

1. *Nucleophilic initiators* that react by the addition of negative ion. Examples of these include metal amides such as $NaNH_2$ and $LiN(C_2H_5)_2$, alkoxides, hydroxides, cyanides, phosphines, amines, and organometallic compounds such as lithium reagents (e.g., n-C_4H_9Li) and Grignard reagents (e.g., PhMgBr). Organometallic compounds of alkali metals are the most common anionic initiators employed commercially in the polymerization of 1,3-butadiene and isoprene. Initiation proceeds by the addition of the metal alkyl to monomer:

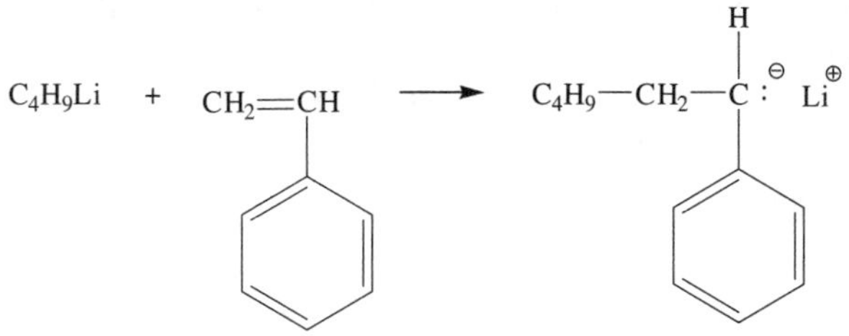

$$C_4H_9Li \quad + \quad CH_2{=}CH \longrightarrow C_4H_9{-}CH_2{-}\overset{\overset{\displaystyle H}{|}}{\underset{}{C}} : ^{\ominus} \; Li^{\oplus}$$

TABLE 14.23   Celing Temperatures of Some Common Polymers [21]

Polymer	Structure	$T_c(°C)$
Polyisobutylene	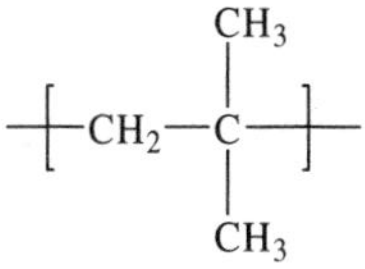 $+CH_2-\overset{\overset{\displaystyle CH_3}{\textstyle\vert}}{\underset{\underset{\displaystyle CH_3}{\textstyle\vert}}{C}}+$	175
Poly(methyl methacrylate)	$+CH_2-\overset{\overset{\displaystyle CH_3}{\textstyle\vert}}{\underset{\underset{\displaystyle O}{\underset{\|}{COCH_3}}}{C}}+$	198
Poly(α -methylstyrene)	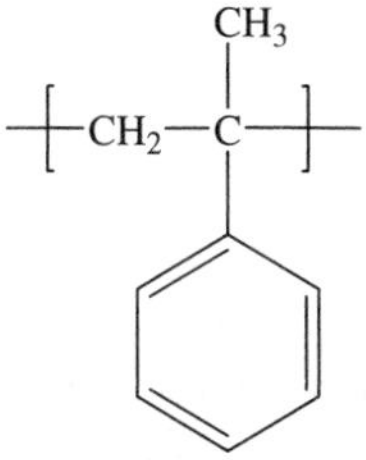 $+CH_2-\underset{\bigcirc}{\overset{\overset{\displaystyle CH_3}{\vert}}{C}}+$	66
Polystyrene	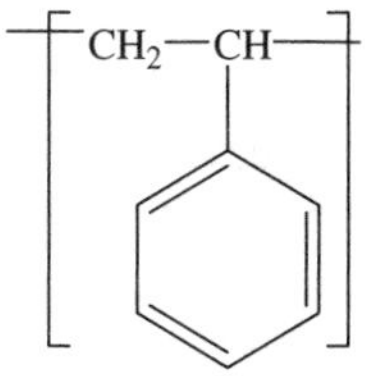 $CH_2-\underset{\bigcirc}{CH}$	395
Polyformaldehyde (Polyoxymethylene, Derlin)	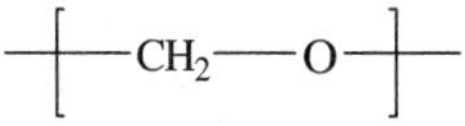 $+CH_2-O+$	116
Polyethylene	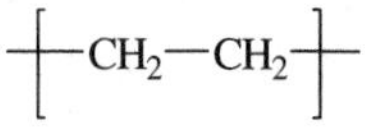 $+CH_2-CH_2+$	610
Polytetrafluoroethylene	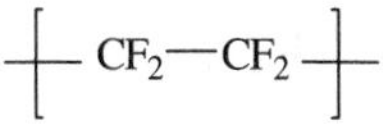 $+CF_2-CF_2+$	1100

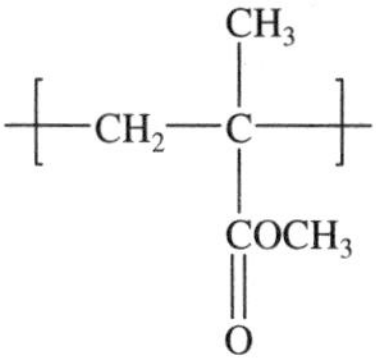

2. *Electron transfer initiators* such as free alkali metals (e.g., Na, Li) or complexes of alkali metals and unsaturated or aromatic compounds (e.g., sodium naphthalene). These bring about initiation as shown in the following scheme:

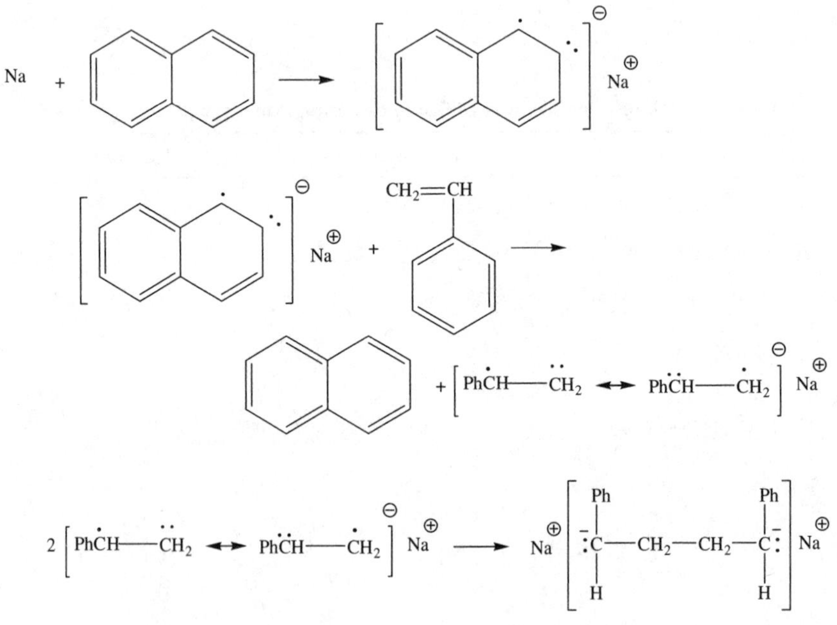

During the initiation process, the addition of the initiator anion to a monomer (e.g., styrene) produces a carbanion at the head end in association with a positively-charged metal counterion.

Propagation. The chain propagates by insertion of additional monomers between the carbanion and counterion.

Termination. Anionic polymerization has no termination associated with it in the time scale of the polymerization reaction. For this reason, anionic polymerization is sometimes called *living* polymerization. As a result, if the starting reagents are pure and if the polymerization is moisture- and oxygen-free, propagation can proceed until all monomer is consumed. In this case, termination occurs only by the deliberate introduction of oxygen, carbon dioxide, methanol or water as follows:

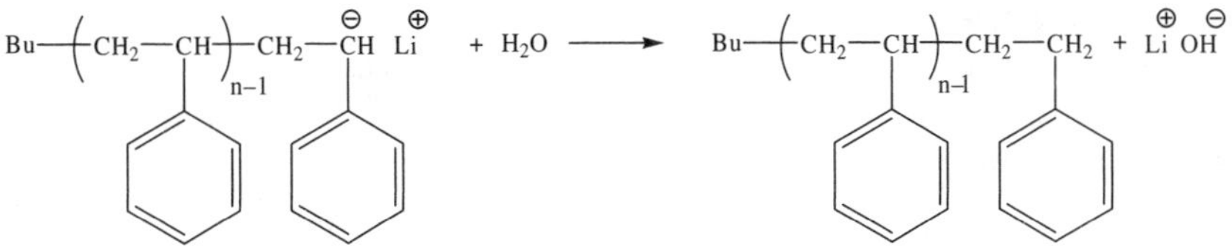

$$\text{Bu}-\left(\text{CH}_2-\text{CH}\right)_{n-1}\text{CH}_2-\overset{\ominus}{\text{CH}}\ \overset{\oplus}{\text{Li}}\ +\ \text{H}_2\text{O}\ \longrightarrow\ \text{Bu}-\left(\text{CH}_2-\text{CH}\right)_{n-1}\text{CH}_2-\text{CH}_2\ +\ \overset{\oplus}{\text{Li}}\ \overset{\ominus}{\text{OH}}$$

In the absence of a termination mechanism, each monomer in an anionic polymerization has an equal probability of attaching to an anion site. Therefore, the number-average degree of polymerization, $\overline{DP}$, is simply equal to the ratio of initial monomer to initial initiator concentration as

$$\overline{DP} = \frac{[M]_o}{[I]_o} \tag{38}$$

The absence of termination during a living polymerization leads to a very narrow molecular-weight distribution with polydispersities as low as 1.06. By comparison, polydispersities above 2 and as high as 20 are typical in free radical polymerization.

Cationic polymerization. In cationic chain polymerization the propagating species is a carbocation. Cationic polymerizations require monomers that have electron-releasing groups such as an alkoxy, phenyl, or a vinyl group (Table 14.20).

Mechanism and kinetics of cationic polymerization initiation. Unlike free-radical and anionic polymerization, initiation in cationic polymerization employs a true catalyst that is restored at the end of the polymerization and does not become incorporated into the terminated polymer chain. Initiation of cationic polymerization is brought about by addition of an electrophile to a monomer molecule. Typical compounds used for cationic polymerization include protonic acids (e.g., H_2SO_4, H_3PO_4), Lewis acids (e.g., $AlCl_3$, BF_3, $TiCl_4$, $SnCl_4$), and stable carbenium-ion salts (e.g., triphenylmethyl halides, tropylium halides):

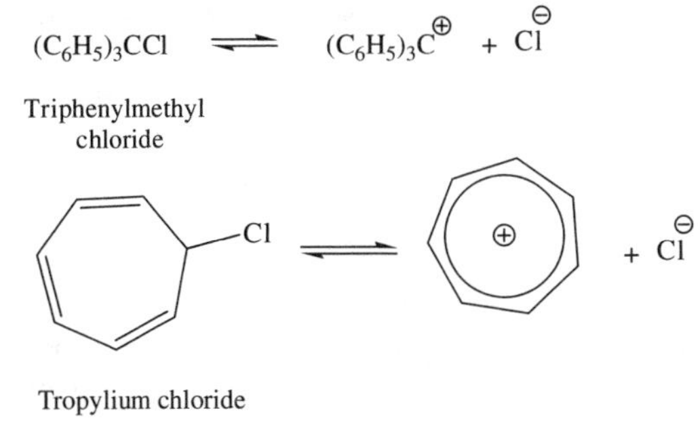

$$(C_6H_5)_3CCl \ \rightleftharpoons\ (C_6H_5)_3\overset{\oplus}{C}\ +\ \overset{\ominus}{Cl}$$

Triphenylmethyl
chloride

Tropylium chloride

Initiation by Lewis acids requires the presence of a trace amount of a *cocatalyst* such as water or other proton or cation source. The Lewis base coordinates with the electrophilic Lewis acid, producing a proton, which is the actual initiator:

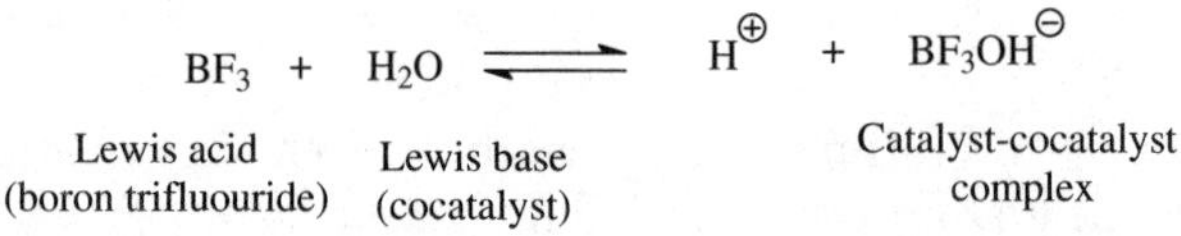

$$BF_3 \; + \; H_2O \; \rightleftharpoons \; H^{\oplus} \; + \; BF_3OH^{\ominus}$$

Lewis acid   Lewis base   Catalyst-cocatalyst
(boron trifluouride)   (cocatalyst)   complex

Initiation of a monomer takes place through the addition of the catalyst ion pair across the double bond, such that the proton adds to the carbon atom bearing the greatest electron density as illustrated with isobutylene in the following reaction. This mode of addition forms the most stable carbonium ion.

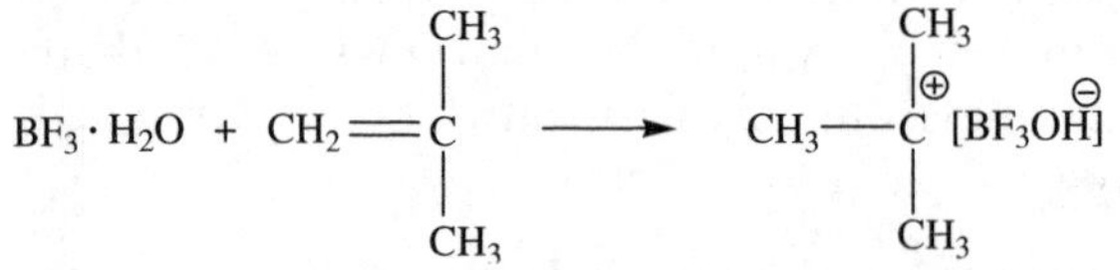

$$BF_3 \cdot H_2O \; + \; CH_2 = \underset{\underset{\displaystyle CH_3}{|}}{\overset{\overset{\displaystyle CH_3}{|}}{C}} \; \longrightarrow \; CH_3 - \underset{\underset{\displaystyle CH_3}{|}}{\overset{\overset{\displaystyle CH_3}{|}}{C}}^{\oplus} [BF_3OH]^{\ominus}$$

The rate of initiation (R_i) is proportional to the concentration of the monomer $[M]$ and the concentration of the catalyst-cocatalyst complex $[C]$.

$$R_i = k_i[M][C] \tag{39}$$

Propagation. Propagation or chain growth takes place by successive addition of monomer molecules in a head-to-tail configuration. At low temperatures, the chain growth takes place rapidly, and the rate constant (k_p) is essentially the same for all propagation steps.

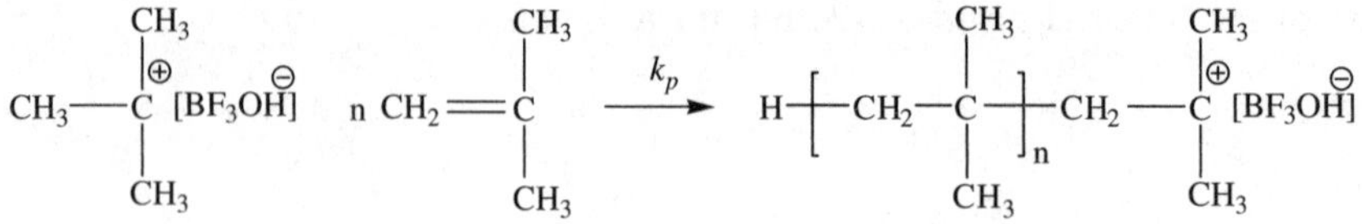

$$CH_3 - \underset{\underset{\displaystyle CH_3}{|}}{\overset{\overset{\displaystyle CH_3}{|}}{C}}^{\oplus}[BF_3OH]^{\ominus} \; n\,CH_2 = \underset{\underset{\displaystyle CH_3}{|}}{\overset{\overset{\displaystyle CH_3}{|}}{C}} \; \overset{k_p}{\longrightarrow} \; H \left[CH_2 - \underset{\underset{\displaystyle CH_3}{|}}{\overset{\overset{\displaystyle CH_3}{|}}{C}} \right]_n CH_2 - \underset{\underset{\displaystyle CH_3}{|}}{\overset{\overset{\displaystyle CH_3}{|}}{C}}^{\oplus}[BF_3OH]^{\ominus}$$

$$R_p = k_p[M][M^+] \tag{40}$$

The rate of ionic chain polymerization is dependent on the dielectric constant of the solvent, the resonance stability of the carbonium ion, the stability of the counterion, and the electropositivity of the initiator. The rate constant is affected by the polarity of the solvent—the rate is fastest in solvents with high dielectric constants as a result of better separation of the carbocation-counterion pair.

Termination. Termination of the polymer chain can occur by chain transfer reaction where a proton is transferred from a terminal side group to a monomer molecule. The newly initiated monomer molecule can generate a new chain.

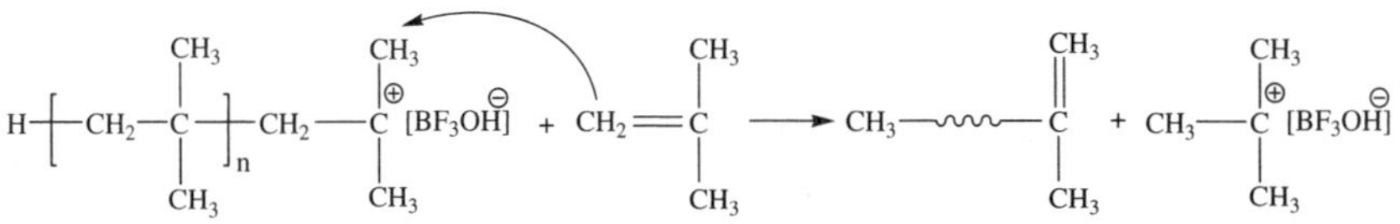

$$R_t = k_{tr}[M][M^+] \tag{41}$$

Termination may also take place by dissociation of the macrocarbocation-counterion complex where a proton is lost to the counter ion. Termination can also take place by the reaction of a growing chain end with traces of water or other protonic reagents.

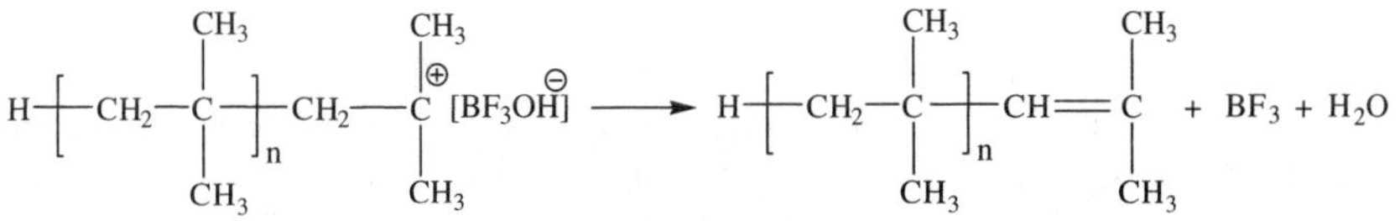

Termination reactions regenerate the catalyst complex, therefore, the complex is a true catalyst, unlike free-radical initiators.

$$R_t = k_t[M^+] \tag{42}$$

As with free radical polymerization, to express the rate of polymerization in terms of measurable terms, one can approximate a steady state for the growing chain end, which implies that the rate of initiation equals the rate of termination, thus $R_i = R_t$, and

$$k_i[M][C] = k_t[M^+]$$

or

$$[M^+] = \frac{k_i[M][C]}{k_t} \tag{43}$$

Substituting for $[M^+]$ in R_p (Eq. [40]) yields the overall rate for cationic polymerization as

$$R_p = \frac{k_i k_p [C][M]^2}{k_t} = k'[C][M]^2 \qquad (44)$$

The value for the average degree of polymerization, $\overline{DP}$, can be expressed as

$$\overline{DP} = \frac{R_p}{R_t} = \frac{k_p[M][M^+]}{k_t[M^+]} = \frac{k_p[M]}{k_t} = k''[M] \qquad (45)$$

when termination occurs via internal dissociation. However, if termination occurs predominantly via chain transfer, then

$$\overline{DP} = \frac{R_p}{R_{tr}} = \frac{k_p[M][M^+]}{k_{tr}[M][M^+]} = k''' \qquad (46)$$

The above kinetic expressions illustrate some basic differences between cationic and free radical processes. In the cationic polymerization, the propagation rate is of first order with respect to the initiator concentration, whereas in free radical polymerization it is proportional to the square root of initiator concentration (Eq. [34]). Furthermore, the molecular weight (or $\overline{DP}$) of the polymer synthesized by the cationic process is independent of the concentration of the initiator, regardless of how termination takes place, unlike free radical polymerization where $\overline{DP}$ is inversely proportional to $[I]^{1/2}$ in the absence of chain transfer (Eq. [35]).

Cationic polymerization can produce polymers with stereoregular structures. It has been observed that in cationic polymerization processes:

1. The amount of stereoregularity is dependent on the nature of the initiator,

2. Stereoregularity increases with a decrease in temperature,

3. The amount and type of polymer (isotactic or syndiotactic) is dependent on the polarity of the solvent; for instance, t-butyl vinyl ether has the

isotactic form preferred in nonpolar solvents, but the syndiotactic form is preferred in polar solvents.

$$\left[\begin{array}{c} CH_2-CH \\ | \\ OC(CH_3)_3 \end{array}\right]$$

Poly(*t*-butyl vinyl ether)

Coordination polymerization. Prior to 1950, the only commercial polymer of ethylene was a highly branched polymer called high-pressure polyethylene (extremely high pressures were used in the polymerization process). This polymer had a $-CH_2-CH_2-$ backbone with some short and long alkane branches. This grade of polyethylene is called low-density polyethylene (LDPE). Similarly, free radical polymerization of propylene yields an amorphous polymer that is a tacky gum at room temperature and has no commercial use. Marvel and Hogan, in the 1940s, and Nobel laureate Karl Ziegler in the early 1950s, developed techniques for making a linear polyethylene. Ziegler prepared high-density polyethylene by polymerizing ethylene at low pressure and ambient temperatures using mixtures of triethylaluminum and titanium tetrachloride. This polyethylene—high-density polyethylene (HDPE) had fewer branches and, therefore, could obtain a higher degree of crystallinity than LDPE. Another Noble laureate, Giulio Natta, used Ziegler's complex coordination catalyst to produce isotactic polypropylene. This form of polypropylene had a level of crystallinity comparable to LDPE and exhibited good mechanical properties over a wide range of temperatures. Ziegler and Natta were awarded the Nobel Prize in 1963.

In general, a Ziegler-Natta catalyst is a metal-organic complex. The catalyst may be described as a combination of a transition metal compound from groups IV–VIII and an organometallic compound of a metal cation from groups I–III of the periodic table. It is customary to refer to the transition metal compounds, such as $TiCl_4$, as the catalyst, and the organometallic compound, such as diethylaluminum chloride, as the cocatalyst. The processes used in polymerization of both HDPE and *i*-PP use a coordination- or insertion-type mechanism during polymerization.

The exact mechanism of coordination polymerization is still unclear. Stereoregularity of the resulting polymers could be explained based on reactions involving either monometallic or bimetallic sites of the catalyst-cocatalyst system. However, it is generally agreed that the growing polymer chain is bound to the metal atom of the catalyst and that the monomer is inserted between the transition metal atom and the terminal carbon atom in the growing chain. That is, the monomer molecule is always the terminal group on the chain. Coordination polymerization can be terminated upon introducing water, hydrogen, aromatic alcohol, or metals like zinc in the polymerization reactor. The following schemes outline the reactions involved in coordination polymerization based on the monometallic and bimetallic approaches, respectively.

Bimetallic Mechanism

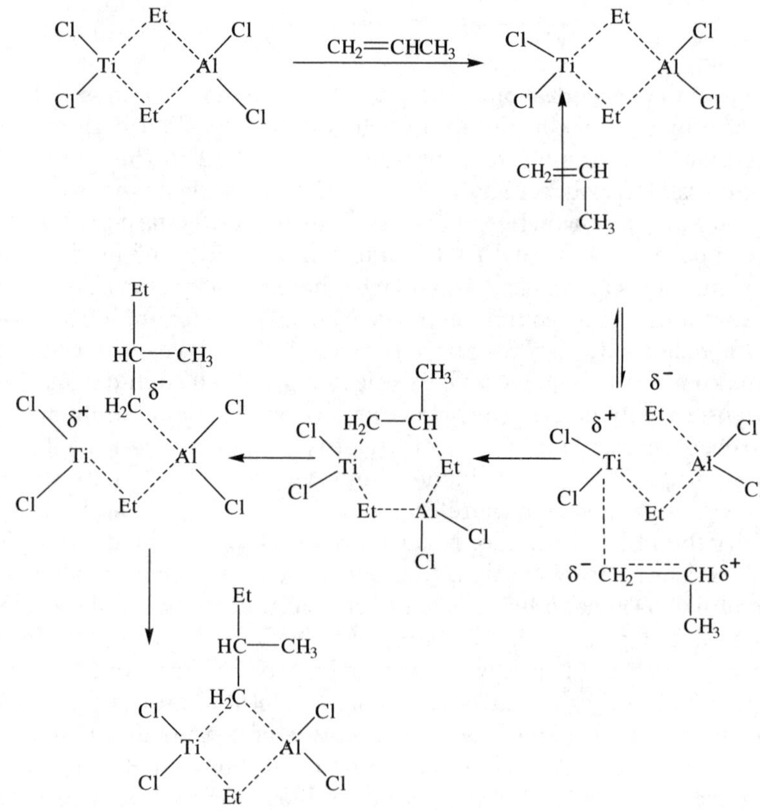

Monometallic Mechanism

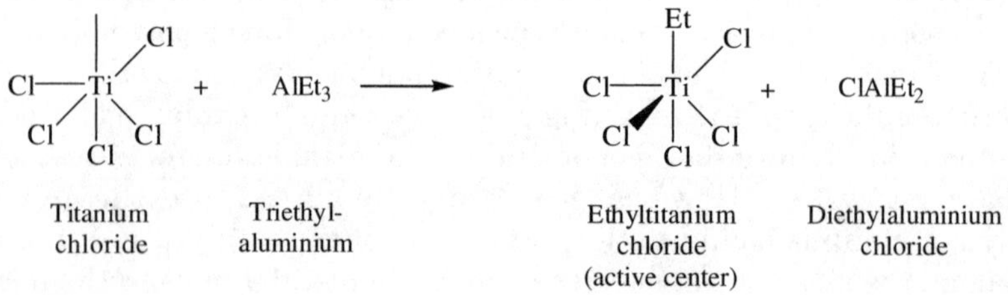

| Titanium chloride | Triethyl-aluminium | Ethyltitanium chloride (active center) | Diethylaluminium chloride |

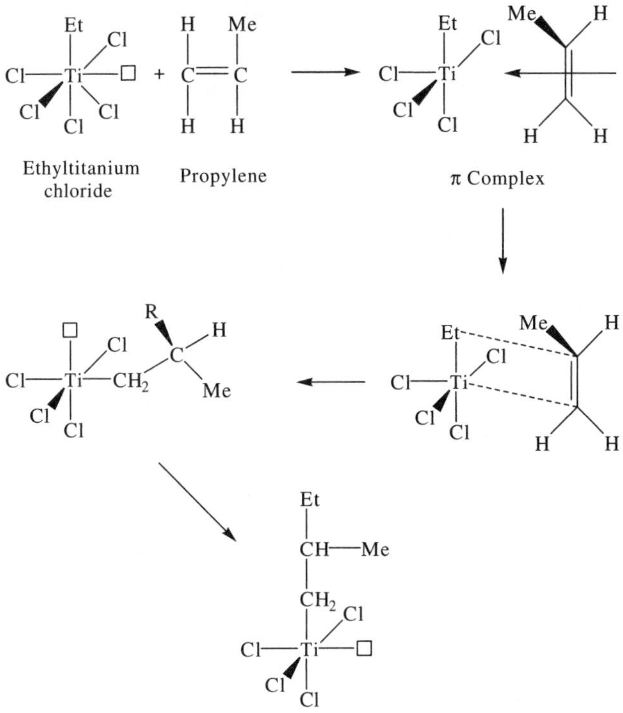

Metathesis polymerization. Metathesis reaction is a catalytically induced reaction in which cyclic olefins such as cyclopentene undergo bond reorganization that results in the formation of so-called polyalkenamers. The resulting polymers contain double bonds that can be subsequently used to introduce cross-linking. Metathesis reactions are induced with transition metal catalysts of the Ziegler-Natta types or similar catalyst-cocatalyst combinations. The catalyst systems that induce metathesis polymerizations include species derived from WCl_6, $WOCl_4$, MoO_3, and ruthenium or rhenium halides. The reactions proceed with good rates at room temperature and the steroregularity can be controlled through choice of reaction conditions and catalysts. A single monomer like cyclopentene can be polymerized to give the *cis* product with the use of molybdenum-based catalyst, or the *trans* product with the use of a tungsten-based catalyst [20].

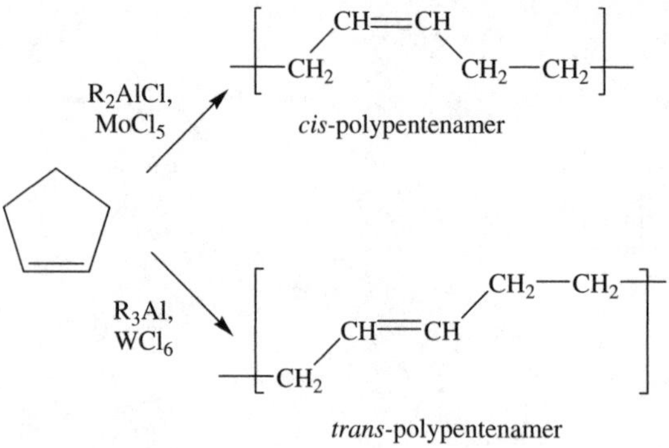

The polymerization probably occurs via complete scission of the carbon-carbon double bond through the reaction with metal carbene precursors giving an active carbene species as shown in the following scheme:

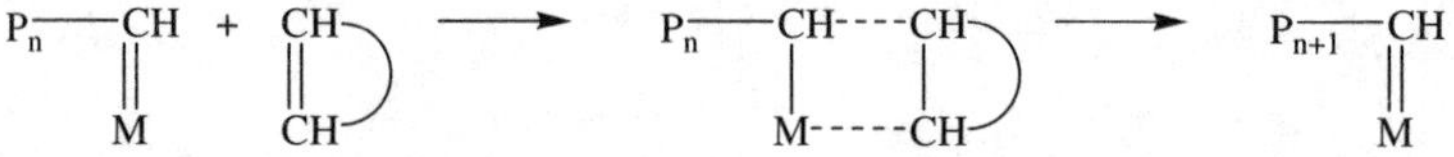

where P represent repeat unit, n is the degree of polymerization, and M represents a metal complex.

Ring-opening polymerization.   Some polymers that are traditionally recognized as belonging to the addition class are polymerized, not by addition to an ethylene double bond, but through a ring-opening polymerization of a sterically strained cyclic monomer. Examples of commercially important ring-opening polymerizations are listed in Table 14.24. The driving force for ring-opening is the relief of the bond-angle strain, or steric repulsions, or both between atoms crowded into the center of the ring. Ring-opening polymerization can happen by both anionic and cationic mechanisms.

Group-transfer polymerization.   Group-transfer polymerization (GTP) is a silicon-mediated Michael addition reaction. It allows the polymerization of α,β-unsaturated esters, ketones, amides, or nitriles through the use of silyl ketenes. The initiator is activated by suitable nucleophilic catalysts such as soluble Lewis acids, fluorides, cyanides, azides, and bifluorides, HF_2^-. During polymerization, the initiating ketene silyl acetal functionality is transferred to the head of each new monomer molecule as it adds to the chain. As in anionic polymerization, molecular weight is determined by the ratio of monomer to initiator concentration. Reactions are generally conducted at low temperatures (about 0 to 50°C)

TABLE 14.24   Commercially Important Polymers Prepared by Ring-Opening Polymerization

Polymer type	Polymer repeating unit	Monomer structure	Monomer type
Polyalkene	$-\!\!\left[\!-CH\!=\!CH(CH_2)_x\!-\!\right]\!-$	$(CH_2)_x \ \substack{CH \\ \| \\ CH}$	Cycloalkene
Polyether	$-\!\!\left[\!-CH_2O\!-\!\right]\!-$	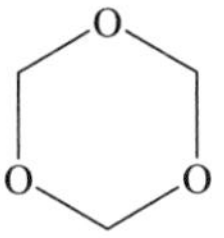 (trioxane ring structure)	Trioxane
Polyether	$-\!\!\left[\!-(CH_2O)_x\!\right]\!-$	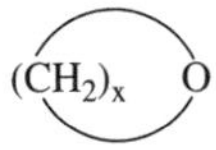 $(CH_2)_x \quad O$	Cyclic ether
Polyester	$-\!\!\left[\!-(CH_2)_x\overset{O}{\overset{\|}{C}}O\!-\!\right]\!-$	$(CH_2)_x \ \substack{C=O \\ \| \\ O}$	Lactone
Polyamide	$-\!\!\left[\!-(CH_2)_x\overset{O}{\overset{\|}{C}}NH\!-\!\right]\!-$	$(CH_2)_x \ \substack{C=O \\ \| \\ NH}$	Lactam
Polysiloxane	$-\!\!\left[\substack{CH_3 \\ \| \\ -SiO- \\ \| \\ CH_3}\right]\!-$	(cyclic siloxane ring with Si, O, CH₃ groups)	Cyclic siloxane
Polyphosphazene	$-\!\!\left[\substack{Cl \\ \| \\ -P\!=\!N- \\ \| \\ Cl}\right]\!-$	(hexachlorocyclotriphosphazene ring with P, N, Cl)	Hexaxhlorcyclotriphosphazene
Polyamine	$-\!\!\left[\!-CH_2CH_2NH\!-\!\right]\!-$	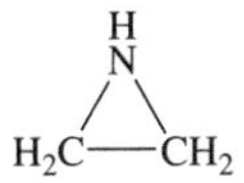 $\substack{H \\ N \\ H_2C\!-\!CH_2}$	Aziridine

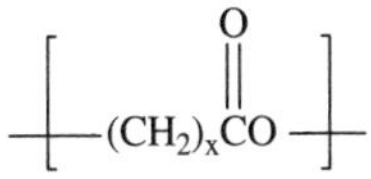

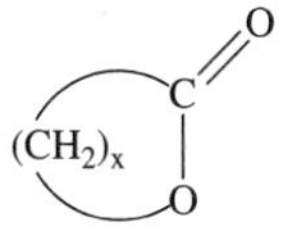

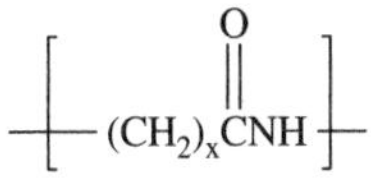

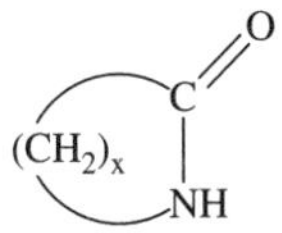

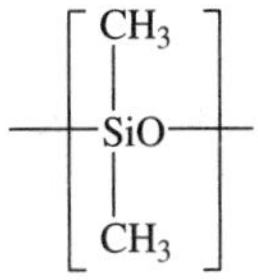

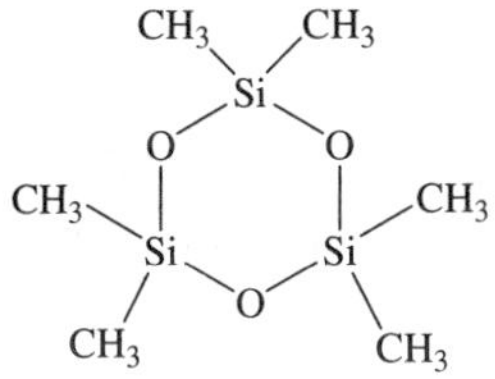

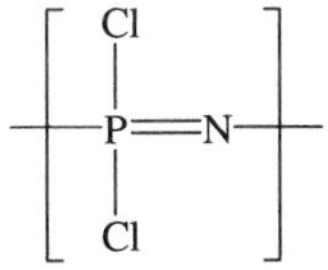

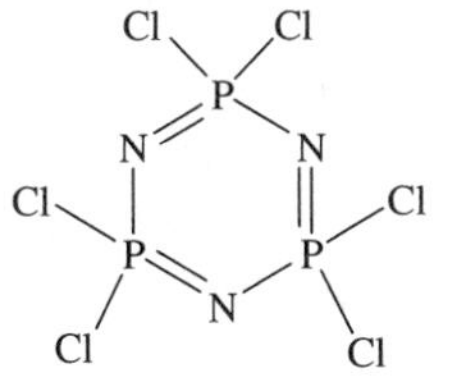

in organic nonprotonic solvent such as toluene and thetrahydrofuran. Compounds with *active* hydrogens such as water and alcohols will stop the polymerization. Under the right conditions, polymerization will continue until all of the monomer has been consumed. An important example is group transfer polymerization of methyl methacrylate shown in the following reaction scheme:

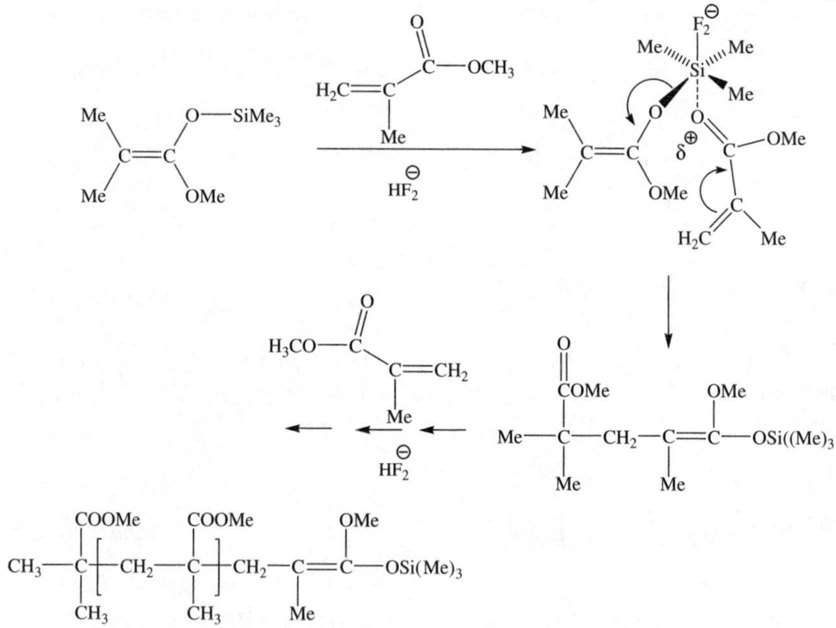

Potential applications of GTP include high-performance automotive finishes, the fabrication of silicon chips, and coatings for optical fibers.

14.9   Polymerization Techniques

Polymers can be prepared by many different processes. Free radical polymerization can be accomplished in bulk, suspension, solution, or emulsion. Ionic and other nonradical polymerizations are usually produced in solution polymerizations. Each technique has characteristic advantages and disadvantages.

Bulk polymerization.   Bulk polymerization is the simplest and most direct method (from the standpoint of formulation and equipment) for converting monomer to polymer. It requires only monomer (and possibly monomer-soluble initiator or catalyst), and perhaps a chain transfer agent for molecular weight control, and as such gives the highest-purity polymer. However, extra care must be taken to control the process when the polymerization reaction is very exothermic and particularly when it is run on a large scale. Poly(methyl methacrylate), polystyrene, or low-density (high pressure) polyethylene, for example, can be produced from heating the respected monomer in the presence of an initiator and the absence of oxygen.

In polymerization, viscosity increases and termination reaction progressively becomes more hindered because the macroradicals are unable to diffuse readily and get together in the viscous medium. In contrast, the small monomer molecules continue to diffuse readily to a growing chain end. This means that the termination rate decreases more rapidly than the propagation rate. As a result, the overall polymerization rate increases with accompanying additional heat production. This leads to the production of high molecular weight macroradicals as a result of propagation in the absence of termination. The vinyl monomers have relatively large exothermic heat of polymerization, typically between −10 and −21 kcal/mol. The tremendous viscosities prevent effective convective (mixing) heat dissipation. An increase in temperature will increase the polymerization rate, and, therefore, generate additional heat to dissipate. This leads to a rapid increase in the rate of polymerization and the amount of heat generated. This phenomenon is known as autoacceleration, the Norris-Trommsdroff or gel effect. It leads to the formation of unusually high-molecular-weight polymers, and releases a massive amount of heat. Therefore, special design of equipment is necessary for large-scale bulk polymerizations. In practice, heat dissipation during bulk polymerization can be removed by providing special baffles for improved heat transfer or by performing the bulk polymerization in separate steps of low-to-moderate conversion.

Another example of bulk (or melt) polymerization is the synthesis of polyamides through the direct interaction between a dicarboxylic acid and a diamine. Nylon 66, for example, can be produced from the reaction between hexamethylenediamine and adipic acid. In practice, it is preferable to ensure the existence of a 1:1 ratio of the two reactants by prior isolation of a 1:1 salt of the two. The overall procedure is summarized by the reaction scheme:

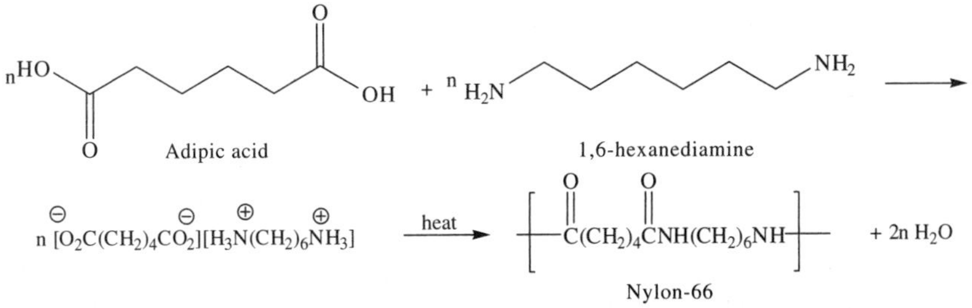

The major commercial uses of bulk vinyl polymerization are in casting formulations and low-molecular-weight polymers for use as adhesives, plasticizers, and lubricant adhesives.

Solution polymerization.   Solution polymerization involves polymerization of a monomer in a solvent in which both the monomer (reactant) and polymer (product) are soluble. Monomers are polymerized in a solution that can be homogeneous or heterogeneous. Many free radical polymerizations are conducted in solution. Ionic polymerizations are almost exclusively solution processes along with many Ziegler-Natta polymerizations. Important water-soluble polymers that can be prepared in aqueous solution include poly(acrylic acid), polyacrylamide, poly(vinyl alcohol), and poly(N-vinylpyrrolidinone). Poly(methyl methacrylate), polystyrene, polybutadiene, poly(vinyl chloride), and poly(vinylidene fluoride) can be polymerized in organic solvents.

The addition of solvent allows minimizing many of the difficulties encountered in bulk polymerization. The solvent acts as diluent that reduces the tendency toward autoacceleration. The requirements for selection of the solvent are that both the initiator and monomer be soluble in it, and that the solvent has acceptable chain-transfer characteristics and suitable melting and boiling points for the conditions of the polymerization and subsequent solvent-removal step. The viscosity of the solution continues to increase until the polymerization is complete, but the concentration of the solution is usually too dilute to exhibit autoacceleration because of the gel effect. Also the solvent aids in the transfer of heat of the bulk process. In addition, the heat of polymerization may be conveniently and efficiently removed by refluxing the solvent. The solvent also allows easier stirring, because the viscosity of the reaction mixture is decreased. On the other hand, the presence of solvent may present new difficulties. Chain transfer to solvent can be a problem that limits the molecular weight. Furthermore, the purity of the polymer may be affected if there are difficulties in the removal of the solvent. The polymer may be recovered by pouring the solution into an agitated poor solvent or nonsolvent. Because of problems usually encountered in removing solvent completely from the resultant polymer, the method is best suited to applications where the solution may be used directly, as with certain adhesives or solvent-based paints.

Precipitation polymerization.   In precipitation polymerization, monomer is polymerized either in bulk or in solution (aqueous or organic), however, the polymer formed is insoluble in the reaction media. As such, the forming polymer precipitates and the viscosity of the medium does not change appreciably. This polymerization is often referred to as *powder* or *granular* polymerization because of the forms in which the polymers are produced. Solution polymerization of acrylonitrile in water, and bulk polymerization of vinyl chloride are examples of precipitation polymerization.

Suspension polymerization. In this method of polymerization, a liquid monomer is suspended in the form of droplets (50 to 500 μm in diameter) in an inert, nonsolvent liquid (almost always water). The monomer:water weight ratio may vary from 1:1 to 1:4. The suspension is maintained by mechanical agitation and the addition of stabilizers. Small quantities (approximately, 0.1 percent) of protective colloid; water-soluble polymers (e.g., poly(vinyl alcohol), hydroxylpropyl cellulose, sodium poly(styrene sulfonate)), or finely divided insoluble inorganic substances (e.g., barium sulfate, calcium phosphate, magnesium phosphate, or magnesium carbonate) are added to prevent both the coalescence of the monomer droplets, and in the later stages of polymerization, coagulation of the polymer particles swollen by monomer. A pH buffer is sometimes also used to help stability. Polymerization takes place in the monomer droplet using monomer-soluble initiator. Each droplet can be looked at as an individual bulk reactor. Thus, from the standpoint of kinetics and mechanism, suspension polymerization is identical to bulk polymerization. The heat can easily be soaked up by and removed from the low-viscosity, inert suspension medium, and the reaction is therefore easily controlled.

During reaction, there is a change in aggregation. If the process is carefully controlled, polymer is obtained in the form of granular beads, hence the method is also called *pearl* or *bead* polymerization. The size of the product beads depends on the strength of agitation, as well as the nature and quantity of the monomer and suspending system. In general, suspension polymerization cannot be used for tacky polymers such as elastomers because of the tendency for agglomeration of polymer particles. However, suspension polymerizations in the presence of high concentration (>1 percent) of the water-soluble stabilizers (and usually water-soluble initiators) produce latex-like dispersions of particles having small particle size in the range 0.5 to 10 μm. This type of suspension polymerization is sometimes referred to as *dispersion polymerization.*

Commercially, suspension polymerizations have been limited to the free radical polymerization of water-insoluble liquid monomers to prepare a number of granular polymers, including polystyrene, poly(vinyl acetate), poly(methyl methacrylate), polytetrafluoroethylene, extrusion and injection-molding grades of poly(vinyl chloride), poly(styrene-*co*-acrylonitrile) (SAN), and extrusion-grade poly(vinylidene chloride-*co*-vinyl chloride). It is possible, however, to perform *inverse suspension polymerizations*, where water-soluble monomer (e.g., acrylamide) is dispersed in a continuous hydrophobic organic solvent.

Emulsion polymerization. The technological origins of emulsion polymerization go back to the 1920s when first developed at Goodyear Tire and Rubber Company. And before World War II, there was a well-established industry for the production of synthetic rubbers and plastics by emulsion techniques. Emulsion polymerization involves the polymerization of monomers that are in the form of emulsions. Emulsion polymerization involves a colloidal dispersion, and resembles suspension polymerization in that water is used as dispersing medium, and heat transfer is efficient. However, it differs from suspension in the type and size of the particles in which the polymerization occurs and in the kind of initiator employed.

Emulsion polymerization system consists of a hydrophobic monomer (e.g., styrene), dispersant (water), water-soluble initiator (e.g., $K_2S_2O_8$), and emulsifier (e.g., surfactant such as soap; sodium stearate; sodium lauryl sulfate). In the water (dispersant) various components are dispersed in an emulsion state by means of the emulsifier which prevents the emulsion from separating into two layers once stirring had stopped. Emulsion system is kept in a well-agitated state during reaction. The ratio of water to monomer is in the range 70:30 to 40:60 by weight. The size of monomer droplets depends upon the polymerization temperature and the rate of agitation. Above certain surfactant concentration, called *critical micelle concentration* (CMC), the excess surfactant molecules aggregate to form small colloidal clusters known as micelle. Surfactant concentration (2 to 3 percent) exceeds CMC by 1 to 3 orders of magnitude; hence the bulk of the surfactant is in micelles.

A simplified representation of an emulsion polymerization system is illustrated in Fig. 14.23. Initiator radicals are generated in the aqueous phase and diffuse into soap micelles swollen with monomer molecules. Polymerization takes place almost exclusively in the interior of the micelles that are present in very high concentration; typically 10^{18} per mL, compared to that of the monomer droplets (10^{10} to 10^{11} per mL). Also, micelles have very high surface to volume ratio compared to droplets. Polymerization starts either by entry of primary radicals or oligomeric radicals (for $n = 3–5$, the oligomer is no longer soluble in water) formed by solution polymerization. As polymerization proceeds, the active micelles (considered as polymer particles) grow by addition of monomer from water solution that in turn gets the replenishment from the monomer droplets. Termination of polymerization occurs by radical combination when a new radical diffuses into the micelle.

The emulsion polymerization process has several distinct advantages of providing a polymer of exceptionally high molecular weight, and narrow molecular weight distribution, while permitting efficient control over the exothermic polymerization reaction because the aqueous phase absorbs the heat of reaction.

Emulsion polymerization is widely used to prepare acrylic polymers, poly(vinyl chloride), poly(vinyl acetate), and a large number of copolymers.

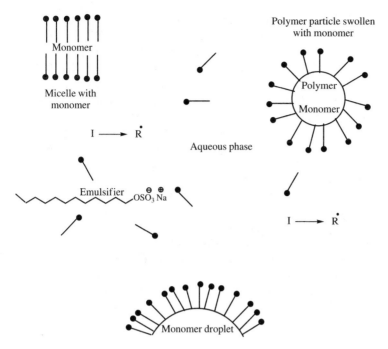

Figure 14.23 A simplified representation of an emulsion polymerization system.

The final product of an emulsion polymerization is referred to as latex. Emulsion polymerization products can in some instances be employed directly without further separations but with appropriate blending operations. Such applications involve coatings, finishes, floor polishes, and paints. Solid polymer can be recovered from the latex by various techniques such as spray drying, coagulation by adding an acid, usually sulfuric acid, or by adding electrolyte salts.

When the monomer is hydrophilic, emulsion polymerization may proceed through what's called an *inverse* emulsion process. In this case, the monomer (usually in aqueous solution) is dispersed in an organic solvent using a water-in-oil emulsifier. The initiator may be either water-soluble or oil-soluble. The final product in an inverse emulsion polymerization is a colloidal dispersion of a water-swollen polymer in the organic phase.

Interfacial polycondensation. A variation of solution polymerization known as interfacial polymerization takes place when the two monomers are present in two immiscible solvents. Reaction then takes place at the interface between the two liquids, and is soluble in neither. Generally, one of the phases also contains an agent that reacts with the condensation byproducts to drive the reaction to completion. This process is especially effective if the rate of polymerization is rapid at moderate temperature (0 to 50°C). The polymerization rate is diffusion-controlled, because the rate of diffusion of the monomers to the interface is slower than the rate of polymerization. Monomer molecules tend to react more rapidly with growing polymer chains than with other monomer molecules because the reaction is too rapid to allow the monomer to diffuse through the layer of polymer. This is why molar mass of the polymers is generally higher than that obtained by the melt method. Stoichiometry automatically exists at the interface. In order to produce long chains and speed up the kinetics of the reaction, the system can be stirred vigorously to ensure a constantly changing interface.

This technique can be used effectively to prepare polyesters, polyamides, and polycarbonates. The process of interfacial polymerization can best be illustrated by the reaction between a diamine and a diacid chloride to produce polyamide. The word *Nylon* is used to represent synthetic polyamides. The various nylons are described by a numbering system that indicates the number of carbon atoms in the monomer chains. Nylons from diamines and dibasic acids are designated by two numbers; the first representing the diamine and the second the dibasic acid. Thus, nylon-6,10 is formed by the reaction of hexamethylenediamine and sebacoyl chloride:

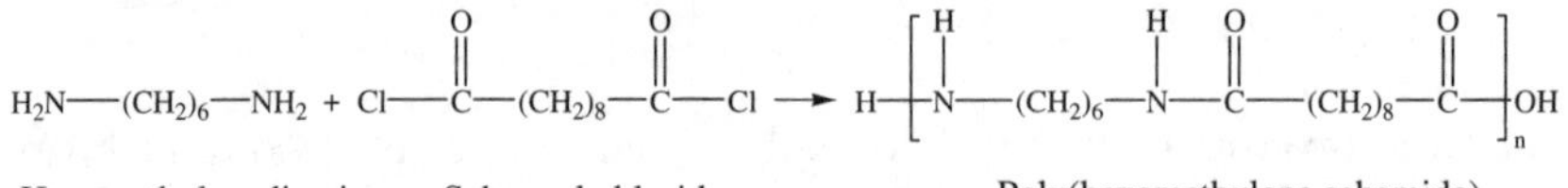

$$H_2N\text{---}(CH_2)_6\text{---}NH_2 \ + \ Cl\text{---}\overset{O}{\overset{\|}{C}}\text{---}(CH_2)_8\text{---}\overset{O}{\overset{\|}{C}}\text{---}Cl \ \longrightarrow \ H\text{---}\left[\overset{H}{\overset{|}{N}}\text{---}(CH_2)_6\text{---}\overset{H}{\overset{|}{N}}\text{---}\overset{O}{\overset{\|}{C}}\text{---}(CH_2)_8\text{---}\overset{O}{\overset{\|}{C}}\right]_n\text{---}OH$$

Hexamethylenediamine   Sebacoyl chloride   Poly(hexamethylene sebamide)
                                                                          (nylon-6,10)

The acid chloride is dissolved, for example, in hexane, and the diamine in water along with some NaOH to soak up the HCl. The aqueous layer is gently poured on top of the diamine solution. The reactants diffuse to the interface, where they react rapidly to form a polymer film. The resulting polymer is insoluble in both phases and can be drawn off in the form of a rope. The continuous thread or *rope* can be wound on a windlass until one or the other of the two reactants is exhausted (Fig. 14.24). This polyamide has found applications in sport equipment and bristles for brushes.

14.10   Copolymerization

Copolymer is a macromolecule consisting of two or more different types of repeat units. The following scheme shows an example of copolymer

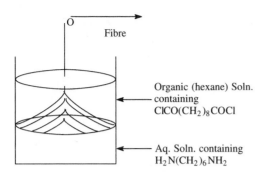

Figure 14.24   Schematic illustration of interfacial polymerization.

formed when styrene and acrylonitrile are polymerized in the same reactor. The polymer with three chemically-different repeating units is termed *terpolymer*. Copolymerization provides a means of producing polymers with new and desirable properties by linking two or three different monomers or repeat units.

Styrene   Acrylonitrile   Poly[styrene-*co*-(acrylo nitrile)]
                                                   styrene-acrylo nitrile copolymer

The exact sequence of monomer units along the chain can vary widely depending upon the relative reactivities of each monomer during the polymerization process. At the extremes, monomer placement may be totally random or may be perfectly alternating. If repeating units are represented by A and B, then the *random* copolymer might have the structure shown as:

AABBABABBAAABAABBAB

An example is the random copolymer made by free radical copolymerization of vinyl chloride and vinyl acetate:

Vinyl chloride   Vinyl acetate   Poly[(vinyl chloride)-*co*-(vnyl acetate)]

In *alternating* copolymer, each monomer of one type is joined to a monomer of a second type. Therefore, there is an ordered (alternating) arrangement of the two repeating units along the polymer chain as shown in the following sequence:

ABABABABABABABABABABABABABABABABABAB

An example is the product made by free radical polymerization of equimolar quantities of styrene and maleic anhydride:

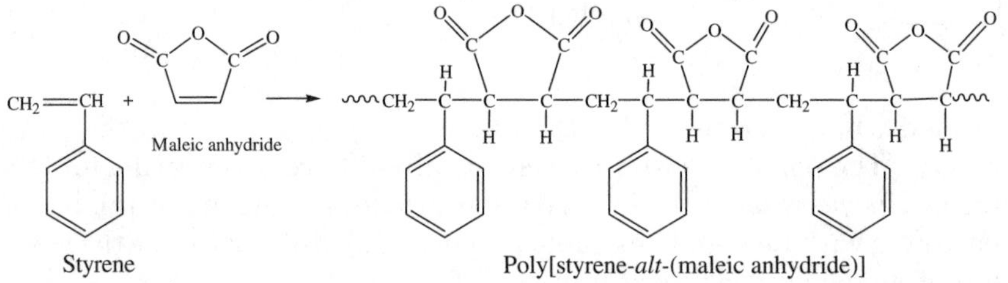

Styrene   Maleic anhydride   Poly[styrene-*alt*-(maleic anhydride)]

Under special circumstances, it is possible to prepare copolymers that contain a long block of one monomer (A) followed by a block of another monomer (B). This type of copolymers is called *block copolymer*, and will have a structure like:

AAAAAAAAABBBBBBBBBBBBAAAAAAAAAAABBBBBBBBBB

Triblock copolymers have a central B block joined by A blocks at the end. A commercially important ABA-triblock copolymer is polystyrene-*block*-polybutadiene-*block*-polystyrne (SBS); a thermoplastic elastomer.

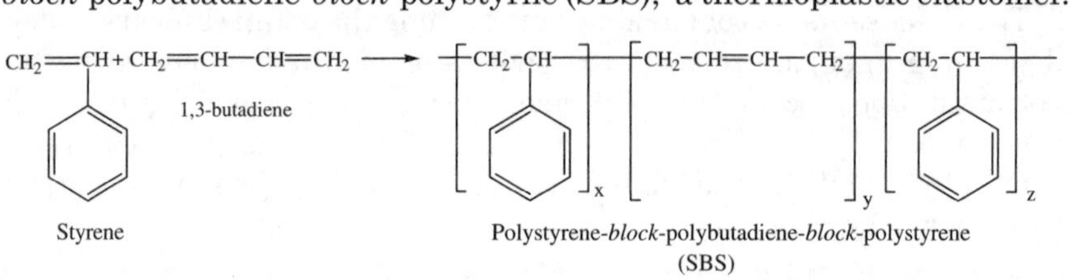

Styrene   1,3-butadiene   Polystyrene-*block*-polybutadiene-*block*-polystyrene
(SBS)

In addition to the above copolymer structures, *graft copolymers* can be prepared, in which sequences of one monomer are grafted onto a backbone of another monomer type:

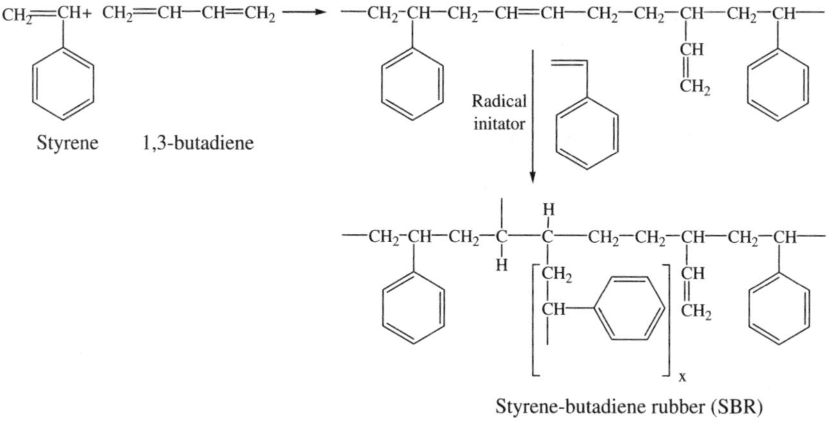

CH₂=CH+ CH₂=CH—CH=CH₂ ⟶ —CH₂-CH—CH₂—CH=CH—CH₂—CH₂—CH—CH₂—CH—

Styrene   1,3-butadiene

Radical initator

Styrene-butadiene rubber (SBR)

Figure 14.25 Synthesis of styrene-butadiene rubber (SBR) by *grafting from* copolymerization.

Graft copolymers are important as elastomeric (e.g., styrene-butadiene rubber (SBR)) and high-impact polymers (e.g., high-impact polystyrene and acrylonitrile-butadiene-styrene (ABS)).

A number of techniques have been developed for the synthesis of graft copolymers. Most commonly, graft copolymers are prepared from pre-polymers that possess groups along the chain that can be activated to initiate polymerization of a second monomer, thus forming branches on the prepolymer. This method is refered to as *grafting from*. Figure 14.25 shows a technique in which polymerization of one monomer is carried out in the presence of a polymer of the other material. Thus, a rubber backbone-styrene graft copolymer results when styrene monomer containing dissolved rubber (SBR) is subjected to polymerization conditions with radical initiators.

Additionally, graft copolymers can be prepared by a *grafting onto* method that involves coupling living polymers to reactive side groups on a prepolymer. An alternative approach to the preparation of graft copolymers involves the use of *macromonomers*. A macromonomer is a prepolymer with terminal polymerizable C=C bond. In this method, graft copolymers are produced by copolymerization of the macromonomer with another olefinic monomer as shown in Fig. 14.26.

Copolymers may be produced by step reaction or by chain reaction polymerization in similar mechanisms to those of homopolymerization. The most widely used synthetic rubber (SBR) is a copolymer of styrene (S) and butadiene (B). Also, ABS, a widely used plastic, is a copolymer or blend of polymers of acrylonitrile, butadiene, and styrene. A special

(*i*) *Synthesis of macromonomer*

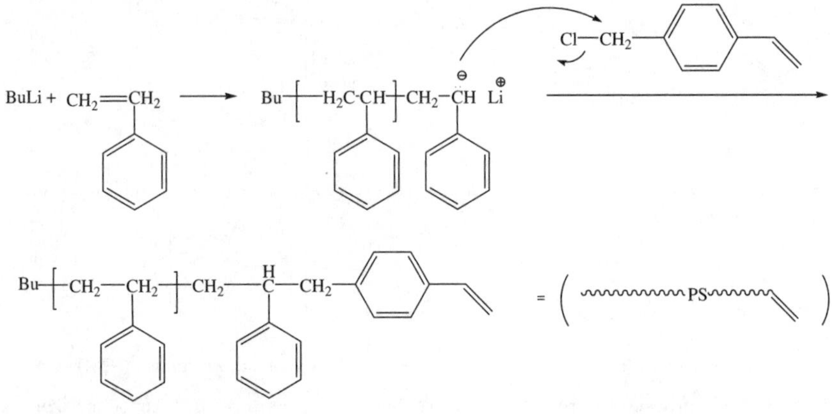

Macromonomer

(*ii*) *Copolymerization of the macromonomer*

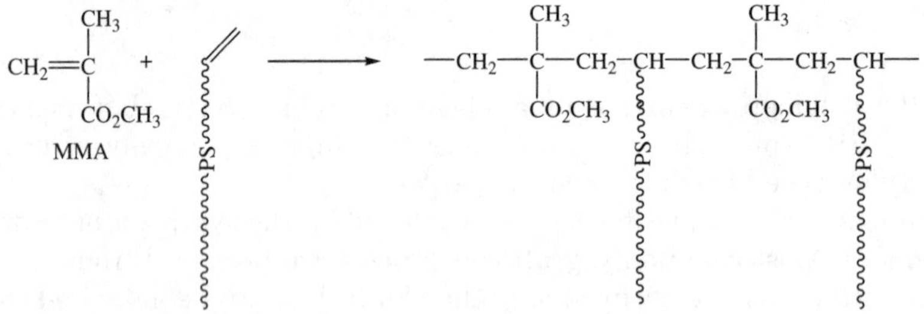

Figure 14.26   Synthesis of graft copolymers via copolymerization of macromonomers.

fiber called Spandex is a block copolymer of stiff polyurethane and flexible polyester.

In the polymerization of a mixture of two or more monomers, the rate at which different monomers add to the growing chain determines the composition and hence the properties of the resulting copolymer. The order as well as the ratio of amounts in which monomers add are determined by their relative reactivities in the chain-growth step, which in turn are influenced by the nature of the end of the growing chain, depending on which monomer added previously. Among the possibilities are random, regular, and alternating additions, as well as block formation.

Kinetics of copolymerization.   With two monomers present, there are four possible propagation reactions, assuming that growth is influenced only by the nature of the end of the growing chain and of the monomer.

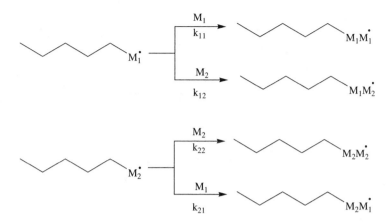

$$r_1 = \frac{k_{11}}{k_{12}} \qquad r_2 = \frac{k_{22}}{k_{21}}$$

k_{11} and k_{22} are called self-propagation rate constants.
k_{12} and k_{21} are called cross-propagation rate constants.

It is experimentally observed that the number of growing chains remains approximately constant throughout the duration of most copolymerizations. In that case, the concentration of $M_1^{\bullet}$ and $M_2^{\bullet}$ are constant (steady state assumption), and the rate of conversion of $M_1^{\bullet}$ to $M_2^{\bullet}$ is equal to the conversion of $M_2^{\bullet}$ to $M_1^{\bullet}$, so

$$k_{12}[M^{\bullet}_1][M_2] = k_{21}[M^{\bullet}_2][M_1]$$

and

$$[M^{\bullet}_1] = \frac{k_{21}[M^{\bullet}_2][M_1]}{k_{12}[M_2]} \tag{47}$$

Rate of disappearance of M_1

$$= -\frac{d[M_1]}{dt} = k_{11}[M^{\bullet}_1][M_1] + k_{21}[M^{\bullet}_2][M_1]$$

Rate of disappearance of M_2

$$= -\frac{d[M_2]}{dt} = k_{22}[M^{\bullet}_2][M_2] + k_{12}[M^{\bullet}_1][M_2]$$

The ratio of disappearance of monomers M_1/M_2 or n is obtained by dividing the two rate equations, followed by substitution of $[M^{\bullet}_1]$, division by k_{21}, and substitution of $r_1 = k_{11}/k_{12}$ and $r_2 = k_{22}/k_{21}$.

$$n = \frac{d[M_1]}{d[M_2]} = \frac{[M_1]}{[M_2]}\left(\frac{r_1[M_1]+[M_2]}{[M_1]+r_2[M_2]}\right)$$

$$= \frac{r_1[M_1]/[M_2]+1}{r_2[M_2]/[M_1]+1} = \frac{r_1 x+1}{(r_2/x)+1} \tag{48}$$

Rearrangement of the above equation will lead to

$$\frac{x(1-n)}{n} = \frac{-r_1 x^2}{n} + r_2 \tag{49}$$

Plotting $x(1-n)/n$ versus x^2/n will give a straight line with a slope of $-r_1$ and an intercept of r_2. The monomer reactivity ratios for some common monomers in radical copolymerization are listed in Table 14.25.

When reactivity ratio is greater than unity, the copolymer contains a larger proportion of the more reactive monomer, and as the difference in reactivity of the two monomers increases, it becomes more and more difficult to produce copolymers containing appreciable amounts of both monomers. In those rare cases when both reactivity ratios are greater than one, there is a tendency to produce block copolymers, but these are better prepared by the anionic *living polymer* techniques. Some specific examples are given below:

1. When $r_1 = \sim 0$, $r_2 = \sim 0$, and $r_1 r_2 = \sim 0$, neither monomer radical will add its own monomer and propagation can continue to produce an alternating copolymer.

2. When $r_1 = 1$, $r_2 = 1$, and $r_1 r_2 = 1$, the copolymerization is said to be ideal; each radical shows the same preference for one of the monomers. The sequence of monomers in the copolymer is completely random, and the polymer composition is the same as the comonomer feed. A plot of mole percent of M_1 in the copolymer against mole percent of M_1 in the corresponding feed will give a straight line with zero intercept.

3. When $r_1 \cdot r_2 = 1$, but neither r_1 nor r_2 is equal to 1 (i.e., $r_1 = 1/r_2$). A plot of mole percent of M_1 in the copolymer against mole percent of M_1 in the corresponding feed will give a curve. The curve will be convex if $r_1 > r_2$ and will be concave if $r_1 < r_2$.

4. When r_1, r_2, and $r_1 \cdot r_2 < 1$, there is a tendency for alternation. The smaller the value of r_1 and r_2, the greater is the tendency for alternation.

TABLE 14.25   Typical Free Radical Chain Copolymerization Reactivity Ratios at 60°C [22]

M_1	M_2	r_1	r_2	$r_1 r_2$
Acrylamide	Acrylic acid	1.38	0.36	0.5
	Methyl acrylate	1.30	0.05	0.07
	Vinylidene chloride	4.9	0.15	0.74
Acrylic acid	Acrylonitrile (50°C)	1.15	0.35	0.40
	Styrene	0.25	0.15	0.04
	Vinyl acetate (70°C)	2	0.1	0.2
Acrylonitrile	Butadiene	0.25	0.33	0.08
	Ethyl acrylate (50°C)	1.17	0.67	0.78
	Maleic anhydride	6	0	0
	Methyl methacrylate	0.13	1.16	0.15
	Styrene	0.04	0.41	0.16
	Vinyl acetate	4.05	0.06	0.24
	Viny chloride	3.28	0.02	0.07
Butadiene	Methyl methacrylate	0.70	0.32	0.22
	Styrene	1.39	0.78	1.08
Chlorotrifluoroethylene	Tetrafluoroethylene	1.0	1.0	1.0
Isoprene	Styrene	1.98	0.44	0.87
Maleic anhydride	Methyl acrylate	0	2.5	0
	Methyl methacrylate	0.03	3.5	0.11
	Styrene	0	0.02	0
	Vinyl acetate (70°C)	0.003	0.055	0.0002
Methyl acrylate	Acrylonitrile	0.67	1.26	0.84
	Styrene	0.18	0.75	0.14
	Vinyl acetate	9.0	0.1	0.90
	Vinyl chloride	5	0	0
Methyl isopropenyl ketone	Styrene (80°C)	0.66	0.32	0.21
Methyl methacrylate	Styrene	0.50	0.50	0.25
	Vinyl acetate	20	0.015	0.30
	Vinyl chloride	12.5	0	0
α–Methylstyrene	Maleic anhydride	0.038	0.08	0.003
	Styrene	0.38	2.3	0.87
Styrene	p-Chlorostyrene	0.74	1.025	0.76
	Fumaronitrile	0.23	0.01	0.002
	p-Methoxystyrene	1.16	0.82	0.95
	Vinyl acetate	55	0.01	0.55
	Vinyl chloride	17	0.02	0.34
	2-Vinylpyridine	0.56	0.9	0.50
Vinyl acetate	Vinyl chloride	0.23	1.68	0.39
	Vinyl laurate	1.4	0.7	0.98
Vinyl chloride	Diethyl maleate	0.77	0.009	0.007
	Vinylidene chloride	0.3	3.2	0.96
N-Vinylpyrrolidone	Styrene(50°C)	0.045	15.7	0.71

NOTE: Temperatures other than 60°C are shown in parentheses.

5. When $r_1 \gg 1$ (or $r_2 \gg 1$) then one obtains homopolymers or block copolymers.

14.11   Modification of Synthetic Polymers

In many cases, a polymer can be modified to improve some property, such as strength, biocompatibility, fire retardancy, adhesion, or to provide a special functional group for certain application by means of postpoly-merization reactions that are similar to those of classical organic chem-ical reactions. Saturated polymeric hydrocarbons such as HDPE may be chlorinated by reaction with chlorine at elevated temperature or in the presence of UV light. Chlorination of PVC yield a product called poly(vinyl dichloride) (PVDC) that has superior heat resistance to that of PVC, and thus finds use in applications like hot water piping systems. Also, chlorination of poly(vinyl chloride) after polymerization is used to increase its softening temperature or to improve its ability to blend with other polymers. Bromination is sometimes used to impart fire retardancy to some polymers.

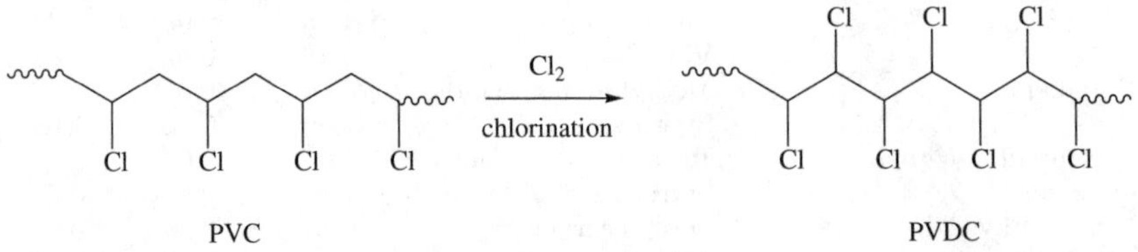

PVC   PVDC

Polyenes (i.e., unsaturated aliphatic polymers) such as polyisoprenes, and polybutadiens may be hydrogenated, halogenated, hydrohalo-genated, cyclized, and epoxidized.

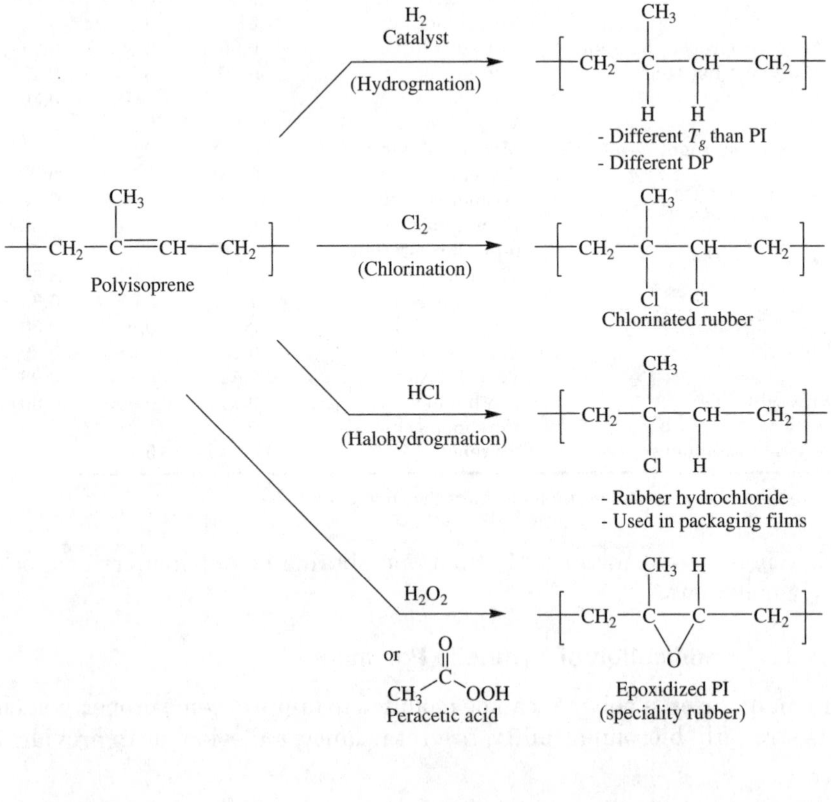

In some cases, important commercial polymers can be produced only by chemical modification of a precursor polymer. Poly(vinyl alcohol) (PVA), which is used as stabilizing agent in emulsion polymerizations and as a thickening and gelling agent, cannot be synthesized directly from its monomer, because vinyl alcohol is isomeric with acetaldehyde. PVA is rather obtained by the direct hydrolysis (or catalyzed alcoholysis) of poly(vinyl acetate). Poly(vinyl acetate) is produced by free radical emulsion or suspension polymerization. Another important polymer, poly(vinyl butyral), which is used as the film between the layers of glass in safety windshields, is obtained by partially reacting poly(vinyl alcohol) with butyraldehyde as shown in the following reaction scheme.

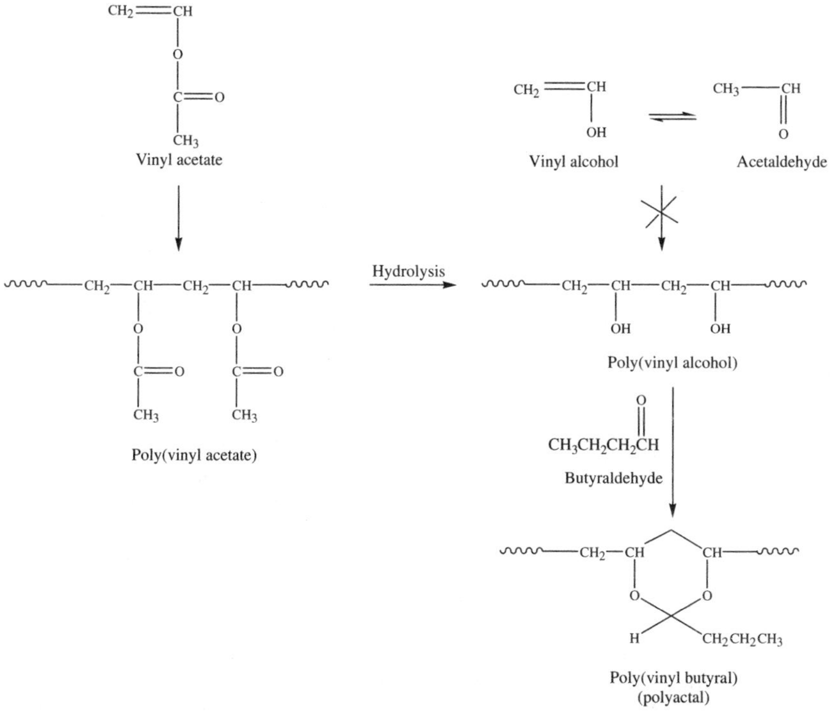

$CH_2 = CH$

$C = O$

CH_3

Vinyl acetate

$CH_2 = CH$ $\rightleftharpoons$ $CH_3 - CH$

OH

O

Vinyl alcohol   Acetaldehyde

Hydrolysis

$CH_2 - CH - CH_2 - CH$

O O

$C = O$ $C = O$

CH_3 CH_3

Poly(vinyl acetate)

$CH_2 - CH - CH_2 - CH$

OH   OH

Poly(vinyl alcohol)

O

$CH_3CH_2CH_2CH$

Butyraldehyde

$CH_2 - CH$ CH

O O

H $CH_2CH_2CH_3$

Poly(vinyl butyral)
(polyactal)

Vinyl amine, like vinyl alcohol, is unstable. Therefore, poly(vinyl amine) is produced by the Hofmann elimination of polyacrylamide.

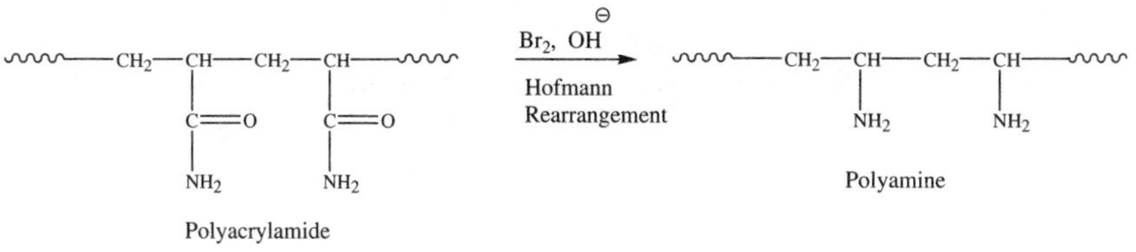

$CH_2 - CH - CH_2 - CH$

$C = O$ $C = O$

NH_2 NH_2

Polyacrylamide

$\overset{\ominus}{Br_2, OH}$

Hofmann
Rearrangement

$CH_2 - CH - CH_2 - CH$

NH_2 NH_2

Polyamine

Polymers with pendant groups that are derivatives of carboxylic acid can be hydrolyzed to yield poly(acrylic acid). This includes polymers like polyacrylamide, polyacrylonitrile, and polyacrylates. When heated, poly(acrylic acid) form polymeric anhydrides, which undergo typical reactions of anhydrides, such as hydrolysis, alcoholysis, and amidation.

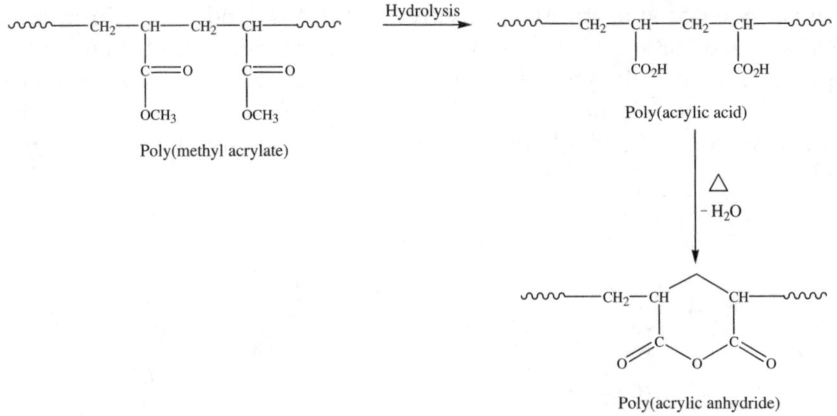

Poly(methyl acrylate)

Poly(acrylic acid)

Poly(acrylic anhydride)

Polyacrylonitrile, upon heating, from a ladder polymer

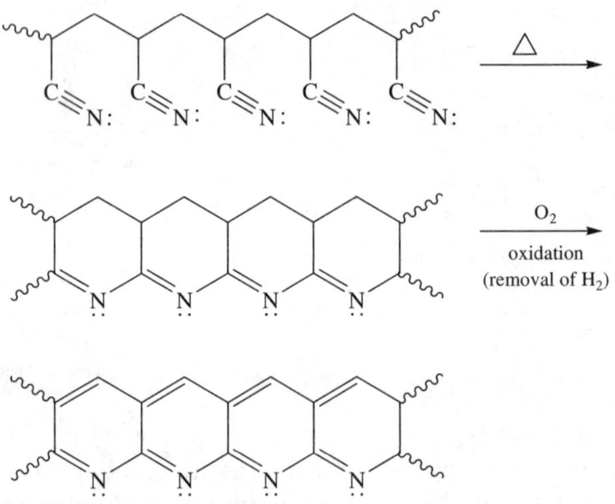

Polymers with phenyl pendant groups such as those present in polystyrene undergo all of the characteristic reactions of benzene, such as alkylation, halogenation, nitration, and sulfonation. Thus, oil-soluble polymers (e.g., poly(vinyl cyclohexylbenzene) used as viscosity improvers in lubricating oils are obtained by the Friedel-Crafts reaction of polystyrene and unsaturated hydrocarbons such as cyclohexene. Also, in the presence of a Lewis acid, halogens such as chlorine react with polystyrene to produce chlorinated polystyrene that has a higher softening point than polystyrene. Polynitrostyrene is produced by the nitration of polystyrene. The latter may be reduced to form polyaminostyrene. Polyaminostyrene may be diazotized to polymeric dyes. Polystyrene and other aromatic polymers could be sulfonated by fuming sulfuric acid. Sulfonated crosslinked polystyrene has been used as an ion exchange resin.

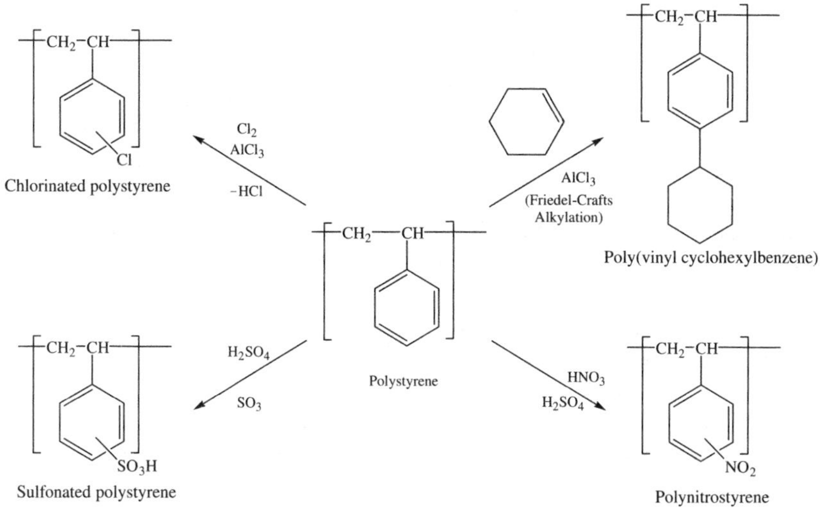

14.12   Degradation, Stability, and Environmental Issues

Most polymers are susceptible to degradation by exposure to high temperature, oxygen and ozone, ultraviolet light, moisture, and chemical agents. Backbone chain scission degradation can occur via depolymerization where monomer is split off from an activated end group in a reaction referred to as *unzipping*. Chain degradation can also occur via random chain breakage where units are split apart in a random manner similar to the opposite of stepwise polycondensation. Chain scission degradation reactions can also occur preferentially at weak links in the polymer backbone. The major means of polymer degradation are given in Table 14.26. Although degradation of polymers might be deleterious, in some cases degradation may be a desirable goal. For example, it is

TABLE 14.26   Major Synthetic Polymer Degradative Agents

Degradation agent	Susceptible polymers	Examples
Acids and bases	Heterochain polymers	Polyesters, polyurethanes
Organic liquids and vapors	Amorphous polymers	Polystyrene, Poly(methyl methacrylate)
Ozone	Unsaturated polymers	Polybutadiene, polyisoprene
Moisture	Heterochain polymers	Polyesters, polyurethanes, polyamides (nylons)
Sunlight	Photosensitive polymers	Polyacetals, polycarbonates
Biodegradation	Heterochain polymers, nitrogen-containing polymers, polyesters	Polyurethanes, polyesters, nylons Polyether-polyurethane
Heat	Vinyl polymers	PVC, poly(α-methylstyrene)
Ionizing radiation	Aliphatic polymers with quaternary carbon atoms	Polypropylene, LDPE, PMMA, poly(α-methylstyrene) polyisobutylene

desired to use polymers that rapidly degrade to environmentally safe by-products in making bottles and packaging films.

Thermal degradation.   Thermal degradation results in a decrease in the degree of polymerization, and generally results in some char and formation of smaller molecules including water, methanol, carbon dioxide, and HCl depending on the structure of the polymer. Polymers lose their mechanical properties and become brittle and break after long-term exposure to sunlight.

Polymers with highly aromatic structures withstand extended exposure to high temperatures. This can be attributed to resonance stabilization (with energies up to 16.7 kJ/mol) that results in high main-chain bond strength and consequently high temperature stability. Thermal stability is further fortified with the presence of heterocyclic rings. Table 14.27 lists examples of high-temperature polymers and their decomposition temperatures. On the other hand, the rate of decomposition of polymers such as PVC at elevated temperatures may be decreased by the addition of heat stabilizers that react with the decomposition products, like HCl. Soluble organic metal compounds, phosphates, and epoxides act as thermal stabilizers or scavengers for HCl.

Oxidative and UV degradation.   Polymers that contain sites of unsaturation, such as polyisoprene and the polybutadienes, are most susceptible to oxygen and ozone oxidation. Figure 14.27 illustrates a typical oxidative degradation of a common elastomer. The figure shows the combined effect of light and oxygen (photolysis) and the action of ozone (ozonolysis).

TABLE 14.27   Examples of Thermally Stable Polymers [6]

Polymer	Structure	Decomposition temperature (°C)
Aromatic polyester	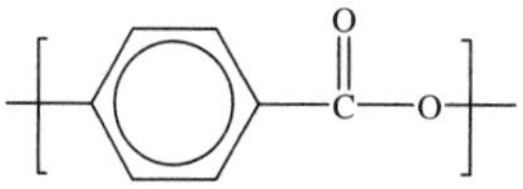	480
Poly(phenylene sulfide)	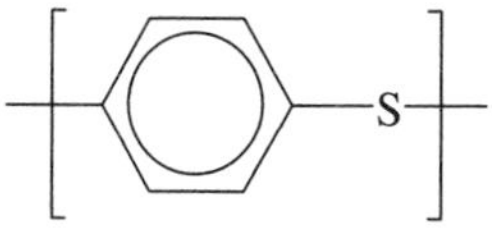	490
Polythiadiazole	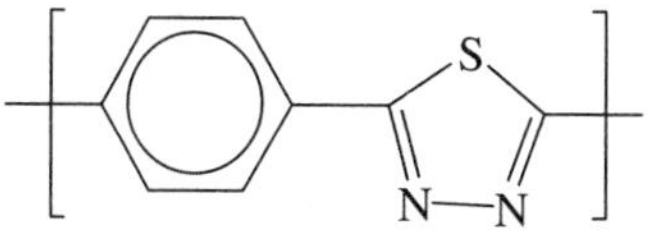	490
Poly(phenylene oxide)	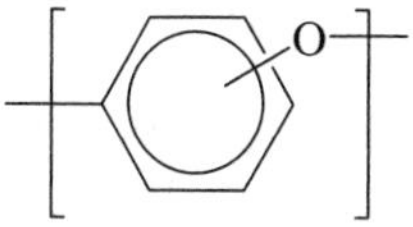	570
Polyimide	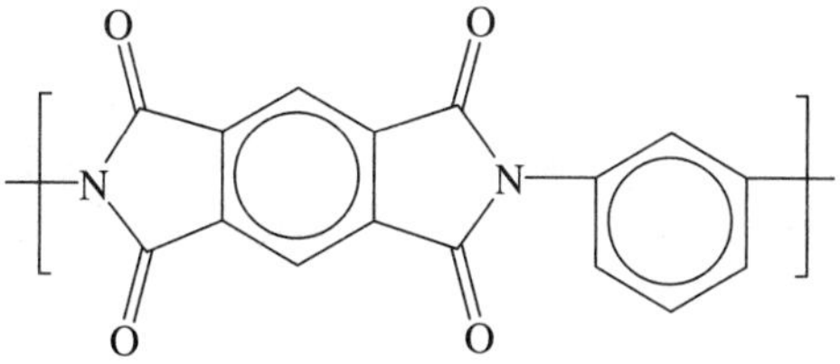	585
Polybenzamide	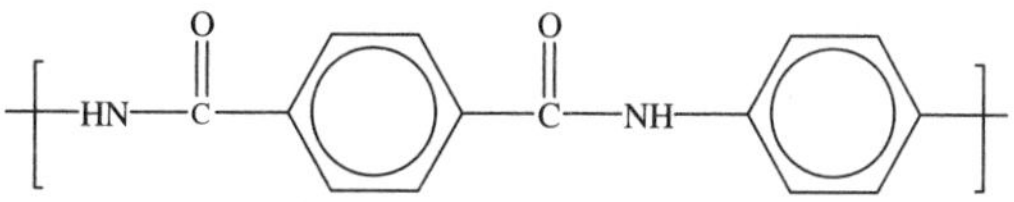	500
Polyoxazole	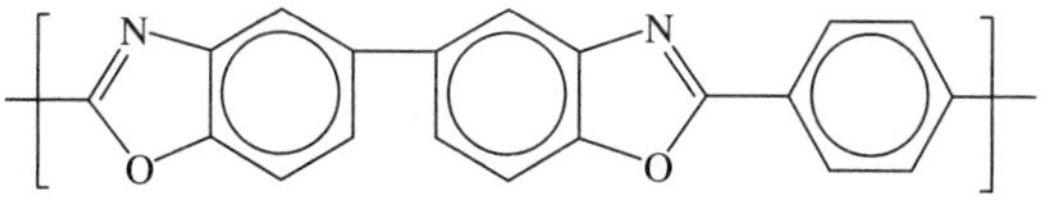	620
Polybenzimidazole	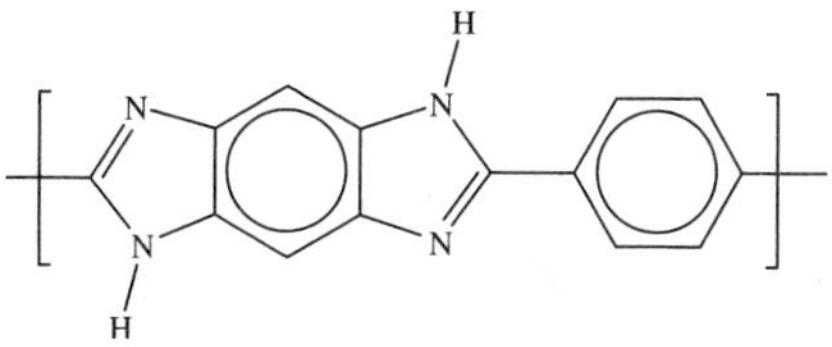	650

(Continued)

TABLE 14.27   Examples of Thermally Stable Polymers [6] (*Continued*)

Polymer	Structure	Decomposition Temperature(C)
Polypyrrole	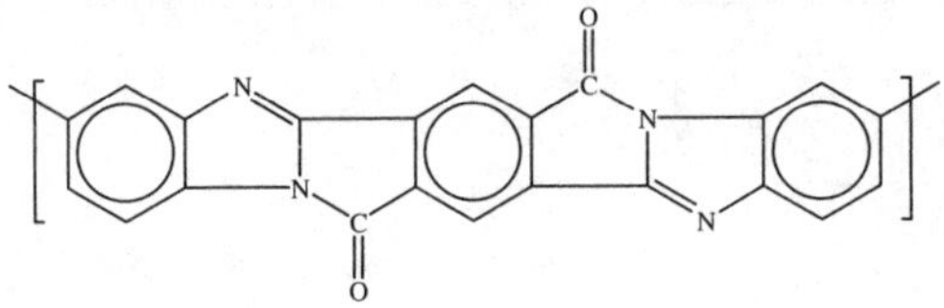	660
Poly(*p*-phenylene)	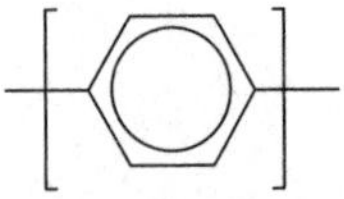	660

Oxidative degradation can also occur in other polymers including natural rubber, polystyrene, polypropylene, nylons, polyurethanes, and most natural and naturally derived polymers. With the exception of fluoropolymers, most polymers are susceptible to oxidation, particularly at elevated temperature or during exposure to ultraviolet light. Oxidation usually leads to increasing brittleness and deterioration in strength.

The rate of degradation of polymers may be retarded by the addition of chain transfer agents called antioxidants. Antioxidants are organic

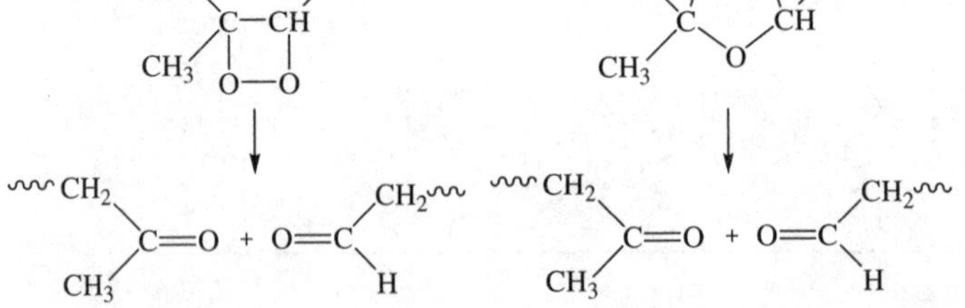

Figure 14.27 Degradation of polyisoprene by photolysis (*a*), and ozonolysis (*b*).

compounds like hindered phenols and aromatic amines that are used as additives to retard oxidative degradation of polymers by acting as free-radical scavengers through producing inactive free radicals. The following equations show two examples of antioxidants derived from hindered phenols that act as chain transfer agents to produce a dead polymer and a stable free radical that does not initiate chain radical degradation.

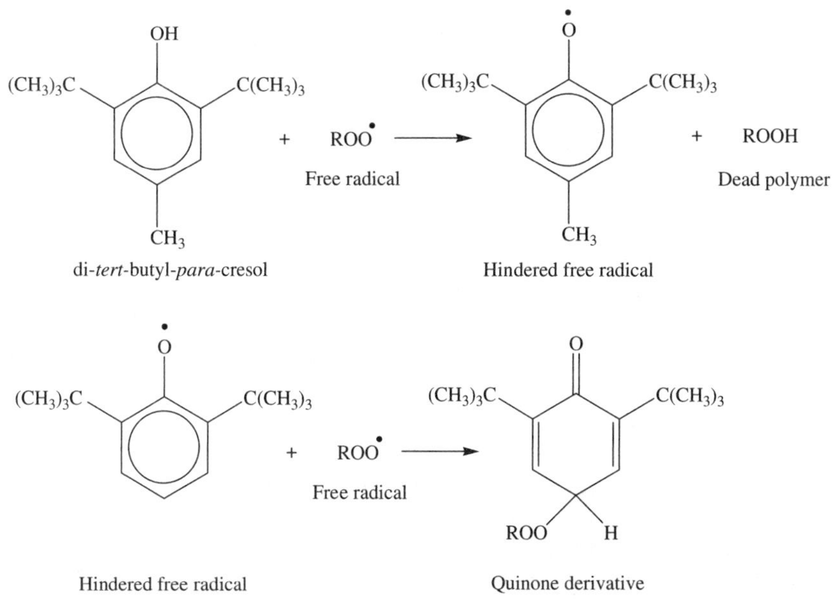

di-*tert*-butyl-*para*-cresol   Hindered free radical

Free radical   Dead polymer

Hindered free radical   Quinone derivative

Free radical

Effects of UV radiation may be reduced by incorporating additives, such as carbon black that screen UV wavelengths (300 to 400 nm). Carbon black has many free electrons and, therefore, can retard free radical degradation of the polymer. Polymer degradation by UV radiation may also be lessened by the addition of compounds such as 2-hydroxybenzophenones that serve as energy transfer agents, that is, they absorb radiation at low wavelengths and radiate it at longer wavelengths (lower energy). Phenyl salicylate rearranges in the presence of high-energy radiation to form 2,2'-dihydroxybenzophenone. The latter and other 2-hydroxybenzophenones act as energy transfer agents, that is, they absorb energy to form chelates that release energy at longer wavelengths by the formation of quinone derivatives. Benzotriazoles such as 2-(o-hydroxyphenyl)benzotriazole are also widely used as UV absorbers.

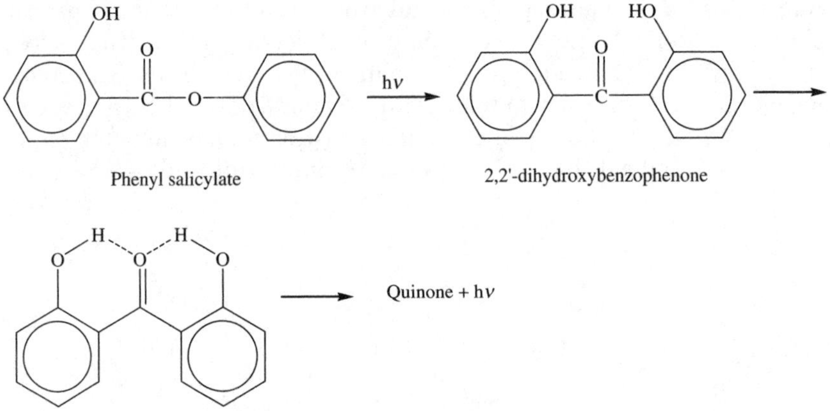

Phenyl salicylate

$h\nu$

2,2'-dihydroxybenzophenone

Quinone + $h\nu$

Chelate

Chemical and hydrolytic degradation. Heteroatomed polymers such as condensation polymers are susceptible to degradation on exposure to aqueous acid or base solutions. These include some naturally occurring polymers, such as polysaccharides and proteins, as well as many synthetic polymers such as polyesters and polyamides. Polymer degradation can also happen by enzymes (produced by microbes) capable of breaking selected bonds such as those that appear naturally, including amide, ester, and ether linkages, and including both natural, naturally derived, and synthetic materials.

Thermoplastics, in contact with organic liquids or vapors, will fail at lower stress or strain even if the interacting chemical is not ordinarily considered to be a solvent for the polymer. The effect of these chemicals is believed to be because of localized plasticization that reduces the effective T_g, and thus increases the localized mobility of polymer chains and promotes craze and crack development.

14.13   Polymer Additives

Polymer properties and performance can be modified through the addition of certain compounds called additives. Additives can be added as solids, liquids, or gases, and cover a wide range of materials. Typical additives include: antimicrobial agents, antioxidants, antistatic agents, coloring agents, flame retardants, impact modifiers, plasticizers, UV/Vis radiation and heat stabilizers, reinforcing agents, viscosity modifiers (such as flow enhancers, thickening agents, or antisag materials), and many more.

Flexibility, for example, can be imparted to stiff polymers through the addition of a compatible liquid or solid that permits spillage of polymer chains and thus reduces the T_g and modulus of the polymer. This will increase not only flexibility, but also workability and distensibility of the plastic.

Additives for the purpose of retarding polymer degradation have been discussed in Sec. 14.12. Many polymers are used as shelter and clothing and in household furnishing, and therefore, it is essential that they have good flame resistance. Polymer resistance to combustion can be increased by adding flame retardants that terminate the free radical combustion reaction. The most common flame-retarding additives for plastics contain large proportions of chlorines and bromines. These elements are believed to quench the free radical flame propagation reactions. Flame-retarding compounds can be simply mixed with the plastics, or they may be reactive monomers that become part of the polymer. Organic phosphates function as flame retardants by forming a char that acts as a barrier to the flame. Hydrated alumina is particulate filler that contains 35 percent water of hydration, the evaporation of which absorbs energy and inhibits flame spread.

A wide variety of organic and inorganic pigments is used as additives to color polymers. Classes of dyes that are used for plastics include azo compounds, anthraquinones, xanthenes, and azines. Among the most important inorganic pigments are ion oxides, cadmium, and chrome yellow. Titanium dioxide is a common pigment where a brilliant, opaque white is desired. Sometimes, pigments perform other functions. For example, calcium carbonate acts as both filler and a pigment in many plastics. Similarly, carbon black is often a UV stabilizer as well as a pigment.

Most polymers are poor electrical conductors, and therefore, electrostatic charge might build up on the surface of polymers. This may cause problems such as dust collection and sparking. The static charges can be dissipated by adding antistatic agents (antistats). Most antistats are hygroscopic (such as fatty-acid amines) and because of their polarity attract a thin film of water to the polymer surface. The antistats reduce the charge by acting as lubricants (moisture film), or they may provide a conductive path for the dissipation of the charge. Examples of antistatic agents are quaternary ammonium compounds, hydroxylalkylamines, organic phosphates, derivatives of polyhydric alcohols such as sorbitol and glycol esters of fatty acids.

Impact modifiers are rubbery additives that improve the resistance of materials. Proper compatibility between the phases is essential. This is often achieved with graft and block copolymers. Most impact modifiers are elastomers such as ABS, BS, methacrylate-butadiene-styrene, acrylic, ethylene-vinyl acetate, and chlorinated polyethylene.

Naturally occurring polymeric materials as well as some synthetic polymers containing linkages like ester or amide are susceptible to attack by microorganisms such as fungi, yeast, and bacteria. This might lead to deterioration of the polymer which, in some cases, can be desired in producing biodegradable polymers. However, it often leads to unwanted growth of microbes on the polymer surface, especially where a polymer will be subjected to a warm, humid environment. The microbial attack can be decreased by the use of biocides as additives to control or destroy bacterial growth. These biocides are generally organic copper, mercury, or tin compounds. It is important that the fungistatic and bacteriostatic additives are nonleaching and are harmless to humans, other animal life, and environment.

Fillers. Fillers are relatively inert material usually added to extend the polymer and thereby reduce the cost of resinous composites. Fillers may also serve in improving processability or dissipating heat in exothermic thermosetting reactions. Reinforcing fillers are used to enhance the structural properties of the polymer and improve some mechanical property or properties such as modulus, tensile or tear strength, abrasion resistance, and fatigue strength. Long glass fibers are used to reinforce epoxy and polyester thermosets. Particulate fillers such as carbon black or silica are widely used to improve the strength and abrasion resistance of commercial elastomers. Fibers in the form of continuous strands, woven fabrics, and chopped (or discontinuous) fibers are used to reinforce thermoplastics and thermosets.

Improvement in cost, processability, and performance of polymers can also be achieved through blending two or more polymers. The blends may be homogeneous, heterogeneous, or a bit of both. Properties of miscible polymer blends may be intermediate between those of the individual components (i.e., additive behavior), as is typically the case for T_g. In other cases, blend properties may exhibit either positive or negative deviation from additivity.

References

1. *American Chemical Society,* Facts & Figures for the Chemical Industry, *Chemical & Engineering News,* 81, 27, 51–61, July 7, 2003.
2. Odian, G., *Principles of Polymerization,* 3rd ed., John Wiley & Sons, Inc., New York, N. Y., 1991.
3. Natta, G., L. Porri, A. Carbonaro, and G. Stoppa, Polymerization of conjugated diolefins by homogeneous aluminum alkyl-titanium alkoxide catalyst systems. I. Cis-1,4 isotactic poly-1,3-pentadiene, *Makromolekulare Chemie,* 77, 114–125, 1964.
4. Natta, G., L. Porri, and A. Carbonaro, Polymerization of conjugated diolefins by homogeneous aluminum alkyl-titanium alkoxide catalyst systems. II., *Makromolekulare Chemie,* 77, 126–138, 1964.
5. Graessley, W. W., *Advances in Polymer Science,* Vol. 16, Springer, New York, N.Y., 1974.
6. Korshak, V. V., *Heat Resistant Polymers,* Halstead Press, New York, N.Y, 1972.
7. Lewis, O. G., *Physical Constants of Linear Homopolymers,* Springer-Verlag, New York, 1968.

8. Andrews, R. J. and E. A. Grulke, *Polymer Handbook*, 4th ed., Brandrup, J., E. H. Immergut, and E. A. Grulke (Eds.), Wiley-Interscience, New York, N.Y., pp. VI/193 ff., 1999.

9. Peyser, P., *Polymer Handbook*, Brandrup, J. and E. H. Immergut (Eds.), 3rd ed., Wiley-Interscience, New York, N.Y., pp. VI/209 ff, 1975.

10. Burfield, D. R. and K. L. Lim, Differential scanning calorimetry analysis of natural rubber and related polyisoprenes. Measurement of the glass transition temperature, *Macromolecules*, 16, 7, 1170–1175, 1983.

11. Stevens, M. P., *Polymer Chemistry: An Introduction*, Oxford University Press, New York, N.Y., 1999.

12. Hildebrand, J. H. and R. L. Scott, *The Solubility of Nonelectrolytes*, 3rd ed., Dover Publication, New York, N.Y., 1964.

13. Small, P. A., Factors affecting the solubility of polymers, *Journal of Applied Chemistry*, 3, 71, 1953.

14. Hoy, K. L., New values of the solubility parameters from vapor pressure data, *Journal of Paint Technology*, 42, 541, 76–78, 115–118, 1970.

15. Grulke, E. A., *Polymer Handbook*, Brandrup, J., E. H. Immergut and E. A. Grulke (Eds.), 4th ed., Wiley-Interscience, New York, N.Y., pp.VII/675 ff, 1999.

16. Bloch, D. R., *Polymer Handbook,* Brandrup, J., E. H. Immergut, and E. A. Grulke (Eds.) 4th ed., Wiley-Interscience, New York, N.Y., pp. VII/497 ff, 1999.

17. Ewart, R. H. Significance of viscosity measurements on dilute solutions of high polymers, *Scientific Progress in the Field of Rubber and Synthetic Elastomers* (*Advances in Colloid Science*) H. Mark and G. S. Whitby (Eds.), Vol. 2, pp. 197–251, Wiley-Interscience, New York, N.Y., 1946.

18. Kurata, M. and Y. Tsunashima, *Polymer Handbook*, 4th ed., Brandrup, J., E. H. Immergut, and E. A. Grulke (Eds.) Wiley-Interscience, New York, N. Y., pp. VII/1 ff., 1999.

19. Winding, C. C. and G. D. Hiatt, *Polymeric Materials*, McGraw-Hill, New York, N.Y., 1961.

20. Carraher, C. E. Jr., *Polymer Chemistry*, 6th ed., Marcel Dekker, Inc., New York, N. Y., 2003.

21. Allcock, H. R. and F. W. Lampe, *Contemporary Polymer Chemistry*, 2nd ed., Prentice Hall, Englewood Cliffs, New Jersy, 1990.

22. Greenley, R. Z., *Polymer Handbook*, Brandrup, J., E. H. Immergut, and E. A. Grulke (Eds.) 4th ed., Wiley-Interscience, New York, N. Y., pp. II/181 ff, 1999.

Index